Biogeoscience and the Ecosystem

Biogeoscience and the Ecosystem

Editor: Acacia Xzavier

RCALLISTO REFERENCE

www.callistoreference.com

Callisto Reference,
118-35 Queens Blvd., Suite 400,
Forest Hills, NY 11375, USA

Visit us on the World Wide Web at:
www.callistoreference.com

ISBN: 978-1-64116-046-9 (Hardback)

Cataloging-in-Publication Data

Biogeoscience and the ecosystem / edited by Acacia Xzavier.
    p. cm.
Includes bibliographical references and index.
ISBN 978-1-64116-046-9
1. Landscape ecology. 2. Geomorphology. 3. Ecosystem management. 4. Biotic communities.
I. Xzavier, Acacia.
QH541.15.L35 B56 2019
577--dc23

Table of Contents

Preface

This book is a valuable compilation of topics, ranging from the basic to the most complex advancements in the field of biogeoscience. This field of study is concerned with the exchange between the biological activities and the chemical and physical environment of Earth's atmosphere or extra-terrestrial space. It integrates the concepts of biology and geosciences. This book studies, analyses and upholds the pillars of biogeoscience and its utmost significance in modern times. While understanding the long term perspectives of the topics, the book makes an effort in highlighting their impact as a modern tool for the growth of the discipline. It is appropriate for students seeking detailed information in this area as well as for experts.

The information contained in this book is the result of intensive hard work done by researchers in this field. All due efforts have been made to make this book serve as a complete guiding source for students and researchers. The topics in this book have been comprehensively explained to help readers understand the growing trends in the field.

I would like to thank the entire group of writers who made sincere efforts in this book and my family who supported me in my efforts of working on this book. I take this opportunity to thank all those who have been a guiding force throughout my life.

Editor

Physical control of interannual variations of the winter chlorophyll bloom in the northern Arabian Sea

Madhavan Girijakumari Keerthi[1], Matthieu Lengaigne[2,3], Marina Levy[2], Jerome Vialard[2],
Vallivattathillam Parvathi[1], Clément de Boyer Montégut[4], Christian Ethé[2], Olivier Aumont[2], Iyyappan Suresh[1],
Valiya Parambil Akhil[1], and Pillathu Moolayil Muraleedharan[1]

[1]Physical Oceanography Division, CSIR-National Institute of Oceanography (CSIR-NIO), Goa, India
[2]Sorbonne Universités (UPMC, Univ Paris 06)-CNRS-IRD-MNHN, LOCEAN Laboratory, IPSL, Paris, France
[3]Indo-French Cell for Water Sciences, IISc-NIO-IITM-IRD Joint International Laboratory, NIO, Goa, India
[4]IFREMER, Univ. Brest, CNRS, IRD, Laboratoire d'Océanographie Physique et Spatiale, IUEM, 29280, Brest, France

Correspondence to: Madhavan Girijakumari Keerthi (keerthanaamg@gmail.com)

Abstract. The northern Arabian Sea hosts a winter chlorophyll bloom, triggered by convective overturning in response to cold and dry northeasterly monsoon winds. Previous studies of interannual variations of this bloom only relied on a couple of years of data and reached no consensus on the associated processes. The current study aims at identifying these processes using both ~ 10 years of observations (including remotely sensed chlorophyll data and physical parameters derived from Argo data) and a 20-year-long coupled biophysical ocean model simulation. Despite discrepancies in the estimated bloom amplitude, the six different remotely sensed chlorophyll products analysed in this study display a good phase agreement at seasonal and interannual timescales. The model and observations both indicate that the interannual winter bloom fluctuations are strongly tied to interannual mixed layer depth anomalies (~ 0.6 to 0.7 correlation), which are themselves controlled by the net heat flux at the air–sea interface. Our modelling results suggest that the mixed layer depth control of the bloom amplitude ensues from the modulation of nutrient entrainment into the euphotic layer. In contrast, the model and observations both display insignificant correlations between the bloom amplitude and thermocline depth, which precludes a control of the bloom amplitude by daily dilution down to the thermocline depth, as suggested in a previous study.

1 Introduction

Located in the western arm of the northern Indian Ocean, the Arabian Sea (AS) is forced by energetic seasonally reversing monsoon winds, which largely control its physical properties. During boreal summer, strong southwesterly winds blow over the western AS (Findlater, 1969) and cause intense upwelling along the coasts of Somalia and Oman and downwelling in the central AS (e.g. Schott and McCreary, 2001). During boreal winter, the Eurasian continent cools, and a high-pressure region develops on the Tibetan Plateau, resulting in cold and dry northerly or northeasterly winds (e.g. Smith and Madhupratap, 2005) and leading to strong evaporative cooling (Dickey et al., 1998). These diverse physical processes cause substantial variations in marine biogeochemical and ecosystem response. Being one of the most productive regions in the world ocean (Satya Prakash and Ramesh, 2007; Prasanna Kumar et al., 2000) and being home to the second-most-intense oxygen minimum zone in the world ocean (Kamykowski and Zentara, 1990), the AS provides an excellent test bed for studying coupled biophysical processes (McCreary et al., 2009).

Previous studies have extensively described the seasonal variability of surface chlorophyll (hereafter, SChl) in the AS. The AS biogeochemical properties vary from stratified oligotrophic conditions during inter-monsoon periods to eutrophic conditions during monsoons (Smith et al., 1998; McCreary et al., 2009). Neither surface irradiance nor temperature limits the biological productivity in this tropical basin:

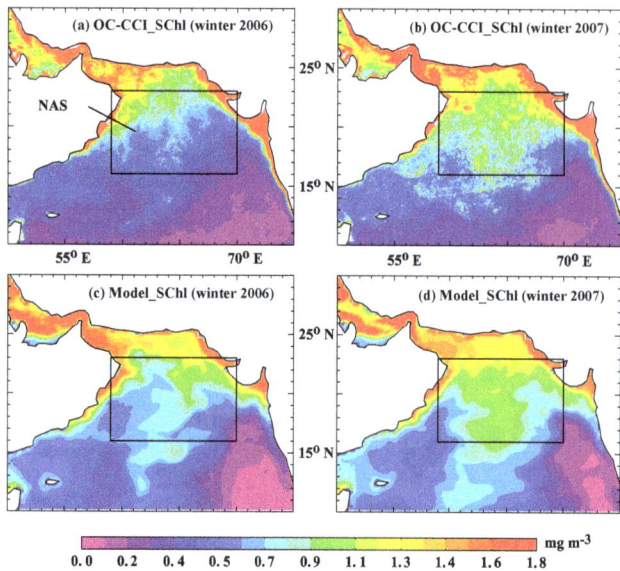

Figure 1. Arabian Sea average SChl for the winter (DJFM) of **(a, c)** 2006 and **(b, d)** 2007 in **(a, b)** OC-CCI product and **(c, d)** model. The NAS (North Arabian Sea) box (59–70° E, 16–23° N) is indicated by a black frame for future reference.

instead, it is mostly attributed to dynamical processes in response to the monsoonal forcing (e.g. Barber et al., 2001; Marra and Barber, 2005). During boreal summer, the largest seasonal blooms are found along the coasts of the Arabian Peninsula (e.g. Banzon et al., 2004; Lévy et al., 2007; Wiggert et al., 2005) and are exported offshore by mesoscale eddy stirring (e.g. Resplandy et al., 2011). In boreal winter, convective overturning allows entrainment of nutrients into the mixed layer and leads to a prominent bloom in the northern AS (Banse and English, 2000; Madhupratab et al., 1996; Prasanna Kumar et al., 2001; Wiggert et al., 2002). In addition to these seasonal variations, several studies have revealed large interannual variations in the AS winter chlorophyll from either satellite (Banse and McClain, 1986; Banse and English, 1993; Sarma et al., 2006, 2012; Wiggert et al., 2002) or in situ measurements (Bauer et al., 1991; Madhupratap et al., 1996; Gundersen et al., 1998; Prasanna Kumar et al., 2001). This strong interannual variability of the northern AS winter bloom is illustrated in Fig. 1a and b for two consecutive winters. A particularly intense bloom developed in the northern AS during winter 2007 (Fig. 1b), with high SChl concentration ($> 1.0 \, \mathrm{mg\,m^{-3}}$) extending southward down to 14° N. In contrast, the winter 2006 bloom remained confined to the northern AS (Fig. 1a), with high chlorophyll concentration ($> 1.0 \, \mathrm{mg\,m^{-3}}$) limited to north of 20° N. The difference in the amplitude of the bloom between winter 2006 and 2007 averaged over the northern AS box (hereafter NAS; region shown in Fig. 1) reaches $0.22 \, \mathrm{mg\,m^{-3}}$, which is approximately 30 % of the climatological winter chlorophyll value.

Understanding the mechanisms driving these interannual chlorophyll variations is important, as this might have a profound influence on the variations of the fish stocks and of the oxygen minimum zone in the AS. To date, only a few studies have discussed the mechanisms that could be responsible for the interannual winter bloom fluctuations (Banse and McClain, 1986; Prasanna Kumar et al., 2001; Wiggert et al., 2002), and no consensus has been reached so far. Comparing in situ time series in February 1995 and 1997, Prasanna Kumar et al. (2001) suggested that increased convective cooling resulted in an intense convective mixing, a deeper mixed layer depth (hereafter, MLD), enhanced nutrients injection through entrainment, and ultimately a stronger bloom in winter 1997 than in winter 1995. Such a mechanism hence implies that the interannual MLD variability should be positively correlated with the interannual SChl variability (the so-called "Bermuda paradigm"). Keerthi et al. (2016) found large winter MLD variations in the NAS over the past 2 decades – largely driven by fluctuations in the advection of dry, cold air from the continent – but did not investigate their biogeochemical consequences. Comparing three consecutive winters from 1998 to 2000, and using a simple one-dimensional model, Wiggert et al. (2002) suggested that interannual variations of the bloom intensity were controlled by the night-time penetration of diurnal mixing, whose maximum downward penetration is constrained by the thermocline depth (hereafter, TCD). In this paradigm, a deeper TCD allows for a deeper night-time mixing, a greater dilution of phytoplankton biomass, and stronger inhibition of the bloom development. This alternative scenario hence implies that the interannual TCD variability should be negatively correlated with the interannual SChl variability in the northern AS during winter, a relationship that directly contradicts the Bermuda paradigm as pointed out by Wiggert et al. (2005). Prasanna Kumar et al. (2001) and Wiggert et al. (2002) hence proposed two conflicting mechanisms, which respectively imply a positive correlation between interannual MLD and SChl variations and a negative correlation between interannual TCD and SChl variations. These papers however based their conclusions on the analysis of rather small samples, i.e. two 1-month-long in situ time series for Prasanna Kumar et al. (2001) and three consecutive winters from satellite chlorophyll data for Wiggert et al. (2002). The absence of consensus on the processes responsible for the interannual NAS winter bloom variations hence pleads for additional studies, especially now that longer remotely sensed and in situ time series are available.

In the present paper, we hence aim at better assessing and understanding the interannual variability of the NAS winter bloom. On the observational side, this study benefits from the extended temporal coverage of the satellite chlorophyll data (~ 15 years) and the advent of the Argo program, which allows monitoring in situ MLD and TCD variations from 2002 onwards. Performing a combined analysis of these datasets allowed us to perform a direct comparison between these

Table 1. Main characteristics of ocean colour satellite products used in the present study.

Name	Satellite	Algorithm	Period	Grid	Winter/summer NAS coverage
SeaWiFS	GeoEye OrbView-2	OC4v6 (O'Reily et al., 2000)	1997–2010	9 km	92 %/25 %
MERIS	ESA ENVISAT	OC4Me (O'Reily et al., 2000)	2002–2012	9 km	96 %/30 %
MODIS	Terra and Aqua	OC3v5 (O'Reily et al., 2000)	2002–present	9 km	96 %/20 %
GSM	Combine SeaWiFs–MODIS–MERIS	GSM (Maritorena and Siegel, 2005)	1997–2012	4 km	98 %/45 %
AVW	Combine SeaWiFs–MODIS–MERIS		2002–2012	4 km	98 %/53 %
OC-CCI	Combine SeaWiFs–MODIS–MERIS	OC-CCI v2.0 (Grant et al., 2015)	1997–2013	4 km	98 %/58 %

physical parameters and the chlorophyll variability. In addition to these satellite and in situ observations, we analyse outputs from a $\sim$ 20-year-long coupled biophysical model simulation. The analysis of this simulation, which accurately simulates interannual NAS winter chlorophyll variations, allows us to investigate the subsurface processes not readily available from observations. Section 2 describes the observational products (satellite chlorophyll estimates and Argo-derived MLD and TCD; Sect. 2.1) and the numerical experiment (Sect. 2.2). Section 3 provides an intercomparison of the available satellite chlorophyll products over the NAS (Sect. 3.1) and the model evaluation (Sect. 3.2). Section 4 provides a description of the interannual chlorophyll variability and its relationship with physical parameters, and discusses the mechanisms driving these fluctuations. Section 5 finally provides a summary and discussion of our results.

2 Data and method

2.1 Observations

The SChl estimates analysed in the present study are derived from different instruments (Sea-Viewing Wide Field-of-View Sensor (SeaWiFS), Medium-Resolution Imaging Spectrometer (MERIS), and Moderate-Resolution Imaging Spectroradiometer (MODIS)) and retrieval algorithms, and they span different periods (Table 1). We will compare these different retrievals in Sect. 3.1, in order to assess the robustness of remotely sensed data to investigate the NAS winter bloom. We used the Level 3 standard mapped images with a 9×9 km spatial and a monthly temporal resolution downloaded from http://oceancolor.gsfc.nasa.gov for all of these single-mission products. In addition, we also used three Level 3 merged ocean colour products downloaded from http://globcolour.info and http://www.oceancolour.org/ at 4×4 km and monthly resolution: the weighted average empirical (AVW) product, the semi-analytical Garver–Siegel–Maritorena (GSM) product, and the Ocean Colour Climate Change Initiative (OC-CCI) product. The longest observational period is provided by the OC-CCI product and spans from October 1997 to December 2013.

The ocean physical parameters are derived from an updated version of the dataset described in Keerthi et al. (2013), with an extended temporal coverage (2002 to 2013) and an estimate of the TCD in addition to that of the MLD. These MLD and TCD datasets are built from a combination of Argo and historical temperature and salinity profiles. To assess whether the mechanism proposed by Prasanna Kumar et al. (2001; i.e. the Bermuda paradigm) dominates the interannual variability of the winter bloom in the northern AS over this extended period, we will investigate if there is a correlation between in-situ-derived interannual MLD and the satellite-derived interannual SChl anomalies. We will test the alternative mechanism proposed by Wiggert et al. (2002) by investigating if there is a negative correlation between interannual in-situ-derived TCD anomalies and interannual satellite-derived SChl anomalies in the northern AS during winter. MLDs were estimated using a temperature criterion and are defined as the depth where the temperature decreases by 0.2 °C with respect to the temperature at 10 m. The reference depth was taken at 10 m to avoid aliasing by the diurnal cycle. The TCD was defined as the depth of the maximal vertical temperature gradient. MLDs and TCDs were estimated from individual temperature profiles at their native vertical resolution. The resolution of the data was then degraded to a regular 2° monthly grid, by taking the median of all MLDs and TCDs in each grid mesh. A more detailed description of this procedure can be found in Keerthi et al. (2013). An overview of the spatio-temporal coverage of this dataset over the NAS is provided in Fig. 2. While the data coverage is particularly sparse in winter before 2002 in our targeted region (e.g. less than 10 data per month are available in winter 2000 in the NAS region), the data density increased considerably after 2002, with the development of the Argo program (Fig. 2a). After 2002, the NAS box winter data density ranged from 25 profiles per month during 2005 to nearly 120 profiles per month during 2012. This implies that the interannual MLD/TCD values averaged over the NAS box during winter 2002 to 2013 are built from an average of 100 to 500 individual values, giving us confidence in the robustness of the interannual MLD/TCD variability derived from this in situ dataset. It should be noticed that the data coverage is

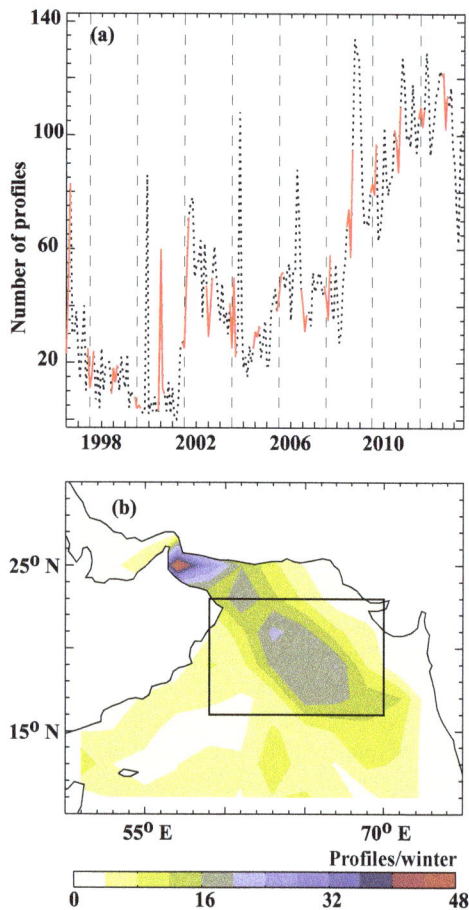

Figure 2. (a) Time series of the number of in situ profiles per month over the NAS box, from 1997 to 2013. The curve is highlighted in red for winter (DJFM). **(b)** Average winter in situ profile density (per $2° \times 2°$ box and per season) for 2002–2012. The NAS region is indicated by a black frame in **(b)**.

however not spatially homogeneous, with the highest coverage along a shipping line crossing the NAS box (Fig. 2b).

We also use the World Ocean Atlas (WOA13) climatology (Boyer et al., 2013) to derive climatologies of the thermocline and nitracline depths, calculated as the depths of maximum temperature and nitrate gradients, respectively. Wind speeds, surface air temperatures, and net heat fluxes are derived from the TropFlux product (Praveen Kumar et al., 2012). In order to assess the variability associated with various interannual climate modes, we have used standard climate indices. The El Niño–Southern Oscillation (ENSO) is represented using the Niño3.4 index, which is the averaged sea surface temperature (SST) anomalies over the Niño3.4 (120–170° W, 5° N–5° S) region during November–January, available from http://www.cpc.ncep.noaa.gov/products/analysis_monitoring/ensostuff/detrend.nino34.ascii.txt. The Indian Ocean Dipole (IOD) is represented by the dipole mode index (DMI; Saji et al., 1999), computed as the difference

between interannual SST anomalies in the western (50–70° E, 10° N– 10° S) and eastern (90–110° E, 10–0° S) equatorial Indian Ocean during September–November, available from http://www.jamstec.go.jp/frcgc/research/d1/iod/DATA/dmi.monthly.txt.

2.2 Model configuration and numerical experiments

These observational products are complemented by a biophysical model simulation, which allows extending our analysis over a longer time period and analysing depth-integrated biogeochemical properties that are not captured by satellites. We use the NEMO (Nucleus for European Modelling of the Ocean; see Madec (2008) for an exhaustive description) ocean general circulation model coupled with the latest version of PISCES (Pelagic Interaction Scheme for Carbon and Ecosystem Studies; see Aumont et al. (2015) for an exhaustive description of the model) biogeochemical component. Briefly, PISCES includes two sizes of sinking particles and four "living" biological pools, which represent two phytoplankton (nano-phytoplankton and diatoms) and two zooplankton (micro-zooplankton and meso-zooplankton) size classes. Phytoplankton growth is limited by five nutrients: NO_3, NH_4, PO_4, SiO_4, and Fe. The ratios among C, N, and P are kept constant for the living compartments, at values proposed by Takahashi et al. (1985). On the other hand, the iron, silicon, and calcite pools of the particles are explicitly modelled. As a consequence, their ratios are allowed to vary. Nutrients are supplied to the ocean from five different sources: atmospheric dust deposition, rivers, sea ice, sediment mobilization, and hydrothermal vents. An interannually varying dust deposition dataset is not available to date. Dust deposition from the atmosphere is hence estimated from climatological monthly deposition maps simulated by the National Center for Atmospheric Research model (Mahowald et al., 2005), assuming constant values for the iron content and solubility (Moore et al., 2004). This choice is further justified by the modelling results of Aumont et al. (2008), who demonstrated that the variability of SChl induced by the interannual variability of aerial iron deposition is likely to be very small everywhere especially relative to the impact of the ocean dynamics, because the largest fluctuations of surface iron produced by dust occur in oligotrophic regions where phytoplankton growth is not primarily controlled by iron availability. The internal Fe contents of both phytoplankton groups and Si contents of diatoms are prognostically simulated as a function of ambient concentrations in nutrients and light level. Details on the red–green–blue model by which light penetration profiles are calculated are given in Lengaigne et al. (2007). The Chl / C ratio is modelled using a modified version of the photo-adaptation model by Geider et al. (1998). For a more detailed description, manuals for NEMO and PISCES are available online at http://www.nemo-ocean.eu/About-NEMO/Reference-manuals.

The regional configuration used in this study is an Indian Ocean sub-domain of the global 1/4° resolution (i.e. cell size ~ 25 km) configuration described by Barnier et al. (2006). It has 46 vertical levels, with a resolution ranging from 5 m at the surface to 250 m at the bottom. The African continent closes the western boundary of the domain. The oceanic portions of the eastern, northern, and southern boundaries use radiative open boundaries (Treguier et al., 2001), constrained with a 150-day relaxation timescale to outputs from a global simulation (Dussin et al., 2009). The circulation and thermodynamics of this regional configuration have been extensively evaluated and reproduce observed variations of key physical parameters well in several Indian Ocean regions (Vialard et al., 2013; Akhil et al., 2014, 2016; Praveen Kumar et al., 2014), including the AS (Nisha et al., 2013; Keerthi et al., 2016).

The simulation starts from rest, with temperature and salinity initialized from the WOA13. PISCES biogeochemical tracers are also initialized from the WOA13 database for nutrients and from the climatology of a global simulation for the other tracers (Aumont and Bopp, 2006). After 5 years of spin-up with a climatological forcing, the model is forced with the Drakkar Forcing Set #4.4 (DFS4.4; Brodeau et al., 2010) from 1980 to 2012. This forcing is a modified version of the CORE dataset (Large and Yeager, 2004), with atmospheric parameters derived from ERA-40 reanalysis (Uppala et al., 2005) and ECMWF analysis after 2002 for latent and sensible heat flux computation. Radiative fluxes are taken from the corrected International Satellite Cloud Climatology Project – Flux Dataset (ISCCP-FD) surface radiations (Zhang et al., 2004), while precipitation forcing is a blend of satellite products described in Large and Yeager (2004). All atmospheric fields are corrected to avoid temporal discontinuities and to remove known biases (see Brodeau et al. (2010) for details). In the following, we will analyse the 1993–2012 period.

3 Evaluation of the interannual variability in the northern Arabian Sea

In this section, we provide an intercomparison of the six ocean colour products described above (Sect. 3.1) and a brief description on model performance at seasonal and interannual timescales in our targeted region (Sect. 3.2).

3.1 Satellite SChl product intercomparison

One of the major limitations of ocean colour imagery is the inability to perform accurate retrievals under clouds and in the presence of aerosols. In the AS, this is particularly challenging during the summer monsoon but not during winter, when the data coverage is larger than 90 % for all datasets considered and reaches 98 % in the case of the merged products (Table 1). Figure 3a shows the SChl climatological sea-

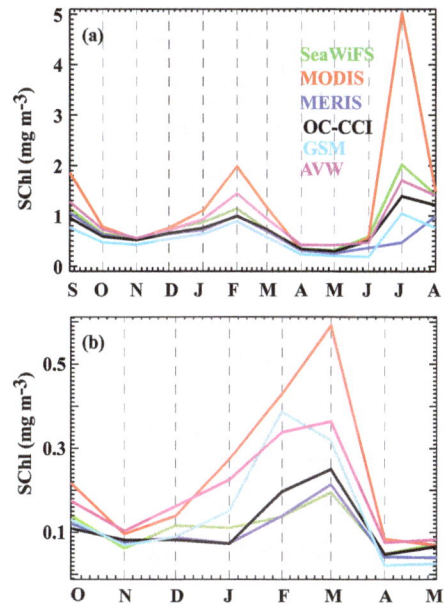

Figure 3. (a) Climatological monthly seasonal cycle of NAS-averaged SChl for all satellite products. **(b)** Standard deviation of monthly (October–May) NAS-averaged SChl for all satellite products.

sonal cycle averaged over the NAS region for the six available satellite products. This figure reveals a good agreement between the different products in terms of seasonal phasing, with a semi-annual cycle associated with two seasonal blooms, one in summer (maximum in July, except for MERIS) and the other in winter (maximum in February for all products). The amplitudes of these seasonal blooms clearly differ depending on the product, with the largest blooms in MODIS (up to $2 \, \mathrm{mg \, m^{-3}}$ in winter and $5 \, \mathrm{mg \, m^{-3}}$ in summer), while seasonal blooms hardly reach $1 \, \mathrm{mg \, m^{-3}}$ in GSM, OC-CCI, and MERIS. However, the amplitude of the summer chlorophyll bloom is uncertain, given the smaller amount of data coverage during summer, especially in July, where there is the least data coverage.

Figure 3b allows assessing the amplitude of the interannual winter bloom variations in each product by computing the standard deviation of the interannual chlorophyll variations for each calendar month from October to May. This figure shows that the largest interannual deviations from the climatological evolution depicted in Fig. 3a occur during February and March, while a minimum in the amplitude of interannual chlorophyll variability is generally found during November and April. Based on this seasonality, we define winter in the following as the period encompassing the large climatological and interannual SChl signals, i.e. from December to March. For instance, we will refer to winter 2002 as the period averaged from December 2002 to March 2003. Figure 3b however also illustrates that the amplitude of the interannual winter bloom variations considerably varies

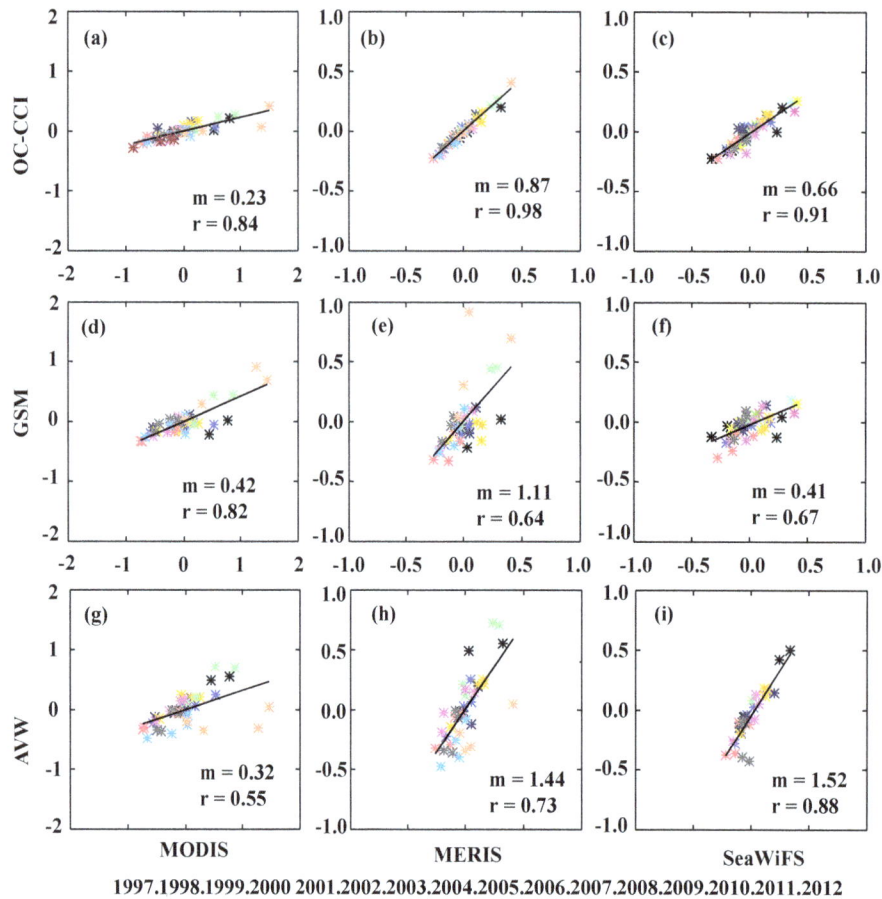

Figure 4. (a, b, c) Scatter plot of interannual anomalies in NAS-averaged monthly winter (DJFM) SChl (mg m^{-3}) in the OC-CCI product against MODIS, MERIS, and SeaWiFS products. **(d, e, f)** Idem for GSM product. **(g, h, i)** Idem for AVW product. All the correlations (r) and regression coefficients (m) indicated in each panel are significantly different from zero at the 90 % confidence level.

amongst SChl products. MODIS displays the largest winter standard deviation (up to 0.6 mg m^{-3} in March), and SeaWiFS the weakest one (up to 0.2 mg m^{-3} in March). This indicates that the analysis of interannual SChl fluctuations may heavily depend on the product considered.

Scatter plots of interannual anomalies in monthly SChl from the different products are shown in Fig. 4. Despite varying amplitudes amongst products, there is generally a good phase agreement between the monthly anomalies from the different products, with correlation ranging from 0.55 between AVW and MODIS to 0.98 between OC-CCI and MERIS. Amongst the three merged products, OC-CCI displays the best match with the three individual satellite products, with a correlation of 0.84, 0.98, and 0.91 with MODIS, MERIS, and SeaWiFS, respectively. The amplitude of anomalies from the merged products generally matches that of MERIS (with regression coefficients ranging from 0.87 to 1.44) but are considerably lower than the ones estimated by MODIS (regression coefficients ranging between 0.23 and 0.42). In the following, most results will be illustrated with the OC-CCI product that displays the best

phase agreement with the three individual satellite products (Fig. 4), offers a very good coverage (Table 1), and spans the longest period (1997–2013). It must however be kept in mind that the amplitude of the interannual SChl anomalies remains uncertain, given the large discrepancies amongst products. In the following, we however show that the interannual relationships existing between SChl and ocean physical parameters are generally robust amongst ocean colour products.

3.2 Model evaluation

A brief evaluation of the seasonal cycle of SChl in the model simulation follows. The model accurately captures the large-scale SChl patterns in the AS for both summer (Fig. 5a, c) and winter (Fig. 5b, d). As for observations, the largest SChl bloom occurs in summer along the Oman coast, while the winter bloom is maximum in the northernmost part of the AS (Fig. 5b, d). The modelled seasonal SChl evolution agrees well with the OC-CCI data in terms of amplitude and timing (Fig. 6a, d), with a clear semi-annual cycle characterized by a larger SChl bloom during summer (up to 1.5 mg m^{-3}) than in

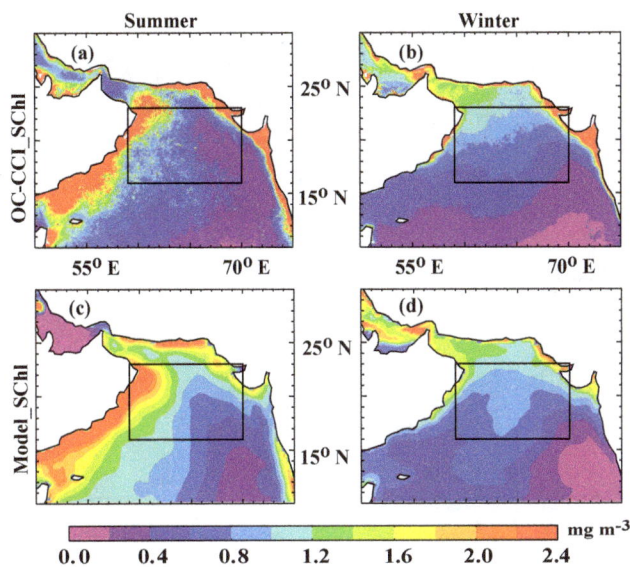

Figure 5. Arabian Sea climatology of **(a, c)** summer (JJAS) and **(b, d)** winter (DJFM, **a**, **b**) OC-CCI and **(c, d)** modelled SChl (mg m^{-3}).

winter (up to $1 \, \text{mg m}^{-3}$) and minimum SChl concentrations (less than $0.5 \, \text{mg m}^{-3}$) during inter-monsoons. Even though the model reasonably simulates the amplitude and timing of the SChl bloom, the summer bloom maximum in the NAS occurs 2 months later in the model (September) than in observations (July). The seasonal timing of the winter bloom in the NAS is however very accurately captured, with a maximum bloom occurring in February in both the model and observations (Fig. 6a, d). This winter bloom occurs in response to convective vertical mixing and related MLD deepening (Fig. 6a, b) driven by the cold, dry northeasterly winds (McCreary and Kundu, 1989; Madhupratab et al., 1996; Prasanna Kumar et al., 2001; Wiggert et al., 2000; Lévy et al., 2007; Kone et al., 2009). The model captures the main features of subsurface physical and biogeochemical variations (Fig. 6b, e): the winter bloom is triggered by the deepening of the mixed layer associated with a strong cooling by surface heat fluxes (Fig. 6c, f), accompanied by a deepening of the thermocline and nitracline. The maximum MLD deepening occurs in January in the model and observations (Fig. 6b, e), 1 month before the SChl peak (Fig. 6a, d). From then on, the upper ocean restratifies, and the MLD shoals in response to increased net heat fluxes into the ocean (Fig. 6c). This MLD shoaling combined with a nitracline that remains deep (Fig. 6b) limits further nitrate supply to the MLD. This analysis briefly illustrates the ability of the model to capture the main biogeochemical features and related mechanisms in the NAS in winter.

This simulation reasonably captures not only the SChl seasonal cycle in the NAS region but also its interannual winter variability. Figure 1 provides a first illustration of the model's ability to capture the amplitude of the contrasted surface blooms during the 2006 and 2007 winters as discussed in the Introduction. In agreement with observations, the simulation displays a winter bloom that extends further south in 2007 than in 2006, resulting in larger mean SChl concentrations over the NAS region in 2007. Figure 7a provides a more thorough validation of the modelled interannual SChl variations in this region. Observed interannual winter SChl anomalies range from $+0.4$ (winter 2011) to $-0.3 \, \text{mg m}^{-3}$ (winter 2012; Fig. 7a). The largest observed winter positive anomalies are found during 2011, 2007, 2002, 2000, and 1999 (Fig. 7a), while the strongest negative anomalies are found in 1997, 2003, 2006, 2008, 2010, and 2012 (Fig. 7a). The modelled interannual NAS winter SChl anomalies agree generally well with those from the OC-CCI dataset in terms of both phase and amplitude (Fig. 7a), with a correlation between the two time series reaching 0.69 over the 2001–2011 period and 0.52 over the 1997–2011 period, both significant at the 99 % confidence level. The main mismatch is found during 1997 and 2002, when the model and observational datasets display opposite anomalies. The observed MLD also exhibits large fluctuations, ranging from $-23 \, \text{m}$ in winter 2006 to around $\sim +14 \, \text{m}$ in winter 2001, 2007, and 2011 (Fig. 7b). The model is also able to capture these observed interannual MLD variations (Fig. 7b), with a 0.65 correlation over the 2001–2011 period, significant at the 95 % confidence level. The main disagreement between the model and observations occurs during the winters of 2002 and 2008, where the observed signals are not well captured by the model. Finally, the observed TCD also exhibits large year-to-year variations in winter, ranging from $-15 \, \text{m}$ in winter 2007 and 2009 to $\sim +15 \, \text{m}$ in winter 2001. In contrast to SChl and MLD, the model does not capture the observed TCD variability well (0.3 correlation), although some major events such as the thermocline shoaling in 2007 and 2009 and the deepening in 2011 are properly simulated. However, the good agreement between the modelled and observed interannual SChl variability in NAS allows us to confidently use the model over a longer period (1993–2012) to further investigate interannual chlorophyll variability and its driving mechanisms.

4 Physical drivers of the interannual SChl variability

In this section, we describe how the main characteristics of the interannual chlorophyll variations in the NAS relate to ocean physical properties (MLD, TCD). The hypotheses of Wiggert et al. (2002) and Prasanna Kumar et al. (2001) for the mechanisms that control interannual SChl variations imply a correlation of SChl anomalies with TCD and MLD anomalies, respectively. In order to test those hypotheses, we compared the time evolution of the OC-CCI SChl, MLD, and TCD anomalies in the NAS box from 2002 to 2013 in Fig. 8. This figure illustrates that observed interannual SChl anoma-

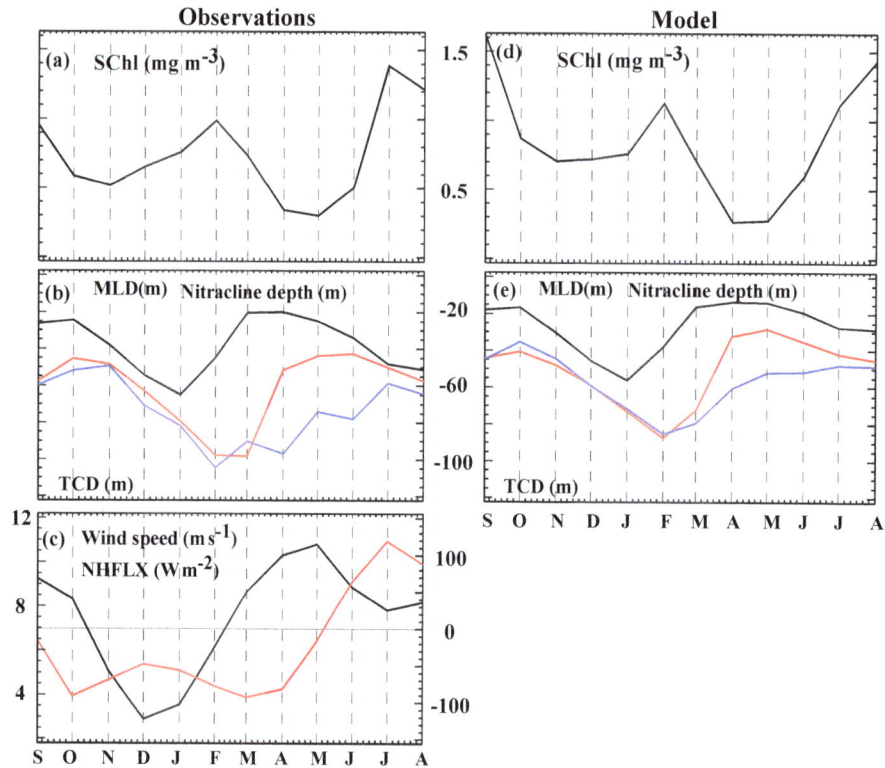

Figure 6. Mean seasonal cycle of NAS-box-averaged monthly **(a)** SChl; **(b)** MLD (black line), TCD (red line), and nitracline depth (blue line); and **(c)** surface net heat flux (NHFLX – black line) and wind speed (red line) in observations. **(d–f)** Same for model. The NAS box is outlined in Fig. 1. Note that the model average is based on the entire NAS box, while observations subsample this box: the good agreement between the two averages however suggests that observational subsampling does not introduce large biases.

Table 2. Correlation between average interannual NAS box winter (DJFM) SChl anomalies derived from satellite products and interannual in situ MLD and TCD anomalies. Bold typeface indicates correlations which are statistically different from zero at the 90 % confidence level.

SChla data	Cor (MLDa)	Cor (TCDa)
SeaWiFS (2003–2010)	0.46	−0.006
MERIS (2003–2012)	**0.69**	0.05
MODIS (2003–2012)	**0.86**	−0.02
OC-CCI (2003–2012)	**0.72**	0.05
GSM (2003–2012)	**0.77**	−0.19
AVW (2003–2012)	0.38	−0.17

lies are closely related to interannual MLD fluctuations, with deeper MLDs generally associated with a positive chlorophyll anomaly, and vice versa. This is verified for most winters, except for 2002 and 2004, where positive chlorophyll anomalies are concomitant with a modest shoaling. In addition, there is a consistent time lag between the MLD and SChl anomalies, with MLD anomalies usually peaking in February and chlorophyll anomalies peaking 1 month later. In contrast, there is no obvious connection between TCD and

SChl anomalies (Fig. 8b): positive SChl anomalies can be associated with either thermocline deepening such as in 2011 or a thermocline shoaling as in 2002 and 2007. Similarly, negative SChl anomalies can be associated with a thermocline deepening as in 2006 or to a shoaling as in 2008 and 2012.

A more quantitative examination of the relationship between interannual SChl anomalies and MLD/TCD anomalies is provided in Fig. 9. As shown in Fig. 9a, c, the winter-averaged SChl and MLD anomalies are strongly correlated in both observations (0.72, statistically significant at the 99 % confidence level; Fig. 9a) and the model (0.59 correlation significant at the 99 % confidence level; Fig. 9c). In contrast, there is no statistically significant correlation at the 90 % confidence level between SChl and TCD variations over the same period for the two datasets (Fig. 9b, d). The dependency of these relationships to the ocean colour product is shown in Table 2. This table indicates that all observational products display larger correlations between SChl and MLD than between SChl and TCD anomalies. The strength of the MLD–SChl relationship however varies depending on the product considered, with the largest correlation for MODIS (0.86) and the weakest for the AVW product (0.38, not significant at the 90 % significance level). None of the SChl prod-

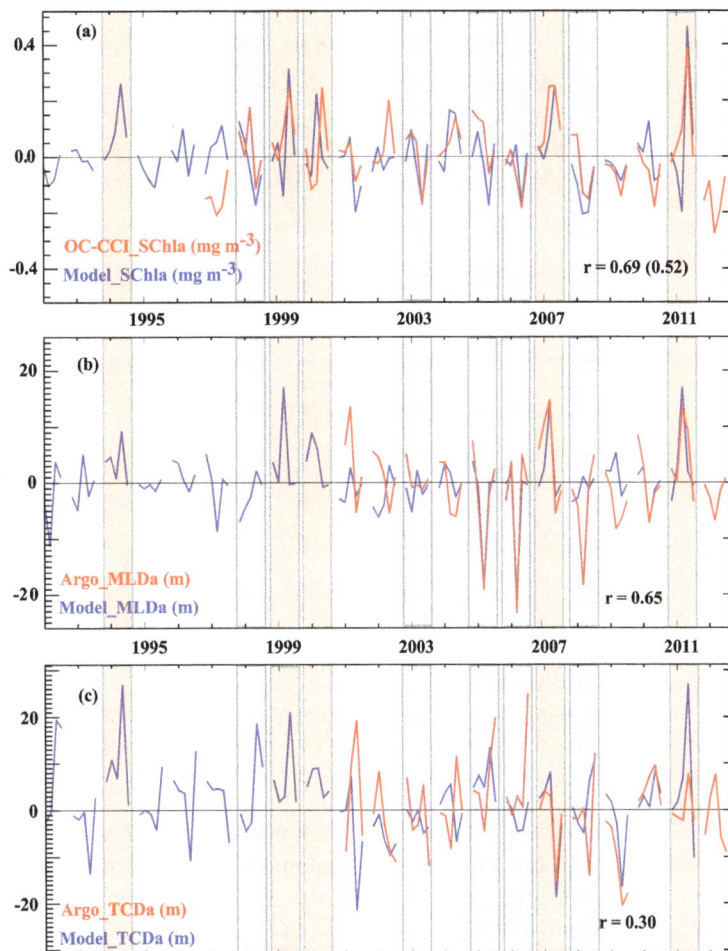

Figure 7. Monthly time series of NAS-box-averaged modelled and observed interannual anomalies in winter for (**a**) SChl, (**b**) MLD, and (**c**) TCD. The correlation between the model and observations from 2002 to 2012 is indicated in (**a**)–(**c**). Red and blue shadings respectively indicate winters for strong and weak blooms in the model, considered for the composite plots of Fig. 10. Note that the model average is based on the entire NAS box, while observations subsample this box: the good agreement between the two averages however suggests that observational subsampling does not introduce large biases.

ucts exhibits a significant relationship with TCD at the 90 % confidence level (Table 2). These results are a strong indication that interannual SChl variations are controlled by MLD rather than by TCD variability.

The spatial distribution of the typical SChl, MLD, and TCD anomalies during an anomalously strong bloom event is shown in Fig. 10 for both observations and the model. This composite pattern is constructed from the half-difference between positive and negative events highlighted in Fig. 7 for the model and Fig. 8 for observations. Observations and the model exhibit very similar spatial patterns: the maximum SChl anomaly signal (exceeding 0.5 mg m^{-3}) occupies the northern part of the box around 21° N, 64° E (Fig. 10a, e), with weaker but still significant SChl signals found everywhere in the NAS box. This SChl pattern matches well with the MLD pattern, with maximum MLD positive anomalies (exceeding 16 m) occurring at the northern boundary of the

NAS box (Fig. 10b, f) and significant positive MLD anomalies everywhere in the NAS box. In contrast, the TCD composite hardly shows any significant anomaly within the NAS box during an anomalously strong bloom (Fig. 10c, g). This composite analysis hence confirms that the relation between interannual SChl and MLD variations (and absence of relation with TCD) deduced from Fig. 8 NAS-averaged values holds over the entire region.

The availability of chlorophyll and nitrate data at depth from the model allows going a step further in the description of the processes driving the chlorophyll variability. Figure 11a–d show the chlorophyll and nitrate evolution between 100 m and the surface, averaged over the NAS box for the two contrasted winters of 2006 and 2007, already discussed in Fig. 1. Both years exhibit a chlorophyll bloom in winter, with maximum chlorophyll concentration in the surface layers in February. The absence of winter deep chloro-

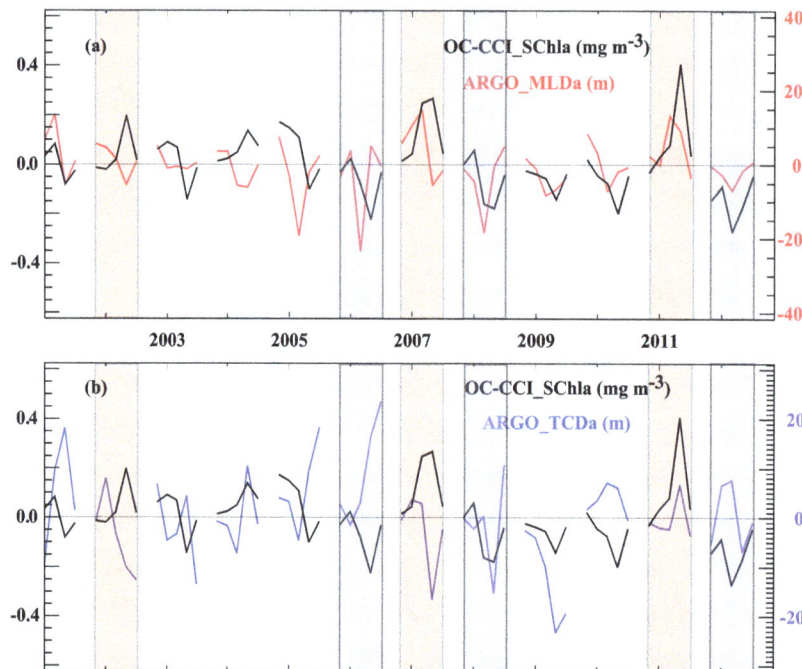

Figure 8. Observed monthly time series of NAS-box-averaged **(a)** SChl and MLD and **(b)** SChl and TCD anomalies in winter over the 2002–2013 period. The red (blue) shadings highlight the winters of strong (weak) blooms used in the composite plots of Fig. 10.

phyll maximum (DCM) precludes the entrainment of chlorophyll from below being responsible for the SChl bloom during this season. This is clearly visible from Fig. 11a and b, which show that the increase in chlorophyll at the surface layer is associated not with a vertical redistribution of chlorophyll but with an increase in the vertically integrated biomass. Let us now investigate what caused this larger phytoplankton concentration during winter 2007. The MLD is deepest in January and reaches $\sim 50\,\mathrm{m}$ during both winters. The similar maximum winter MLDs during the 2 years induce a similar supply of subsurface nutrients (the nitrate concentration $10\,\mathrm{m}$ below the MLD, a proxy for the nitrate content of the water entrained or mixed into the mixed layer, is very similar for both years until January: Fig. 11g, h). This yields a very similar nitrate concentration in January 2007 ($7.23\,\mathrm{mmol\,m^{-3}}$) and January 2008 ($7.29\,\mathrm{mmol\,m^{-3}}$). Consequently, SChl concentrations are very close in January (Fig. 11e, f). The main difference between the two winter blooms is their duration: the 2006–2007 bloom was over in March, while the 2007–2008 bloom still persisted (Fig. 11). This is associated with a MLD that remained deep until February ($\sim 50\,\mathrm{m}$) in 2008, whereas it started shoaling 1 month earlier in 2007. The bottom of the deeper MLD in February 2008 is closer to the nutrient-rich subsurface layer, sustaining a larger nutrient input through turbulent fluxes. As a result, the mixed layer nitrate concentration reaches $\sim 8\,\mathrm{mmol\,m^{-3}}$ in February 2008 against $\sim 4\,\mathrm{mmol\,m^{-3}}$ in 2007 (Fig. 11g, h). In February of both years, those nitrate concentrations are high enough so that phytoplankton growth

is not nutrient-limited, which explains the similar SChl concentrations ($\sim 1.2\,\mathrm{mg\,m^{-3}}$) during that month (Fig. 11e, f). Nitrate becomes limiting in March during both years, yielding no further biomass production after March. It however takes more time for phytoplankton to exhaust the larger February 2008 mixed layer nitrate content, allowing for the 2008 bloom to persist until March. It must finally also be noticed that differences in nitracline depths cannot explain the bloom differences between these two winters: the nitracline is indeed slightly deeper in 2007–2008 than in 2006–2007 (red lines in Fig. 11a–d). Overall, the comparison of these 2 years supports the important role of the February–March MLD variations in setting the near-surface nutrient content and chlorophyll value.

Figure 12 allows exploring if the processes observed for the 2006 and 2007 contrasted winters also operate to explain SChl winter anomalies over the entire period. As the largest winter MLD and SChl variability occurs in February–March (see Figs. 7 and 8), the analysis in Fig. 12 is restricted to the February–March period. Figure 12a shows that the 0–200 m integrated chlorophyll anomalies exhibit an even stronger relationship with MLD fluctuations (0.84 correlation) than with SChl (Fig. 9c; 0.6 correlation), demonstrating that SChl variability does not arise from a vertical redistribution of chlorophyll within the water column but mainly results from phytoplankton growth. In addition, larger interannual MLD anomalies are associated with more nutrients in the mixed layer over the 20-year period analysed (Fig. 12b), with a 0.63 correlation between the two parameters. This can occur ei-

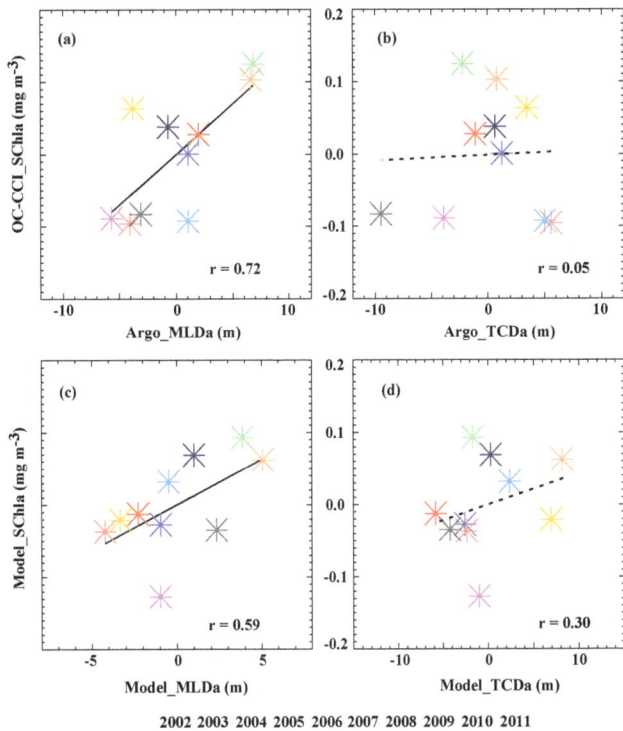

Figure 9. Scatter plot of winter NAS-averaged OC-CCI SChl anomalies against observed **(a)** MLD and **(b)** TCD anomalies over the 2002–2011 period. **(c–d)** Idem for model. Plain (dashed) lines indicate the linear regression that is (not) significantly different from zero at the 90 % confidence level.

h). This is consistent with a MLD deepening controlled by convective overturning, which in turn is controlled by surface heat fluxes. Interannual net heat flux variations in this region are strongly related to 2 m air temperature anomalies (Fig. 12d, correlation of 0.63), indicative of southward advection of anomalously cold/warm air from the continent driving anomalous blooms, as already suggested by Keerthi et al. (2016). A significant relationship also exists between the modelled interannual MLD variability and SST variability in winter (Fig. 12e). A deeper MLD is associated with a cooler SST (-0.69 correlation), which in turn is driven by net heat flux variability (Fig. 12f; 0.66 correlation). This finally results in a large -0.79 negative correlation between the interannual SST and SChl winter anomalies, because they are both initially driven by the same surface heat flux anomalies. Observations exhibit a similar correlation (-0.73), reinforcing the above conclusions from the model results.

5 Summary and discussion

5.1 Summary

The AS is one of the most productive regions in the world ocean, with a strong monsoon-driven seasonal cycle in SChl. The largest SChl bloom occurs during the summer monsoon in the western AS, in response to coastal and offshore upwelling driven by the Findlater jet. There is however also a prominent SChl bloom in winter in the northern AS, which exhibits large year-to-year fluctuations in its extent and intensity. These variations have not yet been described in detail, and there is no consensus on their driving mechanism. In this paper, we described the interannual NAS winter bloom variability and the mechanism driving this variability. To reach that goal, we combined the analysis of several observational datasets (remotely sensed chlorophyll products from various satellites and physical oceanic parameters derived from Argo in situ profiles) and a biogeochemical model simulation.

Our results reveal that SChl anomalies from the various satellite products exhibit a good phase agreement but large amplitude discrepancies. There is a strong ($\sim \pm 50\%$ of the climatological value) year-to-year variability of the NAS winter SChl bloom. These fluctuations of the bloom amplitude are much better correlated ($r \sim 0.4$ to 0.9 depending on the satellite product) with interannual fluctuations in MLD than with interannual fluctuations in TCD ($r \sim -0.2$ to 0.1). As a result, correlations with interannual MLD anomalies are significant at the 90 % confidence level in four out of six chlorophyll satellite products but are insignificant for TCD anomalies irrespective of the product.

The above analysis is based on a limited number of years in observations, due to the in situ data temporal coverage, which only becomes sufficient after 2002 thanks to Argo profilers. Using a biogeochemical model allows us to extend our analyses over a longer period (1993–2012) and to analyse

ther through a modulation of the maximum MLD and hence of the amount of nutrients entrained into the mixed layer or through the period when the MLD is deep (as for 2006–2007) and hence through turbulent fluxes of nutrients into the MLD. As the interannual nitrate anomalies averaged over the mixed layer are correlated with both the interannual MLD anomalies (Fig. 12b; 0.63 correlation) and the maximum absolute MLD (not shown; 0.60 correlation), it is not possible to discriminate between the two processes. In any case, the above results suggest that interannual winter chlorophyll variations largely result from phytoplankton growth through nutrient input to the MLD through turbulent processes. Although these results are consistent with the Prasanna Kumar et al. (2001) hypothesis, the hypothesis of Marra and Barber (2005) could also be valid, with the MLD variations controlling the bloom amplitude through the modulation of the grazing pressure.

Figure 12 also allows discussing the processes driving the interannual MLD variability in winter. There is a significant relationship between the anomalies of modelled interannual MLD and of net surface heat fluxes (Fig. 12c, -0.83 correlation) during winter. The typical spatial pattern of anomalous net heat flux displays a broad heat flux cooling over the entire northern AS, with maximum anomalies located at the northern end of the AS for both model and observation (Fig. 10d,

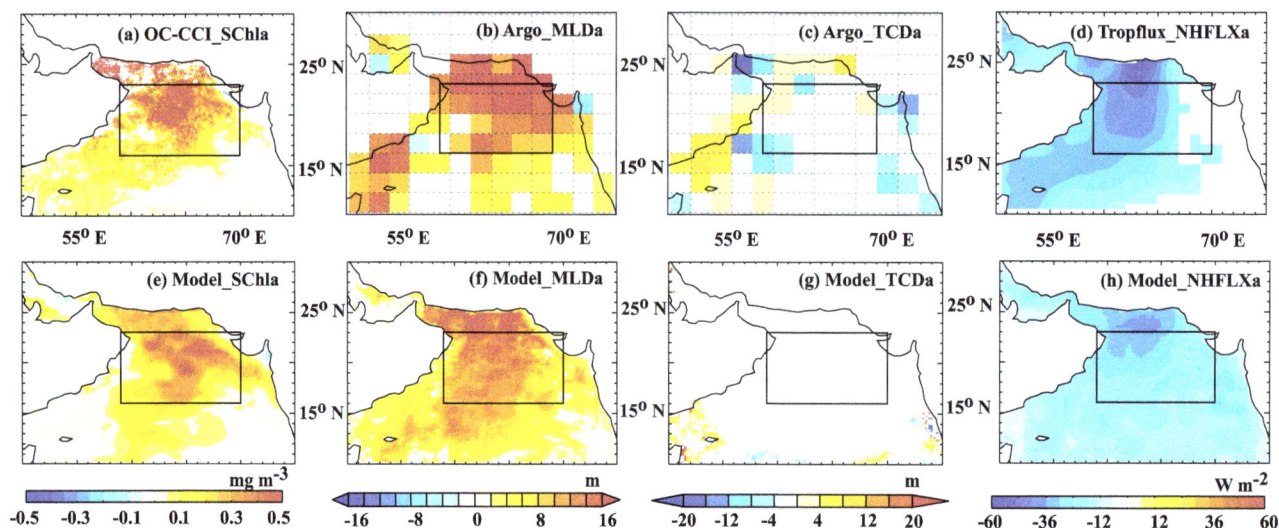

Figure 10. Observed interannual anomalies of **(a)** SChl (OC-CCI), **(b)** MLD, **(c)** TCD (Argo-derived), and **(d)** net surface heat flux (TropFlux) for composite SChl blooms (built from half of the difference between positive and negative events highlighted in Fig. 8). The SChl composites are built from the months of max SChl anomaly: March 2003, March 2008, and March 2012 for positive events; March 2007, March 2009, and February 2013 for negative events. The MLD and TCD composites are built from the months of February of the same year. **(e–h)** Idem for the model. For the model, composites are built from the positive and negative events highlighted in Fig. 7 (SChl is composited using March 1995, 2000, 2008, and 2012 and February 2001 for positive events and March 1999, 2004, 2006, 2007, and 2009 for negative events; MLD and TCD are composited using February of the same years). Regions where composite values are less than the standard error are displayed in white.

subsurface chlorophyll data (which are not available from observations). The model agrees well with observations in terms of interannual winter anomalies in both MLD and SChl averaged over the NAS (typically $r \sim 0.7$). As in observations, we find no relationship between the winter NAS SChl anomalies and the TCD anomalies in this simulation, contrary to what would be expected in the Wiggert et al. (2002) mechanism. Rather, we find a strong relationship ($r \sim 0.6$) between MLD and SChl anomalies, as in the observations ($r \sim 0.7$). The analysis of the model vertical structure indicates that the increase in SChl is not the result of the upward mixing of a pre-existing subsurface chlorophyll maximum. Rather, enhanced surface heat losses due to the advection of cold air by northerly winds result in a more convective overturning, an anomalously deep seasonal MLD, and more turbulent fluxes of nutrients into the MLD. This promotes new production in the surface layer. Our study therefore demonstrates that the mechanisms controlling chlorophyll variations at seasonal timescales (Prasanna Kumar et al., 2001; Lévy et al., 2007, Koné et al., 2009) also operate at the interannual timescale. Despite this convincing evidence on the dominant role played by MLD variations in driving the year-to-year fluctuations of the winter NAS bloom, other oceanic processes such as the Ekman pumping or offshore advection could also play some role. Performing composite analyses similar to those displayed in Fig. 10 but for wind stress curl and surface current anomalies does not reveal any significant relationship between these variables and interannual winter

chlorophyll variations (not shown), suggesting that Ekman pumping or the advection of chlorophyll and/or nutrients is unlikely to play a strong role on the interannual fluctuations of the winter bloom in the NAS.

5.2 Discussion

The present study hence brings new insights on the interannual variability of the NAS winter bloom. As described in the Introduction, there have indeed been to date only two studies addressing the physical mechanisms controlling this interannual variability (Prasanna Kumar et al., 2001; Wiggert et al., 2002), which proposed different mechanisms relying on the analysis of a very limited number of winters. Our study allows demonstrating which mechanism dominates based on the analysis of much longer and various datasets (12 winters in observations and 20 winters in the model). Our observational and modelling results are both inconsistent with the hypothesis proposed by Wiggert et al. (2002), i.e. that the TCD in winter controls the bloom amplitude through a daily dilution effect. In contrast, it is consistent with the hypothesis that interannual MLD variations largely control the amplitude of the bloom (Prasanna Kumar et al., 2001; Marra and Barber, 2005).

On the observational front, our results rely on a comparison of interannual variations in satellite-derived SChl with in-situ-derived MLD and TCD variations. Those datasets are subject to uncertainties arising from both measurements ac-

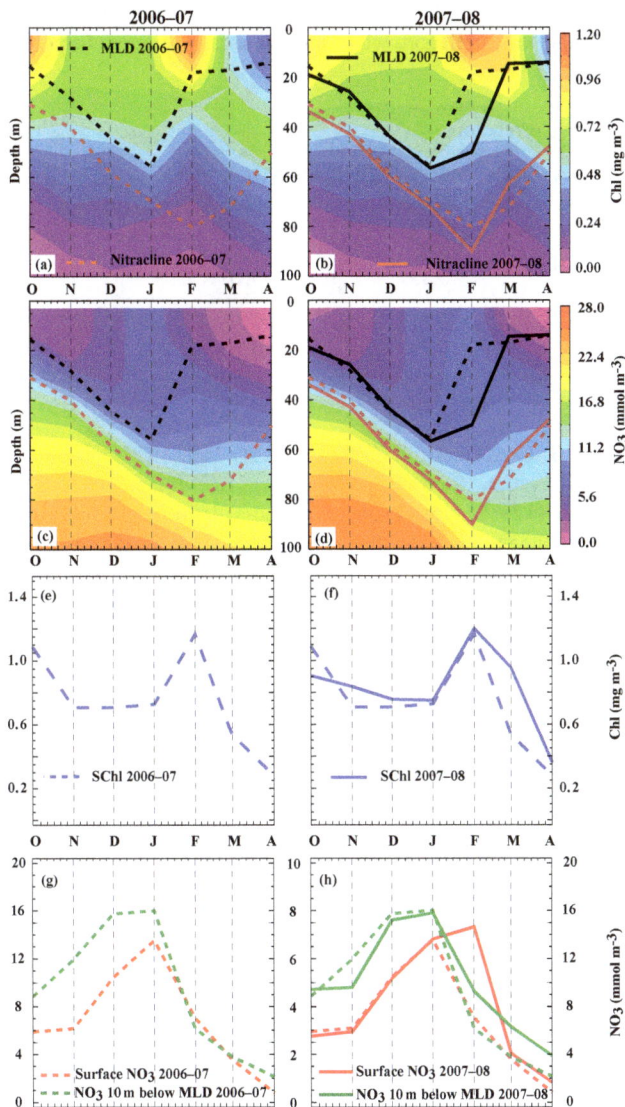

Figure 11. Depth–time section of NAS-averaged **(a, b)** chlorophyll (Chl) and **(c, d)** nitrate (NO_3) for **(a, c)** 2006–2007 and **(b, d)** 2007–2008. The black lines indicate the MLD, thick for 2007–2008 and dashed for 2006–2007. The red lines similarly indicate the nitracline depth. Time series of NAS-averaged **(e, f)** SChl (blue curve) and **(g, h)** surface nitrate (red curve) and nitrate concentration 10 m below the bottom of the MLD (green curve) for **(e, g)** 2006–2007 and **(f, h)** 2007–2008. In **(f, h)**, the 2006–2007 values have been reported as dashed curves to ease comparisons between the 2 years.

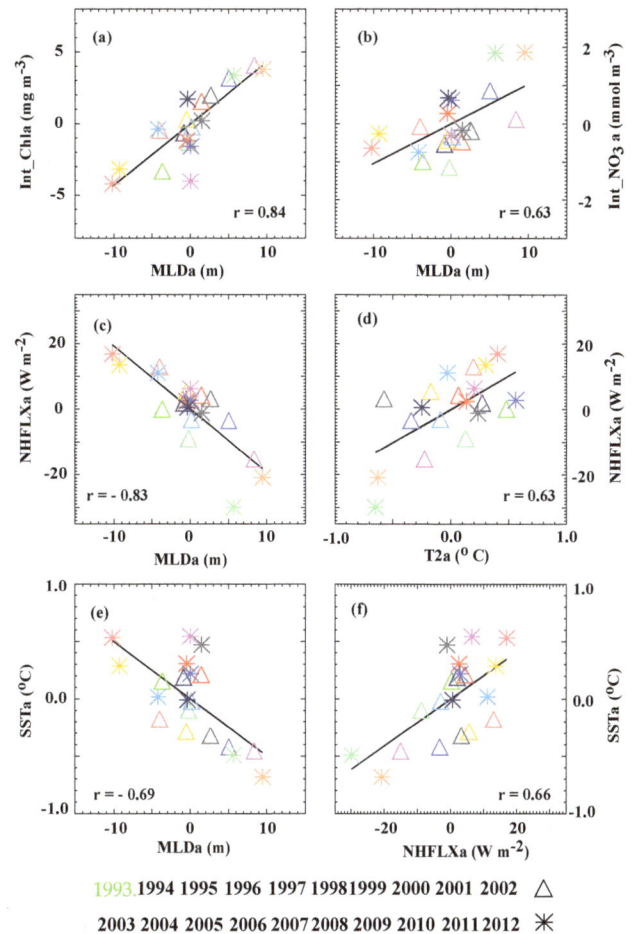

Figure 12. Scatter plot of modelled interannual NAS-box-averaged winter (February–March) anomalies of **(a)** 0–200 m total chlorophyll content against MLD. **(b)** Average MLD nitrate vs. MLD. **(c)** Surface net heat flux vs. MLD. **(d)** Surface net heat flux vs. 2 m air temperature. **(e)** SST vs. MLD. **(f)** SST vs. surface net heat flux. Plain lines indicate regression coefficients that are significantly different from zero at the 90 % confidence level.

curacy and to sampling issues (i.e. the data density used to derive interannual NAS-averaged anomalies). The interannual satellite-derived SChl variations are likely to be very robust as the different satellite products in the NAS exhibit a very good phase agreement (Fig. 4) and a very good data coverage during the winter season (Table 1). Regarding the in-situ-derived MLD and TCD products, the large number of individual measurements used to build seasonal anomalies (i.e. from 100 to 500 depending on the years consid-

ered) yields a good accuracy on the estimate of these seasonal anomalies (~ 2 m uncertainty on MLD and ~ 3 m on TCD estimated from a Monte Carlo approach by subsampling available data, which is relatively small compared to the ~ 20 m peak-to-peak amplitude of observed interannual variations of those fields). One of the major limitation of this in situ dataset may hence only be the inhomogeneous spatial sampling in the region considered, with a higher data density along a shipping line crossing the NAS box. The very good agreement between the in situ data and the totally independent model (for which averages are obtained over the entire NAS region, Fig. 7) however suggests that this is not the case. The fact that similar conclusions can be drawn from the model and the independent in situ dataset also strengthens the trust in each of those datasets.

Table 3. Correlation between interannual SChl anomalies and interannual MLD and TCD anomalies averaged over the NAS box for December–March, December–January, and February–March. Bold typeface indicates correlations which are statistically different from zero at the 90 % confidence level.

		Correlation		
		December–March (DJFM)	December–January (DJ)	February–March (FM)
OC-CCI	MLDa vs. SChla	0.72	−0.05	0.80
	TCDa vs. SChla	0.05	0.43	−0.10
Model	MLDa vs. SChla	0.59	0.21	0.51
	TCDa vs. SChla	0.3	0.37	0.19

The diurnal cycle plays a key role in the hypothesis proposed by Wiggert et al. (2002). Although the spatio-temporal coverage of our in situ dataset is sufficient to accurately sample interannual MLD variations, it does not allow monitoring the diurnal MLD variability. A proper investigation of the impact of diurnal variability on interannual chlorophyll variations from observations would hence require continuous and long-term temperature and chlorophyll profiles from a fixed location, which are not available to date. If the Wiggert et al. (2002) mechanism was dominating, there should however be a negative correlation between interannual variations of chlorophyll and TCD (i.e. a deeper thermocline leading to lower chlorophyll concentrations operating through daily dilution). Table 2 and Figs. 8, 9, and 10 clearly demonstrate that it is not the case. In addition, despite the absence of a diurnal cycle in the model forcing (i.e. by construction, the model cannot reproduce the Wiggert et al. (2002) mechanism), the model displays a good agreement with observed interannual chlorophyll variability in winter in the NAS (see Figs. 7a and 10), which is indirect evidence that the bulk of interannual chlorophyll variations are not linked to a modulation of night-time penetration of diurnal mixing by the TCD.

Wiggert et al. (2005) further argued that the inconsistency between Prasanna Kumar et al. (2001) and Wiggert et al. (2002) may be due to the different seasonal window considered in the two studies: December–January for Wiggert et al. (2002) and February for Prasanna Kumar et al. (2001). Wiggert et al. (2005) argued that during these two periods two distinct processes drive the phytoplankton growth. On one hand, the transition from the winter monsoon to the spring inter-monsoon is characterized by detrainment blooms stimulated by increased irradiance received by phytoplankton due to mixed layer shoaling that follows the relaxation of monsoon winds. On the other hand, the beginning of the northeast monsoon is characterized by entrainment blooms that are stimulated by an increase in nutrients resulting from a deepening of the mixed layer in that period. To revisit this argument, we repeated our analysis over these two periods (Table 3). Our analysis indicates that interannual SChl anomalies during the beginning of the winter monsoon period do not exhibit any significant relationship either with

Table 4. Correlation of IOD and ENSO index with interannual anomalies of NAS-box-averaged winter (DJFM) 2 m surface temperature (T2a) and interannual SChl anomalies derived from different satellite products and the model. Bold typeface indicates correlations which are statistically different from zero at the 90 % confidence level.

	Correlation	
	IOD (SON)	ENSO (NDJ)
T2a (1993–2011)	−0.06	0.27
T2a (2002–2011)	0.05	0.77
SChl_OC-CCI (2002–2011)	0.07	−0.34
SChl_MODIS (2002–2011)	0.26	−0.39
SChl_MERIS (2002–2011)	0.04	−0.37
SChl_GSM (2002–2011)	0.24	−0.44
SChl_AVW (2002–2011)	0.05	−0.11
SChl_Model (2002–2011)	0.02	−0.30

MLD or with TCD anomalies (correlation below 0.30 not significant at the 90 % significance level); i.e. neither the Wiggert et al. (2002) mechanism nor the Prasanna Kumar et al. (2001) mechanism is at work during the bloom initiation period. However, a significant relationship exists between SChl anomalies and MLD anomalies during post-winter period (February–March) with a correlation of 0.80 in the observations and 0.51 for models (Table 3), suggesting that interannual MLD variations control the amplitude of the bloom during the period of peak interannual variability in winter (February–March).

Even though interannual MLD and SChl variations covary for most winters, the winters of 2002 and 2004 behave inconsistently relative to other years in both the model and observations (Figs. 7a, b and 8a), suggesting that another mechanism could be at work during these years. Banerjee and Prasanna Kumar (2014) demonstrated that episodic dust storms could contribute to the interannual variability of the winter bloom in the central AS, away from the region of active winter convection. However, no episodic dust storms were reported during winters of 2002 and 2004 (Banerjee and Prasanna Kumar, 2014), indicating that iron inputs from dust storms are not responsible for the peculiar behaviour observed during these two winters. In any case, our simulation captures most of the observed interannual variability of the chlorophyll bloom in winter (Fig. 7a) despite the use of a climatological iron aerial deposition forcing (i.e. interannual iron deposition variations are not accounted for). This suggests that interannual dust storm variations do not play a dominant role in driving the interannual variability of the bloom in the northern AS. The apparent contradiction between our results and those of Banerjee and Prasanna Kumar (2014) may arise from the different regional focus: the northern AS, where winter convection occurs for our study, and the central AS, where no convective overturning occurs for Banerjee and Prasanna Kumar (2014). The different vertical physics in the two regions may imply a different role of micronutrients.

Our results point toward a strong control of the MLD variability in the NAS through anomalous heat flux perturbations. Although we generally refer to interannual MLD variability in this study, the largest MLD fluctuations in NAS are observed over a single month (see Fig. 7b). These variations hence rather occur at intraseasonal timescales but translate into interannual anomalies when averaging over the entire winter season. Keerthi et al. (2016) already provided a detailed description of these intraseasonal MLD fluctuations in this region in winter, relating them to the advection of continental temperature anomalies from the northern end of the basin. The climate variability behind these heat flux fluctuations is however currently unknown. Results from Wiggert et al. (2009) point to a contrasted biological signature of the western AS during the 1997–1998 and 2006–2007 El Niño events, with an overall decrease of productivity during the former and a slight increase during the latter. The extended analysis of a 1961–2001 model hindcast (Currie et al., 2013) indicates that El Niño generally results in an anomalously low winter and autumn SChl over the AS (as seen in 1997–1998), and a negligible impact of the IOD. In line with Currie et al. (2013), we find that interannual variations of the winter 2 m air temperature in the NAS box over the 2002–2011 period are strongly correlated with ENSO (correlation of 0.77) and weakly correlated with IOD (correlation of 0.05), suggesting that the surface forcing that drives interannual NAS MLD variability may be ENSO-driven (Table 4). Regarding the impact on SChl, all observed and modelled chlorophyll products do exhibit a modest negative correlation between ENSO and interannual fluctuations of the winter bloom over the 2002–2011 period (Table 4). This is consistent with the hypothesis that an El Niño event drives a weaker winter monsoon, warmer surface air, less convective overturning, and hence a weaker bloom. The level of correlation is however modest and not statistically significant in all datasets (ranging from −0.11 to −0.44, depending on the SChl product), indicating that ENSO is not the only driver of the interannual winter bloom variations in this region. In addition, this influence of ENSO is not very stable in time (Table 4), with

a larger correlation with 2 m air temperature over the recent 2002–2011 period (0.77) than over the extended 1993–2011 period (0.27). Further detailed analyses are required to ascertain and better understand the influence of ENSO on the AS winter monsoon. Finally, while interannual SChl variations are rather consistent amongst products, linear SChl trend are not consistent amongst products, some of them showing a decreasing trend and some others showing an increasing trend. These discrepancies hence prevented us performing a robust assessment of these trends in the present study. We however performed the analyses in the paper using detrended data and found that our results regarding the interannual variability are robust.

An obvious perspective of this study is to investigate the processes controlling the interannual chlorophyll variations in summer, which are far larger than in winter. As compared to the winter season, an analysis based on observations during the summer monsoon is complicated by the poorer satellite data coverage (Table 1) and the fewer Argo profiles in the Oman–Somalia upwelling region (Fig. 2b). The model analysis may however provide further insights on the mechanisms that control the upwelling productivity during summer. Long-term variations also deserve further analysis. By analysing a 7-year-long satellite dataset, Goes et al. (2005) suggested that the southwest monsoon intensifies as a result of climate change, driving increased upwelling, primary production, and ecosystem changes in the AS. However, the shortness and discontinuity of the data call into question the reliability of these results (Beaulieu et al., 2013). Recent studies based on longer datasets (Roxy et al., 2015, 2016) rather point towards a reduction of the summer monsoon winds and an AS summer bloom reduction due to enhanced upper-ocean stratification in response to climate change. Climate change projections from coupled experiments however exhibit a large range of responses in terms of changes in the southwest monsoon (e.g. Turner and Annamalai, 2012) and of productivity in the AS (e.g. Bopp et al., 2013). Those large uncertainties call for more targeted studies of the impact of climate change on oceanic productivity in the AS.

Competing interests. The authors declare that they have no conflict of interest.

Acknowledgements. This study was supported by the Centre National d'Etudes Spatiales (CNES) CLIMCOLOR project. M. G. Keerthi is funded by the IFCPAR (Indo-French Centre for Promotion of Advanced Research) proposal 4907-1. Model experiments were performed at HPC Pravah at CSIR-NIO. Matthieu Lengaigne, Christian Ethé, Jerome Vialard, Marina Levy, Clément de Boyer Montegut, and Olivier Aumont are funded by Institut de Recherche pour le Développement (IRD). V. Parvathi is funded by CSIR under SRF. I. Suresh and V. P. Akhil acknowledge the financial support from CSIR, New Delhi. We thank Goddard GSFC/NASA for providing the SeaWiFs, MERIS, and MODIS chlorophyll data; ESA-GlobColour for providing the GSM and AVW chlorophyll data; and ESA Ocean Colour for providing the OC-CCI chlorophyll data. The authors would like to thank Mathew Koll Roxy and one anonymous reviewer as well as the editor and the *Biogeosciences* team for helpful comments that led to a significant improvement of the manuscript. This has NIO contribution number 6068.

Edited by: Marilaure Grégoire

References

Akhil, V. P., Durand, F., Lengaigne, M., Vialard, J., Keerthi, M. G., Gopalakrishna, V. V., Deltel, C., Papa, F., and de Boyer Montégut, C.: A modelling study of the processes of surface salinity seasonal cycle in the Bay of Bengal, J. Geophys. Res., 116, 3926–3947, https://doi.org/10.1002/2013JC009632, 2014.

Akhil, V. P., Lengaigne, M., Vialard, J., Durand, F., Keerthi, M. G., Chaitanya, A. V. S., Papa, F., Gopalakrishna, V. V., and de Boyer Montégut, C.: A modeling study of processes controlling the Bay of Bengal sea surface salinity interannual variability, J. Geophys. Res.-Oceans, 121, 8471–8495, https://doi.org/10.1002/2016JC011662, 2016.

Aumont, O. and Bopp, L.: Globalizing results from ocean in-situ iron fertilization studies, Global Biogeochem. Cy., 20, GB2017, https://doi.org/10.1029/2005GB002591, 2006.

Aumont, O., Ethé, C., Tagliabue, A., Bopp, L., and Gehlen, M.: PISCES-v2: an ocean biogeochemical model for carbon and ecosystem studies, Geosci. Model Dev., 8, 2465–2513, https://doi.org/10.5194/gmd-8-2465-2015, 2015.

Aumont, O., Bopp, L., and Schulz, M.: "What does temporal variability in aeolian dust deposition contribute to sea-surface iron and chlorophyll distributions?", Geophys. Res. Lett., 35, L07607, https://doi.org/10.1029/2007GL031131, 2008.

Banerjee, P. and Prasanna Kumar, S.: ENSO modulation of interannual variability of dust aerosols over the northwest Indian Ocean, J. Climate, 29, 1287–1303, 2014.

Banse, K. and English, D. C.: Revision of satellite-based phytoplankton pigment data from the Arabian Sea during the northeast monsoon, Mar. Res. Pakistan, 2, 83–103, 1993.

Banse, K. and English, D. C.: Geographical differences in seasonality of CZCS-derived phytoplankton pigment in the Arabian Sea for 1978–1986, Deep-Sea Res. Pt. II, 47, 1623–1677, 2000.

Banse, K. and McClain, C. R.: Winter blooms of phytoplankton in the Arabian Sea as observed by Coastal Zone Color Scanner, Mar. Ecol.-Prog. Ser., 34, 201–211, 1986.

Banzon, V. F., Evans, R. E., Gordon, H. R., and Chomko, R. M.: SeaWiFS observations of the Arabian Sea southwest monsoon bloom for the year 2000, Deep-Sea Res. Pt. II, 51, 189–208, 2004.

Barber, R. T., Marra, J., Bidigare, R. C., Codispoti, L. A., Halpern, D., Johnson, Z., Latasa, M., Goericke, R., and Smith, S. L.: Primary productivity and its regulation in the Arabian Sea during 1995, Deep-Sea Res. Pt. II, 48, 1127–1172, 2001.

Barnier, B., Madec, G., Penduff, T., Molines, J., Treguier, A., Sommer, J. L., Beckmann, A., Biastoch, A., Boning, C., Dengg, J.,Derval, C., Durand, E., Gulev, S., Remy, E., Talandier, C., Theetten, S., Maltrud, M., McClean, J., and Cuevas, B. D.: Impact of partial steps and momentum advection schemes in a

global ocean circulation model at eddy permitting resolution, Ocean Dynam., 56, 543–567, https://doi.org/10.1007/s10236-006-0082-1, 2006.

Bauer, S., Hitchcock, G. L., and Olson, D. B.: Influence of monsoonally-forced Ekman dynamics upon the surface-layer depth and plankton biomass distribution in the Arabian Sea, Deep-Sea Res., 38, 531–553, 1991.

Beaulieu, C., Henson, S. A., Sarmiento, J. L., Dunne, J. P., Doney, S. C., Rykaczewski, R. R., and Bopp, L.: Factors challenging our ability to detect long-term trends in ocean chlorophyll, Biogeosciences, 10, 2711–2724, https://doi.org/10.5194/bg-10-2711-2013, 2013.

Bopp, L., Resplandy, L., Orr, J. C., Doney, S. C., Dunne, J. P., Gehlen, M., Halloran, P., Heinze, C., Ilyina, T., Séférian, R., Tjiputra, J., and Vichi, M.: Multiple stressors of ocean ecosystems in the 21st century: projections with CMIP5 models, Biogeosciences, 10, 6225–6245, https://doi.org/10.5194/bg-10-6225-2013, 2013.

Boyer, T. P., Antonov, J. I., Baranova, O. K., Coleman, C., Garcia, H. E., Grodsky, A., Johnson, D. R., Locarnini, R. A., Mishonov, A. V., O'Brien, T. D., Paver, C. R., Reagan, J. R., Seidov, D., Smolyar, I. V., and Zweng, M. M.: World Ocean Database 2013, NOAA Atlas NESDIS 72, edited by: Levitus, S. and Mishonov, A., Silver Spring, MD, 209 pp., 2013.

Brodeau, L., Barnier, B., Treguier, A. M., Penduff, T., and Gulev, S.: An ERA 40-based atmospheric forcing for global ocean circulation models, Sci. Direct, 31, 88–104, https://doi.org/10.1016/j.ocemod.2009.10.005, 2010.

Currie, J. C., Lengaigne, M., Vialard, J., Kaplan, D. M., Aumont, O., Naqvi, S. W. A., and Maury, O.: Indian Ocean Dipole and El Niño/Southern Oscillation impacts on regional chlorophyll anomalies in the Indian Ocean, Biogeosciences, 10, 6677–6698, https://doi.org/10.5194/bg-10-6677-2013, 2013.

Dickey, T., Marra, J., Sigurdson, D. E., Weller, R. A., Kinkade, C. S., Zedler, S. E., Wiggert, J. D., and Langdon, C.: Seasonal variability of biooptical and physical properties in the Arabian Sea: October 1994–October 1995, Deep-Sea Res. Pt. II, 45, 2001–2025, 1998.

Dussin, R., Treguier A.-M., Molines, J. M., Barnier, B., Penduff, T., Brodeau, L., and Madec, G.: Definition of the interannual experiment ORCA025-B83, 1958–2007, LPO Report 902, 2009.

Findlater, J.: A major low-level air current near the Indian Ocean during the northern summer, Q. J. Roy. Meteor. Soc., 95, 362–380, 1969.

Geider, R. J., MacIntyre, H. L., and Kana, T. M.: A dynamic regulatory model of phytoplanktonic acclimation to light, nutrients, and temperature, Limnol. Oceanogr., 43, 679–694, 1998.

Goes, J. I., Thoppil, P. G., Gomes, H. D., and Fasullo, J. T.: Warming of the Eurasian landmass is making the Arabian Sea more productive, Science, 308, 545–547, 2005.

Grant, M., Jackson, T., Chuprin, A., Sathyendranath, S., Zühlke, M., Storm, T., Boettcher, M., and Fomferra, N.: Ocean Colour Climate Change Initiative (OC_CCI) – Phase Two, Product User Guide, Plymouth Marine Laboratory, available at: http://www.esa-oceancolour-cci.org/?q=webfm_send/496, last access: 27 May 2015.

Gundersen, J. S., Gardner, W. D., Richardson, M. J., and Walsh, I. D.: Effects of monsoons on the seasonal and spatial distributions of POC and chlorophyll in the Arabian Sea, Deep-Sea Res. Pt. II, 45, 2103–2132, 1998.

Kamykowski, D. and Zentara, S. J.: Hypoxia in the world ocean as recorded in the historical data set, Deep-Sea Res. Pt. I, 37, 1861–1874, 1990.

Keerthi, M. G., Lengaigne, M., Vialard, J., de Boyer Montegut, C., and Muraleedharan, P. M.: Interannual variability of the Tropical Indian Ocean mixed layer depth, Clim. Dynam., 40, 743–759, 2013.

Keerthi M.G., Lengaigne, M., Drushka, K., Vialard, J., de Boyer Montégut, C., Pous, S., Levy, M., and Muraleedharan, P. M.: Intraseasonal variability of mixed layer depth in the tropical Indian, Clim. Dynam., 46, 2633–2655, 2016.

Koné, V., Aumont, O., Lévy, M., and Resplandy, L.: Physical and Bio-geochemical controls of the Phytoplankton Seasonal Cycle in the Indian Ocean: a modeling study, edited by: Wiggert J. D., Hood, R. R., Naqvi, S. W. A., Brink, K. H., and Smith, S. L., American Geophysical Union, Washington DC, USA, 185, 147–166, 2009.

Large, W. G. and Yeager, S. G.: Diurnal to decadal global forcing for ocean and sea-ice models: the data sets and flux climatologies, NCAR/TN-460 STR, 111 pp., 2004.

Lengaigne, M., Menkes, C., Aumont, O., Gorgues, T., Bopp, L., Andre, J. M., and Madec, G.: Influence of the oceanic biology on the tropical Pacific climate in a coupled general circulation model, Clim. Dynam., 28, 503–516, 2007.

Levy, M., Shankar, D., Andre, J. M., Shenoi S. S. C., Durand, F., de Boyer, C., and Montegut, C.: Basinwide seasonal evolution of the Indian Ocean's phytoplankton blooms, J. Geophys. Res., 112, C12014, https://doi.org/10.1029/2007JC004090, 2007.

Madec, G.: NEMO, the ocean engine, Note du Pole de modelisation, Institut Pierre-Simon Laplace (IPSL), France, No 27 ISSN No 1288–1619, available at: http://www.nemo-ocean.eu/About-www.nemo-ocean.eu/About-NEMO/Reference-manuals (last access: 10 January 2016), 2008.

Madhupratap, M., Kumar, S. P., Bhattathiri, P. M. A., Kumar, M. D., Raghukumar, S., Nair, K. K. C., and Ramaiah, N.: Mechanism of the biological response to winter cooling in the northeastern Arabian Sea, Nature, 384, 549–552, 1996.

Mahowald, N., Baker, A., Bergametti, G., Brooks, N., Duce, R.,Jickells, T., Kubilay, N., Prospero, J., and Tegen, I.: The atmospheric global dust cycle and iron inputs to the ocean, Global Biogeochem. Cy., 19, GB4025, https://doi.org/10.1029/2004GB002402, 2005.

Maritorena, S. and Siegel, D. A.: Consistent Merging of Satellite Ocean Colour Data Sets Using a Bio-Optical Model, Remote Sens. Environ., 94, 429–440, 2005.

Marra, J. and Barber, D.: Primary production in the Arabian Sea: A synthesis of JGOFS data, Prog. Oceanogr., 65, 159–175, 2005.

McCreary, J. P., Murtugudde, R., Vialard, J., Vinayachandran, P. N., Wiggert, J. D., Hood, R. R., Shankar, D., and Shetye, S.: Biophysical processes in the Indian Ocean, Indian Ocean Biogeochemical Processes and Ecological Variability, 9–32, 2009.

McCreary, J. P. and Kundu, P. K.: A numerical investigation of sea surface temperature variability in the Arabian Sea, J. Geophys. Res., 94, 16097–16114, 1989.

Moore, J. K., Doney, S. C., and Lindsay, K.: Upper ocean ecosys- tem dynamics and iron cycling in a global 3-D model, Global Biogeochem. Cy., 18, GB4028, https://doi.org/10.1029/2004GB002220, 2004.

Nisha, K., Lengaigne, M., Gopalakrishna, V. V., Vialard, J., Pous, S., Peter, A-C., Durand, F., and Naik, S.: Processes of summer intraseasonal sea surface temperature variability along the coasts of India, Ocean Dynam., 63, 329–346, 2013.

O'Reilly, J. E., Maritorena, S., Siegel, D. A., O'Brien, M. C., Tool, D., Mitchell, B. G., Karhu, M., Chavez, F. P., Strutton, P., Cota, G., Hooker, S. B., McClain, C. R., Carder, K. L., Muller-Karger, F., Harding, L., Magnusson, A., Phinney, D., Moore, G. F., Aiken, J., Arrigo, K. R., Letelier, R., and Culver, M.: Ocean color chlorophyll-a algorithms for SeaWiFS, OC2 and OC4: Version 4, edited by: Hooker, S. B. and Firestone, E. R., SeaWiFS postlaunch technical report series 11 SeaWiFS postlaunch calibration and validation analyses: Part 3, Greenbelt, MD: NASA Goddard Space Flight Center, 9–23, 2000.

Prasanna Kumar, S., Madhupratap, M., Dileep Kumar, M., Gauns, M., Muraleedharan, P. M., Sarma, V. V. S., and De Souza, S. N.: Physical control of primary productivity on a seasonal scale in central and eastern Arabian Sea, J. Earth Syst. Sci., 109, 433–441, https://doi.org/10.1007/BF02708331, 2000.

Prasanna Kumar, S., Ramaiah, N., Gauns, M., Sarma, V., Muraleedharan, P. M., Raghukumar, S., Kumar, M. D., and Madhupratap, M.: Physical forcing of biological productivity in the northern Arabian Sea during the Northeast Monsoon, Deep-Sea Res. Pt. II, 48, 1115–1126, 2001.

Praveen Kumar, B., Vialard, J., Lengaigne, M., Murty, V. S. N., and McPhaden, M. J.: TropFlux: Air-Sea Fluxes for the Global Tropical Oceans – Description and evaluation, Clim. Dynam., 38, 1521–1543, https://doi.org/10.1007/s00382-011-1115-0, 2012.

Praveen Kumar B., Vialard, J., Lengaigne, M., Murty, V. S. N., Foltz, G., McPhaden, M. J., Pous, S., and de Boyer Montégut, C.: Processes of interannual mixed layer temperature variability in the Thermocline Ridge of the Indian Ocean, Clim. Dynam., 43, 2377–2397, 2014.

Resplandy, L., Levy, M., Madec, G., Pous, S., Aumont, O., and Kumar, D.: Contribution of mesoscale processes to nutrient budgets in the Arabian Sea, J. Geophys. Res., 116, C11007, https://doi.org/10.1029/2011JC007006, 2011.

Roxy, M. K., Ritika, K., Terray, P., Murtugudde, R., Ashok, K., and Goswami, B. N.: Drying of Indian subcontinent by rapid Indian Ocean warming and a weakening land-sea thermal gradient, Nat. Commun., 6, 7423, https://doi.org/10.1038/ncomms8423, 2015.

Roxy M. K., Modi, A., Murutugudde, R., Valsala, V., Panickal, S., Prasanna Kumar, S., Ravichandran, M., Vichi, M., and Levy, M.: A reduction in marine primary productivity driven by rapid warming over the tropical Indian Ocean, Geophys. Res. Lett., 43, 826–833, 2016.

Saji, N. H., Goswami, B. N., Vinayachandran, P. N., and Yamagata, T.: A dipole mode in the tropical Indian Ocean, Nature, 401, 360–363, 1999.

Sarma, V. V. S. S.: The influence of Indian Ocean Dipole (IOD) on biogeochemistry of carbon in the Arabian Sea during 1997–1998, J. Earth Syst. Sci., 115, 433–450, https://doi.org/10.1007/BF02702872, 2006.

Sarma, Y. V. B., Al Azri, A., and Smith, L. S.: Inter-annual Variability of Chlorophyll-a in the Arabian Sea and its Gulfs, Int. J. Mar. Sci., 2.1, 1–11, https://doi.org/10.5376/ijms.2012.02.0001, 2012.

Satya Prakash, S. and Ramesh, R.: Is the Arabian Sea getting more productive?, Curr. Sci. India, 92, 667–670, 2007.

Schott, F. and McCreary, J. P.: The monsoon circulation of the Indian Ocean, Prog. Oceanogr., 51, 1–123, 2001.

Smith, S. L. and Madhupratap, M.: Mesozooplankton of the Arabian Sea: patterns influenced by seasons, upwelling, and oxygen concentrations, Prog. Oceanogr., 65, 214–239, https://doi.org/10.1016/j.pocean.2005.03.007, 2005.

Smith, S. L., Roman, M., Prusova, I., Wishner, K., Gowing, M., Codispoti, L. A., Barber, R., Marra, J., and Flagg, C.: Seasonal response of zooplankton to monsoonal reversals in the Arabian Sea, Deep-Sea Res. Pt. II, 45, 2369–2403, 1998.

Takahashi, T., Broecker, W. S., and Langer, S.: Redfield ratio based on chemical data from isopycnal surfaces, J. Geophys. Res., 90, 6907–6924, 1985.

Treguier, A., Barnier, B., de Miranda, A., Molines, J. M., Grima, N., Imbard, M., Madec, G., Messager, C., Reynaud, T., and Michel, S.: An eddy-permitting model of the Atlantic circulation: Evaluating open boundary conditions, J. Geophys. Res., 106, 22115–22129, 2001.

Turner, A. G. and Annamalai, H.: Climate Change and the South Asian Monsoon, Nature Climate Change, 2, 587–595, https://doi.org/10.1038/nclimate1495, 2012.

Uppala, S. M., Kållberg, P. W., Simmons, A. J., Andrae, U., Bechtold, V. D. C., Fiorino, M., Gibson, J. K., Haseler, J., Hernandez, A., Kelly, G. A., Li, X., Onogi, K., Saarinen, S., Sokka, N., Allan, R. P., Andersson, E., Arpe, K., Balmaseda, M. A., Beljaars, A. C. M., Berg, L. V. D., Bidlot, J., Bormann, N., Caires, S., Chevallier, F., Dethof, A., Dragosavac, M., Fisher, M., Fuentes, M., Hagemann, S., Hólm, E., Hoskins, B. J., Isaksen, L., Janssen, P. A. E. M., Jenne, R., Mcnally, A. P., Mahfouf, J.-F., Morcrette, J.-J., Rayner, N. A., Saunders, R. W., Simon, P., Sterl, A., Trenberth, K. E., Untch, A., Vasiljevic, D., Viterbo, P. and Woollen, J.: The ERA-40 re-analysis, Q. J. Roy. Meteor. Soc., 131, 2961–3012, 2005.

Vialard, J., Drushka, K., Bellenger, H., Lengaigne, M., Pous, S., and Duvel, J-P.: Understanding Madden–Julian-induced sea surface temperature variations in the North Western Australian Basin, Clim. Dynam., 41, 3203–3218, https://doi.org/10.1007/s00382-012-1541-7, 2013.

Wiggert, J. D., Vialard, J., and Behrenfeld, M. J.: Basinwide modification of dynamical and biogeochemical processes by the positive phase of the Indian Ocean Dipole during the SeaWiFS era, in: Indian Ocean Biogeochemical Processes and Ecological Variability, vol. 185, edited by: Wiggert, J. D., Hood, R. R., Wajih, S., Naqvi, A., Brink, K. H., and Smith, S. L., p. 350, 2009.

Wiggert, J. D., Hood, R., Banse, K., and Kindle, J.: Monsoon driven biogeochemical processes in the Arabian Sea, Progr. Oceanogr., 65, 176–213, https://doi.org/10.1016/j.pocean.2005.03.008, 2005.

Wiggert, J. D., Jones, B. H., Dickey, T. D., Brink, K. H., Weller, R. A., Marra, J., and Codispoti, L. A.: The northeast monsoon's impact on mixing, phytoplankton biomass and nutrient cycling in the Arabian Sea, Deep-Sea Res. Pt. II, 47, 1353–1385, 2000.

Wiggert, J. D., Murtugudde, R., and McClain, C. R.: Processes controlling interannual variations in wintertime (northeast monsoon) primary productivity in the central Arabian Sea, Deep-Sea Res. Pt. II, 47, 2319–2343, 2002.

Zhang, Y. C., Rossow, W. B., Lacis, A. A., Oinas, V., and Mishchenko, M. I.: Calculation of radiative fluxes from the surface to top of atmosphere based on ISCCP and other global data sets: refinments of the radiative transfer model and the input data, J. Geophys. Res., 109, D19105, https://doi.org/10.1029/2003JD004457, 2004.

2

Soil water regulates the control of photosynthesis on diel hysteresis between soil respiration and temperature in a desert shrubland

Ben Wang[1,2,3], **Tian Shan Zha**[1,2], **Xin Jia**[1,2,3], **Jin Nan Gong**[3], **Charles Bourque**[4], **Wei Feng**[1,2], **Yun Tian**[1,2], **Bin Wu**[1,2], **Yu Qing Zhang**[1,2], **and Heli Peltola**[3]

[1]Yanchi Research Station, School of Soil and Water Conservation, Beijing Forestry University, Beijing 100083, PR China
[2]Key Laboratory of State Forestry Administration on Soil and Water Conservation, Beijing Forestry University, Beijing, China
[3]School of Forest Sciences, University of Eastern Finland, P.O. Box 111, 80101 Joensuu, Finland
[4]Faculty of Forestry and Environmental Management, University of New Brunswick, P.O. Box 4400, 28 Dineen Drive, Fredericton, New Brunswick, E3B 5A3, Canada

Correspondence to: Tian Shan Zha (tianshanzha@bjfu.edu.cn)

Abstract. Explanations for the occurrence of hysteresis (asynchronicity) between diel soil respiration (R_s) and soil temperature (T_s) have evoked both biological and physical mechanisms. The specifics of these explanations, however, tend to vary with the particular ecosystem or biome being investigated. So far, the relative degree of control of biological and physical processes on hysteresis is not clear for drylands. This study examined the seasonal variation in diel hysteresis and its biological control in a desert-shrub ecosystem in northwest (NW) China. The study was based on continuous measurements of R_s, air temperature (T_a), temperature at the soil surface and below (T_{surf} and T_s), volumetric soil water content (SWC), and photosynthesis in a dominant desert shrub (i.e., *Artemisia ordosica*) over an entire year in 2013. Trends in diel R_s were observed to vary with SWC over the growing season (April to October). Diel variations in R_s were more closely associated with variations in T_{surf} than with photosynthesis as SWC increased, leading to R_s being in phase with T_{surf}, particularly when SWC > 0.08 m³ m⁻³ (ratio of SWC to soil porosity = 0.26). However, as SWC decreased below 0.08 m³ m⁻³, diel variations in R_s were more closely related to variations in photosynthesis, leading to pronounced hysteresis between R_s and T_{surf}. Incorporating photosynthesis into a Q_{10}-function eliminated 84.2 % of the observed hysteresis, increasing the overall descriptive capability of the function. Our findings highlight a high degree of control by photosynthesis and SWC in regulating seasonal variation in diel hysteresis between R_s and temperature.

1 Introduction

Diel hysteresis (asynchronicity) between soil respiration (R_s) and soil temperature (T_s) is widely documented for forests (Tang et al., 2005; Gaumont-Guay et al., 2006; Riveros-Iregui et al., 2007; Stoy et al., 2007; Vargas and Allen, 2008; Jia et al., 2013), grasslands (Carbone et al., 2008; Barron-Gafford et al., 2011), and desert ecosystems (Wang et al., 2014; Feng et al., 2014). Diel hysteresis, which appears as an elliptical loop in the relationship between R_s and T_s, is difficult to model with theoretical functions, such as the Q_{10}, Lloyd–Taylor, Arrhenius, or van't Hoff functions (Lloyd and Taylor, 1994; Winkler et al., 1996; Davidson et al., 2006; Phillips et al., 2011; Oikawa et al., 2014), leading to an inadequate understanding of temperature sensitivity in R_s (Gaumont-Guay et al., 2008; Phillips et al., 2011; Darenova et al., 2014). Therefore, in order to accurately predict soil carbon dioxide (CO_2) fluxes and their responses to climate change, it is necessary to understand the biophysical mechanisms that have a role in controlling seasonal variation in diel hysteresis.

Over decades of research, two main processes have been reported to relate to diel hysteresis between R_s and T_s. One is associated with the physical processes of heat and gas transport in soils (Vargas and Allen, 2008; Phillips et al., 2011; Zhang et al., 2015). Generally, soil CO_2 fluxes are measured at the soil surface and are related to temperatures in the soil. Transport of CO_2 gas to the soil surface takes time to occur, which may cause delays to appear in observed respiration rates, causing hysteretic loops to form between R_s and T_s (Zhang et al., 2015). The other is associated with the biological process of photosynthate supply (Tang et al., 2005; Kuzyakov and Gavrichkova, 2010; Vargas et al., 2011; Wang et al., 2014). Beyond the control of temperature, soil CO_2 fluxes have been associated with plant photosynthesis. Photosynthesis usually peaks at midday (e.g., 11:00–13:00), providing substrate for belowground roots and rhizosphere-microbe respiration, but oscillates out of phase with T_s, usually peaking in the afternoon (e.g., 14:00–16:00). Such influences of current photosynthesis could lead to the formation of hysteretic loops in the relationship between R_s and T_s. These studies highlight the need to consider the inherent role of photosynthesis for a more accurate interpretation of R_s (Tang et al., 2005; Kuzyakov and Gavrichkova, 2010; Vargas et al., 2011). Physical and biological processes that relate to substrates and production transport of carbon (C) in plants and soils are not mutually exclusive and both likely play crucial roles in affecting diel variation in R_s (Stoy et al., 2007; Phillips et al., 2011; Zhang et al., 2015; Song et al., 2015a, b).

Diel hysteresis between R_s and T_s has been shown to vary seasonally with soil water content (SWC; Tang et al., 2005; Riveros-Iregui et al., 2007; Carbone et al., 2008; Vargas and Allen, 2008; Ruehr et al., 2009; Wang et al., 2014). However, the influences of SWC on diel hysteresis are not uniform. Based on the Millington–Quirk model, high SWC blocks CO_2 gas and thermal diffusion (Millington and Quirk, 1961), resulting in large hysteresis loops (Riveros-Iregui et al., 2007; Zhang et al., 2015). In contrast, other studies have reported that low SWC and high water vapor pressure deficits can promote partial stomata closure, which leads to higher photosynthesis in the morning (e.g., 9:00–10:00) and suppressed photosynthesis in mid-afternoon, leading to pronounced hysteresis during dry periods (Tang et al., 2005; Vargas and Allen, 2008; Carbone et al., 2008; Wang et al., 2014). Clearly to understand the causes of diel hysteresis the role of SWC needs to be closely evaluated.

Drylands cover a quarter of the earth's land surface and play an important role in the global C cycle (Safriel and Adeel, 2005; Austin, 2011; Poulter et al., 2014). Many studies in forest ecosystems are based on the application of physical soil CO_2 and heat transport models and evaluate the influences of SWC on CO_2 gas and thermal diffusion (Riveros-Iregui et al., 2007; Phillips et al., 2011; Zhang et al., 2015). In general, many of these studies conclude that diel hysteresis is the result of physical processes alone. Few studies have evaluated the causes of diel hysteresis in drylands. Currently, it is not clear to what degree physical and biological processes control hysteresis in drylands.

Drylands are characterized with low productivity. As weak organic C-storage pools (West et al., 1994; Lange, 2003), drylands are noted for their large contribution of autotrophic production of CO_2. The autotrophic component of R_s occurs

as a direct consequence of root respiration, which is firmly coupled (within several hours) to recent photosynthesis (Liu et al., 2006; Baldocchi et al., 2006; Högberg and Read, 2006; Bahn et al., 2009; Kuzyakov and Gavrichkova, 2010). Consequently, photosynthesis may govern the level of variation in asynchronicity between R_s and T_s in drylands. In drylands, especially in desert ecosystems characterized by sandy soils with high soil porosity, the influence of SWC on gas diffusion is likely nominal. As a rule, most of the available water is used directly in sustaining biological activity in drylands (Noy-Meir, 1973). Under drought conditions, stomata closure in plants at midday reduces water losses, resulting in a corresponding suppression of photosynthesis (Jia et al., 2014). Such changes in diel patterns of photosynthesis likely result in modifications of patterns in R_s, leading to hysteresis between R_s and T_s. Soil water content likely regulates photosynthesis and, in so doing, causes hysteresis between R_s and T_s to vary over the growing season.

In this study, we hypothesize that (1) photosynthesis has a high degree of control in the formation of hysteretic loops between R_s and T_s and (2) SWC regulates this control and its variation over the growing season. The main objectives of this research were to (1) assess biological controls on diel hysteresis between R_s and T_s, (2) explore the causes that lead to variation in seasonal variation in diel hysteresis, and (3) understand SWC's role in influencing hysteresis. To undertake this work, we measured R_s, SWC, T_s, and photosynthesis in a dominant desert shrub on a continuous basis for 2013.

2 Materials and methods

2.1 Site description

The study was conducted at Yanchi Research Station of Beijing Forestry University, Ningxia, northwest China (37°42′31″ N, 107°13′37″ E; 1550 m a.s.l.). The station is located at the southern edge of the Mu Us desert in the transition between the arid and semi-arid climatic zones. Based on 51 years of data (1954–2004) from the meteorological station at Yanchi, the mean annual air temperature at the station was 8.1 °C and the mean annual total precipitation (PPT, mm) was 292 mm (ranging between 250 and 350 mm), 63 % of which fell in late summer (i.e., July–September; Wang et al., 2014; Jia et al., 2014). Annual potential evaporation was on average 5.5 kg m^{-2} d^{-1} (Gong et al., 2016). The soil at the research station was of a sandy type, with a bulk density of 1.6 g cm^{-3}. The total soil porosity within 0–2 and 5–25 cm depths was 50 and 38 %, respectively. Soil organic matter, soil nitrogen, and pH were 0.21–2.14 g kg^{-1}, 0.08–2.10, and 7.76–9.08, respectively (Wang et al., 2014; Jia et al, 2014). The vegetation was regenerated from aerial seeding applied in 1998 and is currently dominated by a semi-shrub species cover of *Artemisia ordosica*, averaging about 50 cm tall with a canopy size of about 80 cm × 60 cm (for additional site description, consult Jia et al., 2014, and Wang et al., 2014, 2015).

2.2 Soil respiration and photosynthesis measurement

Two permanent polyvinyl chloride soil collars were initially installed on a small fixed sand dune in March 2012. Collar dimensions were 20.3 cm in diameter and 10 cm in height, with 7 cm inserted into the soil. One collar was set on bare land with an opaque chamber (LI-8100-104, Nebraska, USA) and the other over an *Artemisia ordosica* plant (~ 10 cm tall) with a transparent chamber (LI-8100-104C). Soil respiration ($\mu mol\, CO_2\, m^{-2}\, s^{-1}$) was directly estimated from CO_2-flux measurements obtained with the opaque-chamber system. Photosynthetic rates ($\mu mol\, CO_2\, m^{-2}\, s^{-1}$) of the selected plants were determined as the difference in CO_2 fluxes obtained with the transparent and opaque chambers.

Continuous measurements of CO_2 fluxes ($\mu mol\, CO_2\, m^{-2}\, s^{-1}$) were made in situ with a Li-8100 CO_2 gas analyzer and a LI-8150 multiplexer (LI-COR, Nebraska, USA) connected to each chamber. Instrument maintenance was carried out bi-weekly during the growing season, including removing plant regrowth in the opaque-chamber installation and cleaning to avoid blackout conditions associated with the transparent chamber. Measurement time for each chamber was 3 min and 15 s, including a 30-second pre-purge, 45-second post-purge, and 2 min measurement period.

2.3 Measurements of temperatures, soil water content, and other environmental factors

Hourly soil temperature (T_s, °C) and volumetric SWC ($m^3\, m^{-3}$) at a 10 cm depth were measured simultaneously about 10 cm from the chambers using a LI-8150-203 temperature and EC_{H_2O} soil-moisture sensor (LI-COR, Nebraska, USA; see Wang et al., 2014). Other environmental variables were recorded every half hour using sensors mounted on a 6 m tall eddy-covariance tower approximately 800 m from our soil CO_2-flux measurement site. Air temperature (T_a, °C) was measured with a thermohygrometer (HMP155A, Vaisala, Finland). Soil-surface temperature (T_{surf}, °C) was measured with an infrared-emission sensor (model SI-111, Campbell Scientific Inc., USA). Incident photosynthetically active radiation (PAR) was measured with a light-quantum sensor (PAR-LITE, Kipp and Zonen, the Netherlands) and PPT, with three tipping-bucket rain gages (model TE525MM, Campbell Scientific Inc., USA) placed 50 m from the tower (see Jia et al., 2014).

2.4 Data processing and statistical analysis

In this study, CO_2-flux measurements were screened by means of limit checking; i.e., hourly CO_2-flux data < -30 or $> 15\, \mu mol\, CO_2\, m^{-2}\, s^{-1}$ were considered to be anomalous

as a result of, for instance, gas leakage or plant damage by insects and removed from the dataset (Wang et al., 2014, 2015). After limit checking, hourly CO_2 fluxes greater than three times the standard deviation from the calculated mean of 5 days' worth of flux data were likewise removed. Quality control and instrument failure together resulted in 5 % loss of hourly fluxes for all chambers, 4 % for temperatures, and 8 % for SWC (Fig. 1). Differences in mean annual T_s and SWC between the two chambers were 0.01 °C and 0.003 $m^3\, m^{-3}$, respectively.

The Q_{10} function (e.g., Eq. 1) was used here to describe the response of R_s to temperature. Earlier studies have shown strong correlation between basal rate of R_s and photosynthesis (Irvine et al., 2005; Sampson et al., 2007). Response of R_s to changes in photosynthesis was, in turn, characterized as a linear function (Eq. 2). Interaction between photosynthesis and temperature on R_s was conveyed through Eq. (3). The instantaneous relative importance (RI) of photosynthesis and temperature on R_s over the growing season was calculated with a correlation-based ratio (see Eq. 4). The importance of photosynthesis on R_s increases with a corresponding increase in RI:

$$R_s = R_{10} \times Q_{10}^{(T-10)/10}, \tag{1}$$

$$R_s = a \times P + b, \tag{2}$$

$$R_s = (a \times P + b) \times c^{(T-10)/10}, \tag{3}$$

$$RI = \frac{\rho_p}{\rho_t}, \tag{4}$$

where R_{10} is the respiration at 10 °C; Q_{10} is the temperature sensitivity of respiration; T is temperature; P is photosynthesis ($\mu mol\, CO_2\, m^{-2}\, s^{-1}$); a, b, and c are regression coefficients; and ρ_p and ρ_t are the correlation coefficients between photosynthesis and R_s and temperature and R_s, respectively.

Pearson correlation analysis was used to calculate the correlation coefficient between temperature or photosynthesis and R_s. Cross-correlation analysis was used to estimate hysteresis in the relationship between temperature and R_s and photosynthesis and R_s. We used root-mean-square error (RMSE) and the coefficient of determination (R^2) as criteria in evaluating function performance. To evaluate seasonal variation in diel hysteresis, the mean monthly daily cycles of R_s, T_a, T_{surf}, T_s, and photosynthesis were generated by averaging their hourly means at a given hour over a particular month (Table 1). Exponential and linear regression was used to evaluate the influence of SWC on the control of photosynthesis on temperature–R_s hysteresis. Likewise, influences of SWC on diel hysteresis was examined during a wet month with high rainfall and adequate SWC (July, PPT = 117.9 mm) and a dry month with low rainfall and inadequate SWC (August, PPT = 10.9 mm; Wang et al., 2014). In order to evaluate the influence of photosynthesis on diel hysteresis in the temperature–R_s relationship, we compared the time lag (in hours) between measured and modeled R_s by means of Eqs. (1) through (3) with a 1-day moving window

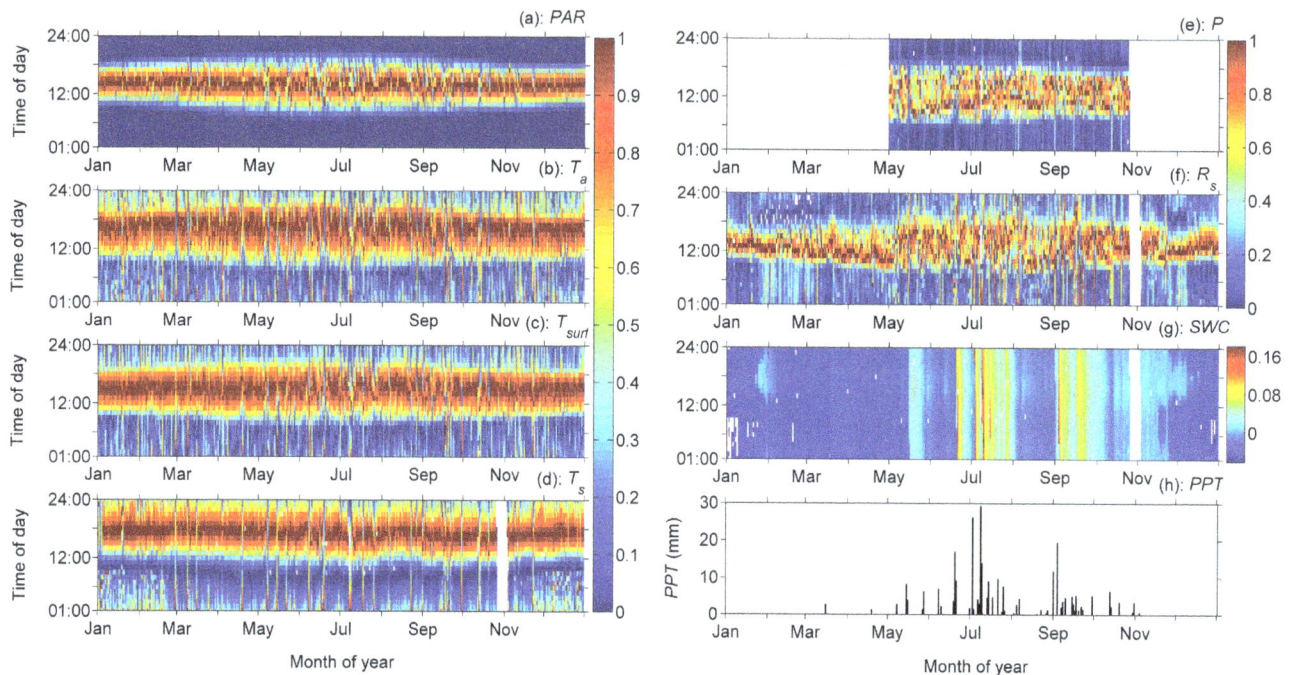

Figure 1. Seasonal variation in incident photosynthetically active radiation (PAR), temperature (i.e., air (T_a), soil-surface (T_{surf}), and soil temperatures (T_s)), photosynthesis (P), and soil respiration (R_s) at an *Artemisia ordosica*-dominated site, and seasonal variation in soil water content (SWC) and precipitation (PPT) for 2013. Hourly PAR, T_a, T_{surf}, T_s, R_s, and P are normalized against all values for each day. Each hourly value (y axis) for each day (x axis) is shown as a value of 1 through 0; 1 denotes the peak value for a given day and 0 is the daily minimum value.

Table 1. Analysis of mean monthly diel cycles of soil respiration (R_s), air temperature (T_a), soil-surface temperature (T_{surf}), soil temperature at a 10 cm depth (T_s), and photosynthesis (P) in a dominant desert-shrub ecosystem, including correlation coefficients and time lag times in R_s vs. T_a, T_{surf}, T_s, and P cycles. Statistically significant Pearson's correlation coefficients (r; $p < 0.05$) are denoted in bold.

		Jan	Feb	Mar	Apr	May	Jun	Jul	Aug	Sep	Oct	Nov	Dec
$R_s - T_a$	Lag	2	4	3	3	1	1	1	2	1	1	1	1
	r	**0.64**	0.25	**0.49**	**0.46**	**0.85**	**0.85**	**0.93**	**0.76**	**0.94**	**0.89**	**0.78**	**0.77**
$R_s - T_{surf}$	Lag	1	2	2	2	0	0	0	1	0	0	1	1
	r	**0.82**	**0.57**	**0.75**	**0.72**	**0.96**	**0.96**	**0.98**	**0.87**	**0.98**	**0.97**	**0.89**	**0.87**
$R_s - T_s$	Lag	4	5	5	5	3	3	2	4	2	2	4	4
	r	−0.06	−0.31	−0.06	−0.07	**0.54**	**0.58**	**0.80**	0.31	**0.77**	**0.65**	0.23	0.12
$R_s - P$	Lag					−1	−1	−2	0	−1	−1		
	r					**0.84**	**0.83**	**0.82**	**0.94**	**0.86**	**0.88**		

and a 1-day time step over the growing season (April to October). Modeled R_s was calculated using the fitted parameters of each function and the measured hourly T_{surf} and photosynthesis for each day. All statistical analyses were performed in MATLAB, with a significance level of 0.05 (R2010b, Mathworks Inc., Natick, MA, USA).

3 Results

3.1 Diel patterns of soil respiration, photosynthesis, and environmental factors

Incident photosynthetically active radiation, T_a, T_{surf}, and T_s exhibited distinctive daily patterns over the year (Fig. 1a–d), peaking at $\sim$ 12:00 (local time, LT), $\sim$ 16:00, $\sim$ 14:00, and $\sim$ 17:00, respectively (Fig. 1a–d). Unlike the environmental factors, daily patterns in R_s remained constant over the non-growing part of the year, peaking at 11:00–13:00, and highly variable during the growing season of the year (April to Oc-

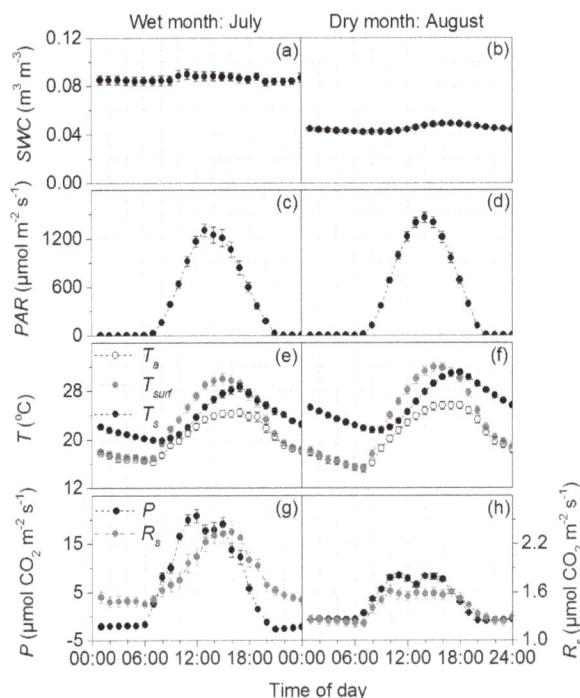

Figure 2. Mean monthly diel cycle of soil water content (SWC), incident photosynthetically active radiation (PAR), temperature (i.e., air (T_a), soil-surface (T_{surf}), and soil temperatures (T_s)), soil respiration (R_s), and photosynthesis (P) at an *Artemisia ordosica*-dominated site during a wet and dry month. Each point is the monthly mean for a particular time of day. Bars represent standard errors.

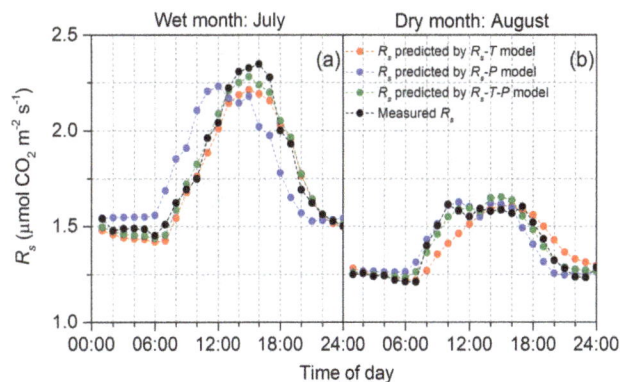

Figure 3. Diel variation of measured soil respiration (R_s) and modeled R_s by using temperature and photosynthesis as input variables in the calculation of R_s for both a wet and dry month (i.e., July and August, respectively); $R_s - T$ function (Eq. 1), $R_s - P$ function (Eq. 2), and $R_s - T - P$ function (Eq. 3).

tober), peaking between 10:00 and 16:00 (Fig. 1f). Similar to R_s during the growing season, diel patterns of photosynthesis were also highly variable, peaking between 10:00 and 16:00 (Fig. 1e).

Diel patterns of monthly mean R_s were similar to those of T_{surf} during the wet month and similar to those of photosynthesis during the dry month (Fig. 2g, h). During the wet month (July), monthly mean diel R_s was out of phase with photosynthesis, but in phase with T_{surf} (Fig. 2g). Soil respiration peaked at 16:00, exhibiting similar timing to T_{surf} (i.e., 15:00), but 4 h later than photosynthesis (peaking at 12:00; Fig. 2g). During the dry month (August), diel R_s was generally in phase with photosynthesis but out of phase with T_{surf} (Fig. 2h). Both photosynthesis and R_s plateaued between 10:00 and 16:00, whereas T_{surf} peaked at 15:00 (Fig. 2h).

3.2 Control of photosynthesis and temperature on diel soil respiration

Among temperatures at the three levels, T_{surf} correlated the strongest with R_s due to the high R^2 with monthly mean diel R_s (Table 1). Over the growing season, monthly mean diel R_s correlated fairly well with photosynthesis (Table 1). The response of R_s to temperature and photosynthesis was

shown to be affected by SWC (Table 2, Fig. 3). During the wet month, T_{surf} alone explained 97 % of the variation in diel R_s (via Eq. 1), whereas photosynthesis explained 67 % of the variation (Table 2, Fig. 3a). However, during the dry month, photosynthesis explained 88 % of the variation in diel R_s (via Eq. 2), whereas T_{surf} explained 76 % of the variation (Fig. 3b, Table 2). Irrespective of dry or wet periods, T_{surf} and photosynthesis together explained over 90 % of the diel variation in R_s (via Eq. 3; see Fig. 3 and Table 2). On the whole, RI varied as a function of SWC, decreasing whenever SWC increased (Fig. 4).

3.3 Effects of soil water content and photosynthesis on diel hysteresis in temperature–R_s relationship

During the wet month, hysteresis was not observed to occur in the monthly mean T_{surf}–R_s relationship, whereas 2-hour lags were found to occur in the photosynthesis–R_s relationship (Table 1; Fig. 3a). During the dry month, the opposite was observed, where 1-hour lags were found to occur in the T_{surf}–R_s relationship (Table 1, Fig. 3b). Over the growing season, T_{surf} lagged behind R_s by about 0–4 h (Fig. 5b), and R_s lagged behind photosynthesis by about the same amount (Fig. 5c). This led to time lags between measured and modeled R_s regardless of the variable, T_{surf}, or photosynthesis, resulting in about 26 % of the days of the growing season (accounting for 184 days, in total) having no time lag (Fig. 5e, f). However, taking into account both T_{surf} and photosynthesis as input variables in the definition of R_s (via Eq. 3), time lags between measured and modeled R_s were mostly eliminated (Fig. 5a, d), with 84 % of the days of the growing season displaying no time lag.

Diel hysteresis in both relationships (i.e., T_{surf}–R_s and photosynthesis–R_s) was shown to be affected by SWC (Fig. 6). Over the growing season, diel hysteresis between R_s and T_{surf} was linearly related to SWC in a downward

Table 2. Regressions based on the Q_{10}, linear, and Q_{10}-linear functions of soil respiration (R_s) for both a wet (July) and dry month (August) in 2013. Variables T_{surf} (°C) refers to the soil-surface temperature, P photosynthesis in the dominant shrub layer, R^2 the coefficient of determination, and RMSE the root-mean-square error.

	Model	Wet month: July	Dry month: August
$R_s - T$	Q_{10}	$R_s = 1.13 \times 1.4^{\frac{T_{surf}-10}{10}}$ $R^2 = 0.97$ RMSE $= 0.0521$	$R_s = 1.12 \times 1.1^{\frac{T_{surf}-10}{10}}$ $R^2 = 0.76$ RMSE $= 0.0796$
$R_s - P$	Linear	$R_s = 0.03 \times P + 1.61$ $R^2 = 0.67$ RMSE $= 0.1889$	$R_s = 0.04 \times P + 1.29$ $R^2 = 0.88$ RMSE $= 0.05752$
$R_s - P - T$	Linear $\times Q_{10}$	$R_s = (0.002 \times P + 1.16) \times 1.38^{\frac{T_{surf}-10}{10}}$ $R^2 = 0.98$ RMSE $= 0.0491$	$R_s = (0.024 \times P + 1.20) \times 1.08^{\frac{T_{surf}-10}{10}}$ $R^2 = 0.94$ RMSE $= 0.0408$

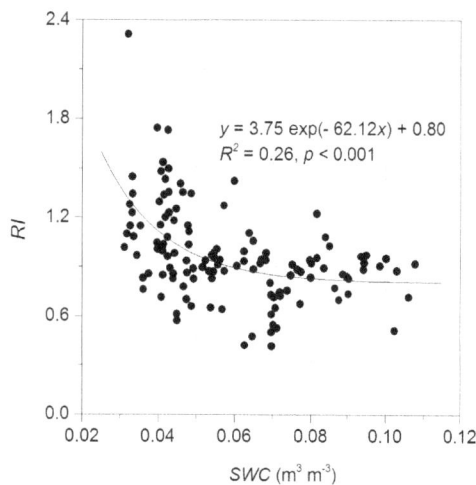

Figure 4. Relationship between soil water content (SWC) and the relative importance (RI) of soil-surface temperature and photosynthesis at an *Artemisia ordosica*-dominated site as a function of soil respiration (R_s).

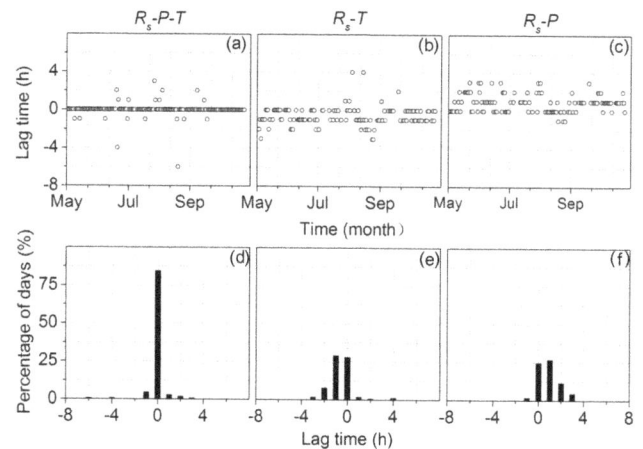

Figure 5. Time lags between measured and modeled soil respiration by means of soil-surface temperature and photosynthesis over the growing season; $R_s - T$ function (Eq. 1), $R_s - P$ function (Eq. 2), and $R_s - P - T$ function (Eq. 3).

manner, when SWC < 0.08 m^3 m^{-3} (ratio of SWC to soil porosity of 0.26; Fig. 6a). Hysteresis was not evident when SWC > 0.08 m^3 m^{-3} (Fig. 6a). In contrast, diel hysteresis between R_s and photosynthesis was linearly related to SWC in an upward manner when SWC < 0.08 m^3 m^{-3} (Fig. 6b), but ceased to be related when SWC > 0.08 m^3 m^{-3} (Fig. 6b).

4 Discussion

4.1 Degree of control of photosynthesis on diel hysteresis

In our study, we found that the diurnal pattern in temperature (T_a, T_{surf}, and T_s) lagged behind R_s by 0–4 h, which resulted in a counterclockwise loop in the relationship between R_s and temperature. Although the magnitude of hys-

teresis between R_s and temperature differed among the three temperature measurements, their seasonal variation was generally uniform. Among the temperature measurements, T_{surf} was more closely related to diel R_s, resulting in weaker hysteresis. Magnitude of hysteresis between R_s and temperature was comparable to those in other plant systems, e.g., 3.5–5 h in a boreal aspen stand (Gaumont-Guay et al., 2006) and 0–5 h in a Chinese pine plantation (Jia et al., 2013). However, the direction of hysteresis was unlike that reported by Phillips et al. (2011), who had reported R_s lagging behind soil temperature.

In general, transfer of heat (downward) and gases (upward) through the soil complex by simple diffusion would take time to occur. Increased SWC would serve to impede this transfer (Millington and Quirk, 1961). If physical processes alone controlled hysteresis, you would expect R_s to lag behind T_{surf} and hysteresis to increase with increasing

SWC. However, such rationalization is not supported by our observations, which show T_{surf} to lag behind R_s and hysteresis to decrease with increasing SWC. As a result, physical processes alone cannot account for the observed patterns in hysteresis between R_s and temperature. Combining photosynthesis and T_{surf} as explanatory variables of R_s (via Eq. 3), we found 84 % of the days over the growing season had no observable lag between measured and modeled R_s, relative to 27 % of the days when T_{surf} alone was used (associated with to Eq. 2), suggesting that photosynthesis has an important role in governing hysteresis in desert shrubland. Along with other studies, including those of Tang et al. (2005), Vargas and Allen (2008), Carbone et al. (2008), Kuzyakov and Gavrichkova (2010), and Wang et al. (2014), our findings provide increasing evidence of the role of photosynthesis in regulating diel hysteresis between R_s and temperature.

4.2 Photosynthesis control of soil respiration and diel hysteresis

The 0–4 h lag between R_s and photosynthesis observed is consistent with those observed in earlier studies, e.g., 0–4 h lag between ecosystem-level photosynthesis and R_s in a coastal wetland ecosystem (Han et al., 2014) and 0–3 h lag between plant photosynthesis and R_s in a steppe ecosystem (Yan et al., 2011). Short time lags suggest rapid response between recent photosynthesis and R_s (Kuzyakov and Gavrichova, 2010). This response is significantly faster than suggested in earlier studies, when approached from an isotopic or canopy/soil flux-based methodology (Howarth et al., 1994; Mikan et al., 2000; Jonson et al., 2002; Högberg et al., 2008; Kuzyakov and Gavrichova, 2010; Kayler et al., 2010; Han et al., 2014).

According to the "goodness of fit" of Eq. (3) to the field data, the time lag between diel photosynthesis and R_s was likely caused by variations in temperature, regardless of SWC. Photosynthesis provide substrates to roots and rhizosphere microbes (Tang et al., 2005; Kuzyakov and Gavrichova, 2010; Vargas et al., 2011; Han et al., 2014). Temperature directly drives enzymatic kinetics of respiratory metabolism in organisms (Van't Hoff, 1898; Lloyd and Taylor, 1994). Photosynthesis is directly driven by radiation (specifically, photosynthetically active radiation). Temperature is also driven by radiation, but through heating of the surface and subsequent air and soil layers. Thus, diel patterns in temperature continuously lagged behind those of photosynthesis by a few hours (as indicated in Fig. 2). The interactions between photosynthesis and temperature led R_s to lag behind photosynthesis, but temperature lagged behind R_s (Fig. 2). This sequence of events may explain the difference in the direction of hysteresis observed here, in contrast to that reported in Phillips et al. (2011). Such explanation is different from the explanations for forest ecosystems, where the transport of photosynthates and influence of turgor and osmotic pressure may be responsible for the specific coupling

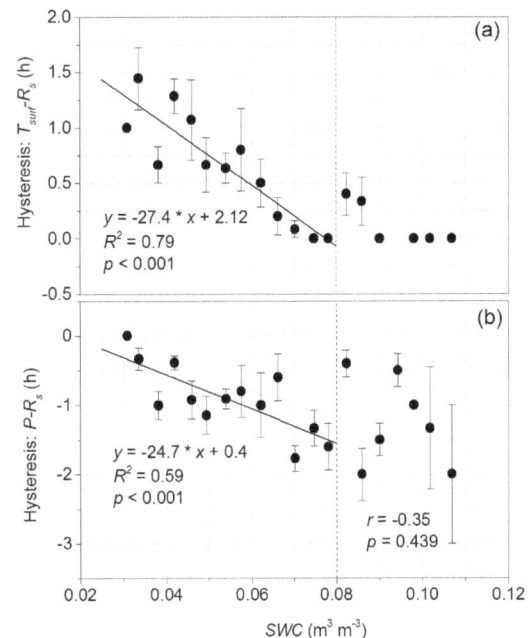

Figure 6. Time lags between soil respiration (R_s) and soil-surface temperature (T_{surf}), R_s, and photosynthesis at an *Artemisia ordosica*-dominated site with respect to soil water content (SWC). Time lags were bin-averaged using SWC intervals of $0.004 \, m^3 \, m^{-3}$.

observed between current photosynthesis and R_s (Steinmann et al., 2004; Högberg et al., 2008; Hölttä et al., 2006, 2009; Mencuccini and Hölttä, 2010). Variations in coupling dynamics may occur because of differences in vegetation height among ecosystems (Kuzyakov and Gavrichova, 2010; Mencuccini and Hölttä, 2010). Unlike forest ecosystems, low-statured vegetation in shrub systems ($\sim 0.5 \, m$) may elicit a few minutes of delay in the transportation of photosynthates and influence of turgor and osmotic pressure (Kuzyakov and Gavrichova, 2010). Such small time lags cannot be easily identified in hourly measurements, resulting in an apparent temperature-dominated control of photosynthesis and R_s.

4.3 Influences of soil water content on seasonal variation in diel hysteresis

Diel R_s varied consistently with T_{surf}, with no observable signs of hysteresis, when SWC $> 0.08 \, m^3 \, m^{-3}$. However, as SWC decreased from this value, diel R_s varied more closely with photosynthesis, leading to increased diel hysteresis between R_s and T_{surf}. These results suggest that SWC played a more important role in regulating the relative control of photosynthesis and temperature on diel R_s over the growing season, supporting our second hypothesis.

A possible explanation for SWC regulating hysteresis might be associated with changes in substrate supply. During the wet period with SWC $> 0.08 \, m^3 \, m^{-3}$, increases in SWC ameliorate diffusion of soil C substrates and its access to soil microbes (Curiel Yuste et al., 2003; Jarvis et al., 2007). Amount of substrate to roots and rhizosphere microbes is also expected to be high as a result of high current photosynthesis (Baldocchi et al., 2006). As a result, diel R_s is not limited by C substrates provided by current photosynthesis and soil organic matter. Consequences of diel R_s may vary repeatedly in synchrony with diel temperature, with no indication of hysteresis when SWC > 0.08 m^3 m^{-3} (Fig. 6a). By contrast, during dry and hot phases, with SWC < 0.08 m^3 m^{-3}, inadequate soil water limits diffusion of soil C substrates and its access to soil microbes (Jassal et al., 2008) and also suppresses photosynthesis (supported by Fig. 2g, h). As a result, R_s may be limited by C substrates under dry conditions. It has been reported that current photosynthesis can account for about 65–70 % of total R_s over the growing season (Ekblad and Högberg et al., 2001; Högberg et al., 2001). Thus, diel R_s may vary more closely to photosynthesis during dry and hot phases over the growing season (Fig. 2h), resulting in increased hysteresis with decreasing SWC below 0.08 m^3 m^{-3} (Fig. 6b).

The 0.08 m^3 m^{-3} SWC threshold of this study was consistent with an earlier study by Wang et al. (2014) that reported that seasonal R_s decoupled from soil temperature as SWC fell below 0.08 m^3 m^{-3}. Earlier studies have reported similar responses of R_s to temperature (Palmroth et al., 2005; Jassal et al., 2008). For example, R_s in an 18-year-old temperate Douglas fir stand decoupled from T_s when SWC fell below 0.11 m^3 m^{-3}. Our results suggest that the decoupling of R_s from temperature for low SWC was due to a shift in control from temperature to photosynthesis. Our work provides urgently needed new knowledge concerning causes/mechanisms involved in defining variation in diel hysteresis in desert shrubland. Based on our work, we suggest that photosynthesis should be considered in simulations of diel R_s in drylands, especially when SWC falls below 0.08 m^3 m^{-3}.

5   Conclusions

Soil water content regulated the relative control between photosynthesis and temperature on diel R_s by changing the relative contribution of autotrophic and heterotrophic respiration to total R_s, causing seasonal variation in diel hysteresis between R_s and temperature. Hysteresis was not observed between R_s and T_{surf}, when SWC > 0.08 m^3 m^{-3}, but the lag hours increased as SWC decreased below this SWC threshold. Incorporating photosynthesis into R_s–temperature-based models reduces diel hysteresis and increases the overall level of goodness of fit. Our findings highlight the importance of biological mechanisms in diel hysteresis between R_s and temperature and the importance of SWC in plant photosynthesis–soil respiration dynamics in dryland ecosystems.

Competing interests. The authors declare that they have no conflict of interest.

Special issue statement. This article is part of the special issue "Ecosystem processes and functioning across current and future dryness gradients in arid and semi-arid lands". It is not associated with a conference.

Acknowledgements. We acknowledge the grants obtained from National Natural Science Foundation of China (NSFC; 31670710, 31670708 and 31361130340), the National Key Research and Development Program of China (2016YFC0500905), the Fundamental Research Funds for the Central Universities (BLYJ201601 and 2015ZCQ-SB-02), and the Finnish–Chinese research collaboration project EXTREME (2013-2016) between Beijing Forestry University and University of Eastern Finland (EXTREME project 14921 funded by Academy of Finland). Also the US–China Carbon Consortium (USCCC) supported this work by way of helpful discussions and exchange of ideas. We acknowledge Paul Stoy, Associate Editor, and anonymous reviewers for their valuable comments and suggestions on this paper. We also acknowledge Huishu Shi, Yuming Zhang, Wei Feng, Yajuan Wu, Peng Liu, Qiang Yang, and Mingyan Zhang for their assistance with the field measurements and instrumentation maintenance.

Edited by: Paul Stoy

References

Austin, A. T.: Has water limited our imagination for aridland biogeochemistry?, Trends Ecol. Evol., 26, 229–235, 2011.

Baldocchi, D., Tang, J., and Xu, L.: How switches and lags in biophysical regulators affect spatial-temporal variation of soil respiration in an oak-grass savanna, J. Geophys. Res.-Biogeo., 111, G02008, https://doi.org/10.1029/2005JG000063, 2006.

Barron-Gafford, G. A., Scott, R. L., Jenerette, G. D., and Huxman, T. E.: The relative controls of temperature, soil moisture, and plant functional group on soil CO$_2$ efflux at diel, seasonal, and annual scales, J. Geophys. Res.-Biogeo., 116, G01023, https://doi.org/10.1029/2010JG001442, 2011.

Bahn, M., Schmitt, M., Siegwolf, R., Richter, A., and Brüggemann, N.: Does photosynthesis affect grassland soil-respired CO$_2$ and its carbon isotope composition on a diurnal timescale?, New Phytol., 182, 451–460, 2009.

Carbone, M. S., Winston, G. C., and Trumbore, S. E.: Soil respiration in perennial grass and shrub ecosystems: Linking environmental controls with plant and microbial sources on seasonal and diel timescales, J. Geophys. Res.-Biogeo., 113, G02022, https://doi.org/10.1029/2007JG000611, 2008.

Curiel Yuste, J. C., Janssens, I. A., Carrara, A., Meiresonne, L., and Ceulemans, R.: Interactive effects of temperature and precipitation on soil respiration in a temperate maritime pine forest, Tree Physiol., 23, 1263–1270, 2003.

Davidson, E. A., Janssens, I. A., and Luo, Y. Q.: On the variability of respiration in terrestrial ecosystems: moving beyond Q_{10}, Glob. Change Biol., 12, 154–164, 2006.

Darenova, E., Pavelka, M., and Acosta, M.: Diurnal deviations in the relationship between CO_2 efflux and temperature: A case study, Catena, 123, 263–269, 2014.

Ekblad, A. and Högberg, P.: Natural abundance of ^{13}C in CO_2 respired from forest soils reveals speed of link between photosynthesis and root respiration, Oecologia, 127, 305–308, 2001.

Feng, W., Zhang, Y., Wu, B., Qin, S., and Lai, Z.: Influence of environmental factors on carbon dioxide exchange in biological soil crusts in desert areas, Arid Land Res. Manag., 28, 186–196, 2014.

Gaumont-Guay, D., Black, T. A., Griffis, T. J., Barr, A. G., Jassal, R. S., and Nesic, Z.: Interpreting the dependence of soil respiration on soil temperature and water content in a boreal aspen stand, Agr. Forest Meteorol., 140, 220–235, 2006.

Gaumont-Guay, D., Black, T. A., Barr, A. G., Jassal, R. S., and Nesic, Z.: Biophysical controls on rhizospheric and heterotrophic components of soil respiration in a boreal black spruce stand, Tree Physiol., 28, 161–171, 2008.

Gong, J., Jia, X., Zha, T., Wang, B., Kellomäki, S., and Peltola, H.: Modeling the effects of plant-interspace heterogeneity on water-energy balances in a semiarid ecosystem, Agr. Forest Meteorol., 221, 189–206, 2016.

Han, G., Luo, Y., Li, D., Xia, J., Xing, Q., and Yu, J.: Ecosystem photosynthesis regulates soil respiration on a diurnal scale with a short-term time lag in a coastal wetland, Soil Biol. Biochem., 68, 85–94, 2014.

Högberg, P., Nordgren, A., Buchmann, N., Taylor, A. F. S., Ekblad, A., Högberg, M. N., Nyberg, G., Ottosson-Löfvenius, M., and Read, D. J.: Large-scale forest girdling shows that current photosynthesis drives soil respiration, Nature, 411, 789–792, 2001.

Högberg, P. and Read, D. J.: Towards a more plant physiological perspective on soil ecology, Trends Ecol. Evol., 21, 548–554, 2006.

Högberg, P., Högberg, M. N., Gottlicher, S. G., Betson, N. R., Keel, S. G., Metcalfe, D. B., Campbell, C., Schindlbacher, A., Hurry, V., Lundmark, T., Linder, S., and Näsholm, T.: High temporal resolution tracing of photosynthate carbon from the tree canopy to forest soil microorganisms, New Phytol., 177, 220–228, 2008.

Hölttä, T., Vesala, T., Sevanto, S., Perämäki, M., and Nikinmaa, E.: Modeling xylem and phloem water flows in trees according to cohesion theory and Münch hypothesis, Trees: Structure and Function, 20, 67–78, 2006.

Hölttä, T., Nikinmaa, E., and Mencuccini, M.: Linking phloem function to structure: analysis with a coupled xylem-phloem transport model, J. Theor. Biol., 259, 325–337, 2009.

Howarth, W. R., Pregitzer, K. S., and Paul, E. A.: ^{14}C allocation in tree soil systems, Tree Physiol., 14, 1163–1176, 1994.

Irvine, J., Law, B. E., and Kurpius, M. R.: Coupling of canopy gas exchange with root and rhizosphere respiration in a semi-arid forest, Biogeochemistry, 73, 271–282, 2005.

Jarvis, P., Rey, A., Petsikos, C., Wingate, L., Rayment, M., Pereira, J., Banza, J., David, J., Miglietta, F., Borghetti, M., Manca, G., and Valentini, R.: Drying and wetting of Mediterranean soil stimulates decomposition and carbon dioxide emission: the "Birch effect", Tree Physiol., 27, 929–940, 2007.

Jassal, R. S., Andrew Black, T., Novak, M. D., Gaumont-Guay, D., and Nesic, Z.: Effect of soil water stress on soil respiration and its temperature sensitivity in an 18-year-old temperate Douglasfir stand, Glob. Change Biol., 14, 1305–1318, 2008.

Jia, X., Zhan, T., Wu, B., Zhang, Y., Chen, W., Wang, X., Yu, H., and He, G.: Temperature response of soil respiration in a Chinese pine plantation: hysteresis and seasonal vs. diel Q_{10}, PLoS ONE, 8, e57858, https://doi.org/10.1371/journal.pone.0057858, 2013.

Jia, X., Zha, T. S., Wu, B., Zhang, Y. Q., Gong, J. N., Qin, S. G., Chen, G. P., Qian, D., Kellomäki, S., and Peltola, H.: Biophysical controls on net ecosystem CO_2 exchange over a semiarid shrubland in northwest China, Biogeosciences, 11, 4679–4693, https://doi.org/10.5194/bg-11-4679-2014, 2014.

Johnson, D., Leake, J. R., Ostle, N., Ineson, P., and Read, D. J.: In situ $^{13}CO_2$ pulse labelling of upland grassland demonstrates a rapid pathway of carbon flux from arbuscular mycorrhizal mycelia to soil, New Phytol., 153, 327–334, 2002.

Kayler, Z., Gessler, A., and Buchmann, N.: What is the speed of link between aboveground and belowground processes? New Phytol., 187, 885–888, 2010.

Kuzyakov, Y. and Gavrichkova, O.: Review: Time lag between photosynthesis and carbon dioxide efflux from soil: a review of mechanisms and controls, Glob. Change Biol., 16, 3386–3406, 2010.

Lange, O. L.: Photosynthesis of soil-crust biota as dependent on environmental factors, in: Biological soil crusts: structure, function, and management, edited by: Belnap, J. and Lange, O. L., Berlin: Springer, 217–40, 2003.

Liu, Q., Edwards, N. T., Post, W. M., Gu, L., Ledford, J., and Lenhart, S.: Temperature independent diel variation in soil respiration observed from a temperate deciduous forest, Glob. Change Biol., 12, 2136–2145, 2006.

Lloyd, J. and Taylor, J. A.: On the temperature dependence of soil respiration, Funct. Ecol., 8, 315–323, 1994.

Mencuccini, M. and Hölttä, T.: The significance of phloem transport for the speed with which canopy photosynthesis and belowground respiration are linked, New Phytol., 185, 189–203, 2010.

Mikan, C. J., Zak, D. R., Kubiske, M. E., and Pregitzer, K. S.: Combined effects of atmospheric CO_2 and N availability on the belowground carbon and nitrogen dynamics of aspen mesocosms, Oecologia, 124, 432–445, 2000.

Millington, R. J. and Quirk, J. P.: Permeability of porous solids, T. Faraday Soc., 57, 1200–1207, 1961.

Noy-Meir, I.: Desert ecosystems: environment and producers, Annu. Rev. Ecol. Syst., 4, 25–51, 1973.

Oikawa, P. Y., Grantz, D. A., Chatterjee, A., Eberwein, J. E., Allsman, L. A., and Jenerette, G. D.: Unifying soil respiration pulses, inhibition, and temperature hysteresis through dynamics of labile soil carbon and O_2, J. Geophys. Res.-Biogeo., 119, 521–536, 2014.

Palmroth, S., Maier, C. A., McCarthy, H. R., Oishi, A. C., Kim, H.-S., Johnsen, K. H., Katul, G. G., and Oren, R.: Contrasting responses to drought of forest floor CO_2 efflux in a Loblolly pine plantation and a nearby Oak–Hickory forest, Glob. Change Biol., 11, 421–434, 2005.

Phillips, C. L., Nickerson, N., Risk, D., and Bond, B. J.: Interpreting diel hysteresis between soil respiration and temperature, Glob. Change Biol., 17, 515–527, 2011.

Poulter, B., Frank, D., Ciais, P., Myneni, R. B., Andela, N., Bi, J., Broquet, G., Canadell, J. G., Chevallier, F., Liu, Y. Y., Running, S. W., Sitch, S., and Van der Werf, G. R.: Contribution of semi-arid ecosystems to interannual variability of the global carbon cycle, Nature, 509, 600–603, 2014.

Riveros-Iregui, D. A., Emanuel, R. E., Muth, D. J., McGlynn, B. L., Epstein, H. E., Welsch, D. L., Pacific, V. J., and Wraith, J. M.: Diurnal hysteresis between soil CO_2 and soil temperature is controlled by soil water content, Geophys. Res. Lett, 34, L17404, https://doi.org/10.1029/2007GL030938, 2007.

Ruehr, N. K., Offermann, C. A., Gessler, A., Winkler, J. B., Ferrio, J. P., Buckmann, N., and Barnard, R. L.: Drought effects on allocation of recent carbon: from beech leaves to soil CO_2 efflux, New Phytol., 184, 950–961, 2009.

Safriel, U. and Adeel, Z.: Dryland ecosystems, II, in: Ecosystems and human well-being: current state and trends, Vol. 1, edited by: Hassan, R., Scholes, R., and Neville, A., Island Press, Washingon, DC, 623–662, 2005.

Sampson, D. A., Janssens, I. A., Curiel Yuste, J., and Ceulemans, R.: Basal rates of soil respiration are correlated with photosynthesis in a mixed temperate forest, Glob. Change Biol., 13, 2008–2017, 2007.

Song, W., Chen, S., Zhou, Y., Wu, B., Zhu, Y., Lu, Q., and Lin, G.: Contrasting diel hysteresis between soil autotrophic and heterotrophic respiration in a desert ecosystem under different rainfall scenarios, Sci. Rep., 5, 16779, https://doi.org/10.1038/srep16779, 2015a.

Song, W., Chen, S., Wu, B., Zhu, Y., Zhou, Y., Lu, Q., and Lin, G.: Simulated rain addition modifies diurnal patterns and temperature sensitivities of autotrophic and heterotrophic soil respiration in an arid desert ecosystem, Soil Biol. Biochem., 82, 143–152, 2015b.

Steinmann, K., Siegwolf, R. T. W., Saurer, M., and Körner, C.: Carbon fluxes to the soil in a mature temperate forest assessed by ^{13}C isotope tracing, Oecologia, 141, 489–501, 2004.

Stoy, P. C., Palmroth, S., Oishi, A. C., Siqueira, M. B. S., Juang, J-Y., Novick, K. A., Ward, E. J., Katul, G. G., and Oren, R.: Are ecosystem carbon inputs and outputs coupled at short time scales? A case study from adjacent pine and hardwood forests using impulse-response analysis, Plant Cell Environ., 30, 700–710, 2007.

Tang, J., Baldocchi, D. D., and Xu, L.: Tree photosynthesis modulates soil respiration on a diurnal time scale, Glob. Change Biol., 11, 1298–1304, 2005.

Van't Hoff, J. H.: Lectures on Theoretical and Physi-cal Chemistry. Part I. Chemical Dynamics (translated by R. A. Lehfeldt), Edward Arnold, London, 224–229, 1898.

Vargas, R. and Allen, M. F.: Environmental controls and the influence of vegetation type, fine roots and rhizomorphs on diel and seasonal variation in soil respiration, New Phytol., 179, 460–471, 2008.

Vargas, R., Baldocchi, D. D., Bahn, M., Hanson, P. J., Hosman, K. P., Kulmala, L., Pumpanen, J., and Yang, B.: On the multi-temporal correlation between photosynthesis and soil CO_2 efflux: reconciling lags and observations, New Phytol., 191, 1006–1017, 2011.

Wang, B., Zha, T. S., Jia, X., Wu, B., Zhang, Y. Q., and Qin, S. G.: Soil moisture modifies the response of soil respiration to temperature in a desert shrub ecosystem, Biogeosciences, 11, 259–268, https://doi.org/10.5194/bg-11-259-2014, 2014.

Wang, B., Zha, T. S., Jia, X., Gong, J. N., Wu, B., Bourque, C. P. A., Zhang, Y., Qin, S. G., Chen, G. P., and Peltola, H.: Microtopographic variation in soil respiration and its controlling factors vary with plant phenophases in a desert-shrub ecosystem, Biogeosciences, 12, 5705–5714, https://doi.org/10.5194/bg-12-5705-2015, 2015.

West, N. E., Stark, J. M., Johnson, D. W., Abrams, M. M., Wight, J. R., Heggem, D., and Peck, S.: Effects of climatic-change on the edaphic features of arid and semiarid lands of western North America, Arid Land Res. Manag., 8, 307–355, 1994.

Winkler, J. P., Cherry, R. S., and Schlesinger, W. H.: The Q_{10} relationship of microbial respiration in a temperate forest soil, Soil Biol. Biochem., 28, 1067–1072, 1996.

Yan, L., Chen, S., Huang, J., and Lin, G.: Water regulated effects of photosynthetic substrate supply on soil respiration in a semiarid steppe, Glob. Change Biol. 17, 1990–2001, 2011.

Zhang, Q., Katul, G. G., Oren, R., Daly, E., Manzoni, S., and Yang, D.: The hysteresis response of soil CO_2 concentration and soil respiration to soil temperature, J. Geophys. Res.-Biogeo., 120, 1605–1618, 2015.

3

Sediment and carbon deposition vary among vegetation assemblages in a coastal salt marsh

Jeffrey J. Kelleway[1,2], **Neil Saintilan**[2], **Peter I. Macreadie**[1,3], **Jeffrey A. Baldock**[4], and **Peter J. Ralph**[1]

[1]Climate Change Cluster, School of Life Sciences, University of Technology Sydney, Ultimo, NSW 2007, Australia
[2]Department of Environmental Sciences, Macquarie University, Sydney, NSW 2109, Australia
[3]School of Life and Environmental Sciences, Centre for Integrative Ecology, Deakin University, Victoria 3216, Australia
[4]CSIRO Agriculture and Food, Glen Osmond, SA 5064, Australia

Correspondence to: Jeffrey J. Kelleway (jeffrey.kelleway@mq.edu.au)

Abstract. Coastal salt marshes are dynamic, intertidal ecosystems that are increasingly being recognised for their contributions to ecosystem services, including carbon (C) accumulation and storage. The survival of salt marshes and their capacity to store C under rising sea levels, however, is partially reliant upon sedimentation rates and influenced by a combination of physical and biological factors. In this study, we use several complementary methods to assess short-term (days) deposition and medium-term (months) accretion dynamics within a single marsh that contains three salt marsh vegetation types common throughout southeastern (SE) Australia.

We found that surface accretion varies among vegetation assemblages, with medium-term (19 months) bulk accretion rates in the upper marsh rush (*Juncus*) assemblage (1.74 ± 0.13 mm yr^{-1}) consistently in excess of estimated local sea-level rise (1.15 mm yr^{-1}). Accretion rates were lower and less consistent in both the succulent (*Sarcocornia*, 0.78 ± 0.18 mm yr^{-1}) and grass (*Sporobolus*, 0.88 ± 0.22 mm yr^{-1}) assemblages located lower in the tidal frame. Short-term (6 days) experiments showed deposition within *Juncus* plots to be dominated by autochthonous organic inputs with C deposition rates ranging from 1.14 ± 0.41 mg C cm^{-2} d^{-1} (neap tidal period) to 2.37 ± 0.44 mg C cm^{-2} d^{-1} (spring tidal period), while minerogenic inputs and lower C deposition dominated *Sarcocornia* (0.10 ± 0.02 to 0.62 ± 0.08 mg C cm^{-2} d^{-1}) and *Sporobolus* (0.17 ± 0.04 to 0.40 ± 0.07 mg C cm^{-2} d^{-1}) assemblages.

Elemental (C : N), isotopic (δ^{13}C), mid-infrared (MIR) and ^{13}C nuclear magnetic resonance (NMR) analyses revealed little difference in either the source or character of materials being deposited among neap versus spring tidal periods. Instead, these analyses point to substantial redistribution of materials within the *Sarcocornia* and *Sporobolus* assemblages, compared to high retention and preservation of organic inputs in the *Juncus* assemblage. By combining medium-term accretion quantification with short-term deposition measurements and chemical analyses, we have gained novel insights into above-ground biophysical processes that may explain previously observed regional differences in surface dynamics among key salt marsh vegetation assemblages. Our results suggest that *Sarcocornia* and *Sporobolus* assemblages may be particularly susceptible to changes in sea level, though quantification of below-ground processes (e.g. root production, compaction) is needed to confirm this.

1 Introduction

1.1 Coastal salt marshes

Coastal salt marshes are dynamic ecosystems, vegetated by herbs, grasses and rushes, that are found in a range of sedimentary settings along low-energy coastlines. Globally, vegetation type and floristic assemblage have been used to classify general types of salt marsh (Adam, 1990, 2002). On the local scale, vegetation zonation is one of the most striking ecological features of many salt marshes, reflecting the elevation requirements of a small number of dominant species, although mosaics of species within a zone are also common (Adam, 2002; Hickey and Bruce, 2010). Whilst the biodiversity values and exceptionally high productivity of salt marshes have long been recognised, increasing attention is now being focused upon ecosystem services such as carbon (C) accumulation and storage (Chmura et al., 2003; Duarte et al., 2013) and the response of these coastal ecosystems to changes in climate (Kirwan and Mudd, 2012) and sea level (Rogers et al., 2014).

Sedimentation dynamics partially determine the survival of salt marshes under rising sea level (Baustian et al., 2012; Kirwan and Megonigal, 2013; Kirwan et al., 2016) and the delivery and storage of organic matter (OM; Duarte et al., 2013; Lovelock et al., 2013). Salt marsh soils may be minerogenic (dominated by mineral inputs) or organogenic (dominated by biomass and litter production and/or allochthonous OM inputs), although most comprise both mineral and organic fractions (Adam, 2002; Baustian et al., 2012). Consequently, soil properties and surface dynamics may be influenced by both physicochemical and biological factors. Physical drivers of accretion (the vertical accumulation of sediments) in intertidal wetlands include the suspended sediment supply of inundating waters (Zhou et al., 2007), the tidal range of a site and position within the tidal range (Ouyang and Lee, 2014; Saintilan et al., 2013; van Proosdij et al., 2006). High tides may play an important role in importing sediment into salt marshes (Rosencranz et al., 2016), while low-tide rainfall may act to redistribute or export materials, including particulate organic carbon (Chen et al., 2015).

Numerous studies have investigated the interactions between vegetation and marsh surface dynamics (Langley et al., 2009; Rooth et al., 2003), although the majority of these studies have focussed on the genus *Spartina* (e.g. Baustian et al., 2012; Mudd et al., 2010, 2009; Nyman et al., 2006). Generally, these studies have shown that the presence of vegetation may have a significant positive influence on surface accretion through (1) accumulating organic matter and (2) facilitating sediment trapping (Morris et al., 2002; Mudd et al., 2010; Nyman et al., 2006). Findings of comparative studies of the effect of vegetation composition on deposition rates, however, vary from no difference in accretion among different vegetation species (e.g. Culberson et al., 2004) to substantial differences among mangroves and different salt marsh species (e.g. Saintilan et al., 2013). Little is known about the extent to which surface materials are redistributed among neighbouring salt marsh species assemblages, though stable isotope approaches have been used to demonstrate the small-scale (i.e. a few metres) movement of OM at mangrove–salt marsh interfaces (Guest et al., 2004, 2006).

1.2 C storage

Average global rates of soil carbon accumulation is extremely high in salt marshes, relative to most terrestrial and coastal ecosystems, with a mean $\pm$ standard error (SE) accumulation rate of $0.024 \pm 0.003\,\mathrm{g\,C\,cm^{-2}\,yr^{-1}}$ (Ouyang and Lee, 2014). While much of this OM is produced belowground by roots and rhizomes, contributions from aboveground sources may be significant (Boschker et al., 1999; Zhou et al., 2006). Sources of above-ground C may include both autochthonous (produced within the community) and allochthonous (deposited from outside the community) OM, although their relative contributions may vary within and among salt marsh settings (Kelleway et al., 2016a). The contribution of C redistributed within the community (i.e. within and among different species assemblages) to surface deposition and longer-term C accumulation remains unquantified and may vary with vegetation structure and geomorphic position. Regardless of OM source, the capacity of salt marshes to store carbon in the long-term remains dependent upon the balance between OM input and its decay (Mueller et al., 2016). While there is considerable debate as to which factors most influence the long-term retention of C in soils, litter quality has long been identified as a key driver of decay rates (Cleveland et al., 2014; Enríquez et al., 1993; Hemminga and Buth, 1991; Josselyn and Mathieson, 1980; Kristensen, 1994) and is of particular relevance to C stock accumulation in surface soils.

1.3 Measuring surface deposition and accretion

A variety of methods have been developed for measuring and monitoring surface dynamics in tidal wetlands (for reviews see Nolte et al., 2013; Thomas and Ridd, 2004). These include techniques relevant to short-term deposition events (of sediments and plant litter) through to medium- and long-term measures of accretion or accumulation (the net effect of multiple deposition and removal events) as well as surface elevation change. Because methods vary in their effectiveness of trapping and retaining different materials, a combination of techniques may be required to identify the different physical and biotic influences on deposition and accretion (Nolte et al., 2013). In this study, we combine several methods to assess short-term (days) deposition and medium-term (months) accretion dynamics within three salt marsh vegetation assemblages common throughout SE Australia. Our aim is to use three different measurement methods to identify the role of vegetation and physical factors in surface deposition and/or accretion. We hypothesise that (1) mineral deposition and accretion will be highest in lower-elevation assemblages but organic deposition and accretion will be highest in the *Juncus* assemblage; (2) the source and character of material deposited will vary temporally according to tidal inundation patterns, with a greater proportion of allochthonous material deposited during times of high inundation frequency; and (3) there will be no difference in biomass–litter–sediment decay patterns among the vegetation assemblages. Together, we expect this information will improve our understanding of how materials (including C) are deposited and accumulate in

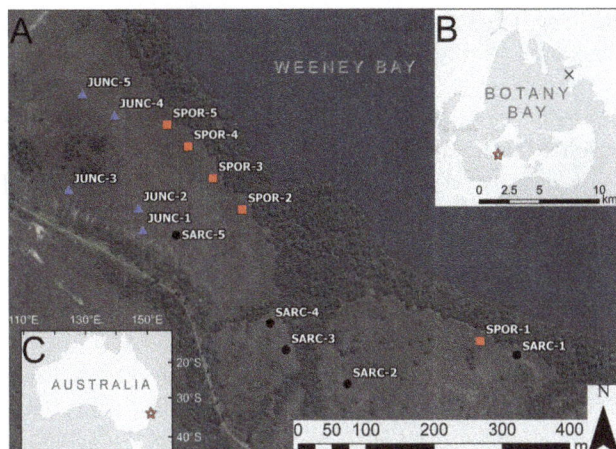

Figure 1. Location of experimental plots within Weeney Bay salt marsh of Towra Point Nature Reserve **(a)**, located along the southern shoreline of Botany Bay **(b)** in SE Australia **(c)**. Location of the nearest tidal gauge is marked by an X in inset B. SARC = *Sarcocornia quinqueflora* assemblage; SPOR = *Sporobolus virginicus* assemblage; JUNC = *Juncus kraussii* assemblage.

coastal wetlands and how these ecosystems might respond under rising sea levels.

2 Methods

2.1 Study setting

Broadly, the salt marshes of SE Australia have been classified within the temperate group of salt marshes, which also includes those of Europe, the Pacific coast of North America, Japan and South Africa (Adam, 1990). These are distinct from the well-studied *Spartina*-dominated marshes of North America's Atlantic coast. Towra Point Nature Reserve is located within the oceanic embayment Botany Bay, approximately 16 km south of central Sydney, Australia's largest city. The intertidal estuarine wetland complex at this site is the largest remaining within the Sydney region and is listed as a Ramsar Wetland of International Importance. Within the site, a large salt marsh area adjacent to Weeney Bay was chosen as a study site as this area exhibits vegetation zonation typical of SE Australian salt marshes (Fig. 1). The lower salt marsh is bordered by the mangrove species *Avicennia marina*, beyond which seagrass meadows (including *Posidonia australis*) occur within the subtidal zone. In some areas, the upslope limit of the salt marsh extends into small patches of the supratidal trees *Casuarina glauca* and *Melaleuca ericifolia*, but for the most part the salt marsh is bordered by a levee, which was constructed between 1947 and 1951. Previous investigation has revealed vegetation zonation across the site coinciding with ranges in elevation and tidal extent (Hickey and Bruce, 2010).

The salt marsh within this site comprises two broad vegetation communities. The lower and middle marsh is characterised by an association of the perennial succulent *Sarcocornia quinqueflora* (C3 photosynthetic pathway) and the perennial grass *Sporobolus virginicus* (C4 photosynthetic pathway). The upper marsh assemblage is dominated by the rush *Juncus kraussii* (C3), with *S. virginicus* (C4) ubiquitous as a subdominant lower stratum across this assemblage.

On the basis of salt marsh vegetation zonation, 15 plots were selected – five plots randomly chosen within the *Juncus*-dominated assemblage and 10 plots strategically selected within the *Sarcocornia–Sporobolus* association (five plots vegetated exclusively by *Sarcocornia*, and five vegetated exclusively by *Sporobolus*). Hereafter, these three assemblages are referred to by genus (*Sarcocornia*, *Sporobolus*, *Juncus*), while reference to the plant species themselves involves the species name (*S. quinqueflora*, *S. virginicus*, *J. kraussii*).

Data previously collected within the study region showed a substantial difference in above-ground biomass of the rush assemblage (*Juncus* mean $= 1116 \, \text{g DW m}^{-2}$; range $= 51$–$4832 \, \text{g DW m}^{-2}$), compared to that of the non-rush assemblages (*Sarcocornia* mean $= 320 \, \text{g DW m}^{-2}$, range $= 52$–$1184 \, \text{g DW m}^{-2}$; *Sporobolus* mean $= 350 \, \text{g DW m}^{-2}$, range $= 148$–$852 \, \text{g DW m}^{-2}$). Moreover, there do not appear to be distinct seasonal patterns of biomass stock for any of these species (Clarke and Jacoby, 1994). Both *Sarcocornia* and *Sporobolus* are perennial species, while *J. kraussii* culms undergo initiation and senescence throughout the year, but with peak culm initiation before and after summer flowering and fruiting (Clarke and Jacoby, 1994). Below-ground biomass data are rare, though on the basis of 0–20 cm depth data presented by Clarke and Jacoby (1994) we have calculated a mean above-ground : below-ground biomass ratio of 1.5 for *Juncus*. No below-ground data have been reported for either *Sarcocornia* and *Sporobolus*.

Tides along the New South Wales coast are semidiurnal (two flood and two ebb periods each lunar day) with a maximum spring tidal range of 2.0 m (Roy et al., 2001). Astronomical (i.e. predicted) maxima occur during the new moon in summer and during the full moon in winter (spring tides). Tidal inundation of and recession from the study area occurs via Weeney Bay, with the causeway acting as a barrier to surface water exchange with the western section of the Nature Reserve and Woolooware Bay. The linear rate of sea-level rise in Botany Bay since local records commenced in 1981 is $1.15 \, \text{mm yr}^{-1}$ (Kelleway et al., 2016b). Rainfall in the region is spread throughout the year, with an annual rainfall of $1084 \, \text{mm yr}^{-1}$ in Botany Bay (Bureau of Meteorology, 2016).

2.2 Surface elevation measurements

Plot elevation was recorded to assess relationships between deposition dynamics and plot position within the tidal frame.

Elevation was measured using a modified version of the tidal inundation method described by English et al. (1994), whereby three vertical rods marked with water-soluble dye were inserted into the ground immediately prior to a summer spring tide (23 January 2015; measured tidal height of 1.897 m above lowest astronomical tide (LAT) datum at nearest tidal gauge). Depth of inundation above the salt marsh surface was measured immediately after the tide receded and was subtracted from the measured tide height to obtain an estimate of surface elevation. Care was taken during the measurement procedure and in the selection of a calm day (to minimise wind and wave effects) to minimise discrepancies between measurements at different plots. Comparison of three replicate rods revealed a SE of the mean < 1.3 cm for each plot.

2.3 Feldspar marker horizons

The feldspar marker horizon (MH) technique (Cahoon and Turner, 1989) has been proposed as a suitable method to investigate the effects of above-ground vegetation structure on the accretion (vertical accumulation) of material on the marsh surface over the medium term (Nolte et al., 2013). The feldspar MH technique was used to record the amount of accretion of bulk materials at each plot on the temporal scale of multiple months. A total of 45 feldspar MHs were installed across the study site on 23 January 2014, comprising three replicates in each of the 15 study plots. Accretion was determined at later dates as the height difference between the marsh surface and the feldspar (i.e. the material accumulated above the MH) and was recorded as the mean of three replicate measurements from within the marker horizon at each sampling event. Measurements were taken 11, 13, 15, 17 and 19 months after installation. During the later sampling events, many MHs in *Sarcocornia* and *Sporobolus* plots became increasingly difficult to discern within the soil, probably due to bioturbation and mixing of sediments (Cahoon and Turner, 1989; Krauss et al., 2003). Consequently, monitoring of all plots was terminated after 19 months.

2.4 Sediment traps

Two complementary types of sediment trap were installed concurrently for the purpose of quantifying short-term (days) deposition of materials among the three vegetation assemblages. These types of traps were selected on the basis of the types of materials that they are most likely to collect, with the aim of providing insights into the processes driving deposition among assemblages. First, pre-weighed 50 mL centrifuge vials (30 mm mouth diameter, 115 mm depth) were placed into the ground, so that the "lip" of each tube was 10 mm above the ground surface. This vial method has a bias towards the collection of non-buoyant materials washing over the mouth of the tube (i.e. mineral matter) and a bias against collection of coarse and/or buoyant materials,

including large fragments of plant litter. Second, a modified version of the filter paper method described by Reed (1989) and Adame et al. (2010) was used to quantify passive sedimentation and litter accretion on the salt marsh surface. Pre-weighed 90 mm hydrophilic nylon filters (pore size 0.45 μm) were placed over 90 mm upturned plastic Petri dishes and attached to the salt marsh surface by two small staples, so that the nylon filter lay level with the surface. The resolution of this method (i.e. the smallest accumulation increment detectable), using a 90 mm filter, has been calculated as $0.0015 \, \mathrm{mg \, cm^{-2}}$ (Thomas and Ridd, 2004).

Three replicates of each short-term trap were installed at the centre of each of the study areas described above during the summer of 2014/15. We chose to base our sampling strategy upon expected tidal inundation patterns rather than capturing seasonal variability for several reasons. First, based on relevant literature (Rogers et al., 2014) we expect tidal inundation patterns to be of primary importance to deposition and accretion dynamics. Second, we do not expect there to be substantial seasonal variability due to factors other than tidal pattern variation. That is, the study region does not experience high seasonal variability in rainfall, nor are there clear seasonal patterns in terms of biomass standing stock or senescence (Clarke and Jacoby, 1994).

Short-term traps were deployed for 6-day (12 high tides) periods on four instances on the basis of tide chart predictions. Two neap ("December neap" and "January neap") periods were selected to reflect periods when high tides were at their lowest. While these neap periods were intended to measure periods without any inundation, higher than predicted tides occurred in both neap periods. Although unconfirmed, inundation of some plots within lower-elevation zones of the study area were expected to have occurred at least once during the December neap (up to 80 % of *Sarcocornia* plots and 100 % of *Sporobolus* plots) and/or the January neap (up to 60 % of *Sarcocornia* plots only; Table S1 in the Supplement). Two other periods ("December spring" and "January spring") were selected as maximum salt marsh inundation events with between five and 10 high tides inundating each plot in each period (Table S1). Although unintended, the fact that a small number of inundations were likely captured during neap tides more accurately reflects the differences in tidal behaviour that naturally occur among the three vegetation assemblages (i.e. lower-elevation assemblages are subject to a greater number of high tides throughout the year than higher-elevation assemblages). Consequently, all results from short-term measures were considered in the context of these varied inundation patterns.

Great care was taken not to disturb sediments or litter collected on or surrounding the removable traps during their installation and collection. Filters with visible crab-excavated sediment ($n = 23/180$) or that were physically upturned during inundation ($n = 3$ January spring inundation only) were excluded from analysis, although all plant (autochthonous and allochthonous) materials were retained for analysis as we considered these to be largely unaffected by crab excavation.

In the laboratory, vials were centrifuged, the supernatant decanted and the vial was placed in an oven for drying. All samples and vessels (filters and centrifuge vials) were dried at 60 °C until constant weight was achieved (≤ 72 h) and subtracted from initial vessel mass to obtain the dry weight of material collected. In addition, all identifiable litter was removed from each filter, identified to the species level and weighed. Litter samples of the main salt marsh species encountered (*S. quinqueflora*, *S. virginicus* and *J. kraussii*), wrack of the seagrass *Posidonia australis* and macroalga *Hormosira banksii*, fresh leaves of the mangrove *Avicennia marina,* and composite samples of all residual deposits (mineral component and unidentified organic matter; referred to hereafter as residues) from filters were also prepared for chemical analyses.

2.5 Elemental and isotopic analysis

Elemental C and N content was measured in order to quantify C deposition rates and infer biomass, litter and soil consumption quality (C : N, by weight). To infer the source of samples relative to reference source material and literature values, δ^{13}C was analysed. Dried above-ground plant biomass, litter and residues were homogenised and ground into a fine powder using a ball mill. The Champagne test (Jaschinski et al., 2008) was used to determine that no residue samples contained inorganic C. Consequently, acidification of samples was deemed unnecessary. Organic %C, %N and δ^{13}C were measured for all samples using an isotope ratio mass spectrometry – elemental analyzer (Thermo Delta V) at the University of Hawaii (HILO).

2.6 MIR

Diffuse reflectance mid-infrared (MIR) spectroscopy was used to assess the composition of biomass, litter and residue samples. MIR spectroscopy characterises the bulk composition and is therefore inclusive of both mineral and organic components. Spectra were acquired using a Nicolet 6700 Fourier transform infrared (FTIR) spectrometer (Thermo Fisher Scientific Inc., Waltham, MA, USA) following the specifications and procedures outlined by Baldock et al. (2013a). Spectra were acquired over 8000–400 cm^{-1} with a resolution of 8 cm^{-1}, but were truncated to 6000–600 cm^{-1}. This spectral range was chosen to include significant signal intensity (including the first near-infrared overtones of the MIR spectra) in the range 4000–6000 cm^{-1} and allow appropriate baseline correction, but also to exclude noise in the acquired signal intensity outside the selected spectral range. Spectra were baseline corrected using a baseline-offset transformation and were then mean centred using the Unscrambler 10.2 software (CAMO Software AS,

Oslo, Norway) before conducting principal component analysis (PCA).

2.7 ^{13}C NMR

Solid-state ^{13}C nuclear magnetic resonance (NMR) spectroscopy was used to quantify the contribution of C functional groups to live plant biomass, litter and residue samples. This was carried out to identify what compositional changes occurred between the different sample types and to what extent this differed between vegetation assemblages and inundation periods. Residue samples were treated with 2 % hydrofluoric acid (HF) according to the method of Skjemstad et al. (1994) to remove paramagnetic materials and concentrate organic C for ^{13}C NMR analyses. Cross-polarisation ^{13}C NMR spectra were acquired using a 200 MHz Avance spectrometer (Bruker Corporation, Billerica, MA, USA) following the instrument specifications, experimental procedures and spectral processing outlined by Baldock et al. (2013b). The ^{13}C NMR data are presented as the proportion of integral area under each of eight chemical shift regions corresponding to the main types of organic functional groupings found in natural organic materials: alkyl C (0–45 ppm), *N*-alkyl/methoxyl (45–60 ppm), O-alkyl (60–95 ppm), di-O-alkyl (95–110 ppm), aryl (110–145 ppm), O-aryl (145–165 ppm), amide/carboxyl (165–190 ppm) and ketone (190–215 ppm; Baldock and Smernik, 2002).

2.8 Statistical analyses

Separate simple linear regression analyses were conducted using all feldspar MH measurements for each of the three vegetation assemblages for the purpose of obtaining accumulation rates over 19 months and to assess the strength of linear fits for these data. We chose to use regression models that do not force a y intercept of 0, as doing so would place an unrealistic weight upon nil accretion values at time t_0. We present a comparison of regression statistics with and without y-intercept forcing in Table S2.

Bulk short-term deposition variables (bulk material collected in vials, bulk material collected on filters) were log-transformed to achieve normality and analysed with separate linear mixed models to test main and interactive effects of vegetation assemblage (*Sarcocornia, Sporobolus, Juncus*) and tidal event (repeated measures: December neap, December spring, January neap, January spring) on the amount of material retained at the end of a deployment period. Elevation was included as a covariate for each of these analyses. Covariance structure was selected for each model through comparison of Akaike's information criterion (AIC) of four covariance structures (unstructured, compound symmetry, diagonal, scaled identity). Where main effects presented significance differences ($P < 0.05$), post hoc tests (with conservative Bonferroni adjustment) were used to determine the difference among levels of vegetation and tidal event factors. Data are presented as the mean ± SE as there was some minor variation in sample numbers among assemblages and events (as described above). Statistical analyses were performed using SPSS v19 (IBM, USA), Origin Pro 2015 (Originlab, USA) and PRIMER v6 (PRIMER-E, UK).

2.8.1 Isotope mixing model

A two-source, single-isotope mixing model (Phillips, 2012) was used to estimate the proportion of C3 (f_1 in Eq. 1) and C4 (f_2 in Eq. 2) plants to the unidentified organic residue (i.e. the material leftover after macrolitter was removed):

$$f_1 = \frac{\delta^{13}C_{residue} - \delta^{13}C_{C4}}{\delta^{13}C_{C3} - \delta^{13}C_{C4}} \tag{1}$$

$$f_2 = 1 - f_1, \tag{2}$$

in which $\delta^{13}C$ denotes the isotopic signal of different sources of organic C: $C_{residue}$ (the residue organic C), C_{C3} (relevant C3 plants – *S. quinqueflora* litter for *Sarcocornia*–*Sporobolus* association residues or *J. kraussii* litter for *Juncus* assemblage residues) and C_{C4} (litter of the C4 species *S. virginicus*).

2.8.2 MIR analysis

PCA was performed using the transformed MIR spectra to (1) identify differences in composition among samples due to sample type and vegetation assemblage and (2) define the MIR spectral components most important to differentiating the samples. Loadings were plotted for the first two principal components to assist in the latter and to guide interpretation of differences in composition among samples.

3 Results

3.1 Feldspar MHs

Net accretion (i.e. vertical surface accumulation) was measured among *Juncus* plots throughout the entire 19 months, reflected in the moderate-strong linear fit ($R^2 = 0.68$, $P < 0.001$) and a mean accumulation rate with relatively low variance (1.74 ± 0.13 mm yr^{-1}). In contrast, accumulation above the feldspar MHs was more varied and slower overall in the *Sarcocornia* ($R^2 = 0.16$, $P < 0.001$, 0.76 ± 0.18 mm yr^{-1}) and *Sporobolus* plots ($R^2 = 0.14$, $P < 0.001$, 0.88 ± 0.22 mm yr^{-1}; Fig. 2). Accretion varied both spatially and temporally within the *Sarcocornia* and *Sporobolus* assemblages. Across *Sporobolus* plots, there was relatively high accretion recorded at the 11-month interval, followed by multiple peaks and troughs in the height of material measured above MHs, with some similarity among replicate plots in the timing of these (Fig. 2b). After modest gains at the 11-month interval, *Sarcocornia* accretion diverged among plots with two plots (*Sarcocornia* 2 and 5) experiencing continued accretion, whilst *Sarcocornia* 3 and 4

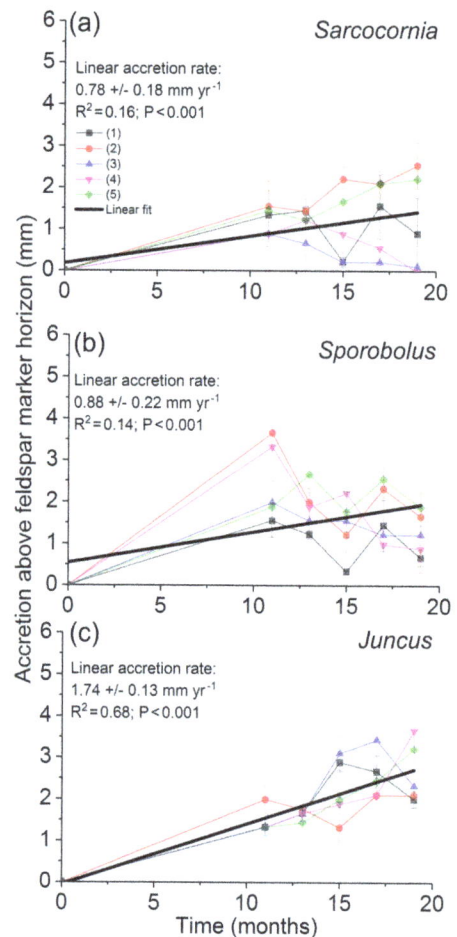

Figure 2. Surface accretion above feldspar marker horizons. Data are presented as the mean ± standard error of three replicate plots at each of five locations for each vegetation assemblage. A linear fit was applied on the basis of all data points ($n = 90$) for each vegetation assemblage.

appeared to lose surface material through the remainder of the study. The pattern of accumulation and loss observed between 13 and 19 months at *Sarcocornia* 1 was mirrored in the nearby *Sporobolus* 1 plot.

3.2 Short-term deposition

3.2.1 Vials

Mean bulk material deposition rates as determined by vials were higher than filter bulk deposition rates across all sampling events and vegetation assemblages (Table 1). Observations of materials retained within vials suggested a dominance of mineral matter and unidentified detritus, except in *Juncus* plots in which *Juncus kraussii* fragments were the dominant material.

Deposition varied significantly among tidal events ($F_{6, 42} = 10.01$, $P < 0.001$), with post hoc tests revealing

Table 1. Summary of sedimentation measurement techniques; mean and standard error values for feldspar marker horizon (MH), vial and filter bulk sediment measures; mean and standard error of organic C deposition rates; and contributions of C3 and C4 vegetation to deposition among vegetation assemblages. Accretion rates slower than local sea-level rise ($1.15\,\mathrm{mm\,yr^{-1}}$) are in red. DN: December neap; DS: December spring; JN: January neap; JS: January spring.

Technique	Parameter / *Measurement* / *Temporal scale* / *Biases/issues*	Period	*Sarcocornia*		*Sporobolus*		*Juncus*	
Feldspar MH	**Bulk measure**				accretion rate (mm y^{-1})			
	sediment accumulation		mean	SE	mean	SE	mean	SE
	mid-term (days to years)	0–19 mo.	0.78	0.18	0.88	0.22	1.74	0.13
	MH may be lost through erosion or bioturbation	95% CI lower	0.42		0.44		1.48	
		95% CI upper	1.14		1.32		2.00	
					see also Figure 2			
Vial	**Bulk measure**				bulk deposition rate (mg cm^{-2} d^{-1})			
	sediment deposition		mean	SE	mean	SE	mean	SE
	short-term (hours to days)	DN	9.33	3.12	7.49	2.48	10.37	3.41
		DS	68.97	10.63	182.11	28.67	49.63	8.94
	Biased towards materials entrained by water. Biased	JN	105.88	25.18	51.59	15.02	13.66	3.65
	against coarse litter larger than vial mouth	JS	275.93	89.62	221.61	24.43	40.85	5.86
Filter					bulk deposition rate (mg cm^{-2} d^{-1})			
	Bulk measure		mean	SE	mean	SE	mean	SE
	sediment deposition	DN	4.03	0.52	2.99	0.72	3.40	2.7E-03
	short-term (hours to days)	DS	5.97	0.95	3.59	0.27	6.49	6.7E-03
		JN	2.18	0.55	1.19	0.13	2.81	5.6E-03
		JS	8.31	2.15	6.84	1.27	6.97	7.7E-03
	Filter + material identification / *composition of material deposited*				*see Figure 3*			
	Filter + elemental analysis / *organic C deposition rate*				organic C deposition rate (mg C cm^{-2} d^{-1}) [a]			
			mean	SE	mean	SE	mean	SE
		DN	0.27	0.04	0.21	0.05	1.21	0.17
		DS	0.25	0.06	0.34	0.06	2.37	0.44
		JN	0.10	0.02	0.17	0.04	1.14	0.41
		JS	0.62	0.08	0.40	0.07	1.99	0.46
	Filter + isotopic analysis / *sources contributing to soil organic C*				isotope mixing model - plant contribution (%)			
			C3	C4	C3	C4	C3	C4
		DN	59.6	40.4	38.3	61.7	80.2	19.8
		DS	77.5	22.5	25.0	75.0	84.6	15.4
		JN	67.9	32.1	19.9	80.1	81.0	19.0
		JS	72.4	27.6	26.0	74.0	78.8	21.2
					see also Table 2			
	Filter + MIR & ^{13}C NMR / *character of deposited materials*				*see Figure 4 & Table 2*			

each event as significantly different from the others. Despite large differences in mean deposition among the three vegetation assemblages during December spring, January neap and January spring events (Table 1), vegetation assemblage was not a significant factor when elevation was included as a covariate ($F_{2,\,45.8} = 1.06$, $P = 0.36$). There was, however, a significant event times assemblage interaction ($F_{6,\,42} = 10.01$, $P < 0.001$), with deposition in *Sarcocornia* vials higher during January neap relative to December spring for *Sarcocornia* plots, but not so for *Sporobolus* and *Juncus* vials (Table 1). Deposition into vials was lowest for all three assemblages during December neap (Table 1) and was highest overall in *Sarcocornia* vials during January spring ($275.93 \pm 89.62\,\mathrm{mg\,cm^{-2}\,d^{-1}}$).

Regression of the log (mass of material retained within vials) versus plot surface elevation revealed no clear relationship between the two variables during the December neap period (Fig. S1a in the Supplement), but negative relationships ($P < 0.001$, $R^2 = 0.35$ to 0.59) existed for all other time periods (Fig. S1b–d). That is, there were broad trends of higher sedimentation at lower-elevation plots than higher-elevation plots during these periods.

3.2.2 Filters

Retention of bulk materials on filters also varied among all four tidal periods ($F_{3,\,109.3} = 48.82$, $P < 0.001$), with overall deposition highest in January spring, followed by December spring (Table 1, Fig. 3). Bulk deposition on filters varied among vegetation assemblages ($F_{2,\,30.85} = 48.82$, $P = 0.004$), with significantly lower deposition in *Sporobolus* plots relative to both *Sarcocornia* (Bonferroni-adjusted P value $= 0.010$) and *Juncus* (Bonferroni-adjusted P value $= 0.023$) plots across all tidal events (Fig. 3, Table 1). In contrast to the vials, there was no clear relationship between bulk material retained on filter papers and plot surface elevation during either of the neap or spring tidal events (Fig. S2).

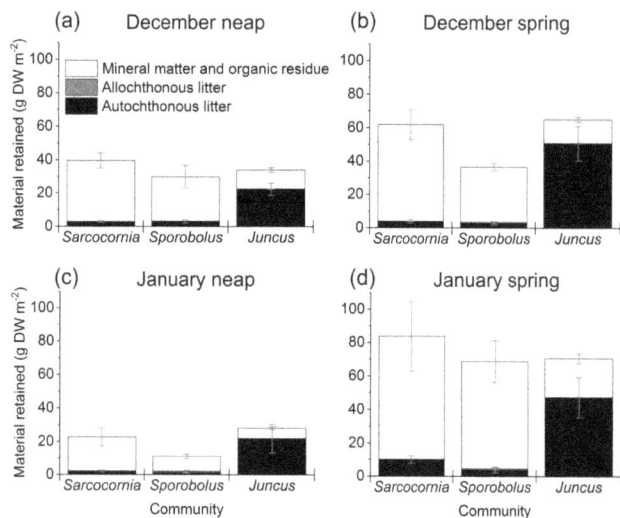

Figure 3. Mean mass of autochthonous litter, allochthonous litter and mineral matter–organic residue retained on filters at the end of a 6-day deployment during December neap (**a**), December spring (**b**), January neap (**c**) and January spring (**d**) tidal periods. Error bars are presented for each component and represent 1 standard error each side of the mean.

Although the mass of bulk material retained on filters was similar across *Sarcocornia* and *Juncus* plots, Fig. 3 demonstrates that different materials were contributing to surface accumulation among the two vegetation assemblages. In *Juncus* plots, autochthonous plant litter (that is, from the dominant species *Juncus* kraussii and the subdominant species *Sporobolus* virinicus) contributed between 66 % (December neap) and 78 % (both December spring and January neap) of all deposited material mass. In contrast, litter contributions were low (≤ 12 % of all deposited material) in both *Sarcocornia* and *Sporobolus* assemblages, regardless of the tidal period. Contributions from identifiable allochthonous materials were low in all cases, with negligible quantities of *Posidonia australis* litter (recorded in five out of all 60 *Sporobolus* filters) and a single large piece of *Hormosira banksii* deposited on a *Sporobolus* filter during December spring – the latter was considered an outlier and was therefore excluded from Fig. 3.

Chemical analysis of the unidentified portion of material deposited on filters also highlights differences between the vegetation assemblages. The organic content (%C, %N) of unidentified material pooled across *Juncus* plots was much higher than for the other assemblages (Table S3), with this difference also apparent in the disparity between C accumulation rates in *Juncus* versus *Sarcocornia* and *Sporobolus* assemblages (Table 1).

3.3 Elemental and isotopic ratios

Elemental C : N ratios and δ^{13}C values of plant biomass, litter and unidentified residues are presented in Table 2. The

biomass and litter samples of the C4 grass *Sporobolus* were more enriched in ^{13}C relative to those of the C3 species *Sarcocornia* and *Juncus* (Table 2). This distinction, however, was not as great for the unidentified residue samples, with δ^{13}C values from all assemblages sitting between the δ^{13}C values of the C3 and C4 salt marsh plants.

Outputs from the isotope mixing model (Table 1) highlighted differences in source contribution among the vegetation assemblages. *Sarcocornia* residues showed a higher contribution of C3 plant material during spring tides relative to the neap tides. Further, similar contribution from the host plant (i.e. C3 in *Sarcocornia* and C4 in *Sporobolus*) to residues was apparent for all tidal periods except January neap, when the C4 contribution to *Sporobolus* residue was higher. Overall, contributions of the host plant ranged from 59.6 to 77.5 % in *Sarcocornia* plots and from 61.7 to 80.1 % in *Sporobolus* plots.

Source contributions across the four tidal periods were most consistent in the *Juncus* assemblage, in which estimates ranged between 78.8–84.6 % for C3 plant material and 15.4–21.2 % for C4 plant material. These contributions aligned well with visual observations of plant cover across plots (in which the C3 plant *J. kraussii* is dominant over the C4 plant *S. virginicus* in approximately an 80 : 20 % biomass mix). Quantification of litter fall onto filters, however, highlighted a skew towards *J. kraussii* litter (85.9–97.0 %) over *S. virginicus* litter (3.0–14.1 %) across the *Juncus* assemblage. Residue C : N ratios were also highest for *Juncus*, followed by *Sporobolus* and then *Sarcocornia*. While *Juncus* litter samples had a higher C : N, relative to all other *Juncus* biomass (Table 2), this difference was not noted for *Sarcocornia* or *Sporobolus*.

3.4 MIR and ^{13}C NMR

Together, the first two principal components explained 96.4 % of the variation in MIR spectra of all samples assessed. A clear separation of residue samples from litter and biomass is apparent along PC1 (Fig. 4a) with inspection of the loadings plot (Fig. 4b) highlighting variation in the range of 600–2000 cm^{-1} (quartz) and distinct troughs at 3400 cm^{-1} (water) and 2900 cm^{-1} (OM-alkyl). Residue samples are separated along PC2, with differentiation among vegetation assemblages, regardless of tidal event. The loadings plot for PC2 (Fig. 4C) also exhibits variation in the range of 600–2000 cm^{-1} (quartz), a peak at 2900 cm^{-1} (OM-alkyl) and 3600–3700 cm^{-1} (kaolinite).

The proportions of C within each of eight organic functional groupings for each sample analysed with ^{13}C NMR are presented in Table 2. For all samples O-alkyl C was the most abundant. O-alkyl C content was higher in live plant biomass than litter for both *Sarcocornia*, but less so for *Sporobolus* and essentially unchanged for *Juncus*. Generally, residues were higher in alkyl C, and amide/carboxyl C, and lower in O-alkyl, di-O-alkyl and aromatics relative to litter and

Table 2. Results of ^{13}C NMR, δ^{13}C and elemental (C : N) analyses for each component of three salt marsh plant assemblages and other potential sources. ^{13}C NMR outputs are the average of two samples for each of the biomass and litter components (except *Sporobolus* litter, for which only one reliable spectrum was obtained). δ^{13}C and C : N values are the mean ($\pm$ SE) of biomass ($n = 3$) and litter ($n = 4$) samples. Residue samples were pooled from 15 filters in each vegetation assemblage. The tidal period was deemed not appropriate (n/a) for live biomass sample collection.

Community	Component	Tide	^{13}C NMR chemical assignment and region (ppm)								δ^{13}C	C : N
			Alkyl (0–45)	N-alkyl/methoxyl (45–60)	O-alkyl (60–95)	Di-O-alkyl (95–110)	Aryl (110–145)	O-aryl (145–165)	Amide/carboxyl (165–190)	Ketone (190–215)		
Sarcocornia	biomass	n/a	14.4	6.6	45.0	10.6	10.5	4.4	7.1	1.4	−26.5 ± 0.5	38.0 ± 4.2
	litter	combined	10.2	5.1	34.0	9.0	16.3	8.6	13.2	3.6	−25.7 ± 0.1	33.0 ± 1.2
	residue	neap	23.1	8.2	33.7	7.3	11.3	4.3	11.0	1.1	−22.1 ± 0.3	14.2 ± 0.4
		spring	25.1	8.1	32.8	7.1	10.5	4.0	11.3	1.1	−23.3 ± 0.2	14.2 ± 0.7
Sporobolus	biomass	n/a	8.6	4.6	51.0	12.0	11.5	4.6	6.5	1.3	−14.9 ± 0.1	59.8 ± 6.3
	litter	spring	5.2	4.5	47.1	12.0	14.6	6.6	8.0	2.0	−15.8 ± 0.2	59.4 ± 5.2
	residue	neap	19.2	7.9	36.5	8.3	12.6	5.0	9.4	1.2	−18.7 ± 0.6	15.5 ± 0.5
		spring	18.5	7.3	36.6	8.4	13.0	5.2	9.5	1.4	−18.4 ± 0.0	16.3 ± 1.1
Juncus	biomass	n/a	8.6	5.4	51.8	12.2	11.2	4.6	5.2	1.1	−24.8 ± 0.3	61.3 ± 5.6
	litter	combined	5.6	5.4	52.2	12.8	11.9	5.4	5.4	1.3	−25.6 ± 0.1	89.9 ± 4.5
	residue	neap	Insufficient material available for ^{13}C NMR analysis								−23.7 ± 0.0	19.6 ± 0.0
		spring	15.4	7.2	35.7	8.8	15.4	6.6	9.2	1.7	−23.8 ± 0.2	18.2 ± 0.4
Other sources	algal mat	n/a	14.3	5.6	37.5	8.7	11.5	4.9	14.8	2.6	−15.0 ± 0.4	13.2 ± 0.1
	mangrove leaf	n/a	16.4	6.4	40.6	9.1	12.8	4.8	8.7	1.1	−28.7 ± 0.3	24.8 ± 1.7
	seagrass wrack	spring	11.6	3.9	33.3	8.0	14.4	7.7	16.4	4.6	−12.3	29.2
	macroalga wrack	spring	6.0	2.6	52.5	14.0	5.5	6.6	11.5	1.3	−17.7	57.8

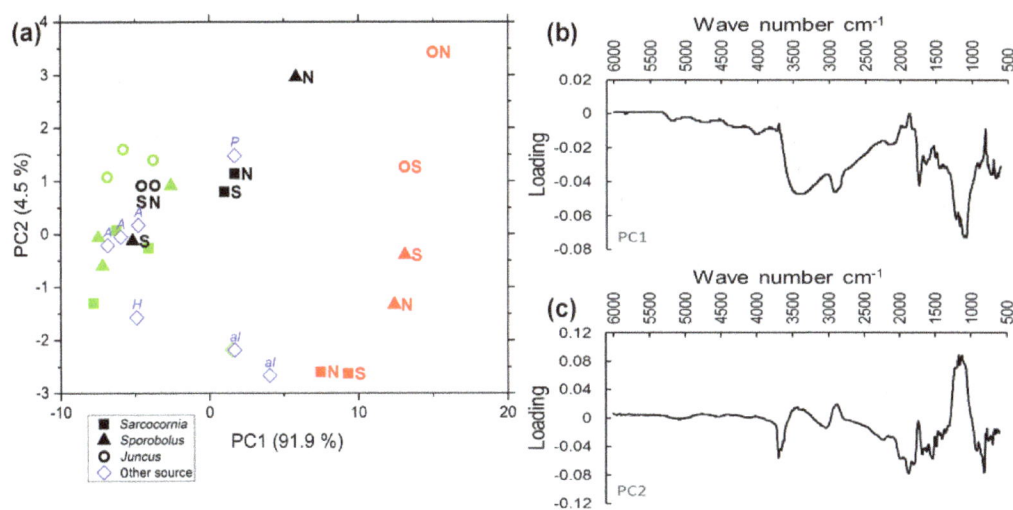

Figure 4. Principal component analysis of MIR spectra with the proportion of variance explained by each component given in parentheses (**a**); spectral loading plots for PC1 (**b**) and PC2 (**c**). S: spring tide samples; N: neap tide samples; H: *Hormosira banksii* (macroalga) wrack; P: *Posidonia australis* (seagrass) wrack; A: *Avicennia marina* (mangrove) leaf; al: benthic algae mat.

biomass samples. There were differences in residue C composition according to which vegetation assemblage they were collected from – aromatics (higher in *Juncus* and *Sporobolus*), alkyl C and amide/carboxyl (higher in *Sarcocornia*). There was high similarity between residues collected under the two different tides, however, for both the *Sarcocornia* and *Sporobolus* assemblages. These similarities among tides are mirrored in the similarity of the residue C : N values. There was insufficient residue material available for analysis from *Juncus* neap tide, even though samples were pooled across a large number of filters, further highlighting the small contribution of unidentifiable sedimentary components within this assemblage.

4 Discussion

In this study we have compared sediment and C accretion dynamics among three vegetation assemblages within a single intertidal wetland complex. Our findings, across a range of methods, showed that there were substantial differences among assemblages in (1) the types of materials deposited on the marsh surface in the short term (days) and (2) the quantities of material accumulated over the medium term (19 months). Here, we first consider the accumulation differences among assemblages over the medium term and then discuss the interactions among vegetation, physical and degradation processes that are likely driving short-term and medium-term differences among assemblages. We conclude with an assessment of the implications for C accumulation and response to relative sea-level rise (RSLR).

4.1 Accretion varies among vegetation assemblages

Surface accretion above feldspar MHs over a period of 19 months and deposition measured with short-term sediment traps provide evidence of the multiple ways in which deposition and accretion dynamics differ between salt marsh vegetation assemblages. First, feldspar MHs highlight a record of continued and consistent accretion across the upper marsh *Juncus* assemblage, amounting to a reliable ($R^2 = 0.68$) accretion rate of 1.74 ± 0.13 mm yr^{-1} (Fig. 2). This value is remarkably similar to the mean accretion rate measured over 10 years above feldspar MHs of 1.76 mm yr^{-1} by Saintilan et al. (2013) for *Juncus* salt marshes across a range of sites in SE Australia. In contrast, accretion above MHs in *Sarcocornia* and *Sporobolus* assemblages varied substantially – both spatially and temporally – in our study (Fig. 2), possibly due in part to the influence of erosion, bioturbation and sediment mixing above MHs (Cahoon and Turner, 1989; Krauss et al., 2003). While our mean accretion estimates for both *Sarcocornia* and *Sporobolus* are lower than the regional estimate for *Sarcocornia–Sporobolus* associations (1.11 ± 0.08 mm yr^{-1}; Saintilan et al., 2013), this regional mean is within the 95 % confidence interval for both species at Towra Point (Table 1). Critically, medium-term accretion rates in the *Juncus* assemblage consistently exceed contemporary rates of sea-level rise within Botany Bay (1.15 mm yr^{-1}), while mean accretion rates for both *Sarcocornia* and *Sporobolus* (including the upper 95 % confidence interval of *Sarcocornia*) are below the contemporary rate of sea-level rise. These patterns of high marsh versus lower marsh accretion are also atypical of results reported outside of the study region (see Sect. 4.2.2), and while in general agreement within previous data from our region, these accretion patterns would benefit from validation across a broader network of sites.

4.2 Processes driving spatial variability in deposition and accretion

One of the key strengths of using short-term deposition methods is the ability to identify and quantify the composition of inputs that may be contributing to differences observed over the medium term. In this study, a distinction was observed between the *Juncus* assemblage (in which medium-term accretion rates were consistently high and organic deposition was higher) and the *Sarcocornia* and *Sporobolus* assemblages (in which medium-term accretion rates were generally lower but more varied and deposition was dominated by mineral material; Table 1, Fig. 3). This distinction was best exemplified by the results of the short-term filter method (Fig. 3), in which differences in the contri-butions of autochthonous litter and the residual sediment (comprising mineral and organic residue components) were stark. There was further evidence of this in the short-term vial results, where mineral-biased deposition was high in the lower-elevation, non-rush assemblages and low in the higher-elevation *Juncus* assemblage during multiple experimental periods (Table 1, Fig. S1). Although higher-than-predicted tides likely influenced some short-term traps during neap experimental periods (Table S1), the fact that deposition into vials was lower during December neap (when up to 80 % of *Sarcocornia* plots and 100 % of *Sporobolus* plots would have been subjected to at least one tidal inundation) relative to January neap (up to 60 % of *Sarcocornia* plots; no inundation of *Sporobolus* plots) suggests that this had a small impact relative to other influences. Non-tidal processes, such as rain- or wind-driven sedimentation and/or bioturbation, are the most likely factors behind sedimentation when inundation was absent. Although filters with visible crab-excavated sediment ($n = 23/180$) were excluded from analysis, such clear identification was not able to be determined for sediments deposited in vials.

In the following sections we interpret the influence of biological, physical and interactive processes on salt marsh surface dynamics on the sub-site scale. We do so by assessing the response of different surface deposition measures (two short term, one medium term) among the three vegetation assemblages studied.

4.2.1 The influence of vegetation on salt marsh surface deposition

The results of this study partially support our first hypothesis. That is, there were broad differences among the vegetation assemblages regarding the amount and type of materials deposited in the short term and rates of accretion in the medium term. The high spatial and temporal variability in marker horizon measurements across *Sarcocornia* and *Sporobolus* plots (Table 1; Fig. 2a, b) limit interpretation of medium-term processes between these two assemblages. In contrast, the relative stability of medium-term accretion patterns in *Juncus* plots (Fig. 2b) and unique short-term deposition results from *Juncus* plots (Table 1, Fig. 3, plus Sect. 4.3.2) allow some interpretation of the potential role of vegetation structure on surface dynamics.

There are fundamental differences in vegetation structure and function that can at least partly account for the variations in the quantity and type of materials being retained in rush (*Juncus*) versus non-rush (*Sarcocornia* and *Sporobolus*) assemblages. First, *Juncus* assemblages have large potential for litter production through the annual replacement of their significant above-ground biomass (1116 g m^{-2}; Clarke and Jacoby, 1994). No clear patterns of annual turnover have been observed in *Sarcocornia* and *Sporobolus* assemblages, in which standing biomass is only about one-third that of the *Juncus* assemblage (Clarke and Jacoby, 1994). There may also be indirect vegetation effects on the deposition and accumulation of surface materials. For instance, the tall (~ 1 m), dense structure of the *Juncus* assemblage is likely to enhance (1) the retention of autochthonous litter (Fig. 3), which may have otherwise been exported during tidal recession, and (2) the capture of mineral particles on plant stems (Morris et al., 2002; Mudd et al., 2004). Dense salt marsh vegetation also has the capacity to enhance sedimentation by reducing the turbulent energy of inundating waters, with Mudd et al. (2010) demonstrating that this phenomenon was responsible for virtually all of the sedimentation increase observed when standing plant biomass of *Spartina alterniflora* was artificially increased. The high short-term litter deposition rates we observed during neap tides (Fig. 3a, c) and the increased contribution of both mineral and litter components during spring tides (Fig. 3b, d) suggested that each of these direct and indirect plant mechanisms may be contributing to the relatively high medium-term accretion rates observed within *Juncus* assemblages.

4.2.2 The influence of physical factors on salt marsh surface deposition

Differences in suspended sediment supply and tidal flooding characteristics (tidal range, position within the tidal prism) have been identified as key physical drivers of salt marsh accretion (Chmura and Hung, 2004; Rogers et al., 2014). Generally, lower elevation within the tidal frame and closer proximity to the source of tidal inundation result in higher sedimentation rates. This is because (1) greater flooding depth allows for greater suspended sediment volume and higher sedimentation, (2) the increase in flooding duration increases the time for sediment deposition to occur and (3) flooding frequency is higher at lower elevations (Baustian et al., 2012; Harter and Mitsch, 2003; Morris, 2007; Oenema and DeLaune, 1988). If these processes were operating at our site, we would have expected to observe higher sedimentation rates in the *Sarcocornia* and *Sporobolus* assemblages, which were generally both lower in the tidal frame (Table S1) and nearer to tidal sources (Fig. 1). Indeed, when measurements relevant to the mineral component were considered, our results appeared to be consistent with this. First, overall mineral retention on filters (Fig. 3) was highest in the *Sarcocornia* and *Sporobolus* assemblages. Second, short-term vials (i.e. the method that is biased toward retention of heavy minerals over organic matter) showed a significant log-linear relationship between deposition and elevation during the periods of greatest tidal inundation (December spring, January spring) and during January neap when significant rainfall (Fig. S3) as well as some inundation of low-elevation sites likely occurred (Table S3). Further, observations made during the first measurement of feldspar MHs at 11 months suggested a high mineral contribution in all *Sporobolus* and *Sarcocornia* plots, though it is unclear why this accretion trend reversed in many plots after this sampling event. Overall,

mineral-specific deposition results are largely supportive of the role of the physical position within the salt marsh towards differences among assemblages.

Importantly, the deposition–elevation relationship expressed by the mineral component did not apply when bulk results of the passive short-term filter method were considered. With the mineral bias effectively removed, no clear relationship between elevation and bulk deposition (i.e. mineral plus organic matter) was observed across any of the tidal periods (see Fig. S2). Instead, total deposition was similar between the minerogenic, lower-elevation *Sarcocornia*–*Sporobolus* plots, and the organogenic higher-elevation *Juncus* plots.

The lack of an elevation relationship in terms of bulk material deposition is somewhat contrary to spatial patterns expected on the basis of physical sedimentary processes in the tidal zone. This disparity extended to medium-term accretion results, in which lower marsh (*Sarcocornia* and *Sporobolus*) assemblages accrete at a slower rate than upper marsh (*Juncus*), both in our study and regionally (Saintilan et al., 2013). This relationship does not necessarily downplay the importance of tidal influence on surface dynamics in SE Australian salt marshes. An alternative explanation is that these physical processes, in interaction with biological factors, instead remobilise and redistribute materials across the lower marsh assemblages, rather than depositing significant amounts of new allochthonous material.

4.2.3 Redistribution of surface materials

The second hypothesis of our study was that the source and character of materials deposited would vary temporally with tidal inundation patterns. For the most part, however, this was not observed, with high degrees of within-assemblage similarity for neap and spring tide samples across the various analyses undertaken (Table 2, Fig. 4). Instead, our results provide multiple lines of evidence that suggest a redistribution of surface materials across the salt marsh, mediated by a range of biological, physical and interactive processes.

The first indication of redistribution of surface materials was the mismatch between rates of short-term bulk deposition and patterns of medium-term accretion among vegetation assemblages. This was best exhibited in the *Sarcocornia* plots, in which short-term measures showed deposition to be as high or higher in *Sarcocornia* plots relative to the other assemblages (Table 1, Fig. 3), while medium-term accretion was actually lowest here (Fig. 2). This suggests that short-term measures in this assemblage were capturing materials that were being moved or redistributed across the salt marsh but not necessarily retained in a given location over longer time periods (i.e. months). While the short- versus medium-term discrepancy was not as large for the *Sporobolus* assemblage, the temporal variability in medium-term feldspar MH measurements (i.e. multiple peaks and troughs across the 19-month period for most plots) also suggested significant redistribution of materials over time in this assemblage. Such movement of materials within the *Sarcocornia* and *Sporobolus* assemblages also fits with the expectation that hydrodynamic energy, and therefore potential for sediment redistribution, would be highest in the salt marsh zones lower in the tidal frame and located closer to tidal sources (Fig. 1, Table S1). We also attribute the fading of feldspar horizons in many *Sarcocornia* and *Sporobolus* plots over time to mixing of sediments (Cahoon and Turner, 1989) in this active zone, with assistance from bioturbation (Cahoon and Turner, 1989; Krauss et al., 2003). In contrast, these temporal discrepancies and variations (including fading of MHs) were not observed in the *Juncus* assemblage, in which hydrodynamic energy is expected to be greatly reduced as a result of both its position within the marsh and the influences of plant biomass (see discussion in Sect. 4.2.2).

Next, it was not expected that tidal inundation would substantially increase salt marsh plant litter production. We therefore interpret the increased concentration of autochthonous litter in *Juncus* plots during spring tides relative to neap tides (Fig. 3) as evidence of the redistribution and trapping of autochthonous material within this assemblage. That is, the extra spring tide litter was material that had been remobilised by inundating water and redistributed within the same community, resulting in a larger amount of material being caught on the *Juncus* filters. The fact that no identifiable *Juncus* litter was collected on any of the *Sporobolus* short-term filters, despite their position being within the expected path of receding tides (Fig. 1), further highlights the retaining capacity within the *Juncus* assemblage. While it is not known over what scale the litter redistribution occurs in the *Juncus* assemblage, we expect it to be highly localised, given the dense structure of standing vegetation here and its capacity to impede movement of coarse litter particles.

Finally, by placing our *Sarcocornia* and *Sporobolus* plots within small patches vegetated exclusively by either the C3 species (*S. quinqueflora*) or the C4 species (*S. virginicus*), we are able to estimate the contribution of each resident plant to the short-term residue collected from within its assemblage. While the dominance of resident plant signatures suggested a strong autochthonous contribution in all instances (see mixing model results in Table 1), residue signatures across all tidal periods reveal a mixture of sources both present (i.e. the resident plant) or neighbouring (i.e. the other co-dominant plant in the association) to the plots. The fact that contributions of sources other than the resident plant were of the order of 20–40 % (Table 1) during the neap tides suggests significant mixing across scales greater than the monospecific patches (i.e. several metres or more). While some of this movement of materials may have been due to the creep of the highest neap period tides into the lower elevation plots (though this appears small – see Sect. 4.2), non-tidal agents such as redistribution by rainfall (Chen et al., 2015) and faunal activity (Guest et al., 2004) may have also contributed. It should be noted that our isotopic mixing model does not ac-

count for any degradation-related kinetic fractionation from plants to litter and sediments. Data in Table 2 suggest little to no fractionation between fresh biomass and partially decomposed litter samples, consistent with other studies comparing $\delta^{13}C$ between fresh and decomposing leaves of estuarine plant species (e.g. Zieman et al., 1984; Fry and Ewel, 2003; Saintilan et al., 2013). We cannot rule out the potential for isotopic fractionation occurring in the decay from litter to residue samples, however, and we recommend this as an area of future research.

A two-source (C3 plant vs. C4 plant) mixing model probably presents an overly simplified estimate of source matter contributions. This is because it does not account for other potential sources that have $\delta^{13}C$ values within or near the range of salt marsh plant sources prescribed in the mixing model. These include mangroves ($-28.7 \pm 0.3\,\text{‰}$), seagrass ($-12.3\,\text{‰}$), macroalgae ($-17.7\,\text{‰}$) and benthic algae ($-15.0 \pm 0.4\,\text{‰}$). Of these, benthic algae would have the greatest potential for contributing to *Sarcocornia* residue, as vegetation is sparsest here (and therefore light penetration to benthos the greatest), while the MIR PC plot (Fig. 4) also points to a similarity in chemical composition between the two. However, the fact that *Sporobolus* residues are consistently depleted in ^{13}C, relative to both the resident plant (*S. virginicus*) and benthic algae, shows that our interpretation of mixing between both C3 and C4 sources is warranted at least in that assemblage. In contrast, the constancy of isotope signatures and their overall similarity with the mix of C3- and C4-derived biomass in the *Juncus* plots provide further evidence of the autochthonous nature and trapping capacity of this assemblage.

Together, these findings allow several hypotheses about redistribution of surface materials to be made. First, short-term deposition measures may capture a significant proportion of within-marsh redistribution and therefore may not necessarily equate with longer-term accretion. Second, the capacity of vegetation to retain autochthonous materials appears to vary substantially among species assemblages. Third, redistribution is likely to be greatest in more exposed, lower-biomass assemblages. These findings also highlight the importance of considering redistributed materials in quantifications of wetland surface dynamics and likely shortcomings for studies that attempt to assess surface dynamics using only short-term methods.

4.3 Implications for wetland functioning

Understanding the biological and physical feedbacks that affect surface dynamics is critical to predict the survival of intertidal wetlands and their associated ecosystem services under changing environmental conditions (Kirwan and Megonigal, 2013). To this end, the data collected as part of this study reveal patterns of how C deposition and accumulation, organic matter decomposition and vulnerability to sea-level rise vary among salt marsh assemblages.

4.3.1 C deposition and accumulation rates

The distinction between organogenic and minerogenic deposits, and their respective locations within the tidal frame, has important implications for surface C deposition and accumulation rates. Here we estimate mean C deposition rates ranging from 0.10 to 0.62 mg C cm^{-2} d^{-1} across the four tidal periods for the minerogenic *Sarcocornia* and *Sporobolus* assemblages and from 1.14 to 2.37 mg C cm^{-2} d^{-1} for the organogenic *Juncus* assemblage (Table 1). It should be noted that such short-term C deposition rates inclusive of plant litter will likely represent a massive overestimation of C that is retained and sequestered over longer timescales, due to diagenesis of deposited OM (Duarte and Cebrian, 1996), and the potential for materials to be redistributed or even exported by tidal and non-tidal processes (see Sect. 4.2.3). Therefore, these deposition rates are not directly comparable to C accumulation rates determined by medium-term (e.g. feldspar MH) or longer-term (e.g. radiometric dating) techniques. Notwithstanding this, the magnitude of the differences we report among assemblages above fit broadly with differences in regional estimates of C accumulation over the medium term (10-year MH experiments), which have been estimated as 4.5 times higher in *Juncus* relative to *Sarcocornia–Sporobolus* salt marsh (Saintilan et al., 2013). Similarly, our results are also in agreement with findings further north in Moreton Bay, where Lovelock et al. (2013) reported much higher C sequestration rates on oligotrophic sand island marshes dominated by *J. kraussii* than *S. quinqueflora*-dominated marshes on the western side of that bay.

4.3.2 Chemical composition of deposits varies among assemblages

We have assessed the chemistry of above-ground biomass, litter and unidentified residues through elemental (C : N) and spectrometric (MIR spectroscopy, ^{13}C NMR) methods. Most importantly, our results highlight assemblage differences in the transformation of OM along the biomass–litter–soil decay continuum. Shifts in the bulk composition of materials were best seen in the principal component plots of MIR spectra, in which biomass, litter and soil residue samples varied across PC1 (Fig. 4a). Broadly, the separation of residues from litter and biomass was primarily due to the addition of mineral components in the residues; however, there was also evidence of a shift in alkyl OM. Specifically, the presence of a single peak at ~ 2900 cm^{-1} in the loadings plot (Fig. 4b) was indicative of a declining cellulose content across PC1, that is, in the general order live biomass – litter – residue. Importantly, cellulose also appears to be a factor in the separation of residues from the three different salt marsh assemblages along PC2 (Fig. 4c), suggesting higher content in the two *Juncus* samples, followed by *Sporobolus* and then *Sarcocornia* samples. This finding was confirmed by ^{13}C NMR data, which showed that greater proportions of plant com-

pounds (carbohydrates more broadly, as well as lignin) were retained within the *Juncus* litter and residue relative to the other species (Table 2). In contrast, the higher proportions of alkyl-C and amide/carboxyl-C within *Sarcocornia* and *Sporobolus* residues were indicative of higher protein and lipid contents, consistent with bacterial biomass and marine algae signatures (Dickens et al., 2006). However, they may also be partly explained by the selective retention of resistant plant waxes, such as suberin and cutan. We therefore reject our third hypothesis that decay patterns would show no differences among the vegetation assemblages.

There are multiple mechanisms that may explain the greater retention of plant-derived C along the biomass–litter–residue pathway for *Juncus*, relative to the other assemblages. The simplest explanation is that a high turnover of *Juncus* biomass (and its exclusion of other sources through shading and/or structural impedance) ensures ample supply of plant-derived C to the benthos. Our data, however, reveal an important biomass to litter transformation in *Juncus* that was not observed in either the *Sarcocornia* or *Sporobolus* assemblage. That is, the C : N of *Juncus* litter increased substantially relative to live biomass. Such an increase is commonly observed in terrestrial (McGroddy et al., 2004) and marine (Stapel and Hemminga, 1997) plants and may be explained by the selective resorption of nutrients (but not carbon) by the plant prior to, or during, senescence (McGroddy et al., 2004; Stapel and Hemminga, 1997). Such a mechanism was supported by the constancy of molecular C composition between *Juncus* biomass and litter (Table 2). The selective resorption of N by a plant has important implications for the fate and processing of the resulting litter and residue, as tissue C : N is considered a primary determinant on salt marsh OM decomposition (Minden and Kleyer, 2015). By retaining nutrients within the living tissues, the plant effectively decreases the lability of resulting litter and residual soils and makes them less attractive to the microbial decomposer community (Hessen et al., 2013; Sterner and Hessen, 1994). This will have the effect of lowering OM remineralisation rates in *Juncus* relative to other assemblages, a result that also coincides with the bacterial and/or marine algae signatures (Dickens et al., 2006) inferred for *Sarcocornia* and *Sporobolus* but not apparent within the more recalcitrant *Juncus* residues (Table 2).

Together, these data from SE Australia contribute to a broader pattern of plant assemblage differences in salt marsh surface dynamics and C sequestration potential (Minden and Kleyer, 2015; Saintilan et al., 2013; Wang et al., 2003). They also highlight short-term processes that may contribute to the high capacity of *Juncus* salt marshes to accumulate significant C stocks globally (0.034 g C cm^{-2} yr^{-1} or 0.093 mg C cm^{-2} d^{-1}), relative to most other salt marsh genera (mean C accumulation rate $= 0.024$ g C cm^{-2} yr^{-1} or 0.066 mg C cm^{-2} d^{-1}; Ouyang and Lee, 2014).

4.3.3 Vulnerability to sea-level rise

There is growing evidence of the capacity of coastal wetlands to maintain surface elevation with relative sea-level rise, in certain situations, by increasing surface elevation through below-ground production, enhanced trapping of sediments or a combination of the two (Baustian et al., 2012; Kelleway et al., 2016b; McKee et al., 2007). Where wetland assemblages are unable to maintain a suitable elevation relative to inundating water levels, vegetation shifts may occur, including the loss of marsh vegetation (Day et al., 1999, 2011; Rogers et al., 2006). While wetland surface elevation is a function of multiple factors, including below-ground production and decomposition, groundwater dynamics, and sedimentary and regional subsidence (Cahoon et al., 1999; Rogers and Saintilan, 2008), the retention of above-ground inputs plays a critical role in wetland survival under changing hydrological conditions (Day et al., 2011).

With this in mind, medium-term accretion data suggest that *Sporobolus* and *Sarcocornia* assemblages at our study site may be particularly vulnerable to current RSLR. That is, mean surface accretion rates were either lower (*Sarcocornia* $= 0.92 \, \mathrm{mm \, yr^{-1}}$) or only marginally higher (*Sporobolus* $= 1.30 \, \mathrm{mm \, yr^{-1}}$) than contemporary rates of local sea-level rise within Botany Bay ($1.15 \, \mathrm{mm \, yr^{-1}}$). Across much of the Towra Point Nature Reserve (as well as elsewhere in the region) upslope encroachment of mangrove shrubs into *Sarcocornia–Sporobolus* association is occurring, possibly in response to sea-level rise (Kelleway et al., 2016b). In contrast, vegetation change (either in the form of mangrove encroachment or dieback) has not been widely reported for *Juncus* assemblages across SE Australia over recent decades, suggesting relative stability during a time of changing sea levels. While below-ground biomass production likely plays a role, average *Juncus* surface accretion rates ($1.70 \, \mathrm{mm \, yr^{-1}}$ in this study, $1.76 \, \mathrm{mm \, yr^{-1}}$ regionally) in excess of local RSLR suggest a potential role of above-ground inputs towards maintaining surface elevation. Dependence upon organogenic inputs for accretion, however, also means that the response of *Juncus* assemblages to RSLR may vary with shifts in productivity or decomposition dynamics (e.g. changes in climate and/or nutrient status). Under present conditions, at least, our analyses have shown these organic inputs to be relatively resistant to early decomposition. In all, our findings are also supportive of recent research that suggests organic accretion may be of critical importance in marsh survival under RSLR, particularly in areas most removed from inorganic sediment delivery (D'Alpaos and Marani, 2015). Whether below-ground organic matter production makes substantial contributions to Australian salt marsh surface elevation dynamics and vulnerability to sea-level rise remains unknown and represents an important area for further research. Better understanding of the temporal dynamics of organic and mineral contributions to elevation maintenance is also required, including in relation to expected non-linear increases in sea level.

By combining medium-term accretion quantification with short-term deposition measurements and chemical analyses, we have gained insights into the various processes behind observed differences in accretion among salt marsh vegetation assemblages. While our study highlights assemblage-scale differences in potential response to RSLR, it represents only a small part of the information needed to accurately predict the future of SE Australian salt marsh assemblages. Further measures of short-term deposition and medium-term accretion across a broader range of sites and geographical settings, longer-term studies of surface elevation change among assemblages and modelling of vegetation response thresholds are all required.

Competing interests. The authors declare that they have no conflict of interest.

Acknowledgements. Charlie Hinchcliffe, Mikael Kim, Sarah Meoli and Frederic Cadera assisted with field measurements. Janine McGowan and Bruce Hawke assisted with sample preparation and spectroscopic methods. Field collections were undertaken in accordance with NSW Office of Environment and Heritage scientific licence SL101217 and NSW Department of Primary Industries Scientific permit P13/0058-1.0. We also thank NSW National Parks and Wildlife Service for supporting access to Towra Point Nature Reserve. This research was supported by the CSIRO Coastal Carbon Cluster. Peter I. Macreadie was supported by an Australian Research Council DECRA fellowship (DE130101084).

Edited by: Steven Bouillon

References

Adam, P. (Ed.): Saltmarsh Ecology, Cambridge University Press, Cambridge, UK, 1990.

Adam, P.: Saltmarshes in a time of change, Environ. Conserv., 29, 39–61, 2002.

Adame, M. F., Neil, D., Wright, S. F., and Lovelock, C. E.: Sedimentation within and among mangrove forests along a gradient of geomorphological settings, Estuar. Coast. Shelf Sci., 86, 21–30, 2010.

Baldock, J. A. and Smernik, R. J.: Chemical composition and bioavailability of thermally altered *Pinus resinosa* (Red pine) wood, Org. Geochem., 33, 1093–1109, 2002.

Baldock, J. A., Hawke, B., Sanderman, J., and Macdonald, L. M.: Predicting contents of carbon and its component fractions in Australian soils from diffuse reflectance mid-infrared spectra, Soil Res., 51, 577–595, https://doi.org/10.1071/sr13077, 2013a.

Baldock, J. A., Sanderman, J., Macdonald, L. M., Puccini, A., Hawke, B., Szarvas, S., and McGowan, J.: Quantifying the allocation of soil organic carbon to biologically significant fractions, Soil Res., 51, 561–576, 2013b.

Baustian, J. J., Mendelssohn, I. A., and Hester, M. W.: Vegetation's importance in regulating surface elevation in a coastal salt marsh facing elevated rates of sea level rise, Glob. Change Biol., 18, 3377–3382, 2012.

Boschker, H., De Brouwer, J., and Cappenberg, T.: The contribution of macrophyte-derived organic matter to microbial biomass in salt-marsh sediments: Stable carbon isotope analysis of microbial biomarkers, Limnol. Oceanogr., 44, 309–319, 1999.

Bureau of Meteorology: Climate Data Online: Station 066037 Sydney Airport, available at: http://www.bom.gov.au/climate/data/, last access: 15 June 2016.

Cahoon, D. and Turner, R. E.: Accretion and canal impacts in a rapidly subsiding wetland II. Feldspar marker horizon technique, Estuaries, 12, 260–268, 1989.

Cahoon, D. R., Day Jr, J. W., and Reed, D. J.: The influence of surface and shallow subsurface soil processes on wetland elevation: A synthesis, Current Topics In Wetland Biogeochemistry, 3, 72–88, 1999.

Chen, S., Torres, R., and Goñi, M. A.: Intertidal zone particulate organic carbon redistribution by low-tide rainfall, Limnol. Oceanogr., 60, 1088–1101, 2015.

Chmura, G. L. and Hung, G. A.: Controls on salt marsh accretion: A test in salt marshes of Eastern Canada, Estuaries, 27, 70–81, 2004.

Chmura, G. L., Anisfeld, S. C., Cahoon, D. R., and Lynch, J. C.: Global carbon sequestration in tidal, saline wetland soils, Global Biogeochem. Cy., 17, 22:21–22:12, 2003.

Clarke, P. J. and Jacoby, C. A.: Biomass and above-ground productivity of salt-marsh plants in south-eastern Australia, Mar. Freshwater Res., 45, 1521–1528, 1994.

Cleveland, C. C., Reed, S. C., Keller, A. B., Nemergut, D. R., O'Neill, S. P., Ostertag, R., and Vitousek, P. M.: Litter quality versus soil microbial community controls over decomposition: a quantitative analysis, Oecologia, 174, 283–294, 2014.

Culberson, S. D., Foin, T. C., and Collins, J. N.: The role of sedimentation in estuarine marsh development within the San Francisco Estuary, California, USA, J. Coast. Res., 20, 970–979, 2004.

D'Alpaos, A. and Marani, M.: Reading the signatures of biologic–geomorphic feedbacks in salt-marsh landscapes, Adv. Water Resour., 93, 265–275, https://doi.org/10.1016/j.advwatres.2015.09.004, 2015.

Day Jr, J., Rybczyk, J., Scarton, F., Rismondo, A., Are, D., and Cecconi, G.: Soil accretionary dynamics, sea-level rise and the survival of wetlands in Venice Lagoon: a field and modelling approach, Estuar. Coast. Shelf Sci., 49, 607–628, 1999.

Day, Jr, J., Kemp, G. P., Reed, D. J., Cahoon, D. R., Boumans, R. M., Suhayda, J. M., and Gambrell, R.: Vegetation death and rapid loss of surface elevation in two contrasting Mississippi delta salt marshes: The role of sedimentation, autocompaction and sea-level rise, Ecol. Eng., 37, 229–240, 2011.

Dickens, A. F., Baldock, J. A., Smernik, R. J., Wakeham, S. G., Arnarson, T. S., Gélinas, Y., and Hedges, J. I.: Solid-state 13C NMR analysis of size and density fractions of marine sediments: Insight into organic carbon sources and preservation mechanisms, Geochim. Cosmochim. Ac., 70, 666–686, 2006.

Duarte, C. M. and Cebrian, J.: The fate of marine autotrophic production, Limnol. Oceanogr., 41, 1758–1766, 1996.

Duarte, C. M., Losada, I. J., Hendriks, I. E., Mazarrasa, I., and Marbà, N.: The role of coastal plant communities for climate change mitigation and adaptation, Nature Climate Change, 3, 961–968, 2013.

English, S. S., Wilkinson, C. C., and Baker, V. V.: Survey manual for tropical marine resources, Australian Institute of Marine Science (AIMS), 1994.

Enríquez, S., Duarte, C. M., and Sand-Jensen, K.: Patterns in decomposition rates among photosynthetic organisms: the importance of detritus C : N:P content, Oecologia, 94, 457–471, 1993.

Fry, B. and Ewel, K. C.: Using stable isotopes in mangrove fisheries research-a review and outlook, Isot. Environ. Health S., 39, 191–196, 2003.

Guest, M. A., Connolly, R. M., and Loneragan, N. R.: Carbon movement and assimilation by invertebrates in estuarine habitats at a scale of metres, Mar. Ecol.-Prog. Ser., 278, 27–34, 2004.

Guest, M. A., Connolly, R. M., Lee, S. Y., Loneragan, N. R., and Breitfuss, M. J.: Mechanism for the small-scale movement of carbon among estuarine habitats: organic matter transfer not crab movement, Oecologia, 148, 88–96, 2006.

Harter, S. K. and Mitsch, W. J.: Patterns of short-term sedimentation in a freshwater created marsh, J. Environ. Qual., 32, 325–334, 2003.

Hemminga, M. and Buth, G.: Decomposition in salt marsh ecosystems of the SW Netherlands: the effects of biotic and abiotic factors, Plant Ecol., 92, 73–83, 1991.

Hessen, D. O., Elser, J. J., Sterner, R. W., and Urabe, J.: Ecological stoichiometry: An elementary approach using basic principles, Limnol. Oceanogr., 58, 2219–2236, 2013.

Hickey, D. and Bruce, E.: Examining Tidal Inundation and Salt Marsh Vegetation Distribution Patterns using Spatial Analysis (Botany Bay, Australia), J. Coast. Res., 26, 94–102, https://doi.org/10.2112/08-1089.1, 2010.

Jaschinski, S., Hansen, T., and Sommer, U.: Effects of acidification in multiple stable isotope analyses, Limnol. Oceanogr.-Meth., 6, 12–15, 2008.

Josselyn, M. N. and Mathieson, A. C.: Seasonal influx and decomposition of autochthonous macrophyte litter in a north temperature estuary, Hydrobiologia, 71, 197–207, 1980.

Kelleway, J. J., Saintilan, N., Macreadie, P. I., and Ralph, P. J.: Sedimentary factors are key predictors of carbon storage in SE Australian saltmarshes, Ecosystems, 19, 865–880, https://doi.org/10.1007/s10021-016-9972-3, 2016a.

Kelleway, J. J., Saintilan, N., Macreadie, P. I., Skilbeck, C. G., Zawadzki, A., and Ralph, P. J.: Seventy years of continuous encroachment substantially increases "blue carbon" capacity as mangroves replace intertidal salt marshes, Glob. Change Biol., 22, 1097–1109, 2016b.

Kirwan, M. L. and Megonigal, J. P.: Tidal wetland stability in the face of human impacts and sea-level rise, Nature, 504, 53–60, 2013.

Kirwan, M. L. and Mudd, S. M.: Response of salt-marsh carbon accumulation to climate change, Nature, 489, 550–553, 2012.

Kirwan, M. L., Temmerman, S., Skeehan, E. E., Guntenspergen, G. R., and Fagherazzi, S.: Overestimation of marsh vulnerability to sea level rise, Nature Climate Change, 6, 253–260, 2016.

Krauss, K. W., Allen, J. A., and Cahoon, D. R.: Differential rates of vertical accretion and elevation change among aerial root types in Micronesian mangrove forests, Estuar. Coast. Shelf Sci., 56, 251–259, 2003.

Kristensen, E.: Decomposition of macroalgae, vascular plants and sediment detritus in seawater – use of stepwise thermogravimetry, Biogeochemistry, 26, 1–24, 1994.

Langley, J. A., McKee, K. L., Cahoon, D. R., Cherry, J. A., and Megonigal, J. P.: Elevated CO_2 stimulates marsh elevation gain, counterbalancing sea-level rise, P. Natl. Acad. Sci. USA, 106, 6182–6186, 2009.

Lovelock, C. E., Adame, M. F., Bennion, V., Hayes, M., O'Mara, J., Reef, R., and Santini, N. S.: Contemporary Rates of Carbon Sequestration Through Vertical Accretion of Sediments in Mangrove Forests and Saltmarshes of South East Queensland, Australia, Estuar. Coast., 37, 763–771, 2013.

McGroddy, M. E., Daufresne, T., and Hedin, L. O.: Scaling of C: N: P stoichiometry in forests worldwide: implications of terrestrial Redfield-type ratios, Ecology, 85, 2390–2401, 2004.

McKee, K. L., Cahoon, D. R., and Feller, I. C.: Caribbean mangroves adjust to rising sea level through biotic controls on change in soil elevation, Global Ecol. Biogeogr., 16, 545–556, 2007.

Minden, V. and Kleyer, M.: Ecosystem multifunctionality of coastal marshes is determined by key plant traits, J. Veg. Sci., 2015.

Morris, J. T.: Ecological engineering in intertidial saltmarshes, Hydrobiologia, 577, 161–168, 2007.

Morris, J. T., Sundareshwar, P., Nietch, C. T., Kjerfve, B., and Cahoon, D.: Responses of coastal wetlands to rising sea level, Ecology, 83, 2869–2877, 2002.

Mudd, S. M., Fagherazzi, S., Morris, J. T., and Furbish, D. J.: Flow, sedimentation, and biomass production on a vegetated salt marsh in South Carolina: Toward a predictive model of marsh morphologic and ecologic evolution, Coast. Estuar. Stud., 59, 165–188, 2004.

Mudd, S. M., Howell, S. M., and Morris, J. T.: Impact of dynamic feedbacks between sedimentation, sea-level rise, and biomass production on near-surface marsh stratigraphy and carbon accumulation, Estuar. Coast. Shelf Sci., 82, 377–389, 2009.

Mudd, S. M., D'Alpaos, A., and Morris, J. T.: How does vegetation affect sedimentation on tidal marshes? Investigating particle capture and hydrodynamic controls on biologically mediated sedimentation, J. Geophys. Res.-Earth, 115, https://doi.org/10.1029/2009JF001566, 2010.

Mueller, P., Jensen, K., and Megonigal, J. P.: Plants Mediate Soil Organic Matter Decomposition In Response To Sea Level Rise, Glob. Change Biol., 22, 404–414, 2016.

Nolte, S., Koppenaal, E. C., Esselink, P., Dijkema, K. S., Schuerch, M., De Groot, A. V., Bakker, J. P., and Temmerman, S.: Measuring sedimentation in tidal marshes: a review on methods and their applicability in biogeomorphological studies, J. Coast. Conserv., 17, 301–325, 2013.

Nyman, J. A., Walters, R. J., Delaune, R. D., and Patrick Jr, W. H.: Marsh vertical accretion via vegetative growth, Estuar. Coast. Shelf Sci., 69, 370–380, 2006.

Oenema, O. and DeLaune, R. D.: Accretion rates in salt marshes in the Eastern Scheldt, south-west Netherlands, Estuar. Coast. Shelf Sci., 26, 379–394, 1988.

Ouyang, X. and Lee, S. Y.: Updated estimates of carbon accumulation rates in coastal marsh sediments, Biogeosciences, 11, 5057–5071, https://doi.org/10.5194/bg-11-5057-2014, 2014.

Phillips, D. L.: Converting isotope values to diet composition: the use of mixing models, J. Mammal., 93, 342–352, 2012.

Reed, D. J.: Patterns of sediment deposition in subsiding coastal salt marshes, Terrebonne Bay, Louisiana: the role of winter storms, Estuaries, 12, 222–227, 1989.

Rogers, K. and Saintilan, N.: Relationships between surface elevation and groundwater in mangrove forests of southeast Australia, J. Coast. Res., 24, 63–69, 2008.

Rogers, K., Wilton, K. M., and Saintilan, N.: Vegetation change and surface elevation dynamics in estuarine wetlands of southeast Australia, Estuar. Coast. Shelf Sci., 66, 559–569, 2006.

Rogers, K., Saintilan, N., and Copeland, C.: Managed Retreat of Saline Coastal Wetlands: Challenges and Opportunities Identified from the Hunter River Estuary, Australia, Estuar. Coast., 37, 67–78, 2014.

Rogers, K., Saintilan, N., and Woodroffe, C. D.: Surface elevation change and vegetation distribution dynamics in a subtropical coastal wetland: Implications for coastal wetland response to climate change, Estuar. Coast. Shelf Sci., 149, 46–56, 2014.

Rooth, J. E., Stevenson, J. C., and Cornwell, J. C.: Increased sediment accretion rates following invasion by Phragmites australis: the role of litter, Estuar. Coast., 26, 475–483, 2003.

Rosencranz, J. A., Ganju, N. K., Ambrose, R. F., Brosnahan, S. M., Dickhudt, P. J., Guntenspergen, G. R., MacDonald, G. M., Takekawa, J. Y., and Thorne, K. M.: Balanced Sediment Fluxes in Southern California's Mediterranean-Climate Zone Salt Marshes, Estuar. Coast., 39, 1035–1049, 2016.

Roy, P., Williams, R., Jones, A., Yassini, I., Gibbs, P., Coates, B., West, R., Scanes, P., Hudson, J., and Nichol, S.: Structure and function of south-east Australian estuaries, Estuar. Coast. Shelf Sci., 53, 351–384, 2001.

Saintilan, N., Rogers, K., Mazumder, D., and Woodroffe, C.: Allochthonous and autochthonous contributions to carbon accumulation and carbon store in southeastern Australian coastal wetlands, Estuar. Coast. Shelf Sci., 128, 84–92, 2013.

Skjemstad, J., Clarke, P., Taylor, J., Oades, J., and Newman, R.: The removal of magnetic materials from surface soils-a solid state 13C CP/MAS NMR study, Soil Res., 32, 1215–1229, 1994.

Stapel, J. and Hemminga, M.: Nutrient resorption from seagrass leaves, Mar. Biol., 128, 197–206, 1997.

Sterner, R. W. and Hessen, D. O.: Algal nutrient limitation and the nutrition of aquatic herbivores, Annu. Rev. Ecol. Syst., 25, 1–29, 1994.

Thomas, S. and Ridd, P. V.: Review of methods to measure short time scale sediment accumulation, Mar. Geol., 207, 95–114, 2004.

van Proosdij, D., Davidson-Arnott, R. G. D., and Ollerhead, J.: Controls on spatial patterns of sediment deposition across a macrotidal salt marsh surface over single tidal cycles, Estuar. Coast. Shelf Sci., 69, 64–86, 2006.

Wang, X. C., Chen, R. F., and Berry, A.: Sources and preservation of organic matter in Plum Island salt marsh sediments (MA, USA): long-chain n-alkanes and stable carbon isotope compositions, Estuar. Coast. Shelf Sci., 58, 917–928, 2003.

Zhou, J., Wu, Y., Zhang, J., Kang, Q., and Liu, Z.: Carbon and nitrogen composition and stable isotope as potential indicators of source and fate of organic matter in the salt marsh of the Changjiang Estuary, China, Chemosphere, 65, 310–317, 2006.

Zhou, J., Wu, Y., Kang, Q., and Zhang, J.: Spatial variations of carbon, nitrogen, phosphorous and sulfur in the salt marsh sediments of the Yangtze Estuary in China, Estuar. Coast. Shelf Sci., 71, 47–59, 2007.

4

Changes in the partial pressure of carbon dioxide in the Mauritanian–Cap Vert upwelling region between 2005 and 2012

Melchor González-Dávila[1], J. Magdalena Santana Casiano[1], and Francisco Machín[1,2]

[1]Instituto de Oceanografía y Cambio Global, Grupo QUIMA, Universidad de Las Palmas de Gran Canaria, 35017, Las Palmas de Gran Canaria, Spain
[2]Departamento de Física, Universidad de Las Palmas de Gran Canaria, 35017, Las Palmas de Gran Canaria, Spain

Correspondence to: Melchor González-Dávila (melchor.gonzalez@ulpgc.es)

Abstract. Coastal upwellings along the eastern margins of major ocean basins represent regions of large ecological and economic importance due to the high biological productivity. The role of these regions for the global carbon cycle makes them essential in addressing climate change. The physical forcing of upwelling processes that favor production in these areas are already being affected by global warming, which will modify the intensity of upwelling and, consequently, the carbon dioxide cycle. Here, we present monthly high-resolution surface experimental data for temperature and partial pressure of carbon dioxide in one of the four most important upwelling regions of the planet, the Mauritanian–Cap Vert upwelling region, from 2005 to 2012. This data set provides direct evidence of seasonal and interannual changes in the physical and biochemical processes. Specifically, we show an upwelling intensification and an increase of $0.6\,\mathrm{Tg\,yr^{-1}}$ in CO_2 outgassing due to increased wind speed, despite increased primary productivity. This increase in CO_2 outgassing together with the observed decrease in sea surface temperature at the location of the Mauritanian Cap Blanc, $21°\,\mathrm{N}$, produced a pH rate decrease of $-0.003 \pm 0.001\,\mathrm{yr^{-1}}$.

1 Introduction

The excess of CO_2 in the atmosphere, largely responsible for global climate change, has prompted research on the role of the oceans in the carbon cycle. The aim in recent decades has been to assess how the oceans act as sources or sinks within the carbon cycle. To achieve this goal, highly resolved spatial and temporal observations representative of the distribution of CO_2 fluxes between the ocean and atmosphere are necessary. Automated instruments on volunteer observing ships (VOSs) serve to provide as many observations throughout the global ocean as possible. This is in addition to data collected on scientific cruises and at long-term moorings (i.e., Astor et al., 2005: Lüger et al., 2004, 2006; González-Dávila et al., 2005, 2009; Schuster et al., 2009; Ullman et al., 2009; Watson et al., 2009; Padin et al., 2010; Gruber et al., 2002; Dore et al., 2003; Santana-Casiano et al., 2007; Bates et al., 2014).

With the amount of data already gathered (http://www.socat.info/; Pfeil et al., 2013), climatologies that present average CO_2 fluxes between the atmosphere and the ocean have been developed, identifying areas acting as a source or sink (Key et al., 2004; Takahashi et al., 2009). However, the low spatial resolution of these databases limits the applicability, especially in coastal areas. Upwelling regions are particularly under-represented in such large databases. Upwelling presents a dynamic process that raises nutrient and CO_2-rich water from relatively deep areas to the surface. The nutrients reaching the photic zone promote primary production, which consumes CO_2. This process generates a CO_2 flux into the ocean. On the other hand, upwelling also brings up CO_2 from deep seawater, which generates uncertainty about the actual role of upwelling areas as a source or sink of CO_2 (Michaels et al., 2001). Indeed, upwelling areas may act as a source or sink of CO_2 depending on their location (Cai et al., 2006; Chen et al., 2013), where upwelling regions at low latitudes mainly act as a source of CO_2 (Feely et al., 2002; Astor et al., 2005; Friederich et al., 2008; Santana-Casiano et al., 2009; González-Dávila et al., 2009) and those at midlatitudes mainly act as a sink of CO_2 (Frankignoulle and Borges, 2001; Hales et al., 2005; Borges and Frankignoulle, 2002; Borges et al., 2005; Santana-Casiano et al., 2009; González-Dávila et al., 2009). Several anthropogenic interactive effects strongly influence eastern boundary upwelling systems (EBUSs), including upper ocean warming, ocean acidification, and ocean deoxygenation (Gruber, 2011; Feely et al., 2008; Keeling et al., 2010). Moreover, evidence of increasing wind speed that would favor upwelling (Bakun, 1990; Demarcq, 2009; Oerder et al., 2015) supports the possibility of a change in the dynamics of these highly productive areas. Recently, eddy-resolving regional ocean models have shown how upwelling intensification can cause a major impact on the system's biological productivity and CO_2 outgassing (Lachkar and Gruber, 2013; Oerder et al., 2015). Wind observations and re-analysis products are controversial regarding the Bakun intensification hypothesis (Bakun, 1990). Using different wind databases for the Canary region, Barton et al. (2013) concluded that there was no evidence for a general increase in the upwelling intensity off northwest Africa. Marcello et al. (2011) found an intensification of the upwelling system in the same area during a 20-year period, while the alongshore wind stress remained almost stable. Cropper et al. (2014) found that coastal summer wind speed increased, resulting in an increase in upwelling-favorable wind speeds north of $20°$ N and an increase in downwelling-favorable winds south of $20°$ N. Santos et al. (2005, 2012) showed that sea surface temperature (SST) was not homogeneous either along latitude or longitude and depended on the upwelling index (UI) intensity. Varela et al. (2015) demonstrated opposite results worldwide depending on the length of data, season evaluated, and selected area within the same wind data set or between data sets. For the Mauritanian region, when wind stress data were used (Varela et al., 2015), a more persistent increasing trend in upwelling-favorable winds north of $21°$ N and a decreasing trend south of $19°$ N was determined.

Starting in June 2005, the QUIMA-VOS line visited the Mauritanian–Cap Vert upwelling region northwest of Africa on a monthly basis (Fig. 1 and Table S1 in the Supplement) producing for the first time a high-resolution database of SST and partial pressure of CO_2 expressed as fugacity $f CO_2$. This database shows the variations in the CO_2 system under changes in the upwelling conditions in the Canary ecosystem from 27 to $10°$ N for the period 2005 to 2012. More data for the region from other surveys exist (http://www.socat.info/; Pfeil et al., 2013) but they were not considered in this study as they do not follow the same track as the QUIMA-VOS line. Those data are strongly influenced by the distance to the upwelling cells with the corresponding physical effects in the partial pressure of CO_2.

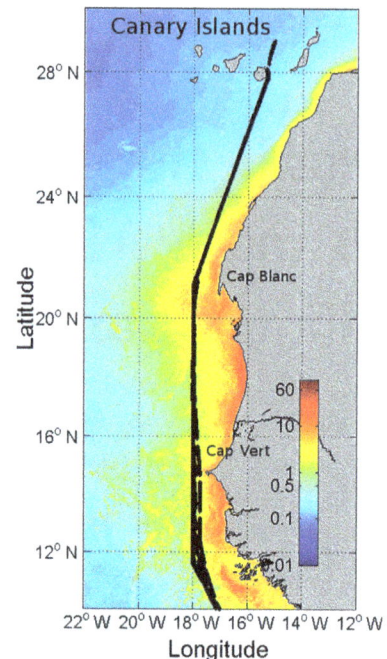

Figure 1. Ship track (black line) in the area from $28°$ N (Gran Canaria, the Canary Islands) to $10°$ N. The locations of Cap Blanc and Cap Vert are indicated. Monthly OceanColor Web (https://oceancolor.gsfc.nasa.gov/) data for average chlorophyll a concentration ($\mathrm{mg\,m^{-3}}$) were included in a MATLAB routine and annually averaged. The map has been generated using MATLAB 7.12 R2011a.

2 Experimental

2.1 Study region

The VOS line crosses the east Atlantic Ocean from the north of Europe (English Channel) to South Africa, calling at Gran Canaria, the Canary Islands, with a periodicity of 2 months, which provides monthly data (southward or northward sections). In this work, the area between Gran Canaria at 27 and $10°$ N has been selected in order to study the Mauritanian–Cap Vert upwelling region. On its route south (Fig. 1), the ship leaves Gran Canaria and goes straight to 100 km off Cap Blanc at $21°$ N, $17°45'$ W. It then follows this longitude, passing at 100 km off Cap Vert until $12°$ N, where it changes direction to Cape Town, reaching $10°$ N, $17°$ W at 330 km off the coast of Guinea. Between 22 and $20°$ N, the ship reaches the 500 m isobath. South of $15°$ N, the ship moves between the 1000 and 500 m isobaths. On its route north, the ship follows the reverse track.

2.2 Experimental data

Experimental data were obtained under the EU projects CARBOOCEAN and CARBOCHANGE (www.CarboOcean.org and https://carbochange.b.uib.no/) and now also available at http://www.socat.info/ (Pfeil et al.,

2013). An autonomous instrument for the determination of the partial pressure of CO_2 developed by Craig Neill following NOAA recommendations was installed on a VOS line. This was operated by the Mediterranean Shipping Company S.A. from 2005 to 2008 and Maersk from 2010 to 2012. This VOS line (QUIMA-VOS) ran between the UK and Cape Town from July 2005 to January 2013 (Table S1 in the Supplement). Temperature was measured at three positions along the sampling circuit: in the intake (Sea-Bird SBE38L), in the equilibrator (Sea-Bird thermosalinograph SBE21 and internal PT100 thermometer), and in the oxygen sensor (Optode 3835, Aanderaa™). After the seawater pump, the intake is divided into two lines, one feeding the CO_2 system and the other feeding the oxygen sensor, the fluorometer, and the Sea-Bird thermosalinometer. Differences between equilibrator and intake temperatures were constant in time due to the high seawater flow but varied among ships due to the different locations of the equipment. Values varied between 0.06 °C when the equipment was placed close to the intake and 0.35 °C when the equipment was one floor above and inside the engine room. The SST was also obtained from the NOAA_OI_SST-V2 data provided by the NOAA/OAR/ESRL PSD from Boulder, Colorado, USA (http://www.esrl.noaa.gov/psd). These data had a spatial resolution of 1° latitude and 1° longitude and monthly averages were used. The correlation between our experimental SST data and satellite data was better than ±1 °C, and improved to ±0.4 °C after removing the most affected upwelling regions (19–22 and 14–16° N), which related to the high variability imposed by the upwelling.

The CO_2 molar fraction, xCO_2, was obtained every 150 s in seawater, while atmospheric xCO_2 data were obtained every 180 min. The seawater intake was located at a 10 m depth. The system was calibrated every 3 hours by measuring four different standard gases with mixing ratios in the ranges of 0.0, 250–290, 380–410, and 490–530 ppm of CO_2 in the air, provided by NOAA and traceable to the World Meteorological Organization scale. The precision of the system is greater than 0.5 µatm and the accuracy estimated with respect to the standard gases is of 1 µatm inside the standards' range. For xCO_2 values higher than the highest standard (532.04 ppm), the accuracy will be reduced, even when linearity was observed in all cases inside the standards range. The fugacity of CO_2 (fCO_2, µatm) was calculated from xCO_2 after correcting for temperature differences between intake and equilibrator, according to the expressions for seawater given by DOE (1994). Normalized fCO_2 ($NfCO_2$) derived from the mean SST for the area (T_{mean}) was computed following Takahashi et al. (1993) as

$$(NfCO_2) = fCO_2 \cdot \exp[0.0423(T_{mean} - SST)]. \quad (1)$$

In order to compute a second carbonate system variable, the surface total alkalinity (A_T) was computed from sea surface salinity (SSS) and SST (Lee et al., 2006). pH_T at the in situ

temperature was computed from fCO_2 and A_T and with average annual surface ocean total phosphate and total silicate concentrations of 0.5 and 4.8 µmol kg^{-1}, respectively, from the World Ocean Atlas 2009, using the carbonic acid acidity constants by Mehrbach et al. (1973) refitted by Dickson and Millero (1987).

Air–sea CO_2 fluxes (FCO$_2$, mmol m^{-2} d^{-1}) were evaluated as

$$FCO_2 = 0.24 \cdot k \cdot s \cdot (fCO_2^{sw} - fCO_2^{atm}), \quad (2)$$

where 0.24 is the scale factor, k is the gas transfer velocity, s is the CO_2 solubility, fCO_2^{sw} is the seawater fugacity of CO_2, and fCO_2^{atm} is the atmospheric fugacity of CO_2. In order to evaluate ($fCO_2^{sw} - fCO_2^{atm}$), fCO_2^{atm} data were linearly interpolated to the fCO_2^{sw} time vector. A positive value for FCO$_2$ corresponds to CO_2 outgassing from the ocean. k (cm h^{-1}) was evaluated with the following parameterization (Nightingale et al., 2000):

$$k = (0.222 \cdot W^2 + 0.333 \cdot w) \cdot (Sc/660)^{-1/2}, \quad (3)$$

where W is the wind speed at 10 m above the sea surface (m s^{-1}) and Sc is the Schmidt number.

The variables involved in estimating FCO$_2$ data (i.e., fCO_2^{sw}, fCO_2^{atm}, SST, and SSS) were fitted to sinusoidal expressions (Lüger et al., 2004) for a given latitude as follows:

$$X(lat)^* = a_0 + a_1(t - 2005) + a_2 \sin(2\pi t) + a_3 \cos(2\pi t)$$
$$+ a_4 \sin(4\pi t) + a_5 \cos(4\pi t), \quad (4)$$

where a_i are the fitting coefficients, t is the sampling time expressed as year fraction, and $X(lat)^*$ represents any of the four fitted variables. This procedure allowed us to reconstruct the series of experimental data for periods without monthly data. The variables were decomposed into an interannual term $X(lat)_t^* = a_0 + a_1(t - 2005)$ plus a periodical term $X(lat)_p^* = a_2 \sin(2\pi t) + a_3 \cos(2\pi t) + a_4 \sin(4\pi t) + a_5 \cos(4\pi t)$, that is, $X(lat)^* = X(lat)_t^* + X(lat)_p^*$. The periodical term accounts for the high-frequency seasonal variability, while the interannual term marks the year-to-year trend. First, observations were grouped in a natural year for a given latitude, as if they had been taken in a single year (no correction was done for interannual variability). The mean seasonal climatology data associated with the periodic coefficients (i.e., a_2, a_3, a_4, and a_5) throughout the sampling period were determined. Next, the interannual coefficient a_1 was calculated by fitting the residuals resulting from subtracting the periodical component, $X(lat)_p^*$, from the original variable $X(lat)$. By fixing these five coefficients (a_1–a_5), new distributions for fCO_2^{sw*}, fCO_2^{atm*}, SST*, and SSS* were constructed with a daily resolution based on the curve fits given for each variable as in Eq. (4), providing the coefficient a_0. The accuracy of this fitting procedure was checked by both computing the correlation between experimental and reconstructed values and by determining the mean residuals. The Pearson coefficients were always over 0.87 for

SST (average 0.94 ± 0.03), over 0.69 for both $f\mathrm{CO}_2^{\mathrm{sw}}$ and $f\mathrm{CO}_2^{\mathrm{atm}}$ (average of 0.79 ± 0.07 and 0.82 ± 0.04, respectively), and over 0.67 for SSS (average 0.79 ± 0.07). The mean residual on the determination of those four variables were $\pm 3.7\,\mu\mathrm{atm}$, $\pm 1.5\,\mu\mathrm{atm}$, $\pm 0.22\,^{\circ}\mathrm{C}$, and ± 0.05 for $f\mathrm{CO}_2^{\mathrm{sw}*}$, $f\mathrm{CO}_2^{\mathrm{atm}*}$, SST*, and SSS*, respectively. When the monthly satellite SST values were considered, the new SST* function averaged for each month produced values within $\pm 0.47\,^{\circ}\mathrm{C}$, confirming that this procedure was able to fit non-sampled periods. It was assumed that the same procedure was valid for non-sampled $f\mathrm{CO}_2$. Finally, daily FCO_2^{*} time series between 10 and $27^{\circ}\,\mathrm{N}$ with a latitudinal resolution of 0.5° were calculated with a standard error of estimation of $0.5\,\mathrm{mmol\,m^{-2}\,d^{-1}}$ (15 % of error) that produced mean residuals (experimental FCO_2–FCO_2^{*}) of $0.4\,\mathrm{mmol\,m^{-2}\,d^{-1}}$ and Pearson correlation coefficients between experimental and computed FCO_2^{*} of $r > 0.6$, $p < 0.01$.

Chlorophyll a was calculated from measurements made by the Moderate-resolution Imaging Spectroradiometer (MODIS) aboard NASA's Aqua satellite. We used monthly averages with a spatial resolution of 9 km supplied by Ocean-Color Web (https://oceancolor.gsfc.nasa.gov).

Wind data were downloaded from the NCEP CFSR database at http://rda.ucar.edu/pub/cfsr.html developed by NOAA and retrieved from the NOAA National Operational Model Archive and Distribution System and maintained by the NOAA National Climatic Data Center. The spatial resolution is approximately $0.3 \times 0.3^{\circ}$ and the temporal resolution is 6 h. The reference height for the wind data is 10 m.

Rainfall data were collected by the precipitation radar installed on the Tropical Rainfall Measuring Mission (TRMM) satellite (http://precip.gsfc.nasa.gov). Monthly averages with a spatial resolution of $0.5 \times 0.5^{\circ}$ (product 3A12, version 07) were used (Fig. S1 in the Supplement) in order to explain changes in seasonal surface salinity distributions.

3 Results and discussion

3.1 Physical properties

The variability of the Mauritanian–Cap Vert upwelling was analyzed in terms of the upwelling index (Nykjaer and Van Camp, 1994) (Fig. 2) using satellite wind data. Negative UI values correspond to upwelling-favorable conditions and positive values to downwelling-favorable conditions. The lowest negative values of the index correspond to more intense upwelling. Results clearly distinguish two main subareas in the upwelling system: (1) north of $20^{\circ}\,\mathrm{N}$, the upwelling conditions were favorable throughout the year, although the highest upwellings were observed from March to September with a northward shift from 20 to $22^{\circ}\,\mathrm{N}$. (2) South of $20^{\circ}\,\mathrm{N}$, a marked seasonality was observed with favorable upwelling conditions during autumn and winter, with the maximum intensity observed during January and February. In this

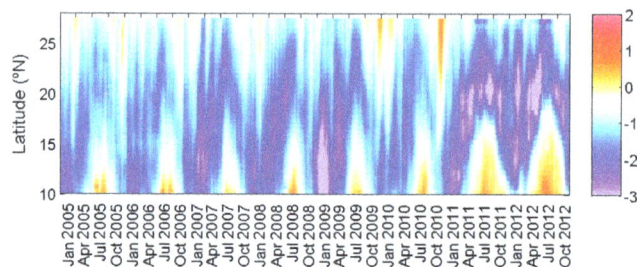

Figure 2. Time series of upwelling index (UI, $\times 10^{-3}\,\mathrm{m^2\,s^{-1}}$) in the Mauritanian–Cap Vert upwelling region along the ship track computed following Nykjaer and Van Camp (1994). Blue colors are related to upwelling events and red colors to downwelling events.

region, a downwelling regime is present between May and November when the summer trade winds are replaced by the monsoonal winds advecting warm water (Fig. 3a) northward along the shore (Nykjaer and Van Camp, 1994). Our results (Fig. 2) are quite consistent with previous research (Nykjaer and Van Camp, 1994; Marcello et al., 2011; Santos et al., 2005, 2012; Cropper et al., 2014) but include the years 2010 to 2012, when the UI at around 20–$21^{\circ}\,\mathrm{N}$ presented a shift of the upwelling intensity from high ($-2000\,\mathrm{m^2\,s^{-1}}$) to strong ($-2800\,\mathrm{m^2\,s^{-1}}$). The analysis of upwelling trends along this route has been controversial since it is highly dependent on the selected region (Santos et al., 2012). The interannual evolution of the UI over the period 2005 to 2012 (Fig. 4, green line) for each degree in latitude indicates an increase in the UI (mean confidence interval of $9\,\mathrm{m^2\,s^{-1}}$) as showed by Santos et al. (2012).

North of $15^{\circ}\,\mathrm{N}$, the upwelling index confirmed the stronger upwelling observed since 1995–1996 in this region after more than a 10-year (from at least 1982 to 1995) period of weaker upwelling (Santos et al., 2012). Local zonal differences between ocean and coastal SST trends determined with satellite data confirmed the intensification of the upwelling regime along the African coast for the period 1982 to 2000 (Santos et al., 2005) and extended by Santos et al. (2012) until 2010 and further extended in this study until 2012 (data not shown). This has been described as a decadal-scale shift of the upwelling regime intensity (Marcello et al., 2011; Santos et al., 2012).

South of $15^{\circ}\,\mathrm{N}$, the annual UI values and trends (Figs. 2 and 4) both for the upwelling (values close to $-2800\,\mathrm{m^2\,s^{-1}}$ in January) and downwelling (values reaching $1850\,\mathrm{m^2\,s^{-1}}$ in July) periods are becoming stronger. At 11–$12^{\circ}\,\mathrm{N}$, where downwelling is becoming stronger, this results in negative annual temperature rates that approach zero. The UI serves as an indication of decadal variability of the summer monsoon winds and associated northward advection of warm water along the coast (Santos et al., 2012).

The highest upwelling intensity along the VOS line was located at the capes, Cap Blanc and Cap Vert. From satellite chlorophyll a data, especially off Cap Blanc, giant fila-

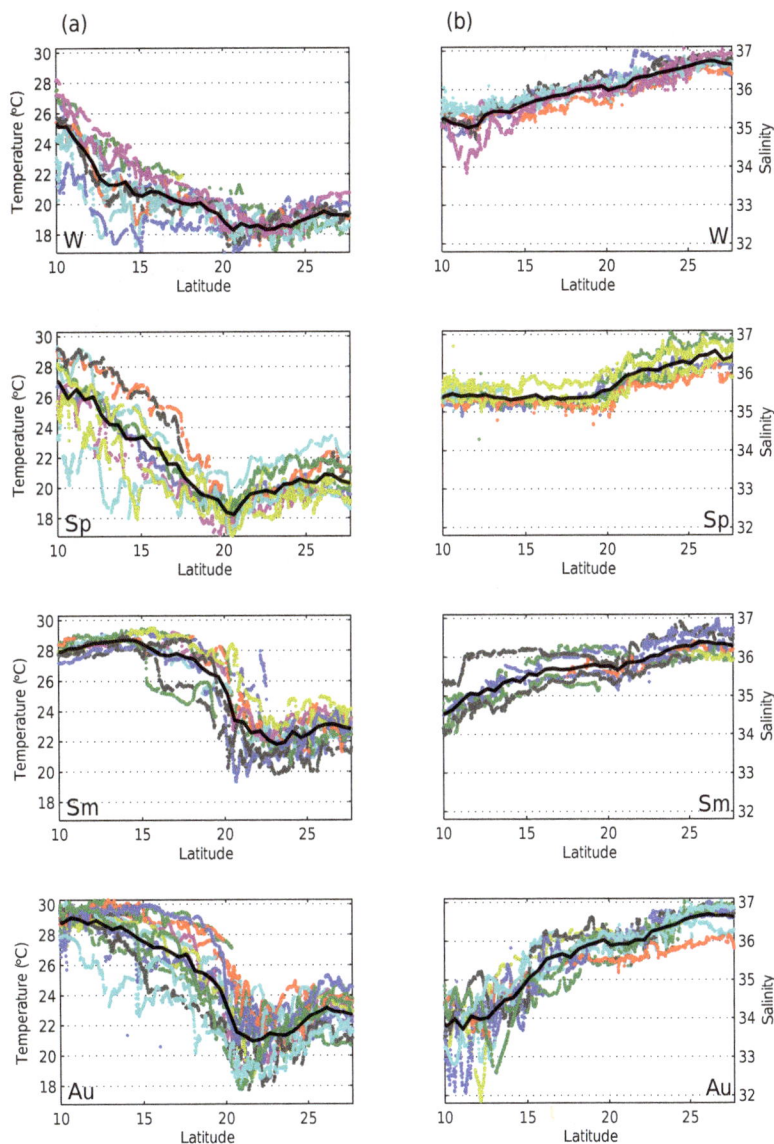

Figure 3. In situ data of column **(a)** SST and column **(b)** SSS in the Mauritanian–Cap Vert coastal region grouped by seasons: winter (W; December, January, and February), spring (Sp; March, April, and May), summer (Sm; June, July, and August), and autumn (Au; September, October, and November). The averaged values for all cruises in Table S1 are shown in black for each season including the 95 % confidence limits. The color code for each cruise is indicated in Table S1.

ments with chlorophyll concentrations above $1 \, \mathrm{mg \, m^{-3}}$ persist year-round, spreading from the coast to several hundred kilometers offshore (Fig. 1). North of Cap Blanc the upwelled water originates from the North Atlantic Central Water, and mixes with South Atlantic Central Water (SACW) towards the south (Mittelstaedt, 1983). South of Cap Blanc, the upwelling of nutrient-rich SACW (Mittelstaedt, 1983) promotes phytoplankton growth between Cap Blanc and Cap Vert. Towards 12° N, upwelling is also fed by the North Equatorial Undercurrent (Hagen and Schemainda, 1984). Moreover, the entire northwest African coast is also influenced by the African desert dust transport by the midtropospheric Harmattan winds originating from the central

Sahara, which supplements the levels of micronutrients (such as iron) to the adjacent marine ecosystem (Mittelstaedt, 1983; Neuer et al., 2004).

The study area is also affected by the migration of the Intertropical Convergence Zone (ITCZ), related to maximum precipitation rates (Hastenrath, 1995). To have a significant satellite precipitation record in our region of interest, precipitation data were integrated longitudinally between 25.25 and 9.75° W. Time series for the latitudinal distribution of integrated precipitation (Fig. S1 in the Supplement) identified the average position of the ITCZ related to maximum precipitation rates. The ITCZ was located at its southernmost position (2° N) during winter, reaching its northernmost position

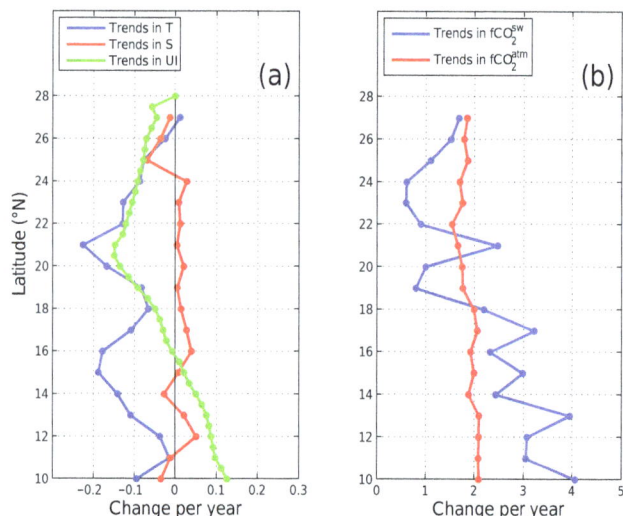

Figure 4. Latitudinal distribution of the interannual trends for the upwelling index (UI) and for the four experimental variables along the QUIMA-VOS line integrated over every degree between 2005 and 2012. Panel **(a)** presents the trends for upwelling index (UI, $\times 10^{-3}$ m^2 s^{-1}, mean confidence interval of 9 m^2 s^{-1}), SST (°C yr^{-1}, confidence interval 0.13 °C), and SSS (yr^{-1}, confidence interval 0.06) and **(b)** the trends for $f\text{CO}_2^{\text{sw}}$ and $f\text{CO}_2^{\text{atm}}$ (confidence intervals 4.23 and 0.44 µatm).

(14–16° N) around summer. The ITCZ reached our area of interest ($> 10°$ N) from late spring to late summer.

The latitudinal distributions of measured SST and SSS along the vessel track are shown in Fig. 3, grouped by seasons (labeled W, Sp, Sm, and Au). The temperature generally decreased from 10 ° N to about 20–21° N, where the ship meets the Mauritanian upwelling. From there to the north, the temperature rises as the ship leaves the upwelling area on its way to the Canary Islands. In situ temperature at 27° N shows temperatures in the range of 18 to 24 °C with the minimum in winter and maximum in late summer to early autumn. The annual temperature range was somewhat higher at 20° N, with a summer maximum of around 26 °C and minimum in spring of about 17 °C. At 10° N, temperatures were the highest throughout the year (> 25 °C), with minimum values in winter and maximum in late spring and late autumn. The low values observed during the end of summer are related to the arrival of the ITCZ (Fig. S1 in the Supplement) at those latitudes. The thermal distribution shows a temperature increase as we move to the Equator and a notable cooling at the upwelled waters off Mauritania. The upwelling of cold water from the Cap Vert area was only detected during winter time and the beginning of spring. Salinity minimum values were normally located at 10° N, increasing to maximum values at the Canaries' latitude. The minimum values of salinity were exceptionally low during autumn from 10 to 16° N by both the freshwater input from rivers that increase their outflow

during this season (Nicholson, 1981) and by the northward shift of the ITCZ during this time of the year.

Anomaly fields for temperature and salinity (data not shown) were calculated as the difference between the observations and the mean values at each season for individual latitudes. For temperature, the largest anomalies in winter and spring were located south of 18° N, with values of ± 2 °C, related to the seasonal cycle of the Cap Vert upwelling. During summer the pattern changed and the largest anomalies were detected in the upwelling area at 18–22 °N, with values of ± 5 °C when the upwelling index for the Mauritanian area was highest (Fig. 2). In autumn the temperature anomalies were shifted slightly to the north, 20–24° N, with values of ± 3 °C related to the observed pulses in upwelling-favorable winds that affected the surface seawater properties. On the other hand, salinity anomalies showed a very homogeneous pattern in all latitudes for winter, spring, and summer, with values generally within ± 0.5. However, during autumn important anomalies south of 18° N were observed, with values in the range of ± 1.5. In this region, the upwelling development, the river discharge, and the rainy season controlled the observed distribution (Yoo and Carton, 1990).

To conclude, the data show a permanent annual upwelling regime observed north of 20° N and a seasonal regime across 10–19° N, in accordance with the climatology of previous studies. The data also confirm an increase in upwelling conditions north of 20° N and an increase in downwelling conditions south of 20° N.

3.2   Carbon dioxide variability

The latitudinal distribution of the seasonal $f\text{CO}_2^{\text{sw}}$ data (Fig. 5a) showed the highest values between 18 and 23° N for all seasons due to the variability imposed by the upwelling off Mauritania. $f\text{CO}_2^{\text{sw}}$ was consistently greater than the $f\text{CO}_2^{\text{atm}}$. During winter, when the Cap Vert upwelling develops (Fig. 2), the 12–15° N region also presented higher $f\text{CO}_2^{\text{sw}}$ values than those in the atmosphere. $f\text{CO}_2^{\text{sw}}$ data showed a latitudinal shift between the seasons following the shift observed in the upwelling index: in winter, the largest values were located between 19 and 24° N; in spring, they were located between 16 and 22° N; and during summer and autumn, the largest $f\text{CO}_2^{\text{sw}}$ values were recorded in the range 20 to 23° N. The difference between $f\text{CO}_2^{\text{sw}}$ normalized to the mean SST of 22 °C for the region (N$f\text{CO}_2^{\text{sw}}$) and $f\text{CO}_2^{\text{sw}}$ ($\Delta f\text{CO}_2 = \text{N}f\text{CO}_2^{\text{sw}} - f\text{CO}_2^{\text{sw}}$, Fig. 5b) reinforced the variability at 20–23° N all year around and at 12–17° N during winter and spring, indicating that upwelling is the major factor contributing to the $f\text{CO}_2$ variability.

According to Takahashi et al. (1993), $f\text{CO}_2^{\text{sw}}$ increases with temperature at a rate of 4.3 % µatm °C^{-1} (between 15 and 26 µatm °C^{-1} in this area) in a thermodynamically controlled system. At 27° N, as SST increases, the rate was only 7.45 µatm °C^{-1} due mainly to biological uptake and also to CO$_2$ outflux. At 20° N the rate became negative with a value

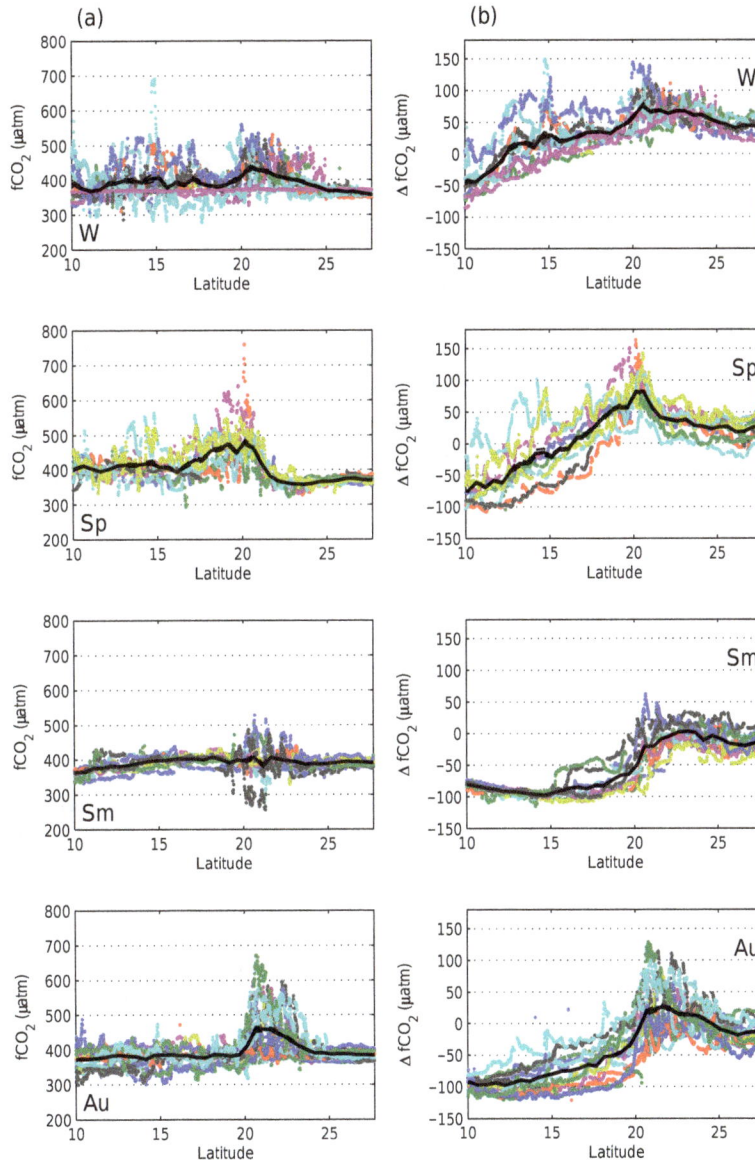

Figure 5. Fugacity of CO_2 data in the Mauritanian–Cap Vert coastal region grouped by seasons: winter (W; December, January, and February), spring (Sp; March, April, and May), summer (Sm; June, July, and August), and autumn (Au; September, October, and November). Column (a) fCO_2^{SW} latitudinal distribution. Column (b), difference between measured and fCO_2^{SW} values normalized to a constant temperature of $22\,°C$. The averaged values for all cruises in Table S1 are shown in black for each season including the 95 % confidence limits. The color code for each cruise is indicated in Table S1.

of $-10.9\,\mu atm\,°C^{-1}$, clearly indicating the important injection of cool and CO_2-rich seawater at the upwelling area. The injection is not being compensated for by the solubility nor by the biological carbon pumps. At $10°\,N$, the rate was still negative but only $-4.3\,\mu atm\,°C^{-1}$ as a result of the seasonal upwelling. $NfCO_2^{SW}$ was related with SST (data not shown) in order to account for effects not removed during normalization. At latitudes 19 to $21°\,N$, in the upwelling vicinity of Cap Blanc, an inverse relationship of 70–$100\,\mu atm\,°C^{-1}$ was found during winter and spring, while in summer and autumn the inverse relationship rate was reduced to 12–$18\,\mu atm\,°C^{-1}$. While the upwelling indexes at those latitudes were quite constant throughout the year, different rates observed should be related to biological consumption of the CO_2 excess. However, during winter and spring the injection of CO_2 in the upwelling is not decreased by the biological activity in the area. But during the Chlorophyll a maximum (late spring and summer), most of the CO_2 was consumed and/or exported and, therefore, the rate was strongly reduced.

Figure 4 depicts the observed interannual trends (a_1 coefficient in Eq. 4) for the four experimentally recorded detrended parameters, together with the UI trend. Confidence intervals of the computed mean annual values for SST, SSS, $f\mathrm{CO}_2^{\mathrm{atm}}$, and $f\mathrm{CO}_2^{\mathrm{sw}}$ were 0.13 °C, 0.06, 0.44, and 4.23 µatm, respectively. There was a clear SST trend whereby seawater along the VOS line track was getting cooler with maximum cooling rates at the location of Cap Blanc (21° N) and Cap Vert upwellings (15° N) with rates higher than $-0.2\,^\circ\mathrm{C}\,\mathrm{yr}^{-1}$. Data from the first 3 years (2005 to 2008) at 21° N showed lower temperatures with higher cooling rates that reached $-0.7\,^\circ\mathrm{C}\,\mathrm{yr}^{-1}$, although 3 years of data are not representative. The area crossed by the VOS line along 17°45′ W from 22 to 10° N is located inside the 1000 m isobath that is well inside the mean frontal activity in the Canary region, about 200 km wide (Wang et al., 2015). The different changes in temperature in the coastal slope and offshore waters are related to the different origins of the waters upwelled from depths of about 100 m to the surface (Mittelstaedt, 1983) that spread off the coastal area. The offshore water SST is less variable owing to longer residence time in the ocean surface. These effects and the fact that the VOS line keeps a track line that crossed the upwelling cells at a distance to the coast that varies among cells contribute to the observed spatial variability. There was no attempt to compare latitudinal and longitudinal effects on the observed values. Our experimental data, however, do not show any positive SST rates in the upwelling affected area, and only when the ship approached the Canary Islands did the trends become less negative, reaching a value of $+0.02\,^\circ\mathrm{C}\,\mathrm{yr}^{-1}$ at 27° N, similar to those obtained for oceanic Atlantic water (Bates et al., 2014).

$f\mathrm{CO}_2^{\mathrm{atm}}$ for the area showed the interannual increase of about $2 \pm 0.3\,\mathrm{µatm}\,\mathrm{yr}^{-1}$ observed in atmospheric stations, while $f\mathrm{CO}_2^{\mathrm{sw}}$ presented a heterogeneous distribution. South of 18° N, the rate of increase was always higher than that in the atmosphere reaching a maximum value of $4.1 \pm 0.4\,\mathrm{µatm}\,\mathrm{yr}^{-1}$ at 10° N. At 27° N, $f\mathrm{CO}_2^{\mathrm{sw}}$ increased at a rate of $1.7 \pm 0.2\,\mathrm{µatm}\,\mathrm{yr}^{-1}$ similar to that determined at the ESTOC time series site (González-Dávila et al., 2010) located at 29°10′ N, 15°30′ W. In the Cap Blanc area, $f\mathrm{CO}_2^{\mathrm{sw}}$ increased at an average rate of $2.5 \pm 0.4\,\mathrm{µatm}\,\mathrm{yr}^{-1}$ with the highest values in the period 2005 to 2008 (a rate of $4.6 \pm 0.5\,\mathrm{µatm}\,\mathrm{yr}^{-1}$ was computed with only those years). Around Cap Blanc, $f\mathrm{CO}_2^{\mathrm{sw}}$ always presented lower rates of increase than in the atmosphere with values well below 1 µatm yr^{-1}. The observed decrease in SST and the trends in $f\mathrm{CO}_2^{\mathrm{sw}}$ can only be explained by a reinforced upwelling. North of 18° N, the lowest rate of increase in $f\mathrm{CO}_2^{\mathrm{sw}}$ compared to $f\mathrm{CO}_2^{\mathrm{atm}}$, together with a decrease in temperature, indicated that upwelling is also favoring an increase in the net community production around the Mauritanian upwelling, consuming and/or exporting the CO_2-rich upwelled waters favored by the lateral transport of the Mauritanian current (Lachkar and Gruber, 2013; Varela et al., 2015). The

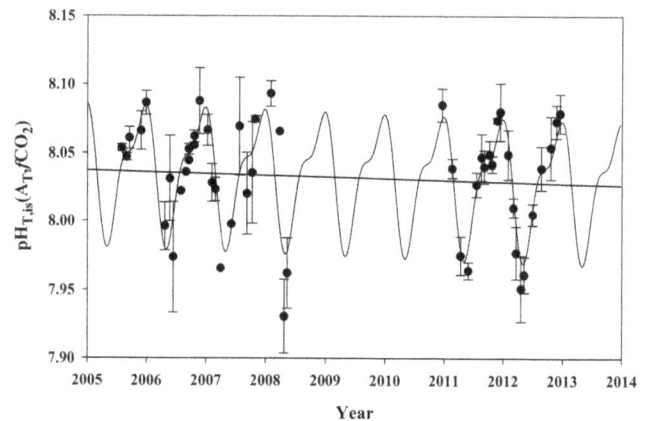

Figure 6. pH of surface waters in total proton scale and at in situ SST computed from total alkalinity (based on regional correlations with SST and SSS; Lee et al., 2006) and $f\mathrm{CO}_2$ at $21 \pm 0.25°$ N. The error bars represent the standard deviation of the computed data for each cruise for the selected latitude. The black curve shows the harmonic fitting of Eq. (4) for the data and the corresponding linear trend is also shown.

upwelling intensification effects observed in the trends of our experimental data support the recent wind stress trends (Cropper et al., 2014; Varela et al., 2015; Santos et al., 2012) of increased upwelling-favorable winds, at least for the period 2005–2012 in the Canary upwelling region (Figs. 2 and 4). The intensification of the upwelling results in a change in the measured upwelled water properties due to either higher upwelling velocities or deeper source upwelled waters. However, what remains unclear from these records is to what extent those changes reflect upwelling variations due to climate change forcing versus natural decadal variability in the upwelling areas occurring over interannual timescales.

Because the upwelling intensity is changing, other variables will also be affected. pH$_{T,\mathrm{is}}$ at $21 \pm 0.25°$ N was computed from $f\mathrm{CO}_2$ and alkalinity pairs of data. Alkalinity was computed from regional correlations with SST and SSS (Lee et al., 2006), which could underrepresent seasonal and interannual variations in upwelling areas. However, pH computed from $f\mathrm{CO}_2$ values are relatively insensitive to errors in A_T, and $f\mathrm{CO}_2$ controls the magnitude and variability of pH (a 60 µmol kg^{-1} change in A_T will affect a 0.1 % in pH, that is, about 0.01 pH units). Figure 6 depicts the computed pH$_{T,\mathrm{is}}(A_T, f\mathrm{CO}_2)$ data and the harmonic fitting of Eq. (4) providing the seasonal variability and interannual trend. Considering the small systematic biases in interannual dynamics, we determined a decrease in pH at a rate of $-0.003 \pm 0.001\,\mathrm{yr}^{-1}$ (Fig. 6). This decrease is one of the highest rate values determined in several time series stations (Bates et al., 2014), where oceanic SST has only slightly increased in the last decades. However, at the Mauritanian upwelling area and at the location where our VOS line approached this region, SST decreased at a rate of $-0.22 \pm 0.06\,^\circ\mathrm{C}\,\mathrm{yr}^{-1}$ (Fig. 4). Solely

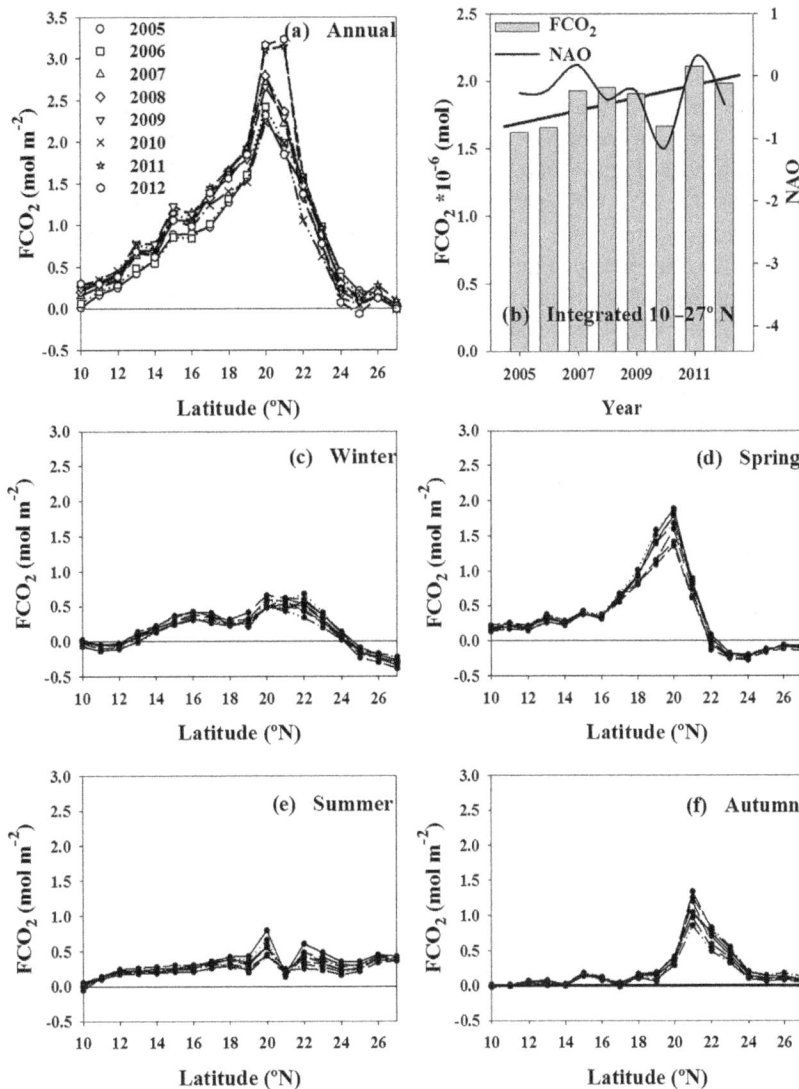

Figure 7. Latitudinal distributions of seasonal and annual CO_2 fluxes (FCO_2, mol m^{-2}). Fluxes of CO_2 were computed using Nightingale et al. (2000) parameterization and satellite winds with a resolution of 6 h. **(a)** Integrated year to year from 2005 to 2012 and **(b)** latitudinally integrated for 2005 to 2012 together with annual values for the North Atlantic Oscillation (NAO) index. Latitudinal distributions of FCO_2 seasonally integrated from 2005 to 2012 are depicted for **(c)** winter (December, January, and February), **(d)** spring (March, April, and May), and **(e)** summer (June, July, and August).

this decrease in temperature would increase the pH by a rate of $+0.004\,\mathrm{yr}^{-1}$ and the fCO_2 would decrease by $4\,\mathrm{\mu atm\,yr}^{-1}$. The net effect of the increase in the amount of rich CO_2 and lower pH upwelled waters in the Mauritanian upwelling would be, therefore, a decrease in the pH rate of over -0.007 ± 0.002 units yr^{-1} and an increase in fCO_2 of $+6.5 \pm 0.7\,\mathrm{\mu atm\,yr}^{-1}$ (with periods where those rates could reach values of $-0.015\,\mathrm{yr}^{-1}$ in pH and $+10.5\,\mathrm{\mu atm\,yr}^{-1}$ in fCO_2 as recorded during 2005–2008). Those values are greatly compensated for by the important decrease in the SST resulting in the determined rates of -0.003 ± 0.001 pH units and $+2.5 \pm 0.4\,\mathrm{\mu atm}$ of fCO_2 per year.

This new data set of experimental values confirmed a decrease in SST and trends in fCO_2^{sw} that can only be explained by reinforced upwelling conditions that favor an increase in the net community production around the Mauritanian upwelling together with a more corrosive environment with pH rates that change by more than $-0.007 \pm 0.002\,\mathrm{yr}^{-1}$ at 21° N. However, the decrease in SST in the upwelling cell buffers this pH rate to values around $-0.003 \pm 0.001\,\mathrm{yr}^{-1}$ and $+2.5 \pm 0.4\,\mathrm{\mu atm\,yr}^{-1}$ in fCO_2, still among the highest observed in other time series.

3.3 Fluxes of CO_2

The annual air–sea CO_2 flux for the full domain was positive (Fig. 7a), with the area off Cap Blanc with values close to $3.3\,mol\,CO_2\,m^{-2}$ (Fig. 7a). North of $24°\,N$, in the area not affected by the coastal upwelling, an average flux of $+0.14 \pm 0.03\,mol\,CO_2\,m^{-2}$ was determined. The ingassing observed during winter and spring of $-0.16 \pm 0.03\,mol\,CO_2\,m^{-2}$ for the full period (Fig. 7) was surpassed by the outgassing during summer and autumn of $0.28 \pm 0.14\,mol\,CO_2\,m^{-2}$. South of $24°\,N$, it was observed that during spring (Fig. 7d) the photosynthetic activity was not intense enough to uptake the CO_2 injected by the strongest upwelling in the surface waters and thus the area acted as a source of CO_2 with values reaching $1.9\,mol\,CO_2\,m^{-2}$ in 2012. During summer (Fig. 7e), primary producers and lateral advection of warm waters by the Mauritanian current could consume and/or export the CO_2-rich waters reaching values of $0.5\,mol\,CO_2\,m^{-2}$. During autumn (Fig. 7f), only the area between 20 and $23°\,N$ acted as a source of $1-1.5\,mol\,CO_2\,m^{-2}$, while the rest was almost in equilibrium. Late autumn–winter upwelling in the 14 to $17°\,N$ region contributed to an increased outgassing with a second annual submaximum of about $0.4\,mol\,CO_2\,m^{-2}$ in winter (Fig. 7c). South of $14°\,N$, annual CO_2 fluxes decreased from about $0.7\,mol\,CO_2\,m^{-2}$ at $14°\,N$ to being roughly in equilibrium at $10°\,N$.

The integrated CO_2 fluxes for the area between 10 and $27°\,N$ along the VOS line section for the years 2005 to 2012 (Fig. 7b) were between 1.6 and 2.1×10^6 mol of CO_2, with an important annual variability. FCO_2 increased during the studied period by $0.05 \pm 0.02 \times 10^6\,mol\,yr^{-1}$. The increase in FCO_2 is related to the observed increase in wind speed (Fig. 4, indicated as UI) north of $16°\,N$. North of $19°\,N$, the influence of wind speed far surpassed the effect of the smaller annual rate of increase in fCO_2^{sw} relative to fCO_2^{atm}, with an exception at $21°\,N$ (Fig. 4). South of $16°\,N$, the decrease in wind speed did not exceed the effect of the incremental change in $(fCO_2^{sw} - fCO_2^{atm})$ associated with the increased downwelling indexes (Fig. 4; Santos et al., 2012), resulting in a slightly increasing FCO_2. The variability observed in the annual integrated CO_2 fluxes (Fig. 7b) was related with the basin-scale oscillations, the North Atlantic Oscillation (NAO) index and the east Atlantic pattern (EA) (http://www.cpc.ncep.noaa. Gov/data/teledoc/telecontents.shtml). Cropper et al. (2014) found winter upwelling variability was strongly correlated with the winter NAO (r values ranged from 0.50 at $12-19°\,N$ to 0.59 at $21-26°\,N$), due to the influence of the Azores semipermanent high-pressure system on the strength of the trade winds. The annual integrated FCO_2 was related with the annual NAO index (Fig. 7b) with a similar $r = 0.54$, even when fluxes are not only controlled by wind strength. However, Fig. 7a clearly indicates that the Mauritanian upwelling area was the most important contributor to FCO_2 in the study

area. The FCO_2 was not significantly correlated with the winter NAO ($r = 0.23$). Also, the EA index, which represents a southward-shifted NAO-like oscillation, presented a lower significant value ($r = 0.48$) (trends not shown), in agreement with the upwelling index (Cropper et al., 2014). Overall, the correlation between fluxes and climate indexes describing the main mode of variability across the Atlantic sector may be directly related to the Azores High and its influence on the trade wind strength.

FCO_2 values along the QUIMA-VOS line were used in order to compute a flux budget for the Mauritanian–Cap Vert region. The observed values were assumed to be valid for at least 100 km on both sides of the QUIMA-VOS line. In this case, the total flux of CO_2 being ejected to the atmosphere would reach a value of 16 Tg of carbon dioxide a year for the period 2005–2012, with a rate of increase of $+0.6\,Tg\,yr^{-1}$. However, it should be considered that the export of the rich fCO_2 upwelled water with high nutrient concentration off the coastal areas would promote a decrease in surface fCO_2 values during productive seasons (as those observed north and south of $21°\,N$) that will result in an ingassing of CO_2. This could balance the observed outgassing increase on a more global scale.

4 Conclusions

The Mauritanian–Cap Vert upwelling area's sensitivity to climatic forcing on upwelling processes strongly affects the CO_2 surface distribution, ocean acidification rates, and air–sea CO_2 exchange.

The experimental SST and carbon dioxide system variable results for the period 2005 to 2012 confirm upwelling intensification at the Mauritanian–Cap Vert upwelling system. Furthermore, we have shown that upwelling regions at low to midlatitudes are important sources of CO_2 for the atmosphere. As a direct result, the pH is decreasing at a rate of $-0.003 \pm 0.001\,yr^{-1}$. Importantly, the amount of emitted CO_2 is increasing annually at a rate of 0.6 Tg due to stronger wind stress, even when primary production seems to also be enhanced in the upwelling area. The monthly record in this EBUS is not yet long enough to determine the extent to which these changes can be attributed to natural decadal variability. These VOS lines must be maintained for years to come and will continue to be one of the most significant contributors to our knowledge of how ocean surface waters are being affected by present and future climate change. The results from VOS lines can provide accurate data for changes in SST, FCO_2, and, consequently, upwelling intensification effects due to global change conditions under decadal natural variability.

Author contributions. MGD and JMSC worked in the equipment installation, data collection and designed the study. FM processed the data and generated figures and results. All of them collaborated in the discussion of the data and the writing of the paper.

Competing interests. The authors declare that they have no conflict of interest.

Special issue statement. This article is part of the special issue "The Ocean in a High-CO_2 World IV". It is a result of the 4th International Symposium on the Ocean in a High-CO_2 World, Hobart, Australia, 3–6 May 2016.

Acknowledgements. Financial support from the European Union through the integrated project FP6 CARBOOCEAN under grant agreement no. 511106-2, FP7 project CARBOCHANGE under grant agreement no. 264879 and H2020 project ATLANTOS under agreement no. 633211 is gratefully acknowledged. Special thanks go to the Mediterranean Shipping Company (MSC) (years 2005–2008) and Maersk (years 2010–2013), who provided the ship platforms and scientific facilities. We thank April Abbott (Macquarie University, Sydney) for her comments and English correction. The MODIS-Aqua Ocean Color Data, 2005–2012 reprocessing, NASA OB.DAAC, Greenbelt, MD, USA, is strongly acknowledged.

Edited by: Kai G. Schulz

References

Astor, Y., Scranton, M., Muller-Karger, F., Bohrer, R. N., and Garcia, J.: CO_2 variability at the CARIACO tropical coastal upwelling time series station, Mar. Chem., 97, 245–261, 2005.

Bakun, A.: Global climate change and intensification of coastal ocean upwelling, Science, 247, 198–201, 1990.

Barton, E. D., Field, D. B., and Roy, C.: Canary current upwelling: More or less?, Prog. Oceanogr., 116, 167–178, 2013.

Bates, N. R., Astor, Y. M., Church, M. J., Currie, K., Dore, J. E., González-Dávila, M., Lorenzoni, L., Muller-Karger, F., Olafsson, J., and Santana-Casiano, J. M.: A time-series view of changing ocean chemistry due to ocean uptake of anthropogenic CO_2 and ocean acidification, Oceanography, 27, 126–141, https://doi.org/10.5670/oceanog.2014.16, 2014.

Borges, A. V. and Frankignoulle, M.: Distribution of surface carbon dioxide and air-sea exchange in the upwelling system off the Galician coast, Global Biogeochem. Cy., 16, 1020, https://doi.org/10.1029/2000GB001385, 2002.

Borges, A. V., Delille, B., and Frankignoulle, M.: Budgeting sinks and sources of CO_2 in the coastal ocean: Diversity of ecosystems counts, Geophys. Res. Lett., 32, L14601, https://doi.org/10.1029/2005GL023053, 2005.

Cai, W.-J., Dai, M., and Wang, Y.: Air–sea exchange of carbon dioxide in oceanmargins: aprovince-based synthesis, Geophys. Res. Lett., 33, L12603, https://doi.org/10.1029/2006GL026219, 2006.

Chen, C.-T. A., Huang, T.-H., Chen, Y.-C., Bai, Y., He, X., and Kang, Y.: Air–sea exchanges of CO_2 in the world's coastal seas, Biogeosciences, 10, 6509–6544, https://doi.org/10.5194/bg-10-6509-2013, 2013.

Cropper, T. E., Hanna, E., and Bigg, G. R.: Spatial and temporal seasonal trends in coastal upwelling off NorthwestAfrica, 1981–2012, Deep-Sea Res. Pt. I, 86, 94–111, 2014.

Demarcq, H.: Trends in primary production, sea surface temperature and wind in upwelling systems (1998–2007), Prog. Oceanogr., 83, 376–385, https://doi.org/10.1016/j.pocean.2009.07.022, 2009.

Dickson, A. G. and Millero, F. J.: A comparison of the equilibrium constants for the dissociation of carbonic acid in seawater media, Deep-Sea Res., 34, 1733–1743, 1987.

DOE: Handbook of methods for the analysis of the various parameters of the carbon dioxide system in sea water, ORNL/CDIAC-74, available at: http://cdiac.ornl.gov/oceans/handbook.html (last access: 7 March 2017), 1994.

Dore, J. E., Lukas, R., Sadler, D. W., and Karl, D. M.: Climate-driven changes to the atmospheric CO_2 sink in the subtropical North Pacific Ocean, Nature, 424, 754–757, 2003.

Feely, R. A., Boutin, J., Cosca, C. E., Dandonneau, Y., Etcheto, J., Inoue, H. Y., Ishii, M., Quéré, C. L., Mackey, D. J., McPhaden, M., Metzl, N., Poisson, A., and Wanninkhof, R.: Seasonal and interannual variability of CO_2 in the equatorial Pacific, Deep Sea Res. Pt. II, 49, 2443–2469, 2002.

Feely, R. A., Sabine, C. L., Hernandez-Ayon, J. M., Ianson, D., and Hales, B.: Evidence for upwelling of corrosive "acidified" water onto the continental shelf, Science, 320, 1490–1492, https://doi.org/10.1126/science.1155676, 2008.

Frankignoulle, M. and Borges, A. V.: European continental shelf as a significant sink for atmospheric carbon dioxide, Global Biogeochem. Cy., 15, 569–576, 2001.

Friederich, G. E., Ledesma, J., Ulloa, O., and Chavez, F. P.: Air-sea carbon dioxide fluxes in the coastal southeastern tropical Pacific, Prog. Oceanogr., 79, 156–166, 2008.

González Dávila, M., Santana-Casiano, M. J., Merlivat, L., Barbero-Munoz, L., and Dafner, E.: Fluxes of CO_2 between the atmosphere and the ocean during the POMME project in the northeast Atlantic Ocean during 2001, J. Geophys. Res., 110, C07S11, https://doi.org/10.1029/2004JC002763, 2005.

González-Dávila, M., Santana-Casiano, J. M., and Ucha, I.: Seasonal variability of fCO_2 in the Angola-Benguela region, Prog. Oceanogr., 83, 124–133, 2009.

González-Dávila, M., Santana-Casiano, J. M., Rueda, M. J., and Llinás, O.: The water column distribution of carbonate system variables at the ESTOC site from 1995 to 2004, Biogeosciences, 7, 3067–3081, https://doi.org/10.5194/bg-7-3067-2010, 2010.

Gruber, N.: Warming up, turning sour, losing breath: ocean biogeochemistry under global change, Philos. T. R. Soc. Lond., 369, 1980–1996, 2011.

Gruber, N., Keeling, C. D., and Bates, N. R.: Interannual variability in the North Atlantic Ocean carbon sink, Science, 298, 2374–2378, 2002.

Hagen, E. and Schemainda, R.: Der Guineadom im ostatlantischen Stromsystem, Beitr. Meereskd., 51, 5–27, 1984.

Hales, B., Takahashi, T., and Bandstra, L.: Atmospheric CO_2 uptake by a coastal upwelling system, Global Biogeochem. Cy., 19, GB1009, https://doi.org/10.1029/2004GB002295, 2005.

Hastenrath, S.: Climate Dynamics of the Tropics, 488 pp., Kluwer Acad., Norwell, Mass., 1995.

Keeling, R. F., Kortzinger, A., and Gruber, N.: Ocean deoxygenation in a warming world, Annu. Rev. Mar. Sci., 2, 199–229, https://doi.org/10.1146/annurev.marine.010908.163855, 2010.

Key, R., Kozyr, A., Sabine, C., Lee, K., Wanninkhof, R., Bullister, J., Feely, R., Millero, F. J., Mordy, C., and Peng, T.-H.: A global ocean carbon climatology: Results from GLODAP, Global Biogeochem. Cy., 18, GB4031, https://doi.org/10.1029/2004GB002247, 2004.

Lachkar, Z. and Gruber, N.: Response of biological production and air–sea CO_2 fluxes to upwelling intensificationin the California and Canary Current Systems, J. Marine Syst., 109–110, 149–160, 2013.

Lee, K., Tong, L. T., Millero, F. J., Sabine, C. L., Dickson, A. G., Goyet, C., Park, G. H., Wanninkhof, R., Feely, R. A., and Key, R. M.: Global relationships of total alkalinity with salinity and temperature in surface waters of the world's oceans, Geophys. Res. Lett., 33, L19605, https://doi.org/10.1029/2006GL027207, 2006.

Lüger, H., Wallace, D. W., Körtzinger, A., and Nojiri, Y.: The $p\mathrm{CO}_2$ variability in the midlatitude North Atlantic Ocean during a full annual cycle, Global Biogeochem. Cy., 18, GB3023, https://doi.org/10.1029/2003GB002200, 2004.

Lüger, H., Wanninkhof, R., Wallace, D. W., and Körtzinger, A.: CO_2 fluxes in the subtropical and subarctic North Atlantic based on measurements from a volunteer observing ship, J. Geophys. Res., 111, C06024, https://doi.org/10.1029/2005JC003101, 2006.

Marcello, J., Alonso, H., Eugenio, F., and Fonte, A.: Seasonal and temporal study of the northwest African upwelling system, Int. J. Remote Sens., 32, 1843–1859, https://doi.org/10.1080/01431161003631576, 2011.

Mehrbach, C., Culberson, C. H., Hawley, J. E., and Pytkowicz, R. N.: Measurement of the apparent dissociation constants of carbonic acid in seawater at atmospheric pressure, Limnol. Oceanogr., 18, 897–907, 1973.

Michaels, A. F., Karl, D. M., and Capone, D. G.: Element stoichiometry, new production and nitrogen fixation, Oceanography, 14, 68–77, 2001.

Mittelstaedt, E.: The upwelling area off Africa – A description of phenomena related to coastal upwelling, Prog. Oceanogr., 12, 307–331, https://doi.org/10.1016/0079-6611(83)90012-5, 1983.

Neuer, S., Torres-Padrón, M. E., Gelado-Caballero, M. D., Rueda, M. J., Hernández-Brito, J. J., Davenport, R., and Wefer, G.: Dust deposition pulses to the eastern subtropical North Atlantic gyre: Does ocean's biogeochemistry respond?, Global Biogeochem. Cy., 18, GB4020, https://doi.org/10.1029/2004GB002228, 2004.

Nicholson, S. E.: Rainfall and atmospheric circulation during drought periods and wetter years in West Africa, Mon. Weather Rev., 109, 2191–2208, 1981.

Nightingale, P. D., Malin, G., Law, C. S., Watson, A. J., Liss, P. S., Liddicoat, M. L., Boutin, J., and Upstill-Goddard, R. C: In situ evaluation of air-sea gas exchange parameterizations using novel conservative and volatile tracers, Global. Biogeochem. Cy., 14, 373–387, 2000.

Nykjaer, L. and Van Camp, L.: Seasonal and interannual variability of coastal upwelling along Northwest Africa and Portugal from 1981 to 1991, J. Geophys. Res., 99, 14197–14207, 1994.

Oerder, V., Colas, F., Echevin, V., Codron, F., Tam, J., and Belmadani, A.: Peru-Chile upwelling dynamics under climate change, J. Geophys. Res.-Oceans, 120, 1152–1172, https://doi.org/10.1002/2014JC010299, 2015.

Padin, X. A., Vázquez-Rodríguez, M., Castaño, M., Velo, A., Alonso-Pérez, F., Gago, J., Gilcoto, M., Álvarez, M., Pardo, P. C., de la Paz, M., Ríos, A. F., and Pérez, F. F.: Air-Sea CO_2 fluxes in the Atlantic as measured during boreal spring and autumn, Biogeosciences, 7, 1587–1606, https://doi.org/10.5194/bg-7-1587-2010, 2010.

Pfeil, B., Olsen, A., Bakker, D. C. E., Hankin, S., Koyuk, H., Kozyr, A., Malczyk, J., Manke, A., Metzl, N., Sabine, C. L., Akl, J., Alin, S. R., Bates, N., Bellerby, R. G. J., Borges, A., Boutin, J., Brown, P. J., Cai, W.-J., Chavez, F. P., Chen, A., Cosca, C., Fassbender, A. J., Feely, R. A., González-Dávila, M., Goyet, C., Hales, B., Hardman-Mountford, N., Heinze, C., Hood, M., Hoppema, M., Hunt, C. W., Hydes, D., Ishii, M., Johannessen, T., Jones, S. D., Key, R. M., Körtzinger, A., Landschützer, P., Lauvset, S. K., Lefèvre, N., Lenton, A., Lourantou, A., Merlivat, L., Midorikawa, T., Mintrop, L., Miyazaki, C., Murata, A., Nakadate, A., Nakano, Y., Nakaoka, S., Nojiri, Y., Omar, A. M., Padin, X. A., Park, G.-H., Paterson, K., Perez, F. F., Pierrot, D., Poisson, A., Ríos, A. F., Santana-Casiano, J. M., Salisbury, J., Sarma, V. V. S. S., Schlitzer, R., Schneider, B., Schuster, U., Sieger, R., Skjelvan, I., Steinhoff, T., Suzuki, T., Takahashi, T., Tedesco, K., Telszewski, M., Thomas, H., Tilbrook, B., Tjiputra, J., Vandemark, D., Veness, T., Wanninkhof, R., Watson, A. J., Weiss, R., Wong, C. S., and Yoshikawa-Inoue, H.: A uniform, quality controlled Surface Ocean CO_2 Atlas (SOCAT), Earth Syst. Sci. Data, 5, 125–143, https://doi.org/10.5194/essd-5-125-2013, 2013.

Santana-Casiano, J., González-Dávila, M., and Ucha, I.: Carbon dioxide fluxes in the Benguela upwelling system during winter and spring: A comparison between 2005 and 2006, Deep-Sea Res. Pt. II, 56, 533–541, 2009.

Santana-Casiano, J. M., González-Dávila, M., Rueda, M., Llinás, O., and González-Dávila, E.-F.: The interannual variability of oceanic CO_2 parameters in the northeast Atlantic subtropical gyre at the ESTOC site, Global Biogeochem. Cy., 21, GB1015, https://doi.org/10.1029/2006GB002788, 2007.

Santos, A. M. P., Kazmin, A. S., and Peliz, A.: Decadal changes in the Canary upwelling system as revealed by satellite observations: Their impact on productivity, J. Mar. Res., 63, 359–379, 2005.

Santos, F., de Castro, M., Gómez-Gesteira, M., and Alvarez, I.: Differences in coastal and oceanic SST warming rates along the Canary upwelling ecosystem from 1982 to 2010, Cont. Shelf Res., 47, 1–6, 2012.

Schuster, U., Watson, A., Bates, N., Corbiere, A., Gonzalez-Davila, M., Metzl, N., Pierrot. D., and Santana-Casiano, J. M.: Trends in North Atlantic sea-surface $f\mathrm{CO}_2$ from 1990 to 2006, Deep-Sea Res. Pt. II, 56, 620–629, 2009.

Takahashi, T., Olafsson, J., Goddard, J. G., Chipman, D. W., and Sutherland, S.: Seasonal variation of CO_2 and nutrients in the high-latitude surface oceans: A comparative study, Glob. Biogeochem. Cy., 7, 843–878, 1993.

Takahashi, T., Sutherland, S., Wanninkhof, R., Sweeney, C., Feely, R., Chipman, D., Hales, B., Friederich, G., Chavez, F., Sabine, C., Watson, A., Bakker, D., Schuster, U., Metzl, N., Yoshikawa-Inoue, H., Ishii, M., Midorikawa, T., Nojiri, Y., Kortzinger, A., Steinhoff, T., Hoppema, M., Olafsson, J., Arnarson, T., Tilbrook, B., Johannessen, T., Olsen, A., Bellerby, A., Wong, C., Delille, B., Bates, N., and de Baar, H.: Climatological mean and decadal change in surface ocean pCO_2, and net sea-air CO_2 flux over the global oceans, Deep-Sea Res. Pt. II, 56, 554–577, 2009.

Ullman, D. J., McKinley, G. A., Bennington, V., and Dutkiewicz, S.: Trends in the North Atlantic carbon sink: 1992–2006, Glob. Biogeochem. Cy., 23, GB4011, https://doi.org/10.1029/2008GB003383, 2009.

Varela, R., Álvarez, I., Santos, F., de Castro, M., and Gómez-Gesteira, M.: Has upwelling strengthened along worldwide over 1982–2010?, Sci. Rep., 5, 10016, https://doi.org/10.1038/srep10016, 2015.

Wang, Y., Castelao, R. M., and Yuan, Y.: Seasonal variability of alongshore winds and sea surface temperature fronts in Eastern Boundary Current Systems, J. Geophys. Res.-Oceans, 120, 2385–2400, https://doi.org/10.1002/2014JC010379, 2015.

Watson, A., Schuster, U., Bakker, D., Bates, N., Corbière, A., González-Dávila, M., Friedrich, T., Hauck, J., Heinze, C., Johannessen, T., Kortzinger, A., Metzl, N., Olafsson, J., Olsen, A., Oschlies, A., Padin, X. A., Pfeil, B., Santana-Casiano, J. M., Steinhoff, T., Telszewski, M., Rios, A. F., Wallace, D. W., and Wanninkhof, R.: Tracking the variable North Atlantic sink for atmospheric CO_2, Science, 326, 1391–1393, https://doi.org/10.1126/science.1177394, 2009.

Yoo, J.-M. and Carton, J. A.: Annual and interannual variation of the freshwater budget in the tropical Atlantic Ocean and the Caribbean Sea, J. Phys. Oceanogr., 20, 831–845, 1990.

5

A biophysical approach using water deficit factor for daily estimations of evapotranspiration and CO$_2$ uptake in Mediterranean environments

David Helman[1], Itamar M. Lensky[1], Yagil Osem[2], Shani Rohatyn[3], Eyal Rotenberg[3], and Dan Yakir[3]

[1]Department of Geography and Environment, Bar-Ilan University, Ramat Gan 52900, Israel
[2]Department of Natural Resources, Agricultural Research Organization, Volcani Center, Bet Dagan 50250, Israel
[3]Earth and Planetary Sciences, Weizmann Institute of Science, Rehovot 76100, Israel

Correspondence to: David Helman (davidhelman.biu@gmail.com, david.helman@biu.ac.il)

Abstract. Estimations of ecosystem-level evapotranspiration (ET) and CO$_2$ uptake in water-limited environments are scarce and scaling up ground-level measurements is not straightforward. A biophysical approach using remote sensing (RS) and meteorological data (RS–Met) is adjusted to extreme high-energy water-limited Mediterranean ecosystems that suffer from continuous stress conditions to provide daily estimations of ET and CO$_2$ uptake (measured as gross primary production, GPP) at a spatial resolution of 250 m. The RS–Met was adjusted using a seasonal water deficit factor (f_{WD}) based on daily rainfall, temperature and radiation data. We validated our adjusted RS–Met with eddy covariance flux measurements using a newly developed mobile lab system and the single active FLUXNET station operating in this region (Yatir pine forest station) at a total of seven forest and non-forest sites across a climatic transect in Israel (280–770 mm yr^{-1}). RS–Met was also compared to the satellite-borne MODIS-based ET and GPP products (MOD16 and MOD17, respectively) at these sites.

Results show that the inclusion of the f_{WD} significantly improved the model, with $R = 0.64$–0.91 for the ET-adjusted model (compared to 0.05–0.80 for the unadjusted model) and $R = 0.72$–0.92 for the adjusted GPP model (compared to $R = 0.56$–0.90 of the non-adjusted model). The RS–Met (with the f_{WD}) successfully tracked observed changes in ET and GPP between dry and wet seasons across the sites. ET and GPP estimates from the adjusted RS–Met also agreed well with eddy covariance estimates on an annual timescale at the FLUXNET station of Yatir (266 ± 61 vs. 257 ± 58 mm yr^{-1} and 765 ± 112 vs.

748 ± 124 gC m^{-2} yr^{-1} for ET and GPP, respectively). Comparison with MODIS products showed consistently lower estimates from the MODIS-based models, particularly at the forest sites. Using the adjusted RS–Met, we show that afforestation significantly increased the water use efficiency (the ratio of carbon uptake to ET) in this region, with the positive effect decreasing when moving from dry to more humid environments, strengthening the importance of drylands afforestation. This simple yet robust biophysical approach shows promise for reliable ecosystem-level estimations of ET and CO$_2$ uptake in extreme high-energy water-limited environments.

1 Introduction

Assessing the water use and carbon uptake in terrestrial ecosystems is important for monitoring biosphere responses to climate change (Ciais et al., 2005; Jung et al., 2010; Reichstein et al., 2013). Accurate estimations of evapotranspiration (ET) and gross primary production (GPP), as a measure of the CO$_2$ uptake, usually require the integration of extensive meteorological, flux and field-based data (e.g., Wang et al., 2014; Kool et al., 2014). However, scaling up field-based measurements to the ecosystem level is not straightforward and requires the use of complex models (Way et al., 2015).

Currently, the eddy covariance (EC) technique is the most direct method for measuring carbon and water vapor fluxes at the ecosystem level (Baldocchi, 2003). The EC approach benefits from continuous temporal coverage; currently (April, 2017), there are more than 560 active EC sites across the globe as part of the FLUXNET program (http://fluxnet.ornl.gov). However, there are also some practical and technical limitations. The EC measurement is representative of a relatively small area ($< 2\,km^2$), and the application of the EC approach is limited to relatively homogeneous and flat terrains. Additionally, most EC towers are concentrated in the US, Europe and Asia, with poor coverage in water-limited regions, such as North Africa and the eastern Mediterranean (Schimel et al., 2015).

Remote-sensing-based models (RS models) have been used to overcome some of the limitations of EC, complementing the information derived from the flux towers. In contrast to process-driven models, RS models benefit from continuous, direct observation of the Earth's surface, acquiring data at a relatively high spatial resolution and with full regional to global coverage. Many RS models for the estimation of ET and GPP exist (see reviews in Kalma et al., 2008, and Hilker et al., 2008), but these algorithms are too complex and most of the models are not provided as accessible products for researchers outside the remote sensing community. Particular exceptions are the satellite-borne MODIS-based ET and GPP products (MOD16 and MOD17), which provide 8-day ET and GPP estimates at 1 km for 2000–2015, globally (Mu et al., 2007, 2011, Running et al., 2000, 2004).

In the past decade, several simple biophysical ET and GPP models based on vegetation indices (from satellite data) have emerged, offering assessment at relatively high to moderate spatial and temporal resolutions with an acceptable accuracy (i.e., daily estimates at 250 m; see, e.g., Veroustraete et al., 2002; Sims et al., 2008; Maselli et al., 2009, 2014; and review of ET models in Glenn et al., 2010). One of those models is the ET model based on the FAO-56 formulation (Allen et al., 1998). The FAO-56 formulation states that the actual ET of irrigated crops can be determined from the reference ET (ET_o) corrected with crop coefficient K_c values (see Eq. 2). The K_c varies mainly with specific plant species characteristics, which enables the transfer of standard K_c values among locations and environments (Allen et al., 2006).

The remote-sensing version of this formulation uses a function of satellite-derived vegetation index, usually the normalized difference vegetation index (NDVI), as a substitute for the crop coefficient. Being a measure of the green plant biomass and the ecosystem leaf area, the NDVI is often used as a surrogate for plant transpiration and rainfall interception capacity (Glenn et al., 2010). Additionally, the NDVI is closely related to the radiation absorbed by the plant and to its photosynthetic capacity (Gamon et al., 1995). However, the direct detection, through NDVI, of the abovementioned parameters on a seasonal timescale is still challenging and usually requires additional meteorological information (Helman et al., 2015a). The RS model based on the FAO-56 formulation combines the two sources of information, satellite and meteorological, providing a daily estimation of actual ET. This model, originally proposed for croplands and other

managed vegetation systems (Allen et al., 1998; Glenn et al., 2010), was recently adjusted for applications in natural vegetation systems (Maselli et al., 2014).

For the estimation of GPP, a simple but robust biophysical GPP model is the one based on the radiation use efficiency (RUE) model proposed by Monteith (1977). The classical Monteith-type model depends on the absorbed radiation and on the efficiency of the vegetation at converting this radiation into carbon-based compounds. Accordingly, this Monteith-based model is driven by radiation and temperature data, acquired from meteorological stations, and by the fraction of absorbed photosynthetically active radiation (fAPAR), which can be calculated from the satellite-derived NDVI. A major challenge in this model, however, is the estimation of the RUE, a key component of the model, which usually depends on plant species type and environmental conditions. Currently, the conventional procedure is to use a plant-species-dependent maximum RUE from a lookup table and adjust it for seasonal changes using some sort of a factor that changes throughout the season based on meteorological data (Running et al., 2004; Zhao and Running, 2010).

Though simple, both ET and GPP models (hereafter RS–Met) were shown to be promising in accurately assessing daily ET and GPP at a relatively high spatial resolution ($< 1\,km$; Helman et al., 2017a; Maselli et al., 2014, 2006; Veroustraete et al., 2002). However, the use of the RS–Met is limited to ecosystems under normally non-stressful conditions because there is no accurate representation of water availability in these models. Recently, the incorporation of a water-deficit factor (f_{WD}) in these models was proposed by Maselli et al. (2009, 2014), adjusting for short-term stress conditions in natural water-limited ecosystems. The proposed f_{WD} is based only on daily rainfall data and daily potential ET calculated from temperature and/or incoming radiation. The RS–Met with the addition of the f_{WD} was successfully validated against EC-derived estimates of ET and GPP at several sites in Italy (Maselli et al., 2014, 2009, 2006).

However, the RS–Met approach has never been tested in extreme high-energy water-limited environments such as those in the eastern Mediterranean. Currently, there is only one active FLUXNET station in the entire eastern Mediterranean (Yatir forest, southern Israel; Fig. 1a) that measures water vapor and carbon fluxes (since 2000), while in this region water is considered to be a valuable resource and the proper management of this resource depends on the accurate assessment of the ET component. Moreover, despite the well-known important contribution of drylands regions to global CO_2 (Ahlström et al., 2015), there are almost no efforts to estimate CO_2 fluxes in forested and non-forested areas in this dry region. This led to the development of the Weizmann mobile lab system (Israel; Fig. 1h), which allows the extension of the permanent FLUXNET measurement sites on campaign basis (e.g., Asaf et al., 2013; for technical detail see http://www.weizmann.ac.il/EPS/Yakir/node/321). Such a

Figure 1. Views of the seven study sites along the climatic gradient **(a–g)** and the newly mobile flux measurement system used in this study **(h)**. Sites include three paired planted pine forests (*Pinus halepensis*) and adjacent non-forest sites (representing the original environment in which these forests were planted): Yatir **(a)** and Wady Attir **(b)**; Eshtaol **(c)** and Modiin **(d)**; Birya **(e)** and Kadita **(f)**. The deciduous oak forest of HaSolelim is shown **(d)**. The three paired sites **(a–f)** represent the geo-climatic transition from xeric to mesic environments in Israel.

system could allow flux and auxiliary analytical measurements across a range of climatic conditions, plant species and ecosystems, as well as addressing land use changes and disturbance. However, to extend these campaign-based measurements in time and space, a model fitted to the high-energy water-limited conditions of this region is required.

Here, we adjusted the RS–Met to the extreme hot and dry conditions of the eastern Mediterranean region. The adjusted RS–Met was examined in a total of seven ecosystems distributed at three precipitation levels along a rainfall gradient (280–770 mm yr^{-1}) in this region (Israel; Fig. 1a–g). Ecosystems included three pairs of planted forests and adjacent non-forest sites (representing the original area in which these forests were planted). Ground-level campaign measurements of ET and net ecosystem CO_2 exchange using the newly developed mobile lab (Fig. 1h) and the continuous flux mea-

surements at the active FLUXNET site in Yatir (Klein et al., 2016; Tatarinov et al., 2016) were used to validate the RS–Met. This combination of model-based estimates and direct flux measurements of ET and CO_2 uptake across a range of climatic conditions and ecosystems provides a unique opportunity to test and validate the RS–Met approach in this high-energy water-limited region. Particularly, we examined the RS–Met with and without the application of the f_{WD}. We also compared the RS–Met with MODIS ET and GPP products at the studied sites.

Our specific goals in this study were to (1) examine the seasonal evolution of the f_{WD} and its role in the RS–Met; (2) compare the model estimates with EC and MODIS ET and GPP products across these high-energy water-limited sites, at a daily and annual basis; and (3) use the RS–Met to estimate changes in water use efficiency (WUE = GPP/ET) following afforestation across the rainfall gradient in Israel, by comparing the three paired forest vs. non-forest sites.

2 Materials and methods

2.1 Study sites

The sites in this study included three pairs of planted pine forests (*Pinus halepensis* Mill.) and adjacent non-forested (dwarf shrublands) sites distributed throughout a climatic range in Israel ($P = 280$–770 mm yr^{-1}), from dry to sub-humid Mediterranean (Table 1 and Fig. 1a–f), which represent the typical Mediterranean vegetation systems in the eastern Mediterranean. The three non-forested sites represent the original natural environment in which the pine forests were planted, while the afforested sites are currently managed by the Jewish National Fund (KKL). The non-forested shrubland sites are mostly dominated by *Sarcopoterium spinosum* (dwarf shrub) in a patchy distribution with a wide variety of herbaceous species, mostly annuals, growing in between the shrub patches during winter to early spring. In addition, we tested the models at one native deciduous forest site dominated by the *Quercus* species. A brief description of the sites is given in the following.

2.1.1 Yatir

The forest of Yatir is an Aleppo pine forest (*Pinus halepensis*) that was planted by the KKL mostly during 1964–1969 in the semiarid region of Israel (31.34° N, 35.05° E; Fig. 1a). It covers a total area of ca. 2800 ha and lies on a predominantly light brown Rendzina soil (79 ± 45.7 cm deep), overlying a chalk and limestone bedrock (Llusia et al., 2016). The average elevation is 650 m. The mean annual rainfall in the forest area is 285 mm yr^{-1} (for the last 40 years) and was 279 mm yr^{-1} at the FLUXNET site during 2001–2015 (Table 1). The mean annual temperature in Yatir is 18.2 °C with 13 and 31 °C for mean winter (November–January) and summer (May–July) temperatures, respectively. Tree den-

Table 1. Site characteristics and locations divided into two groups of forest (top) and non-forest (bottom) sites. In each group, sites are arranged from dry to humid (from top to bottom).

Site	Location (lat, N; long, E)	PFT	Dominant species	Grazing	Altitude	P	AF
Yatir	31.3451; 35.0519	CF	*P. halepensis*	sheep	660	279	0.19
Eshtaol	31.7953; 34.9954	CF	*P. halepensis*	sheep	385	480	0.34
HaSolelim	32.7464; 35.2317	OF	*Q. ithaburensis*	cattle	180	543	0.42
Birya	33.0015; 35.4823	CF	*P. halepensis*	cattle	750	766	0.63
Wady Attir	31.3308; 34.9905	SH	*Phagnalon rupestre*	sheep	490	279	0.11
Modiin	31.8698; 35.0125	SH	*S. spinosum*	cattle	245	480	0.32
Kadita	33.0110; 35.4614	SH	*S. spinosum*	cattle	815	766	0.63

PFT is the plant functional type (CF, coniferous forest; OF, oak forest; SH, shrubland); grazing indicates the main grazing regime at the site; altitude is in meters above sea level; P is the mean annual rainfall ($\mathrm{mm\,yr^{-1}}$); AF is the aridity factor calculated as the P-to-ET_O ratio ($\mathrm{mm\,mm^{-1}}$).

sity in Yatir is ca. 300 $\mathrm{trees\,ha^{-1}}$ (Rotenberg and Yakir, 2011) with a tree average height of ca. 10 m and canopy leaf area index (LAI) of $1.4 \pm 0.4\,\mathrm{m^2\,m^{-2}}$, which displays small fluctuations between winter and summer (Sprintsin et al., 2011). The understory in this forest is mostly comprised of ephemeral herbaceous species (i.e., therophyte, geophytes and hemicryptophytes) growing during the wet season (September–April) and drying out in the beginning of the dry season (May–June). A relatively thin needle litter layer covers the forest floor during the needle senescence period (June–August; Maseyk et al., 2008).

2.1.2 Eshtaol

The forest of Eshtaol was planted in the late 1950s by the KKL with mostly *P. halepensis* trees in the central part of Israel (31.79° N, 34.99° E; Fig. 1c). The current forest area is ca. 1200 ha and lies mainly on Rendzina soils. The average elevation is 330 m. The mean annual rainfall in this area is ca. 500 $\mathrm{mm\,yr^{-1}}$ and was a 480 $\mathrm{mm\,yr^{-1}}$ at the site of the EC measurements during 2012–2015 (Table 1). Tree density in Eshtaol is typically 300–350 $\mathrm{trees\,ha^{-1}}$, with a tree canopy LAI that ranges between 1.9 and 2.6 $\mathrm{m^2\,m^{-2}}$ and a tree average height of 12.5 m (Osem et al., 2012).

2.1.3 Birya

The forest of Birya is a *P. halepensis* forest that was mostly planted during the early 1950s in the northern part of Israel in the Galilee region (33.00° N, 35.48° E; Fig. 1e). The forest covers an area of ca. 2100 ha and lies on Rendzina and Terra rossa soils. The average elevation is 730 m. The average temperature in this area is 16 °C, with an average annual rainfall of 710 and 776 $\mathrm{mm\,yr^{-1}}$ during the years of the EC measurements (2012–2015; Table 1). The average stand density is 375 $\mathrm{trees\,ha^{-1}}$ with an average tree height of 11 m (Llusia et al., 2016).

2.1.4 HaSolelim

The HaSolelim forest is a native deciduous mixed oak forest dominated by *Quercus ithaburensis*, which is accompanied by *Quercus calliprinos* (evergreen) and a few other Mediterranean broadleaved tree and shrub species (Fig. 1g). The forest is located in the northern part of Israel in the Galilee region, 30 km south of the Birya forest (32.74° N, 35.23° E). The forest covers an area of ca. 240 ha and lies on Rendzina and Terra rossa soils. The elevation at the site of the EC measurements is 180 m (Table 1). The average temperature in this area is typically 21 °C, with a mean annual rainfall of 580 $\mathrm{mm\,yr^{-1}}$ and 543 $\mathrm{mm\,yr^{-1}}$ during the years of the EC measurements. The site where the measurements took place is characterized by an average stand density of 280 $\mathrm{trees\,ha^{-1}}$ and an average tree height of 8 m (Llusia et al., 2016).

2.1.5 Wady Attir

This is a xeric shrubland site located southwest from the forest of Yatir (31.33° N, 34.99° E). The average elevation is 490 m. The site is dominated by semi-shrub species, such as *Phagnalon rupestre* L, with *graminae* species, mainly *Stipa capensis* L. (also known as Mediterranean needle grass); *Hordeum spontaneum* K. Koc. (also known as wild barley); and some *Avena* species such as *A. barbata* L. and *A. sterilis* L., appearing shortly after the rainy season (Leu et al., 2014; Fig. 1b). The mean annual rainfall in this area is 230 $\mathrm{mm\,yr^{-1}}$ (Mussery et al., 2016) and was 280 $\mathrm{mm\,yr^{-1}}$ in the years of the EC measurements (2012–2015; Table 1).

2.1.6 Modiin

The shrubland site of Modiin is located a few kilometers from the forest site of Eshtaol and represents the original environment in which this forest was planted (31.87° N, 35.01° E; Fig. 1d). The average elevation is 245 m. The shrubland site is mostly dominated by *Sarcopoterium spinosum* (dwarf shrub) in a patchy distribution with a wide variety of herbaceous species, mostly annuals, growing in between the shrub patches from winter to early spring. The average rainfall amount in this area was $480\,\text{mm}\,\text{yr}^{-1}$ in the years of the EC measurements (Table 1).

2.1.7 Kadita

The shrubland site of Kadita is also dominated by *Sarcopoterium spinosum* (dwarf shrub) in a typical patchy distribution (Fig. 1f). It is located near the forest of Birya at an elevation of 815 m ($33.01°$ N, $35.46°$ E; Table 1). The mean annual rainfall at this site is similar to that recorded in the Birya forest (i.e., $766\,\text{mm}\,\text{yr}^{-1}$ in the years of study).

All shrubland sites have been under continuous livestock grazing for many years, and their vegetation structures are mainly the outcome of both rainfall amount and grazing regime.

2.2 Satellite-derived vegetation index

We used the NDVI from the MODIS onboard NASA's Terra satellite at 250 m spatial resolution (MOD13Q1). The MOD13Q1 NDVI product is a composite of a single day's value selected from 16-day periods based on the maximum value criteria (Huete et al., 2002). Terra's NDVI product is acquired during the morning (10:30 LT) and thus provides a good representation of the peak time of the plants' diurnal activity. The gradual growth of the vegetation enables the interpolation of the 16-day NDVI time series to representative daily values (Glenn et al., 2008; Maselli et al., 2014). We downloaded the 16-day NDVI time series covering the main area of the EC flux measurement for each site from the MODIS subsets (http://daacmodis.ornl.gov/cgi-bin/MODIS/GLBVIZ_1_Glb/modis_subset_order_global_col5.pl) for the period October 2001–October 2015. Then we preprocessed the NDVI time series as described in Helman et al. (2014a, b, 2015b) to remove outliers and uncertainties due to cloud contamination and atmospheric disturbances without removing important information (see Fig. S2 in the Supplement). The processed 16-day NDVI time series were then interpolated on a daily basis using the local scatter plot smoothing technique (LOESS). This technique is suited for eliminating outliers in non-parametric time series and has been shown to be a useful tool in the interpolation of datasets with a seasonal component (Cleveland, 1979).

2.3 The mobile lab system and the FLUXNET station in Yatir

A newly designed mobile flux measurement system was used in all campaigns (Fig. 1h), based on the 28 m pneumatic mast on a 12 tons 4×4 truck that included a laboratory providing an air-conditioned instrument facility (cellular communication, 18 kVA generator, 4200 W UPS). Flux, meteorological and radiation measurements relied on an EC system that provides CO_2 measurements and sensible and latent heat fluxes using a three-dimensional sonic anemometer (R3,

Gill Instruments, Lymington, Hampshire, UK) and enclosed-path CO_2–H_2O infrared gas analyzer (IRGA) (Licor 7200, Li-Cor, Lincoln, NE, USA) using CARBOEUROFLUX methodology (Aubinet et al., 2000), and EddyPro software (www.licor.com). Data were collected using a self-designed program in LabVIEW software. Air temperature and relative humidity (HMP45C probes, Campbell Scientific) and air pressure (Campbell Scientific sensors) were measured at 3 m above the canopy. Energy fluxes relied on radiation sensors, including solar radiation (CMP21, Kipp and Zonen), long-wave radiation (CRG4, Kipp and Zonen) and photosynthetic radiation (PAR, PAR-LITE2) sensors. All sensors were installed in pairs facing both up and down and were connected using the differential mode through a multiplexer to a data logger (Campbell Scientific). GPP for each site was calculated from the measured net ecosystem CO_2 exchange (NEE) after estimating ecosystem respiration, Re, and using the regression of NEE on turbulent nights against temperature, followed by extrapolating the relationship that was found between Re and temperature during the nighttime and daytime periods (Reichstein et al., 2005; modified for our region by Afik, 2009). Flux measurements with the mobile system were carried out on a campaign basis, at six of the seven sites, with each campaign representing approximately 2 weeks at a single site, repeated along the seasonal cycle with mostly two but sometimes only one 2-week sets of measurements per cycle, during the 4 years of measurements, 2012–2015. Continuous flux measurements were carried out at the permanent FLUXNET site of Yatir (xeric forest site). Began in 2000, the EC and supplementary meteorological measurements have been conducted continuously (Rotenberg and Yakir, 2011; Tatarinov et al., 2016), with measurements performed according to the EUROFLUX methodology. Instrumentation is similar to that in the mobile lab except for the use of a closed-path CO_2–H_2O IRGA (LI-7000; Li-Cor, Lincoln, NE) with the inlet placed 18.7 m above the ground. Typical fetch providing 70 % (cumulative) contribution to turbulent fluxes was measured between 100 and 250 m (depending on the site) along the wind distance. This was taken into consideration when using the MOD13Q1 product to derive the modeled fluxes.

During April 2012, at the peak activity season in Yatir forest, for 2 weeks the mobile lab system was deployed 10 m away from the permanent flux measurement tower, where both EC systems were measuring at the same height and fluxes were calculated by the same software (EddyPro 3.0 version; Li-Cor, USA). The linear correlation (R^2) and the slope of the mobile lab vs. the FLUXNET tower was 0.9 and 1.0 for the sensible heat, 0.8 and 0.9 for latent heat, and 0.9 and 1 for the NEE, respectively.

Daily estimates of reference evapotranspiration, i.e., ET_o (mm d^{-1}), for the ET model, the water deficit and the water availability factors were calculated from the mean daily air temperature and the daily total incoming solar radiation, measured at the seven sites following the empirical formulation proposed by Jensen and Haise (1963):

$$\mathrm{ET}_o = \frac{R_\mathrm{g}}{2470}\,(0.078 + 0.0252\,T), \tag{1}$$

in which T is the mean daily air temperature (°C), and R_g is the daily global (total) incoming solar radiation (kJ m^{-2} d^{-1}); ET$_o$ is finally converted into millimeters per day by dividing the R_g by 2470 mm kJ m^{-2} d^{-1} (see in Jensen and Haise, 1963). We decided to use this ET$_o$ formulation of Jensen and Haise (1963) to be consistent with the original RS–Met proposed by Maselli et al. (2014), though we are aware of the large tradition of works devoted to comparing several methods to estimate ET$_o$ and to proving the validity and limitations of these methods under different environmental conditions.

2.4 MODIS ET and GPP products and the PaVI-E model for annual ET

We compared our RS–Met with the products from MODIS-based ET and GPP models, the details of which can be found in Mu et al. (2007, 2011) and Running et al. (2000, 2004) for the ET and GPP models, respectively. These products (MOD16 and MOD17 for ET and GPP, respectively) provide 8-day ET and GPP estimates at 1 km for 2000–2015, globally. MODIS ET and GPP products were compared with RS–Met on seasonal and annual scales at all sites. Importantly, these MODIS products take advantage of the use of vapor pressure information, which was shown to affect the stomatal conductance of plants, whereas our model did not consider this factor directly. We did not use vapor pressure data in the RS–Met because most of the weather stations in this region do not have such information and that would have limited the use of our model. However, the f_WD calculated from radiation, temperature and water supply (rainfall) data is used in the adjusted RS–Met as an indirect proxy for vapor pressure deficit (VPD). To compare with the 8-day MODIS ET and GPP products we averaged the daily RS–Met and EC estimates over the same 8-day periods.

We also compared the RS–Met ET estimates to the annual ET derived from PaVI-E (parameterization of vegetation index for the estimation of the ET model; Helman et al., 2015a), at the six sites on an annual basis. The PaVI-E is an empirical model based on simple exponential relationships found between MODIS-derived enhanced vegetation index (EVI) and NDVI and annual ET estimates from EC at 16 FLUXNET sites, comprising a wide range of plant functional types across Mediterranean climate regions. This simple relationship (PaVI-E) was shown to produce accurate ET estimates on an annual timescale (mm yr^{-1}) and at a moderate spatial resolution of 250 m in this region (Helman et al., 2015a). It was validated against physical-based models (MOD16 and the Land Surface Analysis Satellite Applications Facility (LSA-SAF) product of ET) and ET calculated from water balances across the same study area. PaVI-E was

used for ecohydrological studies in this region, providing insights into the role of climate in altering forest water and carbon cycles (Helman et al., 2017a, b). The advantage of this model is that it does not require any additional meteorological information but is a proper function of the relationship between observed fluxes and satellite-derived vegetation indices. This makes it interesting to compare with the RS–Met model since the RS–Met is highly dependent on meteorological forcing.

3 Description of the models and the use of a water deficit factor

The RS–Met models used here for the daily estimation of ET and GPP are based on the NDVI and the meteorological data. Each model was applied with and without a water deficit factor (f_WD) adjustment (i.e., two model versions for ET and two for the GPP).

3.1 The ET model

The RS–Met of daily ET is based on the FAO-56 formulation (Eq. 2):

$$\mathrm{ET} = \mathrm{ET}_o \times (K_\mathrm{C} + K_\mathrm{S}), \tag{2}$$

in which K_C and K_S stand for the canopy and soil coefficients, respectively (Allen et al., 1998). In the RS–Met a maximum value of K_C ($K_\mathrm{C\_max}$), which depends on the type of the monitored vegetation (Allen et al., 1998, 2006), and a maximum value of K_S ($K_\mathrm{S\_max}$), for soil evaporation, are used as a reference in the model. $K_\mathrm{C\_max}$ and $K_\mathrm{S\_max}$ are then multiplied by a linear transformation of the NDVI (i.e., $f(\mathrm{NDVI})$ and $f(1\text{-NDVI})$, respectively; Maselli et al., 2014) to adjust for the seasonal evolution of the canopy and soil coefficients:

$$K_\mathrm{C} = K_\mathrm{C\_max} \times f(\mathrm{NDVI}) \tag{3}$$
$$K_\mathrm{S} = K_\mathrm{S\_max} \times f(1 - \mathrm{NDVI}). \tag{4}$$

The linear transformation of the NDVI used here is the fractional vegetation cover ($f\mathrm{VC}$) that better represents both ET processes: direct soil evaporation and plant transpiration. The $f\mathrm{VC}$ is a classical two-end member function based on minimum and maximum values of NDVI, corresponding to a typical soil background without vegetation (NDVI$_\mathrm{SOIL}$) and an area fully covered by vegetation (NDVI$_\mathrm{VEG}$):

$$f\mathrm{VC} = (\mathrm{NDVI} - \mathrm{NDVI}_\mathrm{SOIL})/(\mathrm{NDVI}_\mathrm{VEG} - \mathrm{NDVI}_\mathrm{SOIL}). \tag{5}$$

Thus, Eqs. (3) and (4) become

$$K_\mathrm{C} = K_\mathrm{C\_max} \times f\mathrm{VC} \tag{6}$$

and

$$K_\mathrm{S} = K_\mathrm{S\_max} \times (1 - f\mathrm{VC}), \tag{7}$$

respectively.

The fVC in Eq. (5) is calculated on a daily basis from the interpolated NDVI (daily) data. Note that the fVC in Eq. (6) represents the fraction of the area covered by the vegetation, while in Eq. (7) the term 1-fVC represents the fraction of the bare soil area. Both terms, fVC and 1-fVC, in Eqs. (6) and (7), change over the course of a year due to canopy development and/or the appearance of ephemeral herbaceous plants. Here we used the values of 0.1 and 0.8 for the NDVI$_{SOIL}$ and NDVI$_{VEG}$, respectively, which are the values observed for bare ground and dense natural vegetation in this region (Helman et al., 2015b).

Finally, from Eqs. (2) and (5)–(7) we obtain the model without the water deficit factor adjustment (NO f_{WD}):

$$\text{ET} = \text{ET}_o \times \{[f\text{VC} \times K_{C\_max}] \\ + [(1 - f\text{VC}) \times K_{S\_max}]\}. \quad (8)$$

Subsequently, we used water deficit (f_{WD}) and water availability (f_{WA}) factors to adjust the canopy and soil coefficients for water supply conditions in the root zone and top soil in Eqs. (6) and (7), respectively:

$$K_C = K_{C\_max} \times f\text{VC} \times f_{WD} \quad (9)$$

and

$$K_S = K_{S\_max} \times (1 - f\text{VC}) \times f_{WA}. \quad (10)$$

The f_{WD} and f_{WA} in Eqs. (9) and (10) simulate the effects of available water for plant transpiration in the root zone and for surface evaporation in the top soil, respectively, whereas the f_{WD} is defined as follows:

$$f_{WD} = 0.5 + 0.5 \times f_{WA}. \quad (11)$$

The water availability factor (f_{WA}) is calculated as the simple ratio between the daily rainfall amount and the daily ET$_o$, both cumulated over a period of 2 months. Basically, the accumulation period could vary for different ecosystem types and environmental conditions. However, we have taken a period of 2 months for the native shrublands and planted (and native) forests following previous observations that showed that this period is sufficient to maintain the wet topsoil layer for the whole rainy season in ecosystems in this region (Raz-Yaseef et al., 2012). Furthermore, changing the accumulation period did not gave us consistently better results at all sites, as the 2-month period gave us.

The f_{WA} is set to 1 when the cumulated rainfall amount exceeds the atmospheric demand (i.e., the ET$_o$). Note that the f_{WD} would then vary between 0.5 and 1, meaning that ET is reduced to half the potential maximum in the absence of water supply, simulating the basic transpiration levels maintained by evergreen vegetation (Glenn et al., 2011; Maselli et al., 2014). This reduction in the f_{WD} accounts for water deficit in the root zone, which results in reduced plant

transpiration, while short-term effects would be mainly reflected through changes in the NDVI (and consequently in the fVC and fAPAR; Glenn et al., 2010; Running and Nemani, 1988). In contrast to the f_{WD}, the f_{WA} is reduced to zero following a dry period longer than 2 months, making the surface evaporation component null during the dry summer.

The model is adjusted to root zone and surface water deficit conditions (f_{WD} and f_{WA}) by replacing Eqs. (6) and (7) with Eqs. (9) and (10):

$$\text{ET} = \text{ET}_o \times \{[f\text{VC} \times K_{C\_max} \times f_{WD}] \\ + [(1 - f\text{VC}) \times K_{S\_max} \times f_{WA}]\}. \quad (12)$$

Here we used a K_{C_max} value of 0.7 for both forests and non-forest sites, and we used a K_{S_max} value of 0.2 for soil evaporation in both (adjusted and unadjusted for water deficit conditions) models, as in Maselli et al. (2014).

Finally, the model derives daily ET estimates (mm d^{-1}) at the spatial resolution of the MODIS NDVI product, i.e., 250 m.

3.2 The GPP model

For the GPP model, we used the biophysical radiation use efficiency model proposed by Monteith (1977):

$$\text{GPP} = \text{RUE} \times f\text{APAR} \times \text{PAR}, \quad (13)$$

in which PAR is the daily incident photosynthetic active radiation (MJ m^{-2}), calculated as 45.7 % from the incoming measured global solar radiation (Nagaraja Rao, 1984), and fAPAR is the fraction of the PAR that is actually absorbed by the canopy (range from 0 to 1). The fAPAR was derived here from the daily NDVI time series following the linear formulation: fAPAR = 1.1638 NDVI – 0.1426, which was proposed by Myneni and Williams (1994). This linear formulation was successfully applied in similar remote-sensing-based GPP models for similar ecosystems by Veroustraete et al. (2002), Maselli et al. (2006, 2009) and Helman et al. (2017a); RUE is the radiation use efficiency (g C MJ^{-1}), which is the efficiency of the plant for converting the absorbed radiation into carbon-based compounds and which changes over the course of a year (Garbulsky et al., 2008).

The RUE is an important component in the GPP model and is the most challenging parameter to compute. It is usually considered to be related to VPD, water availability, temperature and plant species type (Running et al., 2000), and there have been several recent efforts to directly relate it to the photochemical reflectance index (PRI), which can also be derived from satellites (Garbulsky et al., 2014; Peñuelas et al., 2011; Wu et al., 2015). Currently, the conventional modeling of RUE for Mediterranean ecosystems is not straightforward and is mostly site specific, derived for specific local conditions (Garbulsky et al., 2008). Here, we used the simple approach proposed by Veroustraete et al. (2002) and further developed by Maselli et al. (2009), which states that a potential RUE (RUE_{MAX} in $g\,C\,MJ^{-1}$) can be adjusted for seasonal changes using a function based on temperature and water deficit conditions (f_{WT}):

$$RUE = RUE_{MAX} \times f_{WT}. \qquad (14)$$

The f_{WT} adjusts the RUE_{MAX} for seasonal changes following changes in water availability and temperature conditions:

$$f_{WT} = T_{CORR} \times f_{WD}, \qquad (15)$$

in which T_{CORR} is a temperature correction factor calculated on a daily basis (Veroustraete et al., 2002):

$$T_{CORR} = \frac{e^{\left(a - \frac{\Delta H_{AP}}{G \cdot T}\right)}}{1 + e^{\left(\frac{\Delta S \cdot T - \Delta H_{DP}}{G \cdot T}\right)}}, \qquad (16)$$

in which a is a constant equal to 21.9; ΔH_{AP} and ΔH_{DP} are the activation and deactivation energies ($J\,mol^{-1}$), equal to 52 750 and 211, respectively; G is the gas constant, equal to $8.31\,J\,K^{-1}\,mol^{-1}$; ΔS is the entropy of the denaturation of CO_2 and is equal to $710\,J\,K^{-1}\,mol^{-1}$; T is the mean daily air temperature (in degrees Kelvin); and f_{WD} is the same water deficit factor as in Eq. (11).

The water deficit factor, f_{WD}, is used here only in the model that considers water supply conditions. Thus, in the model without the f_{WD}, the f_{WT} would be only a function of the temperature, and thus $f_{WT} = T_{CORR}$ (in Eq. 15). Following Garbulsky et al. (2008) and Maselli et al. (2009), a constant value of $1.4\,g\,C\,MJ^{-1}$ was used here for RUE_{MAX} at all sites and in model variations (i.e., with and without the f_{WD}). The exclusion of direct measurements of VPD as an input in the model is indeed a limitation; however, we tried to maintain a model with minimal input data that will be available from standard weather stations (VPD information is currently lacking from most of the weather stations in this region). The inclusion of the f_{WD}, which includes radiation, temperature and water supply (rainfall) information, is used as an indirect proxy for VPD in the model.

Finally, daily GPP values were computed from the model at a spatial resolution of 250 m for each of the seven sites and compared with EC estimates and the MODIS GPP product. It should be stated that the use of the EC-derived GPP as a reference in the validation should be taken with caution because GPP by itself is modeled and not directly measured. This may introduce uncertainties in the validation that could be contaminated by self-correlation.

4   Testing the water deficit factor in high-energy water-limited environments

To show the importance of the water deficit factor (f_{WD}) in adjusting the model to seasonal variations in the fluxes, we demonstrate the seasonal evolution of the f_{WD} together with that of the main drivers of the RS–Met at the dryland pine

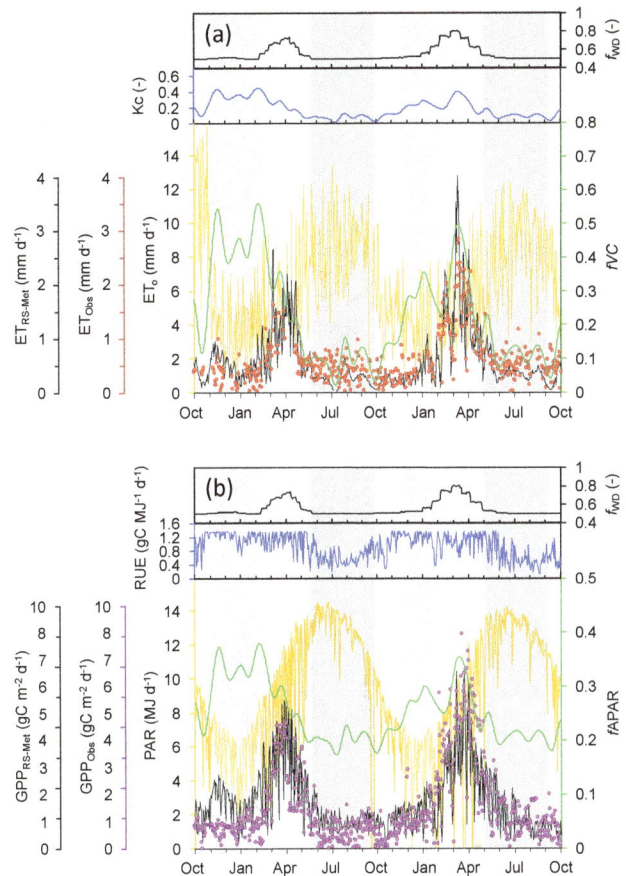

Figure 2. Seasonal evolution of the water deficit factor (f_{WD}; black line in upper panel) and the main drivers of the modeled ET **(a)** and GPP **(b)** in the semiarid pine forest of Yatir (ET_{RS-Met} and GPP_{RS-Met}, respectively; black line in lower panel) for the seasonal years 2008/09 and 2009/10. EC fluxes are also shown (ET_{Obs} and GPP_{Obs}, red and purple dots, respectively). The K_C and the radiation use efficiency (RUE) both without the addition of the f_{WD} (blue in middle panels) are shown together with the potential ET (ET_O, yellow in **a**), the fraction of vegetation cover (fVC, green in **a**), the photosynthetic active radiation (PAR, yellow in **b**) and the fraction of absorbed PAR (fAPAR, green in **b**). Colored vertical bands indicate the critical periods when the addition of f_{WD} is particularly useful.

forest site of Yatir (Fig. 2). Figure 2a shows that the f_{WD} moderates the increase in K_C (blue line in middle panel of Fig. 2a) at the beginning of the rainy season (November–January) even though the fVC (green line in lower panel of Fig. 2a) is relatively high, likely due to the appearance of ephemeral herbaceous vegetation in the forest understory (Helman et al., 2015b). This is a realistic scheme since the herbaceous vegetation has little contribution to the ecosystem fluxes but a significant contribution to the NDVI (and thus to the fVC) signal (Helman, 2017), as observed by the low EC GPP at this time (red dots in lower panel of Fig. 2a). Thus, the f_{WD} has an important role in reducing the K_C to

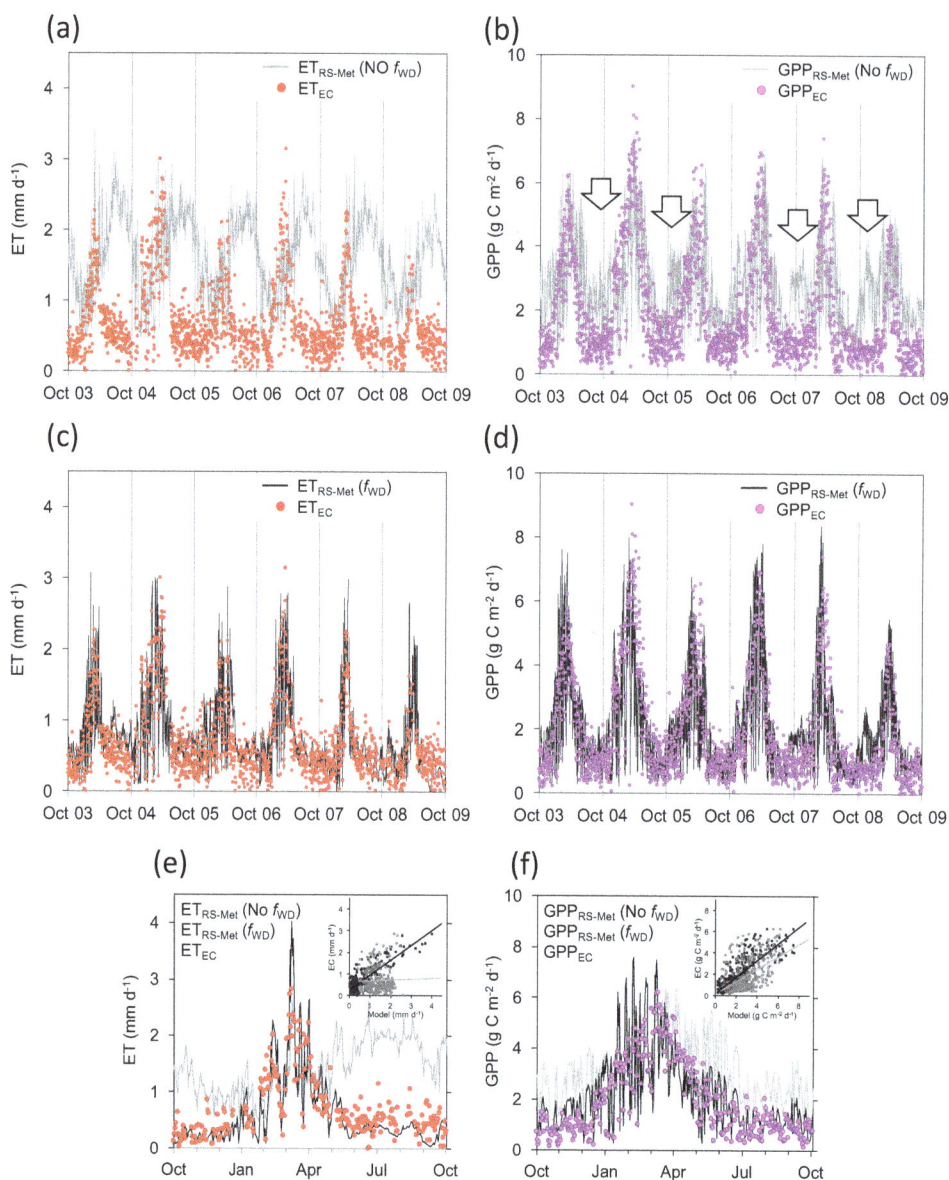

Figure 3. Observed (EC) and modeled (RS–Met) ET and GPP at Yatir. Shown in **(a)**–**(d)** are the RS–Met values with (black) and without (grey) the water deficit factor (f_{WD}). A closer look at the selected years 2009/10 and 2003/04 is shown in **(e)** and **(f)**, respectively. Inserts show the correlations between modeled and observed fluxes with and without the f_{WD} (black and grey dots, respectively).

more realistic low values at this stage of the year when there is less water available for the trees. The same applies for the end of the rainy season and summer, in May–August, when the ET_o is relatively high but there is almost no available water for ET, as implied from the low f_{WD} (black line in upper panel of Fig. 2a).

In the GPP model, the f_{WD} reduces the high RUE at both ends of the rainy season, adjusting the GPP to the water deficit conditions in the root zone during these periods (Fig. 2b). Here again, the low f_{WD} reduces the contribution of the high fAPAR (and the RUE) in the model during the start of the rainy season due to the growth of ephemeral

plants in the understory (green and blue lines in lower and middle panels of Fig. 2b, respectively). This is because there is still not sufficient water in the root zone during this period. Particularly noted, though, is the significant reduction in GPP at the end of the rainy season and during the summer (May–August), when the PAR (yellow line in lower panel of Fig. 2b) is high but less water is available for transpiration and subsequently for photosynthesis.

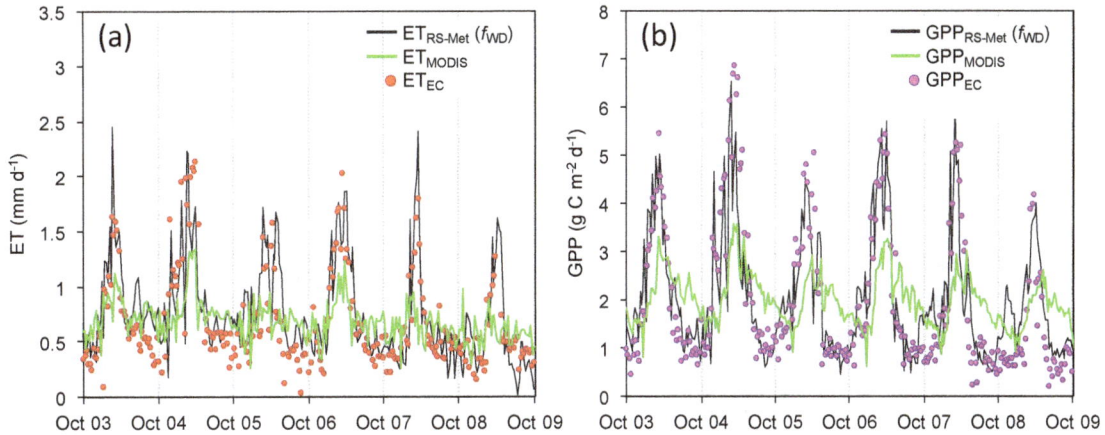

Figure 4. Showing 8-day averaged values of ET (a) and GPP (b) from EC (dots), RS–Met (black) and MODIS (green) at Yatir. The R of the correlation for EC vs. RS–Met is 0.78 and 0.80 for ET and GPP, respectively (slope $= 0.90$ and 0.70; intercept $= 0.19$ and 0.66 for ET and GPP, respectively). The R of the correlation for EC vs. MODIS is 0.47 and 0.60 for ET and GPP, respectively (slope $= 0.21$ and 0.27; intercept $= 0.56$ and 1.47 for ET and GPP, respectively).

Table 2. Statistics of the comparison between the RS–Met with the addition of the water deficit factor (f_{WD}) and without its addition (NO f_{WD}) and the EC-derived measurements.

	ET ($\mathrm{mm\,d^{-1}}$)					GPP ($\mathrm{g\,C\,m^{-1}\,d^{-1}}$)				
	N	Correlation		MAE		N	Correlation		MAE	
		NO f_{WD}	f_{WD}	NO f_{WD}	f_{WD}		NO f_{WD}	f_{WD}	NO f_{WD}	f_{WD}
Yatir	2228	0.05*	0.76	1.9	0.18	2293	0.56	0.77	1.3	0.8
Eshtaol	47	0.16[ns]	0.64	1.3	1.6	54	0.80	0.90	2.3	2.3
HaSolelim	40	0.72	0.79	2.0	0.8	41	0.80	0.88	2.1	2.1
Birya	57	0.72	0.85	1.8	1.8	57	0.64	0.72	4	3
Wady Attir	28	0.80	0.91	0.5	0.7	29	0.90	0.92	0.7	1.0
Modiin	43	0.62	0.64	1.9	1.0	43	0.89	0.90	1.2	1.2
Kadita	28	0.80	0.67	0.8	1.0	28	0.82	0.88	1.8	1.8

The mean absolute error (MAE) is in $\mathrm{mm\,d^{-1}}$ for the ET and in $\mathrm{g\,C\,m^{-2}\,d^{-1}}$ for the GPP. All the correlations were highly statistically significant at $P < 0.001$, except for the ET model without the f_{WD} at the forest site of Yatir (*), which was significant at $P = 0.02$, and the site of Eshtaol, which was not statistically significant (ns). The number of days used for the correlation at each site and flux is indicated (N = days).

5   Comparisons with MODIS and the FLUXNET station in Yatir

5.1   Daily ET and GPP

We compared the daily estimates of the modeled ET with MODIS ET and GPP products and the active FLUXNET station at the dryland pine forest of Yatir for 2002–2012 (Table 1). As expected from the information noted above (Sect. 4), the model without the water deficit factor (NO f_{WD} in Fig. 3a and e) overestimated the ET in comparison to the EC measurements, particularly from mid-spring to the end of the summer (Fig. 3a, e). The peak ET was shifted to late July–early September, while the ET measured from the EC showed an earlier peak, in March. The large overestimation of the model without the f_{WD} was associated with the high ET_o during the spring and summer ($R = 0.91$; $P < 0.001$; see

also Fig. 2a), which is the driver of the ET model (Eqs. 2, 8 and 12), following the low humidity and augmented radiation load at this time of the year (Fig. 2 and Rotenberg and Yakir, 2011; Tatarinov et al., 2016). However, including the f_{WD} in the model helped to correct for this overestimation by linking ET to the available soil water (Fig. 2a), resulting in a good agreement between the model and the EC estimates (Fig. 3c and e; Table 2).

When comparing the modeled GPP with the EC estimates at Yatir, the model without the f_{WD} (NO f_{WD} in Fig. 3b and f) produced higher values during both ends of the rainy season (October–November and May–June). In particular, the model without the f_{WD} overestimated the GPP during the start of the rainy season (indicated by the arrows in Fig. 3b). This was likely due to the increase in the NDVI following the appearance of ephemeral herbaceous plants in the understory of these Mediterranean forests in the beginning of the

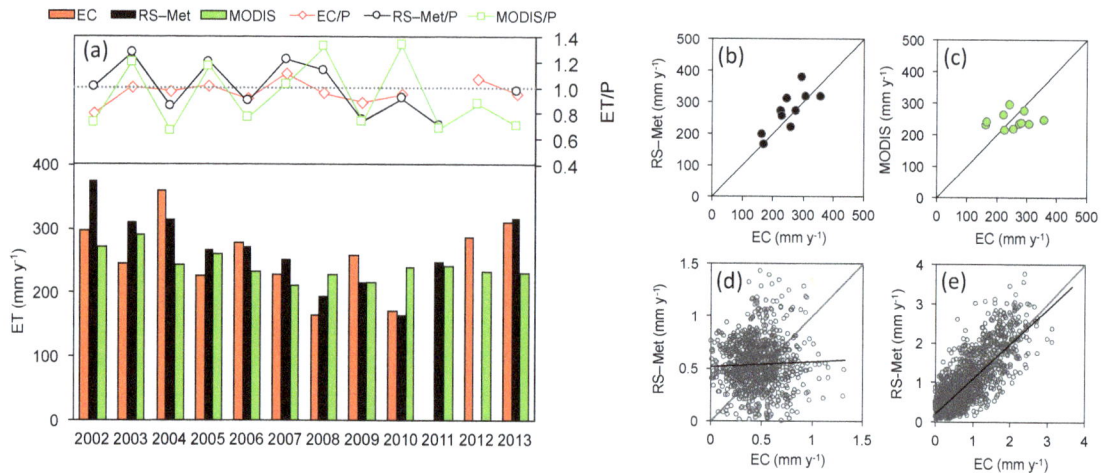

Figure 5. Annual ET (mm yr^{-1}) summed from daily RS–Met (with f_{WD}, black), MODIS (green) and EC (red) at the Yatir forest site for 2003–2014 **(a)**. Linear regressions of the EC vs. RS–Met **(b)** and EC vs. MODIS **(c)**. Daily estimates from RS–Met in dry summer (June–August, **d**) and rainy (October–May, **e**) seasons. The R's of the linear fits for EC vs. RS–Met **(b)** and MODIS **(c)** are 0.78 ($P < 0.05$, $N = 10$) and 0.10 ($P > 0.1$, $N = 11$), respectively. The R's of the linear fits for the daily data in **(d)** and **(e)** are 0.05 ($P > 0.1$, $N = 876$) and 0.80 ($P < 0.0001$, $N = 1570$), respectively. The interannual trends in ET / P from EC, RS–Met and MODIS are presented in the upper panel of **(a)**. Note that the annual sums of ET from EC and RS–Met in 2011 and 2012, respectively, are not displayed due to the scarcity of available data during these years ($> 50\%$ missing data).

rainy season, as already pointed out in the previous section (see also Helman et al., 2015b). The herbaceous vegetation in the understory of Yatir provides a meaningful contribution to the NDVI signal, although it constitutes only a minor component in terms of the biomass and the CO_2 uptake of the forest (Helman et al., 2015b; Rotenberg and Yakir, 2011). Considering f_{WD} in the model thus abridged the RUE, counterbalancing the high contribution of the herbaceous vegetation to the f APAR through the high NDVI. This also better simulated the water deficit conditions experienced by the woody vegetation, which is the main contributor to the GPP in Yatir during the dry period (Fig. 3d and f).

These results explicitly show that the water deficit factor is useful in forcing the model onto the woody vegetation activities (strongly restricted by water shortage at both ends of the rainy season), reducing the impact of other components, such as the peak activities of the understory vegetation that, obviously, does not suffer from water shortage and responds to a small early season moisture input (Helman et al., 2014a, 2017c; Mussery et al., 2016).

Comparison with MODIS ET and GPP products shows a consistent underestimation of the fluxes in the peak season and overestimation in the dry season, implying that these models need to be adjusted to root-zone water deficit conditions at such high-energy water-limited sites (Fig. 4). This is in spite of the use of vapor pressure data in these models (Mu et al., 2007, 2011, Running et al., 2000, 2004). These results suggest that including the f_{WD} in global models, such as the MODIS-based models, might at least reduce the ob-

served dry period overestimations and increase fluxes in the wet season.

5.2 Annual-basis comparisons

We then examined the adjusted RS–Met on an annual scale, first by comparing the inter-annual variation in the modeled ET with that from the EC and with that from the MODIS ET product at Yatir (Fig. 5a). This analysis indicated that RS–Met can also reproduce the annual ET with a fair accuracy, showing a moderate but significant correlation with the total annual ET derived from the daily summed EC estimates ($R = 0.78$; $P < 0.05$; $N = 10$; Fig. 5b) and comparable mean annual ET (266 ± 61 vs. $257 \pm 58 \text{ mm yr}^{-1}$ for ET_{MOD} and ET_{EC}, respectively). MODIS ET, in turn, was not correlated with EC ($R = 0.10$; $P > 0.1$; $N = 11$), showing little year-to-year variation in the annual ET (Fig. 5a, c).

Both the RS–Met and the EC were significantly correlated with P ($R = 0.60$ and 0.93; $P = 0.05$ and < 0.001), showing similar patterns in water use (ET / P ratio), though differing in magnitude in some of the years studied (black and red lines in upper panel of Figs. 5a and S3). The little year-to-year variation in the MODIS ET resulted in a noisier pattern of water use (green line in upper panel of Fig. 5a) compared to that calculated from the RS–Met and EC. A noisy water use pattern was also noted in the RS–Met (compared to that from the EC), particularly in dry years (Fig. S3; e.g., 2003, 2005 and 2008; Fig. 5a). Higher ET in the RS–Met was likely the result of discrepancies in daily estimates during the summer between the RS–Met and EC ($R = 0.05$, $P > 0.1$ for June–August; Fig. 5d). This is supported by the observation of a

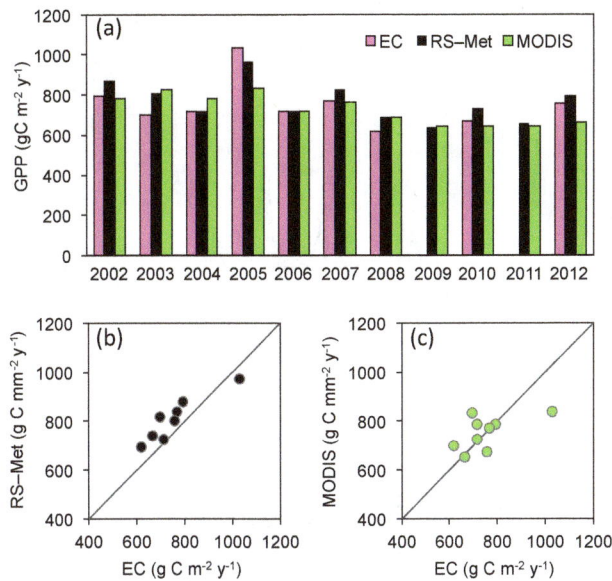

Figure 6. Annual GPP sums ($g\,C\,m^{-2}\,yr^{-1}$) from EC, RS–Met (with f_{WD}) and the 8-day MOD16 product (MODIS) at Yatir **(a)** and the linear regressions of EC vs. RS–Met **(b)** and MODIS **(c)**. The R of the linear fits is 0.91 ($P < 0.05$, $N = 10$) and 0.58 ($P = 0.08$, $N = 10$) for RS–Met and MODIS, respectively. Annual EC GPP for 2009 and 2011 was not calculated due to missing data.

5-fold higher bias between EC and RS–Met summer daily estimates in those dry years (bias $= -0.146\,mm\,d^{-1}$), compared to the remaining years (bias $= -0.029\,mm\,d^{-1}$). These negative biases imply an average overestimation by the RS–Met model during the summer compared to observed (EC) ET estimates.

In contrast, the correlation between the RS–Met and EC was high and significant for daily estimates during the rainy season ($R = 0.80$, $P < 0.0001$ for October–May; Fig. 5e). The relatively large discrepancies between RS–Met and EC during the summer indicate the low sensitivity of the RS–Met model to relatively low ET fluxes (i.e., $< 1.0\,mm\,d^{-1}$). This likely suggests the need to adjust the water availability factor (f_{WA}) to positive values for a longer period, particularly at the end of the rainy season and beginning of the summer.

The annual ET, as estimated from both the RS–Met and EC, was higher than the total rainfall amount in some of the years studied (Fig. S3). A similar pattern was previously reported in forests in water-limited regions (Helman et al., 2017b; Raz-Yaseef et al., 2012; Williams et al., 2012). ET higher than rainwater supply indicates that trees use water stored in deep soil layers during wet years in the subsequent dry years (e.g., 2006 and 2008; Raz-Yaseef et al., 2012; Barbeta et al., 2015). Thus, the transfer of surplus rainwater from previous years should also be taken into consideration when adjusting the model with available water through the f_{WA} and f_{WD}, which are currently calculated only with the seasonal rainfall. Theoretically, this could be done by summing

the available water from the previous year (calculated as $P - ET$) with the 2-month summed P in the calculation of the f_{WA} and f_{WD}. Of course, this would be applied only after completing the ET estimation of the first year.

The adjusted RS–Met GPP (i.e., that with the f_{WD}) was also comparable to the GPP from the EC (765 ± 112 vs. $748 \pm 124\,g\,C\,m^{-2}\,yr^{-1}$, for GPP$_{MOD}$ and GPP$_{EC}$, respectively) and highly correlated on an annual scale (Fig. 6a, b), with an $R = 0.91$ ($P < 0.001$; $N = 9$) and a low mean absolute error (MAE) of $52\,g\,C\,m^{-2}\,yr^{-1}$ (relative error of ca. 7 %). MODIS GPP showed inferior correlation with EC estimates on an annual scale, with an R of 0.58 ($P = 0.08$, $N = 10$; Fig. 6c).

6   Testing the RS–Met across a rainfall gradient

6.1   Comparison with seasonal ET and GPP from EC and MODIS products

Next we compared the ET and GPP estimates from the RS–Met model with the field campaign data and MODIS ET and GPP products across the remaining six ecosystems. Comparison between estimates based on the RS–Met model, with and without the f_{WD}, with those from the EC indicated significantly higher correlations of the adjusted model (i.e., with the f_{WD}) with EC estimates ($P = 0.06$ and $P < 0.01$ for ET and GPP, respectively; Table 2). Only the shrubland site of Kadita showed a higher ET correlation of EC with the unadjusted model. This was likely due to the continuous ET fluxes, which were not captured by the model, throughout the summer period at this relatively moist site.

In general, while using the f_{WD} did not improve (for the ET, $P > 0.1$, as indicated by a two-tailed Student's t test) or only marginally improved (for the GPP, $P = 0.09$, as indicated by a two-tailed Student's t test) RS–Met estimates at the non-forest sites, it significantly improved the ET and GPP estimates at forest sites ($P = 0.05$ and $P = 0.016$ for ET and GPP, respectively, as indicated by a two-tailed Student's t test. The adjusted RS–Met successfully tracked changes in ET and CO_2 fluxes from the dry to wet season at all sites. Similar to those shown in Yatir, MODIS ET and GPP fluxes were much lower than observed fluxes. This underestimation was particularly noted at the forest sites (Figs. 7 and S4).

Overall, the adjusted RS–Met was in good agreement with the EC measurements, with the cross-site regressions producing highly significant linear fits (Fig. 8a, b; $R = 0.82$ and $R = 0.86$; and MAE $= 0.47\,mm\,d^{-1}$ and MAE $= 1.89\,g\,C\,m^{-2}\,d^{-1}$ for ET and GPP, respectively). Comparing between the EC vs. RS–Met regressions and the EC vs. MODIS ET and GPP regressions, using 8-day averaged fluxes values, produced the following linear fits: $ET_{RS-Met} = 1.16\ ET_{EC} - 0.11$ ($R = 0.88$; $P < 0.0001$; $N = 36$), $ET_{MODIS} = 0.38\ ET_{EC} + 0.33$ ($R = 0.65$; $P < 0.0001$; $N = 33$), $GPP_{RS-Met} = 1.09\ GPP_{EC} + 0.21$

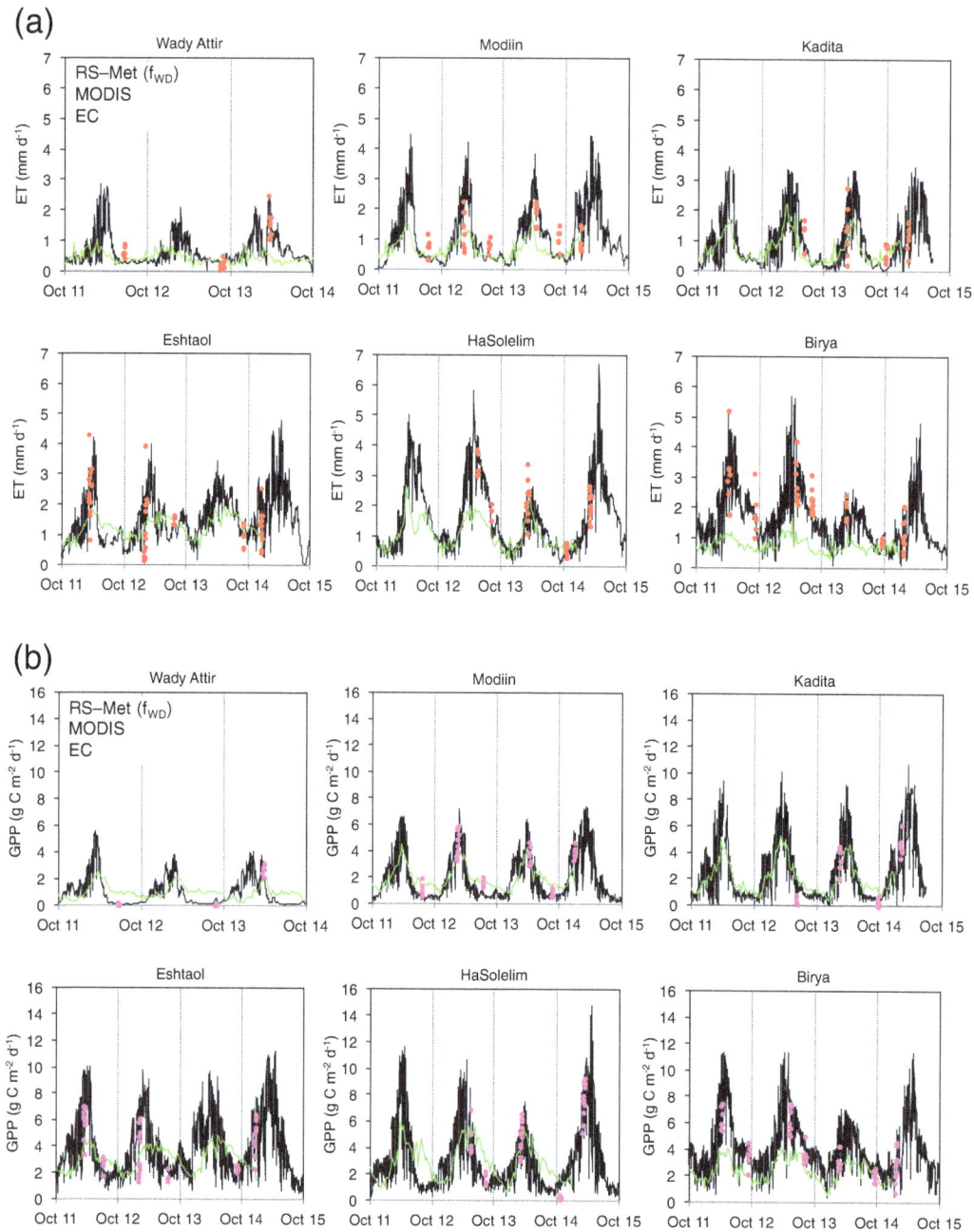

Figure 7. ET (**a**) and GPP (**b**) from EC, RS–Met (with f_{WD}) and MODIS (MOD16A2 and MOD17A2 products) at the six forest and non-forested sites.

$(R = 0.92;$ $P < 0.0001;$ $N = 36)$ and $GPP_{MODIS} = 0.43$ $GPP_{EC} + 1.31$ $(R = 0.77;$ $P < 0.0001;$ $N = 33)$, showing a consistent underestimation of both MODIS products (MOD16 and MOD17) at those sites across sites (Fig. 8c, d).

The WUE (the slope of the regression between ET and GPP in Fig. S5) was slightly higher at $2.32\,\mathrm{g\,C\,kg^{-1}}$ H_2O from the RS–Met compared to the low $1.76\,\mathrm{g\,C\,kg^{-1}}$ H_2O from EC, but it was within the range reported for similar ecosystems in this region (Tang et al., 2014).

6.2 Annual-basis comparisons

To expand our analysis across the rainfall gradient, and because we do not have continuous estimations from the EC at the six sites, we compared the annual ET and GPP from the adjusted RS–Met with those from MODIS ET/GPP products. In the case of ET, we also added annual estimates derived from the empirical PaVI-E model (Helman et al., 2015).

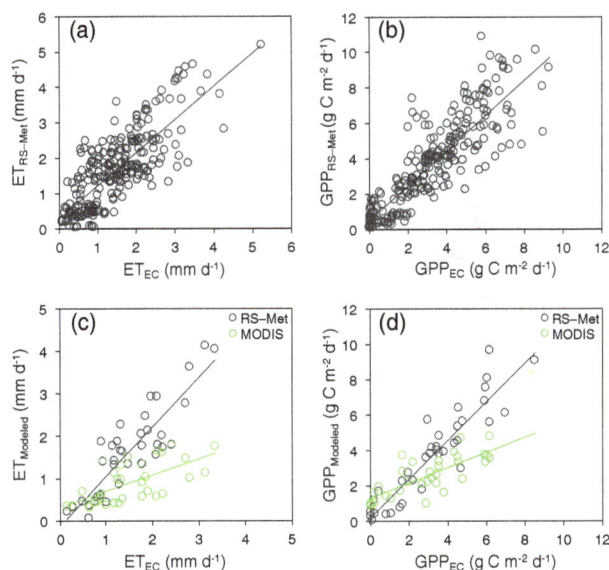

Figure 8. Cross-site EC vs. model correlations of ET **(a, c)** and GPP **(b, d)**. In **(a)** and **(b)** are the EC vs. RS–Met (with f_{WD}) values using all EC data from the six sites (each dot representing a single date), with linear fits of $ET_{RS-Met} = 0.936\,ET_{EC} + 0.281$ ($R = 0.82$, $P < 0.0001$, $N = 243$) and $GPP_{RS-Met} = 0.990\,GPP_{EC} + 0.515$ ($R = 0.86$, $P < 0.0001$, $N = 252$) for ET and GPP, respectively. In **(c)** and **(d)** are the same cross-site correlations but with data averaged over 8-day periods for comparisons with MODIS ET and GPP products (8-day averaged values). Linear fits for EC vs. RS–Met and MODIS in **(c)** are $ET_{RS-Met} = 1.16\,ET_{EC} - 0.11$ ($R = 0.88$, $P < 0.0001$, $N = 36$) and $ET_{MODIS} = 0.38\,ET_{EC} + 0.33$ ($R = 0.65$, $P < 0.0001$, $N = 33$), respectively. In **(d)**, linear fits are $GPP_{RS-Met} = 1.09\,GPP_{EC} + 0.21$ ($R = 0.92$, $P < 0.0001$, $N = 36$) and $GPP_{MODIS} = 0.43\,GPP_{EC} + 1.31$ ($R = 0.77$, $P < 0.0001$, $N = 33$) for EC vs. RS–Met and MODIS, respectively.

The results of our ET comparison showed that the RS–Met and PaVI-E models produced comparable estimates at most of the sites (Fig. 9a), with the only exception being the drylands non-forest site of Wady Attir, which showed higher estimates from RS–Met than from PaVI-E ($P < 0.01$, as indicated by Tukey HSD separation procedure). MODIS ET was in accordance with estimates of RS–Met and PaVI-E at shrubland sites in spite of the underestimation of this product during the wet season likely due to the relatively higher ET at the beginning of the rainy season (Fig. 7a). However, MODIS ET was significantly lower than the other two models at the forest sites and also lower than the shrubland sites (Fig. 9a). The cross-site regression between the annual estimates from RS–Met vs. those from the EC produced a highly significant linear fit ($R = 0.94$, $P < 0.01$), confirming the potential use of the RS–Met in assessing ET on an annual scale across the rainfall gradient at those forest and non-forest sites.

MODIS GPP showed relatively comparable estimates to RS–Met on an annual scale due to its overestimation during the dry season that compensated for its underestimation during the peak growth season (Fig. 9b). Here again, though, underestimation was observed at forest sites, particularly at the sites of Eshtaol and Birya.

6.3   Changes in water use efficiency following afforestation across rainfall gradient

We finally used the adjusted RS–Met to assess the impact of afforestation on the water and carbon budgets across the rainfall gradient in Israel by comparing fluxes in three pine forests (i.e., Yatir, Eshtaol and Birya) with those from the adjacent shrubland sites (i.e., Wady Attir, Modiin and Kadita, respectively). Results showed that the ET significantly increased due to the afforestation of these areas, particularly at the more humid site of Birya (ca. 53 %), but to a lesser extent at the less humid site of Eshtaol (by ca. 20 %) and with almost no change in ET at the dryland site of Yatir (4 %). The GPP also significantly increased at those three paired sites. Overall, afforestation across the rainfall gradient was responsible for a significant increase in the WUE in this region (Fig. 10). Nevertheless, the positive change in the WUE decreased when moving from the dry Yatir–Wady Attir paired site (279 mm yr^{-1}) to the more humid paired site of Birya–Kadita (766 mm yr^{-1}; Fig. 10), strengthening the importance of afforestation efforts in drylands areas.

7   Summary and conclusions

We have tested here a biophysical-based model of ET and CO_2 fluxes driven by satellite-derived vegetation index and meteorological data (RS–Met) and adjusted with a seasonal water deficit factor. The model was validated against direct flux measurements from extensive field campaigns and a fixed FLUXNET station and compared with MODIS ET and GPP products at seven evergreen forest and adjacent non-forested ecosystem sites along a steep rainfall gradient in the high-energy water-limited eastern Mediterranean region. Adjusting the model with the water deficit factor generally improved its performance compared to the model without the use of this factor, particularly at forest sites. The model also outperformed MODIS-based ET and GPP models, which showed generally lower estimates, particularly at the forest sites, suggesting that these models might benefit from the inclusion of the water deficit factor.

Our results show the potential use of this simple biophysical remote-sensing-based model in assessing ET and GPP on a daily basis and at a moderate spatial resolution of 250 m, even in high-energy water-limited Mediterranean environments. The addition of a water deficit factor (based on daily rainfall and radiation and/or temperature data alone) in the RS–Met significantly improved its performance in shrublands and especially in forests in this region and might be used in global vegetation models. Nevertheless, careful at-

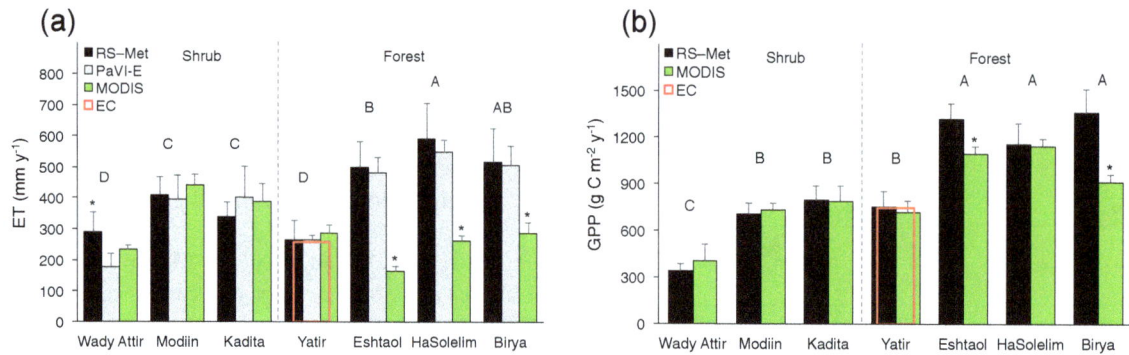

Figure 9. Mean annual (2003–2013) estimates of ET **(a)** and GPP **(b)** from RS–Met (black), MODIS (green) and PaVI-E (grey; only for ET in **a**; Helman et al., 2015a) at the seven sites. Uppercase letters indicate significant differences at $P < 0.05$ between sites from the Tukey HSD separation procedure following two-way ANOVA for the interaction of site × model (using PaVI-E and RS–Met in **a** and MODIS and RS–Met in **b**). Asterisks indicate significantly different values from other models for the specific site, as indicated by Tukey HSD. The EC annual sums at Yatir are also shown (red).

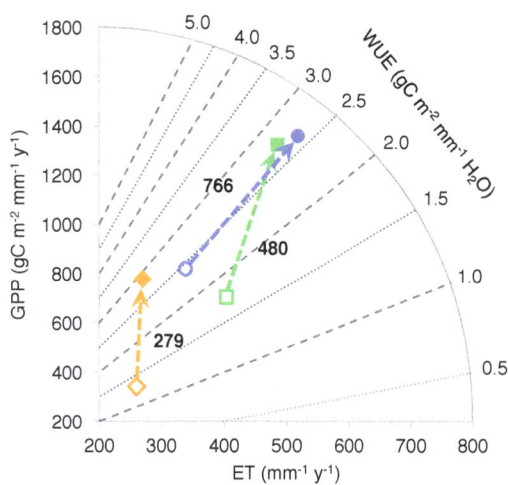

Figure 10. The change in GPP, ET and water use efficiency (WUE; as indicated by the direction of the arrow) attributed to the afforestation (closed symbols) of shrubland areas (open symbols) across a rainfall gradient in Israel (279–766 mm yr^{-1}). The three paired forest and non-forest sites of Yatir–Wady Attir, Eshtaol–Modiin and Birya–Kadita are indicated with yellow, green and blue colors, respectively. The rainfall level at each paired site is indicated near the arrow (mm yr^{-1}). Note the changing slope of the change in ET and GPP, indicating that the gain in WUE due to afforestation decreases from dry to humid areas.

interrelated with VPD, it is commonly suggested that VPD should also be considered in addition to temperature and soil moisture deficit in predicting plant-related biophysical processes such as transpiration and photosynthesis. By including the water deficit factor, we aimed here to indirectly account for these effects of VPD. However, RS–Met still showed a slight overestimation in the fluxes during the peak of the growth season (see results from Yatir site), when VPD is expected to be high. Thus, accounting for VPD effects on stomatal conductance in the RS–Met would have likely reduced these high fluxes during the period of high VPD conditions through the simulation of stomatal closure.

Further work should focus on refining the water deficit factor concept, including the contribution of VPD in the RS–Met. In addition, the contribution of direct surface evaporation from leaves should be accounted for with some sort of a factor adjusted to the seasonal development of the canopy leaf area (likely through the seasonal evolution of satellite-derived fVC and fAPAR). The addition of a soil infiltration factor, adjusted with seasonal fVC and daily rainfall amount, should also probably be considered in the RS–Met. Eventually, a major challenge would be to apply the RS–Met globally, providing a global coverage of daily estimations of ET and CO_2 fluxes at a moderate spatial resolution.

Finally, using the RS–Met, we were able to estimate changes in water use efficiency due to afforestation across the rainfall gradient in Israel. Overall, afforestation across our study area was responsible for a significant increase in the WUE. However, the positive change in the WUE decreased when moving from dry (279 mm yr^{-1}) to more humid (766 mm yr^{-1}) regions, strengthening the importance of drylands afforestation.

The use of this simple approach linked to flexible campaign-based ground validation, as demonstrated in this study, represents a powerful basis for the reliable extension

tention should be paid to adjusting the deficit water factor to local conditions, with its further development particularly required at the end of the rainy season and beginning of the dry period.

We lacked information on VPD at our sites, and thus excluded its simulated effects from the RS–Met model. However, there is much evidence that stomatal conductance is sensitive to VPD, with its effects usually accounted for in global vegetation models. Although temperature is tightly of ET and GPP estimates across spatial and temporal scales in regions with a low density of flux stations.

Competing interests. The authors declare that they have no conflict of interest.

Acknowledgements. We thank the two anonymous referees for thoughtful comments and suggestions that contributed to the improvement of this paper. David Helman acknowledges personal grants provided by the Bar-Ilan University Presidential Office (Milgat Hanasi), the JNF–Rieger Foundation, USA, and the Hydrological Service of Israel, Water Authority. Shani Rohatyn acknowledges scholarships provided by the Ronnie Appleby fund, the Advanced School of Environmental Science of the Hebrew University and the Israel Ministry of Agriculture. We thank Efrat Ramati for helping with field work and data processing, Gerardo Fratini for helping with EddyPro, and Hagai Sagi and Avraham Pelner for technical assistance. We are also grateful to the Meteorological Service of Israel for providing meteorological data and to NASA for making the MODIS NDVI datasets public. This research was partly supported by the Hydrological Service of Israel, Water Authority (grant no. 4500962964). Flux measurements were made possible through financial support from the Israel Science Foundation (ISF), Minerva foundation, JNF–KKL, the Hydrological Service of Israel, Water Authority, and the C. Wills and R. Lewis program in environmental science.

Edited by: Trevor Keenan

References

Afik, T.: Quantitative estimation of CO2 fluxes in a semi-arid forest and their dependence on climatic factors, Thesis submitted to R.H. Smith Faculty of Agriculture, Food and Environment of Hebrew University, Rehovot, Israel (in Hebrew), Thesis submitted to R.H. Smith Faculty of Agriculture, Food and Environment of Hebrew University, Rehovot, Israel, 2009 (in Hebrew).

Ahlström, A., Raupach, M. R., Schurgers, G., Smith, B., Arneth, A., Jung, M., Reichstein, M., Canadell, J. G., Friedlingstein, P., Jain, A. K., Kato, E., Poulter, B., Sitch, S., Stocker, B. D., Viovy, N., Wang, Y. P., Wiltshire, A., Zaehle, S., and Zeng, N.: The dominant role of semi-arid ecosystems in the trend and variability of the land CO2 sink, Science, 80, 348, 895–899, https://doi.org/10.1126/science.aaa1668, 2015.

Allen, R. G., Pereira, L. S., and Raes, D.: Crop evapotranspiration?: guidelines for computing crop water requirements, FAO irrigation and drainage papers, 56, FAO, Rome, 1998.

Allen, R. G., Pruitt, W. O., Wright, J. L., Howell, T. A., Ventura, F., Snyder, R., Itenfisu, D., Steduto, P., Berengena, J., Yrisarry, J. B., Smith, M., Pereira, L. S., Raes, D., Perrier, A., Alves, I., Walter, I., and Elliott, R.: A recommendation on standardized surface resistance for hourly calculation of reference ETo by the FAO56 Penman-Monteith method, Agr. Water Manage., 81, 1–22, https://doi.org/10.1016/j.agwat.2005.03.007, 2006.

Asaf, D., Rotenberg, E., Tatarinov, F., Dicken, U., Montzka, S. A., and Yakir, D.: Ecosystem photosynthesis inferred from measurements of carbonyl sulphide flux, Nat. Geosci., 6, 186–190, 2013.

Aubinet, M., Grelle, A., Ibrom, A., Rannik, S., Moncrieff, J., Foken, T., Kowalski, A. S., Martin, P. H., Berbigier, P., Bernhofer, C., Clement, R., Elbers, J. A., Granier, A., Grünwald, T., Morgenstern, K., Pilegaard, K., Rebmann, C., Snijders, W., Valentini, R., and Vesa, T.: Estimates of the annual net carbon and water exchange of forests: the EUROFLUX methodology, Adv. Ecol. Res., 30, 113–175, 2000.

Baldocchi, D. D.: Assessing the eddy covariance technique for evaluating carbon dioxide exchange rates of ecosystems: past, present and future, Glob. Change Biol., 9, 479–492, 2003.

Barbeta, A., Mejía-Chang, M., Ogaya, R., Voltas, J., Dawson, T. E., and Peñuelas, J.: The combined effects of a long-term experimental drought and an extreme drought on the use of plant-water sources in a Mediterranean forest, Glob. Change Biol., 21, 1213–1225, https://doi.org/10.1111/gcb.12785, 2015.

Ciais, P., Reichstein, M., Viovy, N., Granier, A., Ogée, J., Allard, V., Aubinet, M., Buchmann, N., Bernhofer, C., Carrara, a, Chevallier, F., De Noblet, N., Friend, A. D., Friedlingstein, P., Grünwald, T., Heinesch, B., Keronen, P., Knohl, a, Krinner, G., Loustau, D., Manca, G., Matteucci, G., Miglietta, F., Ourcival, J. M., Papale, D., Pilegaard, K., Rambal, S., Seufert, G., Soussana, J. F., Sanz, M. J., Schulze, E. D., Vesala, T., and Valentini, R.: Europe-wide reduction in primary productivity caused by the heat and drought in 2003, Nature 437, 529–533, https://doi.org/10.1038/nature03972, 2005.

Cleveland, W. S.: Robust locally weighted regression and smoothing scatterplots, J. Am. Stat. Assoc., 74, 829–836, 1979.

Gamon, J. A., Field, C.B., Goulden, M. L., Griffin, K. L., Hartley, A. E., Joel, G., Peñuelas, J., and Valentini, R.: Relationships between NDVI, canopy structure, and photosynthesis in three Californian vegetation types, Ecol. Appl., 28–41, 1995.

Garbulsky, M. F., Penuelas, J., Papale, D., and Filella, I.: Remote estimation of carbon dioxide uptake by a Mediterranean forest, Glob. Change Biol., 14, 2860–2867, https://doi.org/10.1111/j.1365-2486.2008.01684.x, 2008.

Garbulsky, M. F., Filella, I., Verger, A., and Pand eñuelas, J.: Photosynthetic light use efficiency from satellite sensors: From global to Mediterranean vegetation, Environ. Exp. Bot., 103, 3–11, https://doi.org/10.1016/j.envexpbot.2013.10.009, 2014.

Glenn, E., Nagler, P., and Huete, A.: Vegetation Index Methods for Estimating Evapotranspiration by Remote Sensing, Surv. Geophys., 31, 531–555, https://doi.org/10.1007/s10712-010-9102-2, 2010.

Glenn, E. P., Huete, A. R., Nagler, P. L., and Nelson, S. G.: Relationship Between Remotely-sensed Vegetation Indices, Canopy Attributes and Plant Physiological Processes: What Vegetation Indices Can and Cannot Tell Us About the Landscape, Sensors, 8, 2136–2160, https://doi.org/10.3390/s8042136, 2008.

Glenn, E. P., Neale, C. M. U., Hunsaker, D. J., and Nagler, P. L.: Vegetation index-based crop coefficients to estimate evapotranspiration by remote sensing in agricultural and natural ecosystems, Hydrol. Process., 25, 4050–4062, https://doi.org/10.1002/hyp.8392, 2011.

Helman, D.: Land Surface Phenology: What do we really "see" from space?, Total Environ., accepted, https://doi.org/10.1016/j.scitotenv.2017.07.237, 2017.

Helman, D., Lensky, I. M., Mussery, A., and Leu, S.: Rehabilitating degraded drylands by creating woodland islets: Assessing long-term effects on aboveground productivity and soil fertility, Agr. Forest Meteorol., 195–196, 52–60, https://doi.org/10.1016/j.agrformet.2014.05.003, 2014a.

Helman, D., Mussery, A., Lensky, I. M., and Leu, S.: Detecting changes in biomass productivity in a different land management regimes in drylands using satellite-derived vegetation index, Soil Use Manag., 30, 32–39, https://doi.org/10.1111/sum.12099, 2014b.

Helman, D., Givati, A., and Lensky, I. M.: Annual evapotranspiration retrieved from satellite vegetation indices for the eastern Mediterranean at 250 m spatial resolution, Atmos. Chem. Phys., 15, 12567–12579, https://doi.org/10.5194/acp-15-12567-2015, 2015a.

Helman, D., Lensky, I. M., Tessler, N., and Osem, Y.: A phenology-based method for monitoring woody and herbaceous vegetation in Mediterranean forests from NDVI time series, Remote Sens., 7, 12314–12335, https://doi.org/10.3390/rs70912314, 2015b.

Helman, D., Osem, Y., Yakir, D., and Lensky, I. M.: Relationships between climate, topography, water use and productivity in two key Mediterranean forest types with different water-use strategies, Agr. Forest Meteorol., 232, 319–330, https://doi.org/10.1016/j.agrformet.2016.08.018, 2017a.

Helman, D., Lensky, I. M., Yakir, D., and Osem, Y.: Forests growing under dry conditions have higher hydrological resilience to drought than do more humid forests, Glob. Change Biol., 23, 2801–2817, https://doi.org/10.1111/gcb.13551, 2017b.

Helman, D., Leu, S., and Mussery, A.: Contrasting effects of two Acacia species on understorey growth in a drylands environment: Interplay of canopy shading and litter interference, J. Veg. Sci., https://doi.org/10.1111/jvs.12576, 2017c.

Hilker, T., Coops, N. C., Wulder, M. A., Black, T. A., and Guy, R. D.: The use of remote sensing in light use efficiency based models of gross primary production: A review of current status and future requirements, Sci. Total Environ., 404, 411–423, https://doi.org/10.1016/j.scitotenv.2007.11.007, 2008.

Huete, A., Didan, K., Miura, T., Rodriguez, E. P., Gao, X., and Ferreira, L. G.: Overview of the radiometric and biophysical performance of the MODIS vegetation indices, Remote Sens. Environ., 83, 195–213, 2002.

Jensen, M. E. and Haise, H. R.: Estimating evapotranspiration from solar radiation, Proc. Am. Soc. Civ. Eng. J. Irrig. Drain. Div., 89, 15–41, 1963.

Jung, M., Reichstein, M., Ciais, P., Seneviratne, S. I., Sheffield, J., Goulden, M. L., Bonan, G., Cescatti, A., Chen, J., de Jeu, R., Dolman, A. J., Eugster, W., Gerten, D., Gianelle, D., Gobron, N., Heinke, J., Kimball, J., Law, B. E., Montagnani, L., Mu, Q., Mueller, B., Oleson, K., Papale, D., Richardson, A. D., Roupsard, O., Running, S., Tomelleri, E., Viovy, N., Weber, U., Williams, C., Wood, E., Zaehle, S., and Zhang, K.: Recent decline in the global land evapotranspiration trend due to limited moisture supply, Nature, 467, 951–954, https://doi.org/10.1038/nature09396, 2010.

Kalma, J., McVicar, T., and McCabe, M.: Estimating Land Surface Evaporation: A Review of Methods Using Remotely Sensed Surface Temperature Data, Surv. Geophys., 29, 421–469, https://doi.org/10.1007/s10712-008-9037-z, 2008.

Klein, T., Rotenberg, E., Tatarinov, F., and Yakir, D.: Association between sap flow-derived and eddy covariance-derived measurements of forest canopy CO2 uptake, New Phytol., 209, 436–446, https://doi.org/10.1111/nph.13597, 2016.

Kool, D., Agam, N., Lazarovitch, N., Heitman, J. L., Sauer, T. J., and Ben-Gal, A.: A review of approaches for evapotranspiration partitioning, Agr. Forest Meteorol., 184, 56–70, https://doi.org/10.1016/j.agrformet.2013.09.003, 2014.

Leu, S., Mussery, A., and Budovsky, A.: The effects of long time conservation of heavily grazed shrubland: A case study in the Northern Negev, Israel, Environ. Manage., 54, 309–319, 2014.

Llusia, J., Roahtyn, S., Yakir, D., Rotenberg, E., Seco, R., Guenther, A., and Peñuelas, J.: Photosynthesis, stomatal conductance and terpene emission response to water availability in dry and mesic Mediterranean forests, Trees, 30, 749–759, https://doi.org/10.1007/s00468-015-1317-x, 2016.

Maselli, F., Barbati, A., Chiesi, M., Chirici, G., and Corona, P.: Use of remotely sensed and ancillary data for estimating forest gross primary productivity in Italy, Remote Sens. Environ., 100, 563–575, https://doi.org/10.1016/j.rse.2005.11.010, 2006.

Maselli, F., Papale, D., Puletti, N., Chirici, G., and Corona, P.: Combining remote sensing and ancillary data to monitor the gross productivity of water-limited forest ecosystems, Remote Sens. Environ., 113, 657–667, https://doi.org/10.1016/j.rse.2008.11.008, 2009.

Maselli, F., Papale, D., Chiesi, M., Matteucci, G., Angeli, L., Raschi, A., and Seufert, G.: Operational monitoring of daily evapotranspiration by the combination of MODIS NDVI and ground meteorological data: Application and evaluation in Central Italy, Remote Sens. Environ., 152, 279–290, https://doi.org/10.1016/j.rse.2014.06.021, 2014.

Maseyk, K., Hemming, D., Angert, A., Leavitt, S. W., and Yakir, D.: Increase in water-use efficiency and underlying processes in pine forests across a precipitation gradient in the dry Mediterranean region over the past 30 years, Oecologia, 167, 573–585, https://doi.org/10.1007/s00442-011-2010-4, 2011.

Maseyk, K. S., Lin, T., Rotenberg, E., Grünzweig, J. M., Schwartz, A., and Yakir, D.: Physiology-phenology interactions in a productive semi-arid pine forest, New Phytol., 178, 603–616, 2008.

Monteith, J. L.: Climate and the efficiency of crop production in Britain, Philos. T. R. Soc. Lond. B, 281, 277–294, 1977.

Mu, Q., Heinsch, F. A., Zhao, M., and Running, S. W.: Development of a global evapotranspiration algorithm based on MODIS and global meteorology data, Remote Sens. Environ., 111, 519–536, https://doi.org/10.1016/j.rse.2006.07.007, 2007.

Mu, Q., Zhao, M., and Running, S. W.: Improvements to a MODIS global terrestrial evapotranspiration algorithm, Remote Sens. Environ., 115, 1781–1800, https://doi.org/10.1016/j.rse.2011.02.019, 2011.

Mussery, A., Helman, D., Leu, S., and Budovsky, A.: Modeling herbaceous productivity considering tree-grass interactions in drylands savannah: The case study of Yatir farm in the Negev drylands, J. Arid Environ., 124, 160–164, https://doi.org/10.1016/j.jaridenv.2015.08.013, 2016.

Myneni, R. B. and Williams, D. L.: On the relationship between FAPAR and NDVI, Remote Sens. Environ. 49, 200–211, https://doi.org/10.1016/0034-4257(94)90016-7, 1994.

Nagaraja Rao, C. R.: Photosynthetically active components of global solar radiation: Measurements and model computations, Arch. Meteorol. Geophy. B, 34, 353–364, https://doi.org/10.1007/BF02269448, 1984.

Osem, Y., Zangy, E., Bney-Moshe, E., and Moshe, Y.: Understory woody vegetation in manmade Mediterranean pine forests: variation in community structure along a rainfall gradient, Eur. J. Forest Res., 131, 693–704, https://doi.org/10.1007/s10342-011-0542-0, 2012.

Peñuelas, J., Garbulsky, M. F., and Filella, I.: Photochemical reflectance index (PRI) and remote sensing of plant CO2 uptake, New Phytol., 191, 596–599, https://doi.org/10.1111/j.1469-8137.2011.03791.x, 2011.

Raz-Yaseef, N., Yakir, D., Schiller, G., and Cohen, S.: Dynamics of evapotranspiration partitioning in a semi-arid forest as affected by temporal rainfall patterns, Agr. Forest Meteorol., 157, 77–85, https://doi.org/10.1016/j.agrformet.2012.01.015, 2012.

Reichstein, M., Falge, E., Baldocchi, D., Papale, D., Aubinet, M., Berbigier, P., Bernhofer, C., Buchmann, N., Gilmanov, T., Granier, A., Grünwald, T., Havránková, K., Ilvesniemi, H., Janous, D., Knohl, A., Laurila, T., Lohila, A., Loustau, D., Matteucci, G., Meyers, T., Miglietta, F., Ourcival, J.-M., Pumpanen, J., Rambal, S., Rotenberg, E., Sanz, M., Tenhunen, J., Seufert, G., Vaccari, F., Vesala, T., Yakir, D., and Valentini, R.: On the separation of net ecosystem exchange into assimilation and ecosystem respiration: review and improved algorithm, Glob. Change Biol., 11, 1424–1439, https://doi.org/10.1111/j.1365-2486.2005.001002.x, 2005.

Reichstein, M., Bahn, M., Ciais, P., Frank, D., Mahecha, M. D., Seneviratne, S. I., Zscheischler, J., Beer, C., Buchmann, N., Frank, D. C., Papale, D., Rammig, A., Smith, P., Thonicke, K., Velde, M. van der, Vicca, S., Walz, A., and Wattenbach, M.: Climate extremes and the carbon cycle, Nature, 500, 287–295, https://doi.org/10.1038/nature12350, 2013.

Rotenberg, E. and Yakir, D.: Distinct patterns of changes in surface energy budget associated with forestation in the semiarid region, Glob. Change Biol., 17, 1536–1548, https://doi.org/10.1111/j.1365-2486.2010.02320.x, 2011.

Running, S. W. and Nemani, R. R.: Relating seasonal patterns of the AVHRR vegetation index to simulated photosynthesis and transpiration of forests in different climates, Remote Sens. Environ., 24, 347–367, https://doi.org/10.1016/0034-4257(88)90034-X, 1988.

Running, S. W., Thornton, P. E., Nemani, R., and Glassy, J. M.: Global Terrestrial Gross and Net Primary Productivity from the Earth Observing System, in: Methods in Ecosystem Science, edited by: Sala, O. E., Jackson, R. B., Mooney, H. A., and Howarth, R. W., 44–57, 2000.

Running, S. W., Nemani, R. R., Heinsch, F. A., Zhao, M., Reeves, M., and Hashimoto, H.: A Continuous Satellite-Derived Measure of Global Terrestrial Primary Production, Bioscience, 54, 547–560, https://doi.org/10.1641/0006-3568(2004)054[0547:ACSMOG]2.0.CO;2, 2004.

Schimel, D., Pavlick, R., Fisher, J. B., Asner, G. P., Saatchi, S., Townsend, P., Miller, C., Frankenberg, C., Hibbard, K., and Cox, P.: Observing terrestrial ecosystems and the carbon cycle from space, Glob. Change Biol., 21, 1762–1776, https://doi.org/10.1111/gcb.12822, 2015.

Sims, D. A., Rahman, A. F., Cordova, V. D., El-Masri, B. Z., Baldocchi, D. D., Bolstad, P. V, Flanagan, L. B., Goldstein, A. H., Hollinger, D. Y., Misson, L., Monson, R. K., Oechel, W. C., Schmid, H. P., Wofsy, S. C., and Xu, L.: A new model of gross primary productivity for North American ecosystems based solely on the enhanced vegetation index and land surface temperature from MODIS, Remote Sens. Data Assim. Spec. Issue, 112, 1633–1646, https://doi.org/10.1016/j.rse.2007.08.004, 2008.

Sprintsin, M., Cohen, S., Maseyk, K., Rotenberg, E., Grünzweig, J., Karnieli, A., Berliner, P., and Yakir, D.: Long term and seasonal courses of leaf area index in a semi-arid forest plantation, Agr. Forest Meteorol., 151, 565–574, https://doi.org/10.1016/j.agrformet.2011.01.001, 2011.

Tang, X., Li, H., Desai, A. R., Nagy, Z., Luo, J., Kolb, T. E., Olioso, A., Xu, X., Yao, L., Kutsch, W., Pilegaard, K., Köstner, B., and Ammann, C.: How is water-use efficiency of terrestrial ecosystems distributed and changing on Earth?, Sci. Rep. 4, 7483, https://doi.org/10.1038/srep07483, 2014.

Tatarinov, F., Rotenberg, E., Maseyk, K., Ogée, J., Klein, T., and Yakir, D.: Resilience to seasonal heat wave episodes in a Mediterranean pine forest, New Phytol., 210, 485–496, https://doi.org/10.1111/nph.13791, 2016.

Veroustraete, F., Sabbe, H., and Eerens, H.: Estimation of carbon mass fluxes over Europe using the C-Fix model and Euroflux data, Remote Sens. Environ., 83, 376–399, https://doi.org/10.1016/j.ecolmodel.2006.06.008, 2002.

Wang, H., Zhao, P., Zou, L. L., McCarthy, H. R., Zeng, X. P., Ni, G. Y., and Rao, X. Q.: CO2 uptake of a mature Acacia mangium plantation estimated from sap flow measurements and stable carbon isotope discrimination, Biogeosciences, 11, 1393–1411, https://doi.org/10.5194/bg-11-1393-2014, 2014.

Way, D. A., Oren, R., and Kroner, Y.: The space-time continuum: the effects of elevated CO2 and temperature on trees and the importance of scaling, Plant. Cell Environ., 38, 991–1007, https://doi.org/10.1111/pce.12527, 2015.

Williams, C. A., Reichstein, M., Buchmann, N., Baldocchi, D., Beer, C., Schwalm, C., Wohlfahrt, G., Hasler, N., Bernhofer, C., Foken, T., Papale, D., Schymanski, S., and Schaefer, K.: Climate and vegetation controls on the surface water balance: Synthesis of evapotranspiration measured across a global network of flux towers, Water Resour. Res., 48, W06523, https://doi.org/10.1029/2011WR011586, 2012.

Wu, C., Huang, W., Yang, Q., and Xie, Q.: Improved estimation of light use efficiency by removal of canopy structural effect from the photochemical reflectance index (PRI), Agr. Ecosyst. Environ., 199, 333–338, https://doi.org/10.1016/j.agee.2014.10.017, 2015.

Zhao, M. and Running, S. W.: Drought-Induced Reduction in Global Terrestrial Net Primary Production from 2000 Through 2009, Science, 329, 940–943, 2010.

6

Seasonal variability in methane and nitrous oxide fluxes from tropical peatlands in the western Amazon basin

Yit Arn Teh[1], Wayne A. Murphy[2], Juan-Carlos Berrio[2], Arnoud Boom[2], and Susan E. Page[2]

[1]Institute of Biological and Environmental Sciences, University of Aberdeen, Aberdeen, UK
[2]Department of Geography, University of Leicester, Leicester, UK

Correspondence to: Yit Arn Teh (yateh@abdn.ac.uk)

Abstract. The Amazon plays a critical role in global atmospheric budgets of methane (CH_4) and nitrous oxide (N_2O). However, while we have a relatively good understanding of the continental-scale flux of these greenhouse gases (GHGs), one of the key gaps in knowledge is the specific contribution of peatland ecosystems to the regional budgets of these GHGs. Here we report CH_4 and N_2O fluxes from lowland tropical peatlands in the Pastaza–Marañón foreland basin (PMFB) in Peru, one of the largest peatland complexes in the Amazon basin. The goal of this research was to quantify the range and magnitude of CH_4 and N_2O fluxes from this region, assess seasonal trends in trace gas exchange, and determine the role of different environmental variables in driving GHG flux. Trace gas fluxes were determined from the most numerically dominant peatland vegetation types in the region: forested vegetation, forested (short pole) vegetation, *Mauritia flexuosa*-dominated palm swamp, and mixed palm swamp. Data were collected in both wet and dry seasons over the course of four field campaigns from 2012 to 2014. Diffusive CH_4 emissions averaged 36.05 ± 3.09 mg CH_4–C m^{-2} day^{-1} across the entire dataset, with diffusive CH_4 flux varying significantly among vegetation types and between seasons. Net ebullition of CH_4 averaged 973.3 ± 161.4 mg CH_4–C m^{-2} day^{-1} and did not vary significantly among vegetation types or between seasons. Diffusive CH_4 flux was greatest for mixed palm swamp (52.0 ± 16.0 mg CH_4–C m^{-2} day^{-1}), followed by *M. flexuosa* palm swamp (36.7 ± 3.9 mg CH_4–C m^{-2} day^{-1}), forested (short pole) vegetation (31.6 ± 6.6 mg CH_4–C m^{-2} day^{-1}), and forested vegetation (29.8 ± 10.0 mg CH_4–C m^{-2} day^{-1}). Diffusive CH_4 flux also showed marked seasonality, with divergent seasonal patterns among ecosystems. Forested vegetation and mixed palm swamp showed significantly higher dry season (47.2 ± 5.4 mg CH_4–C m^{-2} day^{-1} and 85.5 ± 26.4 mg CH_4–C m^{-2} day^{-1}, respectively) compared to wet season emissions (6.8 ± 1.0 mg CH_4–C m^{-2} day^{-1} and 5.2 ± 2.7 mg CH_4–C m^{-2} day^{-1}, respectively). In contrast, forested (short pole) vegetation and *M. flexuosa* palm swamp showed the opposite trend, with dry season flux of 9.6 ± 2.6 and 25.5 ± 2.9 mg CH_4–C m^{-2} day^{-1}, respectively, versus wet season flux of 103.4 ± 13.6 and 53.4 ± 9.8 mg CH_4–C m^{-2} day^{-1}, respectively. These divergent seasonal trends may be linked to very high water tables (> 1 m) in forested vegetation and mixed palm swamp during the wet season, which may have constrained CH_4 transport across the soil–atmosphere interface. Diffusive N_2O flux was very low (0.70 ± 0.34 µg N_2O–N m^{-2} day^{-1}) and did not vary significantly among ecosystems or between seasons. We conclude that peatlands in the PMFB are large and regionally significant sources of atmospheric CH_4 that need to be better accounted for in regional emissions inventories. In contrast, N_2O flux was negligible, suggesting that this region does not make a significant contribution to regional atmospheric budgets of N_2O. The divergent seasonal pattern in CH_4 flux among vegetation types challenges our underlying assumptions of the controls on CH_4 flux in tropical peatlands and emphasizes the need for more process-based measurements during periods of high water table.

1 Introduction

The Amazon basin plays a critical role in the global atmospheric budgets of carbon (C) and greenhouse gases (GHGs) such as methane (CH_4) and nitrous oxide (N_2O). Recent basin-wide studies suggest that the Amazon as a whole accounts for approximately 7 % of global atmospheric CH_4 emissions (Wilson et al., 2016). N_2O emissions are of a similar magnitude, with emissions ranging from 2 to 3 Tg N_2O–N yr^{-1} (or approximately 12–18 % of global atmospheric emissions) (Huang et al., 2008; Saikawa et al., 2013, 2014). While we have a relatively strong understanding of the role that the Amazon plays in regional and global atmospheric budgets of these gases, one of the key gaps in knowledge is the contribution of specific ecosystem types to regional fluxes of GHGs (Huang et al., 2008; Saikawa et al., 2013, 2014). In particular, our understanding of the contribution of Amazonian wetlands to regional C and GHG budgets is weak, as the majority of past ecosystem-scale studies have focused on *terra firme* forests and savannas (D'Amelio et al., 2009; Saikawa et al., 2013; Wilson et al., 2016; Kirschke et al., 2013; Nisbet et al., 2014). Empirical studies of GHG fluxes from Amazonian wetlands are more limited in geographic scope and have focused on three major areas: wetlands in the state of Amazonas near the city of Manaus (Devol et al., 1990; Bartlett et al., 1988, 1990; Keller et al., 1986), the Pantanal region (Melack et al., 2004; Marani and Alvalá, 2007; Liengaard et al., 2013), and the Orinoco River basin (Smith et al., 2000; Lavelle et al., 2014). Critically, none of the ecosystems sampled in the past were peat-forming ones; rather, the habitats investigated were non-peat forming (i.e., mineral or organo-mineral soils), seasonally inundated floodplain forests (i.e., *varzea*), rivers, or lakes.

Peatlands are one of the major wetland habitats absent from current bottom-up GHG inventories for the Amazon basin and are often grouped together with non-peat-forming wetlands in regional atmospheric budgets (Wilson et al., 2016). Unlike their Southeast Asian counterparts, most peatlands in the Amazon basin are unaffected by human activity at the current time (Lahteenoja et al., 2009a, b; Lahteenoja and Page, 2011), except for ecosystems in the Madre de Dios region in southeastern Peru, which are impacted by gold mining (Householder et al., 2012). Because we have little or no data on ecosystem-level land–atmosphere fluxes from Amazonian peatlands (Lahteenoja et al., 2009b, 2012; Kirschke et al., 2013; Nisbet et al., 2014), it is difficult to ascertain if rates of GHG flux from these ecosystems are similar to or different from mineral soil wetlands (e.g., *varzea*). Given that underlying differences in plant community composition and soil properties are known to modulate the cycling and flux of GHGs in wetlands (Limpens et al., 2008; Melton et al., 2013; Belyea and Baird, 2006; Sjögersten et al., 2014), expanding our observations to include a wider range of wetland habitats is critical in order to improve our understanding of regional trace gas exchange and also to determine whether ag-gregating peat and mineral soil wetlands together in bottom-up emissions inventories are appropriate for regional budget calculations. Moreover, Amazonian peatlands are thought to account for a substantial land area (i.e., up to 150 000 km^2) (Schulman et al., 1999; Lahteenoja et al., 2012), and any differences in biogeochemistry among peat and mineral/organo-mineral soil wetlands may therefore have important implications for understanding and modeling the biogeochemical functioning of the Amazon basin as a whole.

Since the identification of extensive peat forming wetlands in the north (Lahteenoja et al., 2009a, b; Lahteenoja and Page 2011) and south (Householder et al., 2012) of the Peruvian Amazon, several studies have been undertaken to better characterize these habitats, investigating vegetation composition and habitat diversity (Draper et al., 2014; Kelly et al., 2014; Householder et al., 2012; Lahteenoja and Page, 2011), vegetation history (Lähteenoja and Roucoux, 2010), C stocks (Lahteenoja et al., 2012; Draper et al., 2014), hydrology (Kelly et al., 2014), and peat chemistry (Lahteenoja et al., 2009a, b). Most of the studies have focused on the Pastaza–Marañón foreland basin (PMFB), where one of the largest stretches of contiguous peatlands has been found (Lahteenoja et al 2009a; Lahteenoja and Page, 2011; Kelly et al., 2014), covering an estimated area of 35 600 ± 2133 km^2 (Draper et al., 2014). Up to 90 % of the peatlands in the PMFB lie in flooded backwater river margins on floodplains and are influenced by large, annual fluctuations in water table caused by the Amazonian flood pulse (Householder et al., 2012; Lahteenoja et al., 2009a). These floodplain systems are dominated by peat deposits that range in depth from ∼ 3.9 m (Lahteenoja et al., 2009a) to ∼ 12.9 m (Householder et al., 2012). The remaining 10 % of these peatlands are not directly influenced by river flow and form domed (i.e., raised) nutrient-poor bogs that likely only receive water and nutrients from rainfall (Lahteenoja et al., 2009b). These nutrient-poor bogs are dominated by large, C-rich forests (termed "pole forests") that represent a very-high-density C store (total pool size of 1391 ± 710 Mg C ha^{-1}, which includes both above- and belowground stocks), exceeding even the C density of nearby floodplain systems (Draper et al., 2014). Even though the peats in these nutrient-poor bogs have a relatively high hydraulic conductivity, they act as natural stores of water because of high rainwater inputs (> 3000 mm per annum), which help to maintain high water tables, even during parts of the dry season (Kelly et al., 2014).

CH_4 flux in tropical soils are regulated by the complex interplay among multiple factors that regulate CH_4 production, oxidation, and transport. Key factors include redox/water table depth (Couwenberg et al., 2010, 2011; Silver et al., 1999; Teh et al., 2005; von Fischer and Hedin, 2007), plant productivity (von Fischer and Hedin, 2007; Whiting and Chanton, 1993), soil organic matter lability (Wright et al., 2011), competition for C substrates among anaerobes (Teh et al., 2008; Teh and Silver, 2006; von Fischer and Hedin, 2007), and presence of plants capable of facilitating atmospheric egress

(Pangala et al., 2013). Of all these factors, fluctuation in soil redox conditions, as mediated by variations in water table depth, is perhaps most critical in regulating CH_4 dynamics (Couwenberg et al., 2010, 2011) because of the underlying physiology of the microbes that produce and consume CH_4. Methanogenic archaea are obligate anaerobes that only produce CH_4 under anoxic conditions (Conrad, 1996); as a consequence, they are only active in stably anoxic soil microsites or soil layers, where they are protected from the effects of strong oxidants such as oxygen or where competition for reducing equivalents (e.g., acetate, H_2) from other anaerobic microorganisms is eliminated (Teh et al., 2005, 2008; Teh and Silver, 2006; von Fischer and Hedin, 2002, 2007). CH_4 oxidation, in contrast, is thought to be driven primarily by aerobic methanotrophic bacteria in tropical soils (Hanson and Hanson, 1996; Teh et al., 2005, 2006; von Fischer and Hedin, 2002, 2007), with anaerobic CH_4 oxidation playing a quantitatively smaller role (Blazewicz et al., 2012). Thus, fluctuations in redox or water table depth play a fundamental role in directing the flow of C among different anaerobic pathways (Teh et al., 2008; Teh and Silver, 2006; von Fischer and Hedin, 2007), and shifting the balance between production and consumption of CH_4 (Teh et al., 2005; von Fischer and Hedin, 2002). Moreover, water table or soil moisture fluctuations are also thought to profoundly influence CH_4 transport dynamics throughout the soil profile, changing the relative partitioning of CH_4 among different transport pathways such as diffusion, ebullition, and plant-facilitated transport (Whalen, 2005; Jungkunst and Fiedler, 2007).

Controls on N_2O flux are also highly complex (Groffman et al., 2009), with N_2O originating from as many as four separate sources (e.g., bacterial ammonia oxidation, archaeal ammonia oxidation, denitrification, dissimilatory nitrate reduction to ammonium), each with different environmental controls (Baggs, 2008; Morley and Baggs, 2010; Firestone and Davidson, 1989; Firestone et al., 1980; Pett-Ridge et al., 2013; Silver et al., 2001; Prosser and Nicol, 2008). Key factors regulating soil N_2O flux include redox, soil moisture content or water table depth, temperature, pH, labile C availability, and labile N availability (Groffman et al., 2009). As is the case for CH_4, variations in redox/water table depth play an especially prominent role in regulating N_2O flux in tropical peatland ecosystems because all of the processes that produce N_2O are redox-sensitive, with bacterial or archaeal ammonia oxidation occurring under aerobic conditions (Prosser and Nicol, 2008; Firestone and Davidson, 1989; Firestone et al., 1980) whereas nitrate-reducing processes (i.e., denitrification, dissimilatory nitrate reduction to ammonium) occur under anaerobic ones (Firestone and Davidson, 1989; Firestone et al., 1980; Morley and Baggs, 2010; Silver et al., 2001). Moreover, for nitrate-reducing processes, which are believed to be the dominant source of N_2O in wet systems, the extent of anaerobiosis also controls the relative proportion of N_2O or N_2 produced during dissimilatory metabolism

(Firestone and Davidson, 1989; Firestone et al., 1980; Morley and Baggs, 2010; Silver et al., 2001).

In order to improve our understanding of the biogeochemistry and rates of GHG exchange from Amazonian peatlands, we conducted a preliminary study of CH_4 and N_2O fluxes from forested peatlands in the PMFB. The main objectives of this are to

1. quantify the magnitude and range of soil CH_4 and N_2O fluxes from a subset of peatlands in the PMFB that represent dominant vegetation types;

2. determine seasonal patterns of trace gas exchange;

3. establish the relationship between trace gas fluxes and environmental variables.

Sampling was concentrated on the four most dominant vegetation types in the area, based on prior work by the investigators (Lahteenoja and Page, 2011). Trace gas fluxes were captured from both floodplain systems and nutrient-poor bogs in order to account for underlying differences in biogeochemistry that may arise from variations in hydrology. Sampling was conducted during four field campaigns (two wet season, two dry season) over a 27-month period, extending from February 2012 to May 2014.

2 Materials and methods

2.1 Study site and sampling design

The study was carried out in the lowland tropical peatland forests of the PMFB, between 2 and 35 km south of the city of Iquitos, Peru (Lahteenoja et al., 2009a, b) (Fig. 1, Table 1). The mean annual temperature is 26 °C, annual precipitation is ca. 3100 mm, relative humidity ranges from 80 to 90 %, and altitude ranges from ca. 90 to 130 m a.s.l. (above sea level) (Marengo 1998). The northwestern Amazon basin near Iquitos experiences pronounced seasonality, which is characterized by consistently high annual temperatures, but marked seasonal variation in precipitation (Tian et al., 1998) and an annual river flood pulse linked to seasonal discharge from the Andes (Junk, 1989). Precipitation events are frequent, intense and of significant duration during the wet season (November to May) and infrequent, intense, and of short duration during the dry season (June to August). September and October represent a transitional period between dry and wet seasons, where rainfall patterns are less predictable. Catchments in this region receive no less than 100 mm of rain per month (Espinoza Villar et al., 2009a, b) and > 3000 mm of rain per year. River discharge varies by season, with the lowest discharge between the dry season months of August and September. Peak discharge from the wet season flood pulse occurs between April and May, as recorded at the Tamshiyacu River gauging station (Espinoza Villar et al., 2009b).

Table 1. Site characteristics including field site location, nutrient status, plot, and flux chamber replication.

Vegetation type	Site name	Nutrient status*	Latitude (S)	Longitude (W)	Plots	Flux chambers
Forested	Buena Vista	Rich	4°14′45.60″ S	73°12′0.20″ W	21	105
Forested (short pole)	San Jorge (center)	Poor	4°03′35.95″ S	73°12′01.13″ W	6	28
Forested (short pole)	Miraflores	Poor	4°28′16.59″ S	74°4′39.95″ W	41	204
M. flexuosa Palm Swamp	Quistococha	Intermediate	3°49′57.61″ S	73°12′01.13″ W	135	668
M. flexuosa Palm Swamp	San Jorge (edge)	Intermediate	4°03′18.83″ S	73°10′16.80″ W	18	86
Mixed palm swamp	Charo	Rich	4°16′21.80″ S	73°15′27.80″ W	18	90

* After Householder et al. (2012) and Lahteenoja et al. (2009a, b).

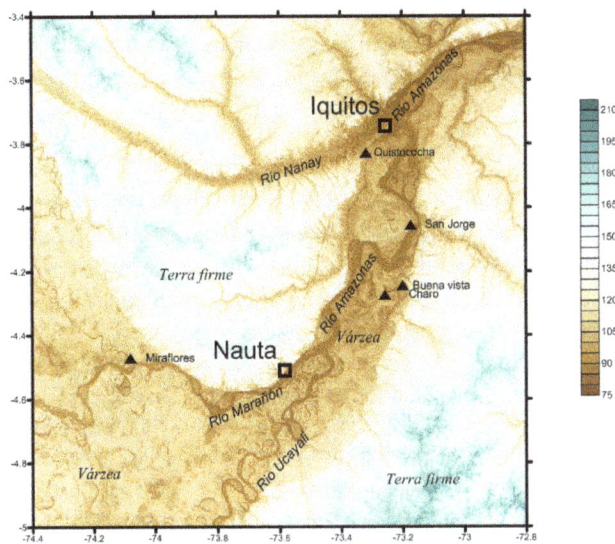

Figure 1. Map of the study region and field sites. The color scale to the right of the map denotes elevation in m a.s.l. Tan and brown tones indicate areas in which peatlands are found; however, not all of these areas are peatland-dominated.

Histosols form the dominant soil type for peatlands in this region (Andriesse, 1988; Lahteenoja and Page, 2011). Study sites are broadly classified as nutrient-rich, intermediate, or nutrient-poor (Lahteenoja and Page, 2011), with pH ranging from 3.5 to 7.2 (Lahteenoja and Page, 2011; Lahteenoja et al., 2009a, b). More specific data on pH for our plots are presented in Table 3. Nutrient-rich (i.e., minerotrophic) sites tend to occur on floodplains and river margins and account for at least 60 % of the peatland cover in the PMFB (Lahteenoja and Page, 2011; Draper et al., 2014). They receive water, sediment, and nutrient inputs from the annual Amazon river flood pulse (Householder et al., 2012; Lahteenoja and Page, 2011), leading to higher inorganic nutrient content, of which Ca and other base cations form major constituents (Lahteenoja and Page, 2011). Many of the soils in these nutrient-rich areas are fluvaquentic Tropofibrists (Andriesse, 1988) and contain thick mineral layers or minerogenic intrusions, reflective of episodic sedimentation events

in the past (Lahteenoja and Page, 2011). In contrast, nutrient-poor (i.e., oligotrophic) sites tend to occur further inland (Lahteenoja and Page, 2011; Draper et al., 2014). They are almost entirely rain-fed and receive low or infrequent inputs of water and nutrients from streams and rivers (Lahteenoja and Page, 2011). These ecosystems account for 10 to 40 % of peatland cover in the PMFB, though precise estimates vary depending on the land classification scheme employed (Lahteenoja and Page, 2011; Draper et al., 2014). Soil Ca and base cation concentrations are significantly lower in these sites compared to nutrient-rich ones, with similar concentrations to that of rainwater (Lahteenoja and Page, 2011). Soils are classified as typic or hydric Tropofibrists (Andriesse, 1988). Even though Ca and base cations themselves play no direct role in modulating CH_4 and N_2O fluxes, underlying differences in soil fertility may indirectly influence CH_4 and N_2O flux by influencing the rate of labile C input to the soil, the decomposability of organic matter, and the overall throughput of C and nutrients through the plant–soil system (Firestone and Davidson, 1989; Groffman et al., 2009; von Fischer and Hedin, 2007; Whiting and Chanton, 1993).

We established 239 sampling plots ($\sim 30\,\text{m}^2$ per plot) within five tropical peatland sites that captured four of the dominant vegetation types in the region (Draper et al., 2014; Householder et al., 2012; Kelly et al., 2014; Lahteenoja and Page, 2011) and which encompassed a range of nutrient availabilities (Fig. 1, Table 1) (Lahteenoja and Page, 2011; Lahteenoja et al., 2009a). These four dominant vegetation types included forested vegetation (nutrient-rich; $n = 21$ plots), forested (short pole) vegetation (nutrient-poor; $n = 47$ plots), *Mauritia flexuosa*-dominated palm swamp (intermediate fertility; $n = 153$ plots), and mixed palm swamp (nutrient-rich; $n = 18$ plots) (Table 1). Four of the study sites (Buena Vista, Charo, Miraflores, and Quistococha) were dominated by only one vegetation type, whereas San Jorge contained a mixture of *M. flexuosa* palm swamp and forested (short pole) vegetation (Table 1). As a consequence, both vegetation types were sampled in San Jorge to develop a more representative picture of GHG fluxes from this location. Sampling efforts were partially constrained by issues of site access; some locations were difficult to access (e.g., centre of the San Jorge peatland) due to water table height and navigability of river channels; as a consequence, sampling patterns were somewhat uneven, with higher sampling densities in some peatlands than in others (Table 1).

In each peatland site, transects were established from the edge of the peatland to its center. Each transect varied in length from 2 to 5 km, depending on the relative size of the peatland. Randomly located sampling plots ($\sim 30 \, \text{m}^2$ per plot) were established at 50 or 200 m intervals along each transect, from which GHG fluxes and environmental variables were measured concomitantly. The sampling interval (i.e., 50 or 200 m) was determined by the length of the transect or size of the peatland, with shorter sampling intervals (50 m) for shorter transects (i.e., smaller peatlands) and longer sampling intervals (200 m) for longer transects (i.e., larger peatlands).

2.2 Quantifying soil–atmosphere exchange

Soil–atmosphere fluxes (CH_4, N_2O) were determined in four campaigns over a 2-year annual water cycle: February 2012 (wet season), June–August 2012 (dry season), June–July 2013 (dry season), and May–June 2014 (wet season). The duration of the campaign for each study site varied depending on its size. Each study site was generally sampled only once for each campaign, except for a subset of plots within each vegetation type where diurnal studies were conducted to determine whether CH_4 and N_2O fluxes varied over daily time steps. Gas exchange was quantified using a floating static chamber approach (Livingston and Hutchinson, 1995; Teh et al., 2011). Static flux measurements were made by enclosing a $0.225 \, \text{m}^2$ area with a dark, single component, vented 10 L flux chamber. No chamber bases (collars) were used due to the highly saturated nature of the soils. In most cases, a standing water table was present at the soil surface, so chambers were placed directly onto the water. In the absence of a standing water table, a weighted skirt was applied to create an airtight seal. Under these drier conditions, chambers were placed carefully on the soil surface. In order to reduce the risk of pressure-induced ebullition or disruption to soil gas concentration profiles caused by the investigators' footfall, flux chambers were lowered from a distance of 2 m away using a 2 m long pole. Gas samples were collected with syringes using > 2 m lengths of Tygon® tubing, after thoroughly purging the dead volumes in the sample lines. To promote even mixing within the headspace, chambers were fitted with small computer fans (Pumpanen et al., 2004). Headspace samples were collected from each flux chamber at five intervals over a 25 min enclosure period using a gas-tight syringe. Gas samples were stored in evacuated Exetainers® (Labco Ltd., Lampeter, UK), shipped to the UK, and subsequently analyzed for CH_4, CO_2, and N_2O concentrations using Thermo TRACE GC Ultra (Thermo Fisher Scientific Inc., Waltham, Massachusetts, USA) at the University of St. Andrews. Chromatographic separation was achieved

using a Porapak-Q column, and gas concentrations were determined using a flame ionization detector (FID) for CH_4, a methanizer-FID for CO_2, and an electron capture detector (ECD) for N_2O. Instrumental precision, determined from repeated analysis of standards, was < 5 % for all detectors.

Diffusive fluxes were determined by using the JMP IN version 11 (SAS Institute, Inc., Cary, North Carolina, USA) statistical package to plot best-fit lines to the data for headspace concentration against time for individual flux chambers, with fluxes calculated from linear or nonlinear regressions depending on the individual concentration trend against time (Teh et al., 2014). Gas mixing ratios (ppm) were converted to areal fluxes by using the ideal gas law to solve for the quantity of gas in the headspace (on a mole or mass basis) and normalized by the surface area of each static flux chamber (Livingston and Hutchinson, 1995). Ebullition-derived CH_4 fluxes were also quantified in our chambers where evidence of ebullition was found. This evidence consisted of either (i) rapid, nonlinear increases in CH_4 concentration over time; (ii) abrupt, stochastic increases in CH_4 concentration over time; or (iii) an abrupt stochastic increase in CH_4 concentration, followed by a linear decline in concentration. For observations following pattern (i), flux was calculated by fitting a quadratic regression equation to the data ($P < 0.05$), and CH_4 flux determined from the initial steep rise in CH_4 concentration. For data following pattern (ii), the ebullition rate was determined by calculating the total CH_4 production over the course of the bubble event, in line with prior work conducted by the investigators (Teh et al., 2011). Last, for data following pattern (iii), a best-fit line was plotted to the CH_4 concentration data after the bubble event and a net rate of CH_4 uptake calculated from the gradient of the line. While observations (i)–(iii) all reflect the effects of ebullition, only observations following patterns (i) and (ii) indicate net emission to the atmosphere, whereas observations following pattern (iii) indicate emission followed by net uptake. As a consequence, patterns (i) and (ii) were categorized as "net ebullition" (i.e., net efflux) whereas observations following pattern (iii) were categorized as "ebullition-driven CH_4 uptake" (i.e., net influx).

2.3 Environmental variables

To investigate the effects of environmental variables on trace gas fluxes, we determined air temperature, soil temperature, chamber headspace temperature, soil pH, soil electrical conductivity (EC; $\mu\text{S m}^{-2}$), dissolved oxygen concentration of the soil pore water (DO; measured as percent saturation, %) in the top 15 cm of the peat column, and water table position concomitant with gas sampling. Air temperature (measured 1.3 m above the soil) and chamber headspace temperature were measured using a Checktemp® probe and meter (Hanna Instruments LTD, Leighton Buzzard, UK). Peat temperature, pH, DO, and EC were measured at a depth of 15 cm below the peat surface and recorded in situ with each gas sample

Table 2. Proportion of observations for each vegetation type that showed evidence of ebullition, mean rates of ebullition, and ebullition-driven CH_4 uptake. Values represent means and standard errors.

Vegetation type	Percentage of observations	Net ebullition ($mg\,CH_4$–$C\,m^{-2}\,day^{-1}$)		Ebullition-driven uptake ($mg\,CH_4$–$C\,m^{-2}\,day^{-1}$)	
	(%)	Wet season	Dry season	Wet season	Dry season
Forested	10.5	0	0	0	-136.4 ± 0.1
Forested (short pole)	6.9	994.6 ± 293.2	512.5 ± 153.0	-95.8 ± 0.0	-245.5 ± 48.9
M. flexuosa palm swamp	16.7	1192.0 ± 305.7	994.3 ± 237.3	-869.4 ± 264.8	-401.4 ± 59.9
Mixed palm swamp	12.2	0	733.6 ± 313.1	0	-464.4 ± 565.9

using a HACH® rugged outdoor HQ30D multimeter and pH, DO, or EC probe. At sites where the water level was above the peat surface, the water depth was measured using a meter rule. Where the water table was at or below the peat surface, the water level was measured by auguring a hole to 1 m depth and measuring water table depth using a meter rule.

2.4 Statistical analyses

Statistical analyses were performed using JMP IN version 11 (SAS Institute, Inc., Cary, North Carolina, USA). Box–Cox transformations were applied where the data failed to meet the assumptions of analysis of variance (ANOVA); otherwise, nonparametric tests were applied (e.g., Wilcoxon signed-rank test). ANOVA and analysis of co-variance (ANCOVA) were used to test for relationships between gas fluxes and vegetation type, season, and environmental variables. When determining the effect of vegetation type on gas flux, data from different study sites (e.g., San Jorge and Miraflores) were pooled together. Means comparisons were tested using Fisher's least significant difference (LSD) test.

3 Results

3.1 Differences in gas fluxes and environmental variables among vegetation types

All vegetation types were net sources of CH_4, with an overall mean ($\pm$ SE – standard error) diffusive flux of $36.1 \pm 3.1\,mg\,CH_4$–$C\,m^{-2}\,day^{-1}$ and a mean net ebullition flux of $973.3 \pm 161.4\,mg\,CH_4$–$C\,m^{-2}\,day^{-1}$ (Fig. 2, Table 2). We also saw examples of ebullition-driven CH_4 uptake (i.e., a sudden or stochastic increase in CH_4 concentration, followed immediately by a rapid linear decline in concentration), with a mean rate of $-504.1 \pm 84.4\,mg\,CH_4$–$C\,m^{-2}\,day^{-1}$ (Table 2). Diffusive fluxes of CH_4 accounted for the majority of observations (83.3 to 93.1 %), while ebullition fluxes accounted for a much smaller proportion of observations (6.9 to 16.7 %; Table 2).

Diffusive CH_4 flux varied significantly among the four vegetation types sampled in this study (two-way ANOVA with vegetation, season, and their interaction, $F_{7,979} = 13.2$,

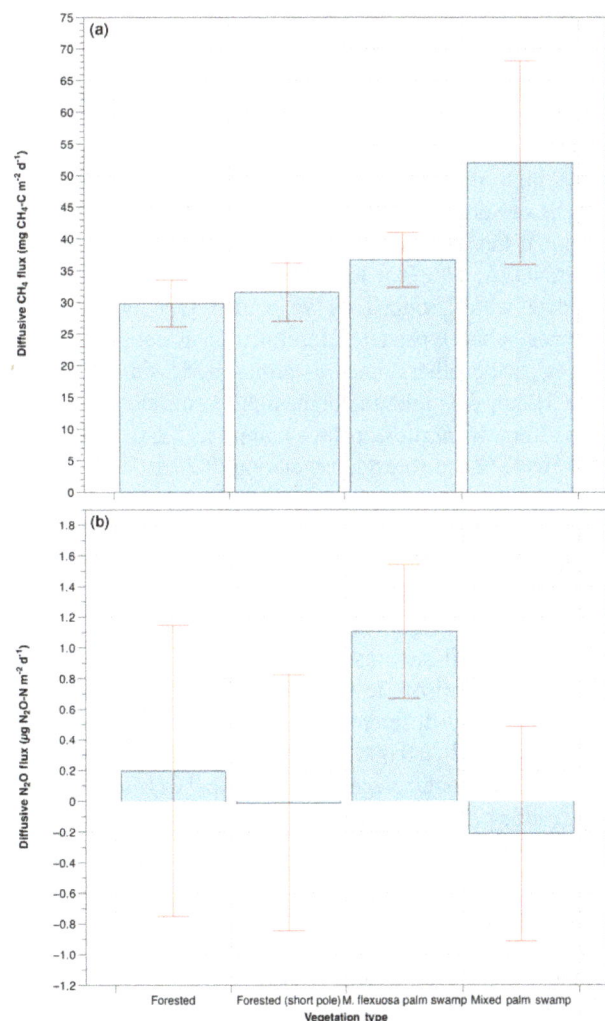

Figure 2. Net diffusive **(a)** methane (CH_4) and **(b)** nitrous oxide (N_2O) fluxes by vegetation type. Error bars denote standard errors.

$P < 0.0001$; Fig. 2a). However, the effect of vegetation was relatively weak (see ANCOVA results in Section 3.3), and a means comparison test on the pooled data was unable to determine which means differed significantly from the others (Fisher's LSD, $P > 0.05$). For the pooled data, the overall numerical trend was that mixed palm swamp showed the highest mean flux (52.0 ± 16.0 mg CH_4–$C\,m^{-2}\,day^{-1}$), followed by *M. flexuosa* palm swamp (36.7 ± 3.9 mg CH_4–$C\,m^{-2}\,day^{-1}$), forested (short pole) vegetation (31.6 ± 6.6 mg CH_4–$C\,m^{-2}\,day^{-1}$), and forested vegetation (29.8 ± 10.0 mg CH_4–$C\,m^{-2}\,day^{-1}$). CH_4 ebullition (i.e., net ebullition and ebullition-driven uptake) did not vary significantly among vegetation types or between seasons (Table 2). Broadly speaking, however, we saw a greater frequency of ebullition in the *M. flexuosa* palm swamp, followed by mixed palm swamp, forested vegetation, and forested short pole vegetation (Table 2).

These study sites were also a weak net source of N_2O, with a mean diffusive flux of $0.70 \pm 0.34\,\mu$g N_2O–$N\,m^{-2}\,day^{-1}$. We saw only limited evidence of ebullition of N_2O, with only three chambers out of 1181 (0.3 % of observations) showing evidence of N_2O ebullition. These data were omitted from the analysis of diffusive flux of N_2O. Because of the high variance in diffusive N_2O flux among plots, analysis of variance indicated that mean diffusive N_2O flux did not differ significantly among vegetation types (two-way ANOVA, $P > 0.5$, Fig. 2b). However, when the N_2O flux data were grouped by vegetation type, we see that some vegetation types tended to function as net atmospheric sources, while others acted as atmospheric sinks (Fig. 2b, Table 3). For example, the highest N_2O emissions were observed from *M. flexuosa* palm swamp ($1.11 \pm 0.44\,\mu$g N_2O–$N\,m^{-2}\,day^{-1}$) and forested vegetation ($0.20 \pm 0.95\,\mu$g N_2O–$N\,m^{-2}\,day^{-1}$). In contrast, forested (short pole) vegetation and mixed palm swamp were weak sinks for N_2O, with a mean flux of -0.01 ± 0.84 and $-0.21 \pm 0.70\,\mu$g N_2O–$N\,m^{-2}\,day^{-1}$, respectively.

Soil pH varied significantly among vegetation types (data pooled across all seasons; ANOVA, $P < 0.0001$, Table 3). Multiple comparison tests indicated that mean soil pH was significantly different for each of the vegetation types (Fisher's LSD, $P < 0.0001$, Table 3), with the lowest pH in forested (short pole) vegetation (4.10 ± 0.04), followed by *M. flexuosa* palm swamp (5.32 ± 0.02), forested vegetation (6.15 ± 0.06), and the mixed palm swamp (6.58 ± 0.04).

Soil DO content varied significantly among vegetation types (data pooled across all seasons; Kruskal–Wallis, $P < 0.0001$, Table 3). Multiple comparison tests indicated that mean DO was significantly different for each of the vegetation types (Fisher's LSD, $P < 0.05$, Table 3), with the highest DO in the forested (short pole) vegetation (25.2 ± 2.1 %), followed by the *M. flexuosa* palm swamp (18.1 ± 1.0 %), forested vegetation (11.8 ± 2.8 %), and the mixed palm swamp (0.0 ± 0.0 %).

EC varied significantly among vegetation types (data pooled across all seasons; Kruskal–Wallis, $P < 0.0001$, Table 3). Multiple comparison tests indicated that mean EC was significantly different for each of the vegetation types (Fisher's LSD, $P < 0.05$; Table 3), with the highest EC in the mixed palm swamp ($170.9 \pm 6.0\,\mu$S m^{-2}), followed by

Table 3. Environmental variables for each vegetation type for the wet and dry season. Values reported here are means and standard errors. Lowercase letters indicate significant differences among vegetation types within the wet or dry season (Fisher's LSD, $P < 0.05$).

Vegetation type	Peat temperature (°C)		Air temperature (°C)		Conductivity (μS m^{-2})		Dissolved oxygen (%)		Water table level (cm)		pH	
	Wet season	Dry season	Wet season	Dry season	Wet season	Dry season	Wet season	Dry season	Wet season	Dry season	Wet season	Dry season
Forested	26.1 ± 0.1a	24.7 ± 0.0a	28.8 ± 0.7a	26.4 ± 0.3a	79.0 ± 5.9a	75.9 ± 5.7a	0.2 ± 0.1a	18.9 ± 4.4a	110.8 ± 9.3a	−13.2 ± 0.7a	5.88 ± 0.15a	6.31 ± 0.04a
Forested (short pole)	25.2 ± 0.0b	24.8 ± 0.1a	27.6 ± 0.1b	27.5 ± 0.1b	21.0 ± 0.0b	48.5 ± 4.8b	4.4 ± 0.0a	33.1 ± 2.6b	26.9 ± 0.5b	−4.7 ± 0.4b	4.88 ± 0.01b	3.8 ± 0.03b
M. flexuosa palm swamp	25.6 ± 0.6c	25.3 ± 0.1b	26.3 ± 0.1c	26.4 ± 0.1a	45.9 ± 2.1c	51.9 ± 1.8b	19.4 ± 1.3b	17.3 ± 1.5a	37.2 ± 1.7c	6.1 ± 1.3c	5.04 ± 0.03c	5.49 ± 0.03c
Mixed palm swamp	26.0 ± 0.0a	25.0 ± 0.1ab	26.1 ± 0.1c	28.2 ± 0.3b	100.0 ± 0.2d	206.4 ± 4.2c	0.0 ± 0.0a	0.0 ± 0.0c	183.7 ± 1.7d	−2.4 ± 0.3b	6.1 ± 0.03a	6.82 ± 0.02d

Table 4. Trace gas fluxes for each vegetation type for the wet and dry season. Values reported here are means and standard errors. Uppercase letters indicate significant differences in gas flux between seasons within a vegetation type, while lowercase letters indicate significant differences among vegetation types within a season (Fisher's LSD, $P < 0.05$).

Vegetation type	Methane flux $(\mathrm{mg\,CH_4\text{–}C\,m^{-2}\,day^{-1}})$		Nitrous oxide flux $(\mathrm{\mu g\,N_2O\text{–}N\,m^{-2}\,day^{-1}})$	
	Wet season	Dry season	Wet season	Dry season
Forested	6.7 ± 1.0Aa	47.2 ± 5.4Ba	2.54 ± 1.48	-1.16 ± 1.20
Forested (short pole)	60.4 ± 9.1Ab	18.8 ± 2.6Bb	1.16 ± 0.54	-0.42 ± 0.90
M. flexuosa palm swamp	46.7 ± 8.4Ac	28.3 ± 2.6Bc	1.14 ± 0.35	0.92 ± 0.61
Mixed palm swamp	6.1 ± 1.3Aa	64.2 ± 12.1Ba	1.45 ± 0.79	-0.80 ± 0.79

forested vegetation ($77.1 \pm 4.2\,\mathrm{\mu S\,m^{-2}}$), *M. flexuosa* palm swamp ($49.7 \pm 1.4\,\mathrm{\mu S\,m^{-2}}$), and the forested (short pole) vegetation ($40.9 \pm 3.5\,\mathrm{\mu S\,m^{-2}}$).

Soil temperature varied significantly among vegetation types (data pooled across all seasons; ANOVA, $P < 0.0001$, Table 3). Multiple comparison tests indicated that soil temperature in forested (short pole) vegetation was significantly lower than in the other vegetation types (Table 3), whereas the other vegetation types did not differ in temperature amongst themselves (Fisher's LSD, $P < 0.05$, Table 3).

Air temperature varied significantly among vegetation types (data pooled across all seasons; ANOVA, $P < 0.0001$, Table 3). Multiple comparison tests indicated that air temperature in *M. flexuosa* palm swamp was significantly lower than in the other vegetation types, whereas the other vegetation types did not differ in temperature amongst themselves (Fisher's LSD, $P < 0.05$, Table 3).

Water table depths varied significantly among vegetation types (data pooled across all seasons; ANOVA, $P < 0.0001$, Table 3). The highest mean water tables were observed in mixed palm swamp (59.6 ± 9.3 cm), followed by forested vegetation (34.0 ± 6.9 cm), *M. flexuosa* palm swamp (17.4 ± 1.2 cm), and forested (short pole) vegetation (3.5 ± 1.0 cm) (Fisher's LSD, $P < 0.0005$).

3.2 Temporal variations in gas fluxes and environmental variables

The peatlands sampled in this study showed pronounced seasonal variability in diffusive CH_4 flux (two-way ANOVA, $F_{7,979} = 13.2$, $P < 0.0001$; Table 4). For ebullition of CH_4 and ebullition-driven uptake of CH_4, mean fluxes varied between seasons, but high variability meant that these differences were not statistically significant (two-way ANOVA, $P > 0.8$; Table 2). Diffusive N_2O flux showed no seasonal trends (two-way ANOVA, $P > 0.5$) and therefore will not be discussed further here. Diurnal studies suggest that diffusive fluxes of neither CH_4 nor N_2O varied over the course of a 24 h period.

For diffusive CH_4 flux, the overall trend was towards significantly higher wet season ($51.1 \pm 7.0\,\mathrm{mg\,CH_4\text{–}}$

$\mathrm{C\,m^{-2}\,day^{-1}}$) compared to dry season ($27.3 \pm 2.7\,\mathrm{mg\,CH_4\text{–}}$ $\mathrm{C\,m^{-2}\,day^{-1}}$) flux (data pooled across all vegetation types; t test, $P < 0.001$, Table 4). However, when diffusive CH_4 flux was disaggregated by vegetation type, very different seasonal trends emerged. For example, both forested vegetation and mixed palm swamp showed significantly greater diffusive CH_4 flux during the *dry season* with net fluxes of 47.2 ± 5.4 and $64.2 \pm 12.1\,\mathrm{mg\,CH_4\text{–}C\,m^{-2}\,day^{-1}}$, respectively (Fisher's LSD, $P < 0.05$, Table 3). In contrast, *wet season* flux was 7–16 times lower, with net fluxes of 6.7 ± 1.0 and $6.1\,\mathrm{mg\,CH_4\text{–}C\,m^{-2}\,day^{-1}}$, respectively (Fisher's LSD, $P < 0.05$, Table 3). In contrast, forested (short pole) vegetation and *M. flexuosa* palm swamp showed seasonal trends consistent with the pooled dataset, i.e., significantly higher flux during the wet season (46.7 ± 8.4 and $60.4 \pm 9.1\,\mathrm{mg\,CH_4\text{–}C\,m^{-2}\,day^{-1}}$, respectively) compared to the dry season (28.3 ± 2.6 and $18.8 \pm 2.6\,\mathrm{mg\,CH_4\text{–}}$ $\mathrm{C\,m^{-2}\,day^{-1}}$, respectively) (Fisher's LSD, $P < 0.05$, Table 3).

Even though seasonal trends in CH_4 ebullition were not statistically significant, we will briefly describe the overall patterns for the different vegetation types as they varied among ecosystems (Table 2). Forested vegetation only showed evidence of ebullition during the dry season, where ebullition-driven uptake was observed. For forested (short pole) vegetation, net ebullition was generally greater during the wet season, while ebullition-driven uptake was higher during the dry season. For *M. flexuosa* palm swamp, both net ebullition and ebullition-driven uptake were greater during the wet season. Lastly, for mixed palm swamp, both net ebullition and ebullition-driven uptake were greater during the dry season.

For the environmental variables, soil pH, DO, EC, water table depth, and soil temperature varied significantly between seasons, whereas air temperature did not. Thus, for sake of brevity, air temperature is not discussed further here. Mean soil pH was significantly lower during the wet season (5.18 ± 0.03) than during the dry season (5.31 ± 0.04) (data pooled across all vegetation types; t test, $P < 0.05$, Table 2). When disaggregated by vegetation type, the overall trend was found to hold true for all vegetation types except forested (short pole) vegetation, which displayed higher pH during the wet season compared to the dry season (Table 2). A two-way ANOVA on Box–Cox transformed data using vegetation type, season, and their interaction as explanatory variables indicated that vegetation type was the best predictor of pH, with season and vegetation type by season playing a lesser role ($F_{7,1166} = 348.9$, $P < 0.0001$).

For DO, the overall trend was towards significantly lower DO during the wet season (13.9 ± 1.0 %) compared to the dry season (19.3 ± 1.2 %) (data pooled across all vegetation types; Wilcoxon test, $P < 0.0001$, Table 2). However, when the data were disaggregated by vegetation type, we found that individual vegetation types showed distinct seasonal trends from each other. Forested vegetation and mixed palm swamp were consistent with the overall trend (i.e., lower wet season compared to dry season DO), whereas forested (short pole) vegetation and *M. flexuosa* palm swamp displayed the reverse trend (i.e., higher *wet season* compared to *dry season* DO) (Table 2). A two-way ANOVA on Box–Cox transformed data using vegetation type, season, and their interaction as explanatory variables indicated that vegetation type was the best predictor of DO, followed by a strong vegetation by season interaction; season itself played a lesser role than either of the other two explanatory variables ($F_{7,1166} = 57.0$, $P < 0.0001$).

For EC, the overall trend was towards lower EC in the wet season ($49.4 \pm 1.8 \, \mu\text{S m}^{-2}$) compared to the dry season ($65.5 \pm 2.2 \, \mu\text{S m}^{-2}$) (data pooled across all vegetation types; Wilcoxon test, $P < 0.05$, Table 2). When the data were disaggregated by vegetation type, this trend was consistent for all the vegetation types except for forested vegetation, where differences between wet and dry season were not statistically significant (Wilcoxon, $P > 0.05$, Table 2).

Water table depths varied significantly between seasons (data pooled across all vegetation types; Wilcoxon test, $P < 0.0001$, Table 2). Mean water table level was significantly higher in the wet (54.1 ± 2.7 cm) than the dry (1.3 ± 0.8 cm) season. When disaggregated by vegetation type, the trend held true for individual vegetation types (Table 2). All vegetation types had negative dry season water tables (i.e., below the soil surface) and positive wet season water tables (i.e., water table above the soil surface), except for *M. flexuosa* palm swamp that had positive water tables in both seasons. Two-way ANOVA on Box–Cox transformed data using vegetation type, season, and their interaction as explanatory variables indicated that all three factors explained water table depth, but that season accounted for the largest proportion of the variance in the model, followed by vegetation by season and lastly by vegetation type ($F_{7,1157} = 440.1$, $P < 0.0001$).

For soil temperature, the overall trend was towards slightly higher temperatures in the wet season ($25.6 \pm 0.0 \, °\text{C}$) compared to the dry season ($25.1 \pm 0.0 \, °\text{C}$) ($t$ test, $P < 0.0001$). Analysis of the disaggregated data indicates this trend was consistent for individual vegetation types (Table 2). Two-

way ANOVA on Box–Cox transformed data using vegetation type, season, and their interaction as explanatory variables indicated that all three variables played a significant role in modulating soil temperature, although season accounted for the largest proportion of the variance whereas the other two factors accounted for a similar proportion of the variance ($F_{7,1166} = 21.3$, $P < 0.0001$).

3.3 Relationships between gas fluxes and environmental variables

To explore the relationships between environmental variables and diffusive gas fluxes, we conducted an ANCOVA on Box–Cox transformed gas flux data, using vegetation type, season, vegetation by season, and environmental variables as explanatory variables. We did not analyze trends between ebullition and environmental variables because of the limitations in the sampling methodology and the limited number of observations.

For diffusive CH_4 flux, ANCOVA revealed that vegetation by season was the strongest predictor of CH_4 flux, followed by a strong season effect ($F_{13,917} = 9.2$, $P < 0.0001$). Other significant drivers included soil temperature, water table depth, and a borderline-significant effect of vegetation type ($P < 0.06$). However, it is important to note that each of these environmental variables was only weakly correlated with CH_4 flux even when the relationships were statistically significant; for example, when individual bivariate regressions were calculated, the r^2 values were less than 0.01 for each plot (see Figs. S1 and S2 in the Supplement).

For diffusive N_2O flux, ANCOVA indicated that the best predictors of flux rates were dissolved oxygen and electrical conductivity ($F_{13,1014} = 2.2$, $P < 0.0082$). As was the case for CH_4, when the relationships between these environmental variables and N_2O flux were explored using individual bivariate regressions, r^2 values were found to be very low (e.g., less than $r^2 < 0.0007$) or not statistically significant (see Figs. S3 and S4).

4 Discussion

4.1 Large and asynchronous CH_4 fluxes from peatlands in the Pastaza–Marañón foreland basin

The ecosystems sampled in this study were strong atmospheric sources of CH_4. Diffusive CH_4 flux, averaged across all vegetation types, was $36.1 \pm 3.1 \, \text{mg CH}_4\text{–C m}^{-2} \text{ day}^{-1}$, spanning a range from -100 to $1510 \, \text{mg CH}_4\text{–C m}^{-2} \text{ day}^{-1}$. This mean falls within the range of other diffusive fluxes observed in Indonesian peatlands ($3.7–87.8 \, \text{mg CH}_4\text{–} \text{C m}^{-2} \text{ day}^{-1}$) (Couwenberg et al., 2010) and other Amazonian wetlands ($7.1–390.0 \, \text{mg CH}_4\text{–C m}^{-2} \text{ day}^{-1}$) (Bartlett et al., 1988, 1990; Devol et al., 1988, 1990). Although the ebullition data must be treated with caution because of the sampling methodology (see below), we observed a mean net ebullition flux of $973.3 \pm 161.4\,\mathrm{mg\,CH_4}$–$\mathrm{C\,m^{-2}\,day^{-1}}$, spanning a range of 27 to $8082\,\mathrm{mg\,CH_4}$–$\mathrm{C\,m^{-2}\,day^{-1}}$. While data on ebullition from Amazonian wetlands are sparse, these values are broadly in line with riverine and lake ecosystems sampled elsewhere (Bastviken et al., 2010; Smith et al., 2000; Sawakuchi et al., 2014). Ebullition-driven CH_4 uptake is not a commonly reported phenomena in other peatland studies because it is likely an artefact of chamber sampling methods; as a consequence, we do not discuss these data further here. To summarize, these data on diffusive CH_4 flux and ebullition suggest that peatlands in the Pastaza–Marañón foreland basin are strong contributors to the regional atmospheric budget of CH_4, given that the four vegetation types sampled here represent the dominant cover types in the PMFB (Draper et al., 2014; Householder et al., 2012; Kelly et al., 2014; Lahteenoja and Page, 2011)

The overall trend in the diffusive flux data was towards greater temporal (i.e., seasonal) variability in diffusive CH_4 flux rather than strong spatial (i.e., inter-site) variability. For the pooled dataset, diffusive CH_4 emissions were significantly greater during the wet season than the dry season, with emissions falling by approximately half from one season to the other (i.e., 51.1 ± 7.0 to $27.3 \pm 2.7\,\mathrm{mg\,CH_4}$–$\mathrm{C\,m^{-2}\,day^{-1}}$). This is in contrast to the data on diffusive CH_4 flux among study sites, where statistical analyses indicate that there was a weak effect of vegetation type on CH_4 flux, that was only on the edge of statistical significance (i.e., ANCOVA; $P < 0.06$ for the vegetation effect term). For the ebullition data, while there was no significant difference among vegetation types or between seasons, it is interesting to note that ebullition was more common for the two vegetation types – mixed palm swamp and *M. flexuosa* palm swamp – that showed the highest rates of diffusive CH_4 flux (Fig. 2, Table 2). In contrast, forested and forested (short pole) vegetation, which showed the lowest rates of diffusive CH_4 flux, also showed the lowest occurrence of ebullition (Fig. 2, Table 2). This is broadly consistent with the notion that mixed palm swamp and *M. flexuosa* palm swamp may produce more CH_4 or possess lower gross CH_4 oxidation rates than the other vegetation types.

At face value, these data on diffusive CH_4 flux suggest two findings: first, the relatively weak effect of vegetation type on diffusive CH_4 flux implies that patterns of CH_4 cycling are broadly similar among study sites. Second, the strong *overall* seasonal pattern suggests that – on the whole – these systems conform to our normative expectations of how peatlands function with respect to seasonal variations in hydrology and redox potential, i.e., enhanced CH_4 emissions during a more anoxic wet season (i.e., when water tables rise) and reduced CH_4 emissions during a more oxic dry season (i.e., when water tables fall). However, closer inspection of the data reveals that different vegetation types showed contrasting seasonal emission patterns (Table 3), challenging our basic assumptions about how these ecosystems function. For example, while forested (short pole) vegetation and

M. flexuosa palm swamp conformed to expected seasonal trends for methanogenic wetlands (i.e., higher wet season compared to dry season emissions), forested vegetation and mixed palm swamp showed the opposite pattern, with significantly greater CH_4 emissions during the dry season. The disaggregated data thus imply that the process-based controls on CH_4 fluxes may vary significantly among these different ecosystems, rather than being similar, leading to a divergence in seasonal flux patterns.

What may explain this pattern of seasonal divergence in CH_4 flux? One explanation is that CH_4 emissions from forested vegetation and mixed palm swamp, compared to the other two ecosystems, may be more strongly transport-limited during the wet season than the dry season. This interpretation is supported by the field data; forested vegetation and mixed palm swamp had the highest wet season water table levels, measuring 110.8 ± 9.3 and $183.7 \pm 1.7\,\mathrm{cm}$, respectively (Table 2). In contrast, water table levels for forested (short pole) vegetation and *M. flexuosa* palm swamp in the wet season were 3–7 times lower, measuring only 26.9 ± 0.5 and $37.2 \pm 1.7\,\mathrm{cm}$, respectively (Table 2). Moreover, a scatter plot of diffusive CH_4 flux against water table depth shows a peak in diffusive CH_4 emissions when water tables are between 30 and $40\,\mathrm{cm}$ above the surface, after which CH_4 emissions decline precipitously (Fig. S2). Thus, the greater depth of overlying water in forested vegetation and mixed palm swamp may have exerted a much greater physical constraint on gas transport compared to the other two ecosystems. This interpretation is broadly consistent with studies from other ecosystems, which indicate that high or positive water tables may suppress CH_4 emissions from wetlands above a system-specific threshold (Couwenberg et al., 2010, 2011).

However, transport limitation alone does not fully explain the difference in dry season CH_4 emissions among vegetation types. Forested vegetation and mixed palm swamp showed substantially higher dry season CH_4 emissions (47.2 ± 5.4 and $85.5 \pm 26.4\,\mathrm{mg\,CH_4}$–$\mathrm{C\,m^{-2}\,day^{-1}}$, respectively) compared to forested (short pole) vegetation and *M. flexuosa* palm swamp (9.6 ± 2.6 and $25.5 \pm 2.9\,\mathrm{mg\,CH_4}$–$\mathrm{C\,m^{-2}\,day^{-1}}$, respectively), pointing to underlying differences in CH_4 production and oxidation among these ecosystems. One possibility is that dry season methanogenesis in forested vegetation and mixed palm swamp was greater than in the other two ecosystems, potentially driven by higher rates of C flow (Whiting and Chanton, 1993). This is plausible given that forested vegetation and mixed palm swamp tend to occur in more nutrient-rich parts of the Pastaza–Marañón foreland basin, whereas forested (short pole) vegetation and *M. flexuosa* palm swamp tend to dominate in more nutrient-poor areas (Lahteenoja et al., 2009a), leading to potential differences in rates of plant productivity and belowground C flow. Moreover, it is possible that the nutrient-rich vegetation may be able to utilize the higher concentration of nutrients, deposited during the flood pulse, during the Amazonian dry season (Morton et al., 2014; Saleska et al., 2016), with implications for overall ecosystem C throughput and CH_4 emissions. Of course, this interpretation does not preclude other explanations, such as differences in CH_4 transport rates among ecosystems (e.g., due to plant-facilitated transport or ebullition) (Panagala et al., 2013) or varying rates of CH_4 oxidation (Teh et al., 2005). However, these other possibilities cannot be explored further without recourse to more detailed process-level experiments. Forthcoming studies on the regulation of GHG fluxes at finer spatial scales (e.g., investigation of environmental gradients within individual study sites) or detailed diurnal studies of GHG exchange (Murphy et al., 2017) will further deepen our understanding of the process controls on soil GHG flux from these peatlands and shed light on these questions.

Finally, while the trends described here are intriguing, it is important to acknowledge some of the potential limitations of our data. First, given the uneven sampling pattern, it is possible that the values reported here do not fully represent the entire range of diffusive flux rates, especially for the more sparsely sampled habitats. However, given the large and statistically significant differences in CH_4 emissions between seasons, it is likely that the main trends that we have identified will hold true with more spatially extensive sampling. Second, the data are a conservative underestimate of CH_4 emissions, because the low-frequency, static chamber sampling approach that we utilized was unable to fully capture erratic ebullition events representatively (McClain et al., 2003). Although we attempted to quantify CH_4 ebullition within our static flux chambers, the sampling approach that we utilized was not the best suited for representatively quantifying ebullition. Given the erratic or stochastic nature of ebullition, automated chamber measurements or an inverted "flux funnel" approach would have provided better estimates of ebullition (Strack et al., 2005). However, we lacked the resources to apply these techniques here. We also did not measure CH_4 emissions from the stems of woody plants, even though woody plants have been recently identified as an important point of atmospheric egress (Pangala et al., 2013). We did not have enough data on floristic composition or individual plant identities within our plots to develop a sampling design that would adequately represent plant-mediated fluxes from our study sites or the resources to implement a separate study of stem fluxes. Third and last, our data probably underestimate net CH_4 fluxes for the PMFB because we chose to include fluxes with strong negative values (i.e., more than $-10\,\mathrm{mg}\,CH_4$–$C\,m^{-2}\,day^{-1}$) in our calculation of mean diffusive flux rates. These observations are more negative than other values typically reported elsewhere in the tropical wetland literature (Bartlett et al., 1988, 1990; Devol et al., 1988, 1990; Couwenberg et al., 2010). However, they represent only a small proportion of our dataset (i.e., 7 %, or only 68 out of 980 measurements), and inspection of our field notes and the data itself did not produce convincing reasons to exclude these observations (e.g., we found no evi-

dence of irregularities during field sampling, and any chambers that showed statistically insignificant changes in concentration over time were removed during our quality control procedures). While headspace concentrations for these measurements were often elevated above mean tropospheric levels ($> 2\,ppm$), this in itself is not unusual in reducing environments that contain strong local sources of CH_4 (Baldocchi et al., 2012). We did not see this as a reason to omit these values as local concentrations of CH_4 are likely to vary naturally in methanogenic forest environments due to poor mixing in the understory and episodic ebullition events. Importantly, exclusion of these data did not alter the overall statistical trends reported above and only produced slightly higher estimates of diffusive CH_4 flux ($41.6 \pm 3.2\,\mathrm{mg}\,CH_4$–$C\,m^{-2}\,day^{-1}$ vs. $36.1 \pm 3.1\,\mathrm{mg}\,CH_4$–$C\,m^{-2}\,day^{-1}$).

4.2 Western Amazonian peatlands as weak atmospheric sources of nitrous oxide

The ecosystems sampled in this study were negligible atmospheric sources of N_2O, emitting only $0.70 \pm 0.34\,\mu g\,N_2O$–$N\,m^{-2}\,day^{-1}$, suggesting that peatlands in the Pastaza–Marañón foreland basin make little or no contribution to regional atmospheric budgets of N_2O. This is consistent with N_2O flux measurements from other forested tropical peatlands, where N_2O emissions were also found to be relatively low (Inubushi et al., 2003; Couwenberg et al., 2010). No statistically significant differences in N_2O flux were observed among study sites or between seasons, suggesting that these different peatlands may have similar patterns of N_2O cycling. Interestingly, differences in N_2O fluxes were not associated with the nutrient status of the peatland; i.e., more nutrient-rich ecosystems, such as forested vegetation and mixed palm swamp, did not show higher N_2O fluxes than their nutrient-poor counterparts, such as forested (short pole) vegetation and *M. flexuosa* palm swamp. This may imply that N availability, one of the principal drivers of nitrification, denitrification, and N_2O production (Groffman et al., 2009; Werner et al., 2007), may not be greater in nutrient-rich versus nutrient-poor ecosystems in this part of the western Amazon. Alternatively, it is possible that even though N availability and N fluxes may differ between nutrient-rich and nutrient-poor systems, N_2O yield may also vary such that net N_2O emissions are not significantly different among study sites (Teh et al., 2014).

One potential source of concern are the negative N_2O fluxes that we documented here. While some investigators have attributed negative fluxes to instrumental error (Cowan et al., 2014; Chapuis-Lardy et al., 2007), others have demonstrated that N_2O consumption – particularly in wetland soils – is not an experimental artifact but occurs due to the complex effects of redox, organic carbon content, nitrate availability, and soil transport processes on denitrification (Ye and Horwath, 2016; Yang et al., 2011; Wen et al., 2016; Schlesinger, 2013; Teh et al., 2014; Chapuis-Lardy et al.,

2007). Given the low redox potential and high carbon content of these soils, it is plausible that microbial N_2O consumption is occurring, because these types of conditions have been found to be conducive for N_2O uptake elsewhere (Ye and Horwath, 2016; Teh et al., 2014; Yang et al., 2011).

5   Conclusions

Our data suggest that peatlands in the Pastaza–Marañón foreland basin are strong sources of atmospheric CH_4 at a regional scale and need to be better accounted for in CH_4 emissions inventories for the Amazon basin as a whole. In contrast, N_2O fluxes were negligible, suggesting that these ecosystems are weak regional sources at best. Divergent or asynchronous seasonal emissions patterns for CH_4 among different vegetation types were intriguing and challenge our underlying expectations of how tropical peatlands function. These data highlight the need for greater wet season sampling, particularly from ecosystems near river margins that may experience very high water tables (i.e., > 40 cm). Moreover, these data also emphasize the need for more spatially extensive sampling across both the Pastaza–Marañón foreland basin and the wider Amazon region as a whole in order to establish if these asynchronous seasonal emission patterns are commonplace or specific to peatlands in the PMFB region. If CH_4 emission patterns for different peatlands in the Amazon are in fact asynchronous and decoupled from rainfall seasonality, then this may partially explain some of the heterogeneity in CH_4 sources and sinks observed at the basin-wide scale (Wilson et al., 2016).

Author contributions. Yit Arn Teh secured the funding for this research, assisted in the planning and design of the experiment, and took the principal role in the analysis of the data and preparation of the manuscript. Wayne A. Murphy planned and designed the experiment, collected the field data, analyzed the samples, and took a secondary role in data preparation, data analysis, and manuscript preparation. Juan-Carlos Berrio, Arnoud Boom, and Susan E. Page supported the planning and design of the experiment and provided substantive input into the writing of the manuscript. Arnoud Boom in particular took a lead role in developing the maps of our study sites in the PMFB.

Competing interests. The authors declare that they have no conflict of interest.

Acknowledgements. The authors would like to acknowledge the UK Natural Environment Research Council for funding this research (NERC award number NE/I015469). We would like to thank MINAG and the Ministerio de Turismo in Iquitos for permits to conduct this research; the Instituto de Investigaciones de la Amazonía Peruana (IIAP) for logistical support; Peruvian rainforest villagers for their warm welcome and acceptance;

Hugo Vasquez, Pierro Vasquez, Gian Carlo Padilla Tenazoa, and Yully Rojas Reátegui for fieldwork assistance; Outi Lahteenoja and Ethan Householder for fieldwork planning; and Paul Beaver of Amazonia Expeditions for lodging and logistical support. Our gratitude also goes to Alex Cumming for fieldwork support and laboratory assistance and Bill Hickin, Gemma Black, Adam Cox, Charlotte Langley, Kerry Allen, and Lisa Barber of the University of Leicester for all of their continued support. Thanks are also owed to Graham Hambley (St Andrews), Angus Calder (St Andrews), Viktoria Oliver (Aberdeen), Torsten Diem (Aberdeen), Tom Kelly (Leeds), and Freddie Draper (Leeds) for their help in the laboratory and with fieldwork planning. Torsten Diem, Viktoria Oliver, and two anonymous referees provided very helpful and constructive comments on earlier drafts of this paper. This publication is a contribution from the Scottish Alliance for Geoscience, Environment and Society (http://www.sages.ac.uk) and the UK Tropical Peatland Working Group (https://tropicalpeat.wordpress.com).

Edited by: Ivonne Trebs

References

Andriesse, J.: Nature and management of tropical peat soils, Food & Agriculture Org., Rome, Italy, 45–59, 1988.

Baggs, E. M.: A review of stable isotope techniques for N_2O source partitioning in soils: Recent progress, remaining challenges and future considerations, Rapid Commun. Mass Spectrom., 22, 1664–1672, 2008.

Baldocchi, D., Detto, M., Sonnentag, O., Verfaillie, J., Teh, Y. A., Silver, W., and Kelly, N. M.: The challenges of measuring methane fluxes and concentrations over a peatland pasture, Agr. Forest Meteorol., 153, 177–187, https://doi.org/10.1016/j.agrformet.2011.04.013, 2012.

Bartlett, K. B., Crill, P. M., Sebacher, D. I., Harriss, R. C., Wilson, J. O., and Melack, J. M.: Methane flux from the Central Amazonian floodplain, J. Geophys. Res.-Atmos., 93, 1571–1582, 1988.

Bartlett, K. B., Crill, P. M., Bonassi, J. A., Richey, J. E., and Harriss, R. C.: Methane flux from the Amazon River floodplain - emissions during rising water, J. Geophys. Res.-Atmos., 95, 16773–16788, https://doi.org/10.1029/JD095iD10p16773, 1990.

Bastviken, D., Santoro, A. L., Marotta, H., Pinho, L. Q., Calheiros, D. F., Crill, P., and Enrich-Prast, A.: Methane Emissions from Pantanal, South America, during the Low Water Season: Toward More Comprehensive Sampling, Environ. Sci. Technol., 44, 5450–5455, https://doi.org/10.1021/es1005048, 2010.

Belyea, L. R. and Baird, A. J.: Beyond "The limits to peat bog growth": Cross-scale feedback in peatland development, Ecol. Monogr., 76, 299–322, 2006.

Blazewicz, S. J., Petersen, D. G., Waldrop, M. P., and Firestone, M. K.: Anaerobic oxidation of methane in tropical and boreal soils: Ecological significance in terrestrial methane cycling, J. Geophys. Res.-Biogeo., 117, 1–9, https://doi.org/10.1029/2011JG001864, 2012.

Chapuis-Lardy, L., Wrage, N., Metay, A., Chotte, J.-L., and Bernoux, M.: Soils, a sink for N_2O? A review, Global Change Biol., 13, 1–17, https://doi.org/10.1111/j.1365-2486.2006.01280.x, 2007.

Conrad, R.: Soil Microorganisms as Controllers of Atmospheric Trace Gases, Microbiol. Rev., 60, 609–640, 1996.

Couwenberg, J., Dommain, R., and Joosten, H.: Greenhouse gas fluxes from tropical peatlands in south-east Asia, Global Change Biol., 16, 1715–1732, https://doi.org/10.1111/j.1365-2486.2009.02016.x, 2010.

Couwenberg, J., Thiele, A., Tanneberger, F., Augustin, J., Bärisch, S., Dubovik, D., Liashchynskaya, N., Michaelis, D., Minke, M., Skuratovich, A., and Joosten, H.: Assessing greenhouse gas emissions from peatlands using vegetation as a proxy, Hydrobiologia, 674, 67–89, https://doi.org/10.1007/s10750-011-0729-x, 2011.

Cowan, N. J., Famulari, D., Levy, P. E., Anderson, M., Reay, D. S., and Skiba, U. M.: Investigating uptake of N_2O in agricultural soils using a high-precision dynamic chamber method, Atmos. Meas. Tech., 7, 4455–4462, https://doi.org/10.5194/amt-7-4455-2014, 2014.

D'Amelio, M. T. S., Gatti, L. V., Miller, J. B., and Tans, P.: Regional N_2O fluxes in Amazonia derived from aircraft vertical profiles, Atmos. Chem. Phys., 9, 8785–8797, https://doi.org/10.5194/acp-9-8785-2009, 2009.

Devol, A. H., Richey, J. E., Clark, W. A., King, S. L., and Martinelli, L. A.: Methane emissions to the troposphere from the Amazon floodplain, J. Geophys. Res.-Atmos., 93, 1583–1592, https://doi.org/10.1029/JD093iD02p01583, 1988.

Devol, A. H., Richey, J. E., Forsberg, B. R., and Martinelli, L. A.: Seasonal dynamics in methane emissions from the Amazon River floodplain to the troposphere, J. Geophys. Res.-Atmos., 95, 16417–16426, https://doi.org/10.1029/JD095iD10p16417, 1990.

Draper, F. C., Roucoux, K. H., Lawson, I. T., Mitchard, E. T. A., Coronado, E. N. H., Lahteenoja, O., Montenegro, L. T., Sandoval, E. V., Zarate, R., and Baker, T. R.: The distribution and amount of carbon in the largest peatland complex in Amazonia, Environ. Res. Lett., 9, 1–12, https://doi.org/10.1088/1748-9326/9/12/124017, 2014.

Espinoza Villar, J. C., Guyot, J. L., Ronchail, J., Cochonneau, G., Filizola, N., Fraizy, P., Labat, D., de Oliveira, E., Ordoñez, J. J., and Vauchel, P.: Contrasting regional discharge evolutions in the Amazon basin (1974–2004), J. Hydrol., 375, 297–311, 2009a.

Espinoza Villar, J. C., Ronchail, J., Guyot, J. L., Cochonneau, G., Naziano, F., Lavado, W., De Oliveira, E., Pombosa, R., and Vauchel, P.: Spatio-temporal rainfall variability in the Amazon basin countries (Brazil, Peru, Bolivia, Colombia, and Ecuador), Int. J. Climatol., 29, 1574–1594, 2009b.

Firestone, M. K. and Davidson, E. A.: Microbiological basis of NO and N_2O production and consumption in soil, in: Exchange of Trace Gases Between Terrestrial Ecosystems and the Atmosphere, edited by: Andrae, M. O. and Schimel, D. S., John Wiley and Sons Ltd., New York, 7–21, 1989.

Firestone, M. K., Firestone, R. B., and Tiedge, J. M.: Nitrous oxide from soil denitrification: Factors controlling its biological production, Science, 208, 749–751, 1980.

Groffman, P. M., Butterbach-Bahl, K., Fulweiler, R. W., Gold, A. J., Morse, J. L., Stander, E. K., Tague, C., Tonitto, C., and Vidon, P.: Challenges to incorporating spatially and temporally explicit phenomena (hotspots and hot moments) in denitrification models, Biogeochemistry, 93, 49–77, https://doi.org/10.1007/s10533-008-9277-5, 2009.

Hanson, R. S. and Hanson, T. E.: Methanotrophic Bacteria, Microbiol. Rev., 60, 439–471, 1996.

Householder, J. E., Janovec, J., Tobler, M., Page, S., and Lähteenoja, O.: Peatlands of the Madre de Dios River of Peru: Distribution, Geomorphology, and Habitat Diversity, Wetlands, 32, 359–368, https://doi.org/10.1007/s13157-012-0271-2, 2012.

Huang, J., Golombek, A., Prinn, R., Weiss, R., Fraser, P., Simmonds, P., Dlugokencky, E. J., Hall, B., Elkins, J., Steele, P., Langenfelds, R., Krummel, P., Dutton, G., and Porter, L.: Estimation of regional emissions of nitrous oxide from 1997 to 2005 using multinetwork measurements, a chemical transport model, and an inverse method, J. Geophys. Res.-Atmos., 113, D17313, https://doi.org/10.1029/2007jd009381, 2008.

Inubushi, K., Furukawa, Y., Hadi, A., Purnomo, E., and Tsuruta, H.: Seasonal changes of CO_2, CH_4 and N_2O fluxes in relation to land-use change in tropical peatlands located in coastal area of South Kalimantan, Chemosphere, 52, 603–608, https://doi.org/10.1016/s0045-6535(03)00242-x, 2003.

Jungkunst, H. F. and Fiedler, S.: Latitudinal differentiated water table control of carbon dioxide, methane and nitrous oxide fluxes from hydromorphic soils: feedbacks to climate change, Global Change Biol., 13, 2668–2683, https://doi.org/10.1111/j.1365-2486.2007.01459.x, 2007.

Junk, W.: Flood tolerance and tree distribution in central Amazonian floodplains, in: Tropical forests: Botanical dynamics, speciation and diversity, Academic Press, London, UK, 47–64, 1989.

Keller, M., Kaplan, W. A., and Wofsy, S. C.: Emissons of N_2O, CH_4 and CO_2 from tropical forest soils, J. Geophys. Res.-Atmos., 91, 1791–1802, https://doi.org/10.1029/JD091iD11p11791, 1986.

Kelly, T. J., Baird, A. J., Roucoux, K. H., Baker, T. R., Honorio Coronado, E. N., Ríos, M., and Lawson, I. T.: The high hydraulic conductivity of three wooded tropical peat swamps in northeast Peru: measurements and implications for hydrological function, Hydrol. Process., 28, 3373–3387, https://doi.org/10.1002/hyp.9884, 2014.

Kirschke, S., Bousquet, P., Ciais, P., Saunois, M., Canadell, J. G., Dlugokencky, E. J., Bergamaschi, P., Bergmann, D., Blake, D. R., Bruhwiler, L., Cameron-Smith, P., Castaldi, S., Chevallier, F., Feng, L., Fraser, A., Heimann, M., Hodson, E. L., Houweling, S., Josse, B., Fraser, P. J., Krummel, P. B., Lamarque, J. F., Langenfelds, R. L., Le Quere, C., Naik, V., O'Doherty, S., Palmer, P. I., Pison, I., Plummer, D., Poulter, B., Prinn, R. G., Rigby, M., Ringeval, B., Santini, M., Schmidt, M., Shindell, D. T., Simpson, I. J., Spahni, R., Steele, L. P., Strode, S. A., Sudo, K., Szopa, S., van der Werf, G. R., Voulgarakis, A., van Weele, M., Weiss, R. F., Williams, J. E., and Zeng, G.: Three decades of global methane sources and sinks, Nat. Geosci., 6, 813–823, https://doi.org/10.1038/ngeo1955, 2013.

Lahteenoja, O. and Page, S.: High diversity of tropical peatland ecosystem types in the Pastaza-Maranon basin, Peruvian Amazonia, J. Geophys. Res.-Biogeo., 116, 1–14, https://doi.org/10.1029/2010jg001508, 2011.

Lähteenoja, O. and Roucoux, K.: Inception, history and development of peatlands in the Amazon Basin, PAGES News, 18, 27–31, 2010.

Lahteenoja, O., Ruokolainen, K., Schulman, L., and Alvarez, J.: Amazonian floodplains harbour minerotrophic and ombrotrophic peatlands, Catena, 79, 140–145, https://doi.org/10.1016/j.catena.2009.06.006, 2009a.

Lahteenoja, O., Ruokolainen, K., Schulman, L., and Oinonen, M.: Amazonian peatlands: an ignored C sink and potential source, Global Change Biol., 15, 2311–2320, https://doi.org/10.1111/j.1365-2486.2009.01920.x, 2009b.

Lahteenoja, O., Reategui, Y. R., Rasanen, M., Torres, D. D., Oinonen, M., and Page, S.: The large Amazonian peatland carbon sink in the subsiding Pastaza-Maranon foreland basin, Peru, Global Change Biol., 18, 164–178, https://doi.org/10.1111/j.1365-2486.2011.02504.x, 2012.

Lavelle, P., Rodriguez, N., Arguello, O., Bernal, J., Botero, C., Chaparro, P., Gomez, Y., Gutierrez, A., Hurtado, M. D., Loaiza, S., Pullido, S. X., Rodriguez, E., Sanabria, C., Velasquez, E., and Fonte, S. J.: Soil ecosystem services and land use in the rapidly changing Orinoco River Basin of Colombia, Agr. Ecosyst. Environ., 185, 106–117, https://doi.org/10.1016/j.agee.2013.12.020, 2014.

Liengaard, L., Nielsen, L. P., Revsbech, N. P., Priem, A., Elberling, B., Enrich-Prast, A., and Kuhl, M.: Extreme emission of N_2O from tropical wetland soil (Pantanal, South America), Front. Microbiol., 3, 1–13, https://doi.org/10.3389/fmicb.2012.00433, 2013.

Limpens, J., Berendse, F., Blodau, C., Canadell, J. G., Freeman, C., Holden, J., Roulet, N., Rydin, H., and Schaepman-Strub, G.: Peatlands and the carbon cycle: from local processes to global implications – a synthesis, Biogeosciences, 5, 1475–1491, https://doi.org/10.5194/bg-5-1475-2008, 2008.

Livingston, G. and Hutchinson, G.: Chapter 2: Enclosure-based measurement of trace gas exchange: applications and sources of error, in: Biogenic Trace Gases: Measuring Emissions from Soil and Water, edited by: Matson, P. and Harriss, R. C., Blackwell Science Ltd, Cambridge, MA, USA, 14–51, 1995.

Marani, L. and Alvalá, P. C.: Methane emissions from lakes and floodplains in Pantanal, Brazil, Atmos. Environ., 41, 1627–1633, https://doi.org/10.1016/j.atmosenv.2006.10.046, 2007.

McClain, M. E., Boyer, E. W., Dent, C. L., Gergel, S. E., Grimm, N. B., Groffman, P. M., Hart, S. C., Harvey, J. W., Johnston, C. A., Mayorga, E., McDowell, W. H., and Pinay, G.: Biogeochemical hot spots and hot moments at the interface of terrestrial and aquatic ecosystems, Ecosystems, 6, 301–312, https://doi.org/10.1007/s10021-003-0161-9, 2003.

Melack, J. M., Hess, L. L., Gastil, M., Forsberg, B. R., Hamilton, S. K., Lima, I. B. T., and Novo, E.: Regionalization of methane emissions in the Amazon Basin with microwave remote sensing, Global Change Biol., 10, 530–544, https://doi.org/10.1111/j.1529-8817.2003.00763.x, 2004.

Melton, J. R., Wania, R., Hodson, E. L., Poulter, B., Ringeval, B., Spahni, R., Bohn, T., Avis, C. A., Beerling, D. J., Chen, G., Eliseev, A. V., Denisov, S. N., Hopcroft, P. O., Lettenmaier, D. P., Riley, W. J., Singarayer, J. S., Subin, Z. M., Tian, H., Zurcher, S., Brovkin, V., van Bodegom, P. M., Kleinen, T., Yu, Z. C., and Kaplan, J. O.: Present state of global wetland extent and wetland methane modelling: conclusions from a model intercomparison project (WETCHIMP), Biogeosciences, 10, 753–788, https://doi.org/10.5194/bg-10-753-2013, 2013.

Morley, N. and Baggs, E. M.: Carbon and oxygen controls on N_2O and N_2 production during nitrate reduction, Soil Biol. Biochem., 42, 1864–1871, https://doi.org/10.1016/j.soilbio.2010.07.008, 2010.

Morton, D. C., Nagol, J., Carabajal, C. C., Rosette, J., Palace, M., Cook, B. D., Vermote, E. F., Harding, D. J., and North, P. R. J.: Amazon forests maintain consistent canopy structure and greenness during the dry season, Nature, 506, 221–224, https://doi.org/10.1038/nature13006, 2014.

Murphy, W. A., Berrio, J. C., Boom, A., Page, S. E., and Teh, Y. A.: Spatial and diurnal trends in methane and nitrous oxide flux within peatland ecosystems, in: Methane emissions from peat swamp forests of differing vegetation types, within the Loreto Region of the Peruvian Amazonas – unpublished PhD thesis, University of Leicester, Leicester, 2017.

Nisbet, E. G., Dlugokencky, E. J., and Bousquet, P.: Methane on the Rise – Again, Science, 343, 493–495, https://doi.org/10.1126/science.1247828, 2014.

Pangala, S. R., Moore, S., Hornibrook, E. R. C., and Gauci, V.: Trees are major conduits for methane egress from tropical forested wetlands, New Phytol., 197, 524–531, https://doi.org/10.1111/nph.12031, 2013.

Pett-Ridge, J., Petersen, D. G., Nuccio, E., and Firestone, M. K.: Influence of oxic/anoxic fluctuations on ammonia oxidizers and nitrification potential in a wet tropical soil, FEMS Microbiol. Ecol., 85, 179–194, https://doi.org/10.1111/1574-6941.12111, 2013.

Prosser, J. I. and Nicol, G. W.: Relative contributions of archaea and bacteria to aerobic ammonia oxidation in the environment, Environ. Microbiol., 10, 2931–2941, https://doi.org/10.1111/j.1462-2920.2008.01775.x, 2008.

Pumpanen, J., Kolari, P., Ilvesniemi, H., Minkkinen, K., Vesala, T., Niinistö, S., Lohila, A., Larmola, T., Morero, M., Pihlatie, M., Janssens, I., Yuste, J. C., Grünzweig, J. M., Reth, S., Subke, J.-A., Savage, K., Kutsch, W., Østreng, G., Ziegler, W., Anthoni, P., Lindroth, A., and Hari, P.: Comparison of different chamber techniques for measuring soil CO_2 efflux, Agr. Forest Meteorol., 123, 159–176, https://doi.org/10.1016/j.agrformet.2003.12.001, 2004.

Saikawa, E., Schlosser, C. A., and Prinn, R. G.: Global modeling of soil nitrous oxide emissions from natural processes, Global Biogeochem. Cy., 27, 972–989, https://doi.org/10.1002/gbc.20087, 2013.

Saikawa, E., Prinn, R. G., Dlugokencky, E., Ishijima, K., Dutton, G. S., Hall, B. D., Langenfelds, R., Tohjima, Y., Machida, T., Manizza, M., Rigby, M., O'Doherty, S., Patra, P. K., Harth, C. M., Weiss, R. F., Krummel, P. B., van der Schoot, M., Fraser, P. J., Steele, L. P., Aoki, S., Nakazawa, T., and Elkins, J. W.: Global and regional emissions estimates for N_2O, Atmos. Chem. Phys., 14, 4617–4641, https://doi.org/10.5194/acp-14-4617-2014, 2014.

Saleska, S. R., Wu, J., Guan, K., Araujo, A. C., Huete, A., Nobre, A. D., and Restrepo-Coupe, N.: Dry-season greening of Amazon forests, Nature, 531, E4–E5, https://doi.org/10.1038/nature16457, 2016.

Sawakuchi, H. O., Bastviken, D., Sawakuchi, A. O., Krusche, A. V., Ballester, M. V. R., and Richey, J. E.: Methane emissions from Amazonian Rivers and their contribution to the global methane budget, Global Change Biol., 20, 2829–2840, https://doi.org/10.1111/gcb.12646, 2014.

Schlesinger, W. H.: An estimate of the global sink for nitrous oxide in soils, Global Change Biol., 19, 2929–2931, https://doi.org/10.1111/gcb.12239, 2013.

Schulman, L., Ruokolainen, K., and Tuomisto, H.: Parameters for global ecosystem models, Nature, 399, 535–536, 1999.

Silver, W. L., Lugo, A., and Keller, M.: Soil oxygen availability and biogeochemistry along rainfall and topographic gradients in upland wet tropical forest soils, Biogeochemistry, 44, 301–328, 1999.

Silver, W. L., Herman, D. J., and Firestone, M. K. S.: Dissimilatory Nitrate Reduction to Ammonium in Upland Tropical Forest Soils, Ecology, 82, 2410–2416, 2001.

Sjögersten, S., Black, C. R., Evers, S., Hoyos-Santillan, J., Wright, E. L., and Turner, B. L.: Tropical wetlands: A missing link in the global carbon cycle?, Global Biogeochem. Cy., 28, 1371–1386, https://doi.org/10.1002/2014GB004844, 2014.

Smith, L. K., Lewis, W. M., Chanton, J. P., Cronin, G., and Hamilton, S. K.: Methane emissions from the Orinoco River floodplain, Venezuela, Biogeochemistry, 51, 113–140, 2000.

Strack, M., Kellner, E., and Waddington, J. M.: Dynamics of biogenic gas bubbles in peat and their effects on peatland biogeochemistry, Global Biogeochem. Cy., 19, 1–9, https://doi.org/10.1029/2004GB002330, 2005.

Teh, Y. A. and Silver, W. L.: Effects of soil structure destruction on methane production and carbon partitioning between methanogenic pathways in tropical rain forest soils, J. Geophys. Res.-Biogeo., 111, 1–8, https://doi.org/10.1029/2005JG000020, 2006.

Teh, Y. A., Silver, W. L., and Conrad, M. E.: Oxygen effects on methane production and oxidation in humid tropical forest soils, Global Change Biol., 11, 1283–1297, https://doi.org/10.1111/j.1365-2486.2005.00983.x, 2005.

Teh, Y. A., Silver, W. L., Conrad, M. E., Borglin, S. E., and Carlson, C. M.: Carbon isotope fractionation by methane-oxidizing bacteria in tropical rain forest soils, J. Geophys. Res.-Biogeo., 111, 1–8, https://doi.org/10.1029/2005jg000053, 2006.

Teh, Y. A., Dubinsky, E. A., Silver, W. L., and Carlson, C. M.: Suppression of methanogenesis by dissimilatory Fe(III)-reducing bacteria in tropical rain forest soils: implications for ecosystem methane flux, Global Change Biol., 14, 413–422, https://doi.org/10.1111/j.1365-2486.2007.01487.x, 2008.

Teh, Y. A., Silver, W. L., Sonnentag, O., Detto, M., Kelly, M., and Baldocchi, D. D.: Large Greenhouse Gas Emissions from a Temperate Peatland Pasture, Ecosystems, 14, 311–325, https://doi.org/10.1007/s10021-011-9411-4, 2011.

Teh, Y. A., Diem, T., Jones, S., Huaraca Quispe, L. P., Baggs, E., Morley, N., Richards, M., Smith, P., and Meir, P.: Methane and nitrous oxide fluxes across an elevation gradient in the tropical Peruvian Andes, Biogeosciences, 11, 2325–2339, https://doi.org/10.5194/bg-11-2325-2014, 2014.

Tian, H., Melillo, J. M., Kicklighter, D. W., McGuire, A. D., Helfrich III, J. V. K., Moore III, B., and Vorosmarty, C. J.: Effect of interannual climate variability on carbon storage in Amazonian ecosystems, Nature, 396, 664–667, 1998.

von Fischer, J. and Hedin, L.: Separating methane production and consumption with a field-based isotope dilution technique, Global Biogeochem. Cy., 16, 1–13, https://doi.org/10.1029/2001GB001448, 2002.

von Fischer, J. C. and Hedin, L. O.: Controls on soil methane fluxes: Tests of biophysical mechanisms using stable isotope tracers, Global Biogeochem. Cy., 21, 9, Gb2007, https://doi.org/10.1029/2006gb002687, 2007.

Wen, Y., Chen, Z., Dannenmann, M., Carminati, A., Willibald, G., Kiese, R., Wolf, B., Veldkamp, E., Butterbach-Bahl, K., and Corre, M. D.: Disentangling gross N_2O production and consumption in soil, Sci. Rep., 6, 8, https://doi.org/10.1038/srep36517, 2016.

Werner, C., Butterbach-Bahl, K., Haas, E., Hickler, T., and Kiese, R.: A global inventory of N_2O emissions from tropical rainforest soils using a detailed biogeochemical model, Global Biogeochem. Cy., 21, Gb3010, https://doi.org/10.1029/2006gb002909, 2007.

Whalen, S. C.: Biogeochemistry of methane exchange between natural wetlands and the atmosphere, Environ. Eng. Sci., 22, 73–94, https://doi.org/10.1089/ees.2005.22.73, 2005.

Whiting, G. J. and Chanton, J. P.: Primary production control of methane emission from wetlands, Nature, 364, 794–795, 1993.

Wilson, C., Gloor, M., Gatti, L. V., Miller, J. B., Monks, S. A., McNorton, J., Bloom, A. A., Basso, L. S., and Chipperfield, M. P.: Contribution of regional sources to atmospheric methane over the Amazon Basin in 2010 and 2011, Global Biogeochem. Cy., 30, 400–420, https://doi.org/10.1002/2015GB005300, 2016.

Wright, E. L., Black, C. R., Cheesman, A. W., Drage, T., Large, D., Turner, B. L., and SjÖGersten, S.: Contribution of subsurface peat to CO_2 and CH_4 fluxes in a neotropical peatland, Global Change Biol., 17, 2867–2881, https://doi.org/10.1111/j.1365-2486.2011.02448.x, 2011.

Yang, W. H., Teh, Y. A., and Silver, W. L.: A test of a field-based N_{15}-nitrous oxide pool dilution technique to measure gross N_2O production in soil, Global Change Biol., 17, 3577–3588, https://doi.org/10.1111/j.1365-2486.2011.02481.x, 2011.

Ye, R. and Horwath, W. R.: Nitrous oxide uptake in rewetted wetlands with contrasting soil organic carbon contents, Soil Biol. Biochem., 100, 110–117, https://doi.org/10.1016/j.soilbio.2016.06.009, 2016.

7

Global high-resolution monthly $p\mathrm{CO_2}$ climatology for the coastal ocean derived from neural network interpolation

Goulven G. Laruelle[1], **Peter Landschützer**[2], **Nicolas Gruber**[3], **Jean-Louis Tison**[1], **Bruno Delille**[4], **and Pierre Regnier**[1]

[1]Department Geoscience, Environment & Society (DGES), Université Libre de Bruxelles, Bruxelles, Belgium
[2]Max Planck Institute for Meteorology, Bundesstr. 53, Hamburg, Germany
[3]Environmental Physics, Institute of Biogeochemistry and Pollutant Dynamics, ETH Zürich, Zürich, Switzerland
[4]Unité d'Oceanographie Chimique, Astrophysics, Geophysics and Oceanography department, University of Liège, Liège, Belgium

Correspondence to: Goulven G. Laruelle (goulven.gildas.laruelle@ulb.ac.be)

Abstract. In spite of the recent strong increase in the number of measurements of the partial pressure of CO_2 in the surface ocean ($p\mathrm{CO_2}$), the air–sea CO_2 balance of the continental shelf seas remains poorly quantified. This is a consequence of these regions remaining strongly under-sampled in both time and space and of surface $p\mathrm{CO_2}$ exhibiting much higher temporal and spatial variability in these regions compared to the open ocean. Here, we use a modified version of a two-step artificial neural network method (SOM-FFN; Landschützer et al., 2013) to interpolate the $p\mathrm{CO_2}$ data along the continental margins with a spatial resolution of $0.25°$ and with monthly resolution from 1998 to 2015. The most important modifications compared to the original SOM-FFN method are (i) the much higher spatial resolution and (ii) the inclusion of sea ice and wind speed as predictors of $p\mathrm{CO_2}$. The SOM-FFN is first trained with $p\mathrm{CO_2}$ measurements extracted from the SO-CATv4 database. Then, the validity of our interpolation, in both space and time, is assessed by comparing the generated $p\mathrm{CO_2}$ field with independent data extracted from the LD-VEO2015 database. The new coastal $p\mathrm{CO_2}$ product confirms a previously suggested general meridional trend of the annual mean $p\mathrm{CO_2}$ in all the continental shelves with high values in the tropics and dropping to values beneath those of the atmosphere at higher latitudes. The monthly resolution of our data product permits us to reveal significant differences in the seasonality of $p\mathrm{CO_2}$ across the ocean basins. The shelves of the western and northern Pacific, as well as the shelves in the temperate northern Atlantic, display particularly pronounced seasonal variations in $p\mathrm{CO_2}$, while the shelves in the south-eastern Atlantic and in the southern Pacific reveal a much smaller seasonality. The calculation of temperature normalized $p\mathrm{CO_2}$ for several latitudes in different oceanic basins confirms that the seasonality in shelf $p\mathrm{CO_2}$ cannot solely be explained by temperature-induced changes in solubility but are also the result of seasonal changes in circulation, mixing and biological productivity. Our results also reveal that the amplitudes of both thermal and nonthermal seasonal variations in $p\mathrm{CO_2}$ are significantly larger at high latitudes. Finally, because this product's spatial extent includes parts of the open ocean as well, it can be readily merged with existing global open-ocean products to produce a true global perspective of the spatial and temporal variability of surface ocean $p\mathrm{CO_2}$.

1 Introduction

The quantitative contribution of the coastal ocean to the global oceanic uptake of atmospheric CO_2 is still being debated (Borges et al., 2005; Chen and Borges, 2009; Cai, 2011; Wanninkhof et al., 2013; Gruber, 2015), but several recent studies have suggested that the flux density, or uptake per unit area, is greater over continental shelf seas than over the open ocean (Chen et al., 2013; Laruelle et al., 2014). Laruelle et al. (2014) used more than 3×10^6 $p\mathrm{CO_2}$ measurements from the SOCATv2 database (Pfeil et al., 2014 Bakker et al., 2016) to demonstrate very strong disparities in air–seawater CO_2 exchange at the regional scale and pronounced seasonal variations, especially at temperate latitudes. Furthermore, it was suggested that, despite the presence of a seasonally varying sea-ice cover, Arctic continental shelves are a regional hotspot of CO_2 uptake (Bates et al., 2006; Laruelle et al., 2014; Yasunaka et al., 2016). However, even with this much larger dataset compared to previous studies, large regions of the global coastal ocean remained either void of data or very poorly monitored in space and time, including the seasonal cycle. These data gaps not only limit our ability to reduce uncertainties in flux estimates and to unravel whether they differ from the adjacent open ocean but also hamper the identification and quantification of the many processes controlling the source–sink nature of the coastal ocean (Bauer et al., 2013). Laruelle et al. (2014) attempted to overcome this limitation by combining various upscaling methods depending on data density in different regions, e.g., resorted to using annual means, wherever the seasonal coverage was deemed to be insufficient. However, they could not overcome the limitation that the data alone are insufficient to assess whether there are any trends in coastal fluxes. This is a serious gap when considering that the influence of human activity on coastal system is increasing rapidly (Doney, 2010; Cai, 2011; Regnier et al., 2013; Gruber, 2015).

In the open ocean, novel statistical methods relying on artificial neural networks (ANNs) have permitted the generation of a series of high-resolution continuous monthly maps for ocean surface CO_2 partial pressures (pCO_2) (e.g., Landschützer et al., 2013; Sasse et al., 2013; Nakaoka et al., 2013; Zeng et al., 2014). Although differing in their details (see, e.g., Rödenbeck et al., 2015, for an overview), these products typically have a nominal spatial resolution of 1° and monthly temporal resolution. By filling in the spatial and temporal gaps, these products greatly facilitate the calculation of the air–sea CO_2 exchange, as they do not require separate assumptions about the surface ocean pCO_2 in areas lacking data. Such methods are well suited to resolve spatial gradients, and they also permit us to determine seasonal and interannual variations and trends in pCO_2 (e.g., Landschützer et al., 2014, 2015, 2016; Zeng et al., 2014). Because of the small relative contribution of the coastal ocean to the total oceanic surface area and the relatively coarse spatial resolution of the ANN-based surface ocean pCO_2 products so far, they are not well suited to resolve the high spatiotemporal variations of the surface ocean pCO_2 fields along the shelves.

Reproducing the complex seasonal dynamics of the CO_2 exchange at the air–water interface in the coastal ocean is of particular importance considering that they often have large intra-annual variability (Signorini et al., 2013). For instance, in temperate climates, it is common for continental shelf waters to turn from CO_2 sinks for the atmosphere during spring to CO_2 sources during summer (Shadwick et al., 2010; Cai, 2011; Laruelle et al., 2014, 2015). Shelf waters are also typically characterized by small-scale physical features such as coastal currents, river plumes and eddies inducing sharp bio-

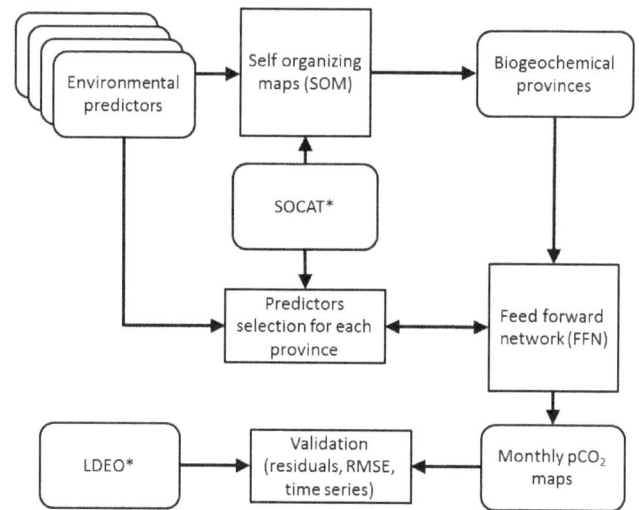

Figure 1. Schematic scheme of the different steps involved in the SOM-FFN artificial neural network calculations leading to continuous monthly pCO_2 maps over the 1998–2015 period.

geochemical fronts (Liu et al., 2010) that markedly influence the spatial patterns of the pCO_2 fields (e.g., Turi et al., 2014).

To resolve the high spatial and temporal variability in air–sea CO_2 exchange over the global shelf region, the two-step ANN method developed by Landschützer et al. (2013) is modified here for the specific conditions that prevail in these environments. Our calculations are performed at a much finer resolution of 0.25° and new environmental drivers such as sea-ice cover are used to account for the potentially significant role of sea ice in the CO_2 exchange (Bates et al., 2006; Vancoppenolle et al., 2013; Parmentier et al., 2013; Moreau et al., 2016; Grimm et al., 2016). The definition of the coastal/open-oceanic boundary varies strongly from one study to the other (Walsh, 1988; Laruelle et al., 2013), with a potentially large impact on the shelf CO_2 budget (Laruelle et al., 2010). Here, we use a very wide definition for this boundary (i.e., 300 km width or 1000 m depth) to secure spatial continuity between our new shelf pCO_2 product and those already existing for the open ocean (Landschützer et al., 2013, 2016; Rödenbeck et al., 2015). Our approach leads to the first continuous and monthly resolved pCO_2 maps over the 1998–2015 period across the global shelf region, permitting us to study the seasonal dynamics of these regions in relationship to that of the adjacent open ocean.

2 Methods

The method used in this study is a modified version of the SOM-FFN method developed by Landschützer et al. (2013) to calculate monthly resolved pCO_2 maps of the Atlantic Ocean at a 1° resolution and later applied to the entire global open ocean (Landschützer et al., 2014). The reconstruction of a continuous pCO_2 field involves establishing

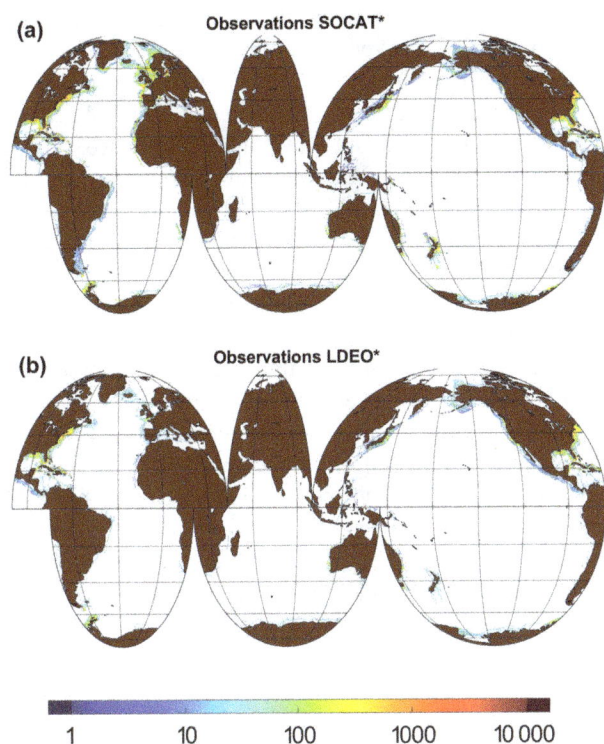

Figure 2. Number of observations contained in each $0.25°$ grid cell of the SOCAT* (a) and LDEO* (b) databases.

numerical relationships between pCO_2 and a number of independent environmental predictors that are known to control its variability in both time and space. The first step of the method relies on the use of a neural network clustering algorithm (self-organizing map, SOM) to define a discrete set of biogeochemical provinces characterized by similar relationships between the independent environmental variables and a monthly resolved pCO_2 field. The second step consists in deriving nonlinear and continuous relationships between pCO_2 and some or all of the aforementioned independent variables using a feed-forward network (FFN) method within each biogeochemical province created by the SOM. The method is extensively documented in Landschützer et al. (2013, 2014) but the specific modifications introduced in this study to better simulate the characteristics of the shelves, the choice of environmental drivers, their data sources and the definition of the geographic extent of this analysis are described in the following sections. Figure 1 summarizes the different steps involved in the calculations of the SOM-FFN.

2.1 Data sources and processing

All the datasets used in our calculations were converted from their original spatial resolutions to a regular $0.25°$ resolution grid. The temporal resolution of all datasets is monthly (i.e., 216 months over the entire period) except for the bathymetry, which is assumed constant over the course of the simulations,

and wind speed, whose original resolution is 6 h. For the latter, monthly averages are calculated for each grid cell to generate monthly values. Sea-surface temperature (SST) and sea-surface salinity (SSS) maps were taken from the Met Office's EN4, which consists of quality-controlled subsurface ocean temperature and salinity profiles and their objective analyses (Good et al., 2013). The bathymetry was extracted from the global ETOPO2 database (US Department of Commerce, 2006). The sea-ice concentrations were taken from the global 25 km resolution monthly data product compiled by the NSIDC (National Snow and Ice Cover Data; Cavalieri et al., 1996). Wind speed data were extracted from ERA-Interim reanalysis (Dee et al., 2011). The chlorophyll surface concentrations were extracted from the monthly 9 km resolution SeaWIFS data product prior to 2010 and from MODIS for later years (NASA, 2016). The list of all data products used in the calculations and the transformations applied to produce monthly $0.25°$ resolution forcing files are summarized in Table 1.

Finally, the surface ocean pCO_2 were taken from the gridded SOCATv4 product (Sabine et al., 2013; Bakker et al., 2016) while those used for the validation stem from the LDEOv2015 database (Takahashi et al., 2016). With our definition of the coastal zone, SOCATv4 contains $\sim 8 \times 10^6$ data points and LDEO $\sim 5.6 \times 10^6$, with over 70 % of the data shared with SOCATv4. Because of this significant overlap between both data products, we created two entirely independent datasets by randomly assigning each of those common data point to either database to ensure that each data only belongs to one dataset. The resulting datasets are named SOCAT* and LDEO*, with the former being used for training and the latter for validation. Prior to the creation of both datasets, all data from SOCAT were converted from fCO_2 (fugacity of CO_2 in water) to pCO_2 using the formulation reported in Takahashi et al. (2012). The data densities of SOCAT* and LDEO* are shown in Fig. 2 and reveal a heterogeneous spatial coverage. Northern temperate shelves are generally well covered, especially in the northern Atlantic. In this region, the data density is better in SOCAT* than LDEO* thanks to the addition of many European cruises in the SOCAT database. In contrast, equatorial regions remain undersampled, especially in the Indian Ocean. Because of the difficulty of sampling in waters seasonally covered in ice, polar regions are very unevenly represented in SOCAT* and LDEO*. Luckily, some areas, such as some parts of Antarctica and Bering Sea do contain enough data to train and validate the SOM-FFN. Overall SOCAT* contains roughly 40 % more data than LDEO*.

2.2 Modifications of the SOM-FFN method

The specific characteristics of the continental shelves motivated a number of modifications of the global ocean SOM-FFN method, including a 16-fold increase in spatial resolution from 1 to $0.25°$, the addition of new environmental variables as biogeochemical predictors and a shortening of the simulation period to the period of 1998 through 2015. All these modifications are detailed here below.

The higher resolution of $0.25° \times 0.25°$ results in over 2 million grid cells that help to better track the global coastline and its complex geomorphological features (Crossland et al., 2005; Liu, 2010). It is also common along eastern and western boundary currents to find continental shelves as narrow as 10–20 km, i.e., an extension that is significantly smaller than a single cell at $1°$ resolution. Additionally, biogeochemical fronts associated with river plumes, coastal currents and upwelling are characterized by spatial scales of the order of tens of kilometers or even smaller (Wijesekera et al., 2003). The chosen resolution is also identical to the gridded coastal pCO_2 product from the SOCAT initiative (Sabine et al., 2013; Bakker et al., 2014).

The definition of the geographic extent of the shelf region excludes estuaries and other inland water bodies but uses a wide limit for the outer continental shelf that encapsulates all current definitions of the coastal ocean. This approach facilitates future integration with existing global ocean data products (e.g., Landschützer et al., 2016; Rödenbeck et al., 2015) and model outputs, which typically struggle to represent the shallowest parts of the ocean (Bourgeois et al., 2016). The outer limit used here is given by whichever point is the furthest from the coast: either 300 km distance from the coastline (which roughly corresponds to the outer edge of territorial waters; Crossland et al., 2005) or the 1000 m isobaths (Laruelle et al., 2013). The resulting domain (Fig. B in the Supplement) covers 77 million km², more than twice the surface area generally attributed to the coastal ocean (Walsh et al., 1998; Liu et al., 2010; Laruelle et al., 2013).

The predictor variables for the SOM-FFN networks were chosen based on a set of trial-and-error experiments with the selection criteria being the quality of fit, i.e., the best reconstruction of the available observations. The first step of the SOM-FFN calculations, i.e., the SOM-based clustering, relies on the assignment of the surface ocean data to biogeochemical provinces sharing common spatiotemporal patterns of SST, SSS, bathymetry, rate of change in sea-ice coverage, wind speed and observed pCO_2. Chlorophyll a is not included in the list of environmental factors used to generate the biogeochemical provinces because of the incomplete data coverage at high latitude in winter due to cloud coverage. Both the use of wind speed and the rate of change in monthly sea-ice concentration are novelties compared to the setup of Landschützer et al. (2013). The latter is calculated from the gridded monthly sea-ice concentration field of Cavalieri et al. (1996). It allows us to account for the complex processes occurring in melting and forming sea ice that are known to strongly influence the dynamics of the carbon within sea-ice-covered areas (Parmentier et al., 2013). This first step is performed without any data normalization of the datasets, as this permits us to give more weight to the pCO_2 data. Based on a series of simulations using different numbers of biogeo-

chemical provinces, we found that a clustering of the data into 10 biogeochemical provinces minimized the average deviation between simulated and observed pCO_2 (see below) while ensuring that at least 1000 different grid cells can be used for validation against LDEO* in each province.

In the second step of the estimation procedure, i.e., the application of the FFN, SST, SSS, bathymetry, sea-ice concentration and chlorophyll a are used as predictors to establish the nonlinear relationships between these predictors and the target pCO_2 (for data sources, see below). Similar to the SOM in step one, the selected variables not only comprise proxies representing the solubility and biological pumps of the coastal ocean but also yield the best fit to the data. These calculations are done iteratively using a sigmoid activation function on an incomplete dataset in order to perform an assessment on the remaining data after each iteration, until an optimal relationship is found. Additionally, as performed in Landschützer et al. (2015), the output pCO_2 data were smoothed using the spatial and temporal mean of each point's neighboring pixels in both time and space within the 3-pixel neighborhood domain. This operation is performed iteratively and does not significantly alter the results, but it ensures smoother transitions in the pCO_2 field at the boundaries between the provinces. This smoothing method yielded good results for the open Southern Ocean where marked pCO_2 fronts are also observed (Landschützer et al., 2015) and was deemed relevant here due to the potentially strong pCO_2 gradients characterizing the shelves.

Another change from the most recent global ocean SOM-FFN application (Landschützer et al., 2016) is the different temporal extension of the simulation period, which covers the period from 1998 to 2015 instead of 1982 to 2011. This overall shortening was necessary because one of environmental driver, i.e., chlorophyll data derived from SeaWIFS, only starts in September 1997 (NASA, 2016). Monthly chlorophyll data throughout the entire simulation period were preferred here over the use of a monthly climatology as done in Landschützer et al. (2016) to better capture interannual variability. At the same time, we have been able to extend the coastal product by 4 years to the end of 2015.

2.3 Model training and evaluation

We evaluated the coastal SOM-FFN product using the root mean square error (RMSE) metric, calculated as the difference between estimated and observed pCO_2. During the early development stage, preliminary simulations were performed using only data from SOCAT v2.0 (Pfeil et al., 2013; Sabine et al., 2013) to train the FFN algorithm. Each simulation was carried out using different subsets of environmental predictors extracted from the complete set (SST, SSS, bathymetry, sea-ice concentration and chlorophyll a). The results obtained were then compared to the more complete dataset of SOCAT*, which contains 40 % more shelf pCO_2 measurements from 1998 to 2015 (Bakker et al., 2016). This

Table 1. Datasets used to create the environmental forcing files. The original spatial and temporal resolution and the main manipulations applied for their use in the SOM-FFN are also reported.

Predictor	Dataset	Resolution	Reference	Manipulation
SST	EN4	0.25°, daily	Good et al. (2013)	Monthly average
SSS	EN4	0.25°, daily	Good et al. (2013)	Monthly average
Bathymetry	ETOPO2	2 min	US Department of Commerce (2006)	Aggregation to 0.25°
Sea ice	NSIDC	0.25°, monthly	Cavalieri et al. (1996)	Monthly rate of change in sea-ice coverage
Chlorophyll a	SeaWIFS, MODIS	9 km, monthly	NASA (2016)	Aggregation to 0.25°
Wind speed	ERA	0.25°, 6 h	Dee et al. (2011)	Monthly average

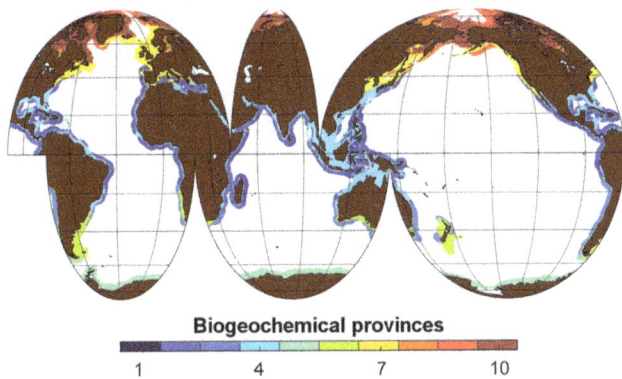

Biogeochemical provinces

Figure 3. Map of the 10 different biogeochemical provinces generated by the artificial neural network method SOM-FFN.

process allowed us, for each province, to calculate the RMSE for several combinations of independent predictor variables for the pCO_2. Next, the combinations of predictors displaying the lowest RMSE were kept for the final simulations, which then used all data from SOCAT*. Thus, the pCO_2 calculations in each province potentially rely on a different set of predictors (Table 1).

The coastal SOM-FFN results are validated through a comparison with the LDEO* dataset through the calculation of residuals and RMSE. Additionally, a model-to-model comparison is also performed with the global ocean results of Landschützer et al. (2016) in the regions where the domains overlap. To perform this latter analysis, the coastal high-resolution coastal pCO_2 product generated here was aggregated to a regular monthly 1° resolution to match the grid used by Landschützer et al. (2016).

Finally, the ability of the coastal SOM-FFN to capture seasonal variations is assessed by comparing the cell-average simulated monthly pCO_2 to monthly means for cells extracted from the LDEO* database. The cells retained for this analysis are all those for which the average for each month could be calculated from measurements performed on at least 3 different years.

3 Results and discussion

3.1 Biogeochemical provinces

Despite the fact that the SOM is not given any prior knowledge regarding space and time, the spatial distribution of the 10 biogeochemical provinces is mostly controlled by latitudinal gradients and distance from the coast (Fig. 3; high-resolution monthly maps are also available in the Supplement). Although the exact spatial extent of each province varies from one month to the next following the seasonal variations of the environmental forcing parameters, each province roughly corresponds to one type of climatological setting. Nevertheless, because of these spatial migrations, most cells belong to different provinces depending on the month (see Fig. B). These seasonal migrations are mostly driven by changes in temperature, sea-ice cover, pCO_2 and, to a lesser degree, salinity. P1, P2 and P4 (Province 1, 2, etc.) are three of the largest provinces, covering a total of 35.7×10^6 km^2 and representing warm tropical regions with bottoms at shallow to intermediate depths. During summer, the spatial coverage of P4 expands north- and southward as a consequence of warming. P2 represents tropical regions with deeper bottom depths. P1 and P2 display less seasonal changes in their spatial distribution than P4 due to weaker seasonal temperature changes. P3 and P6, which cover a combined 9.2×10^6 km^2, are found in the Southern Hemisphere and correspond to sub-polar and temperate regions, respectively. Their spatial distributions are subject to marked latitudinal migrations throughout the year as a result of the large amplitude changes in seasonal temperature observed in midlatitude coastal waters (Laruelle et al., 2014). Similarly, P7 corresponds to temperate northern hemispheric waters and displays marked seasonal changes including the shelves of the Norwegian Basin in summer and most of the Mediterranean Sea in winter. P5, P8, P9 and P10 together cover 22.7×10^6 km^2. These provinces are partly (seasonally) covered by sea ice with an average spatial ice cover over the study period of 57, 39, 54 and 46 % for P5, P8, P9 and P10, respectively. P5 represents the shelves of Antarctica all year round. P8 includes large fractions of the enclosed seas at higher northern latitudes such as the Baltic Sea and Hudson

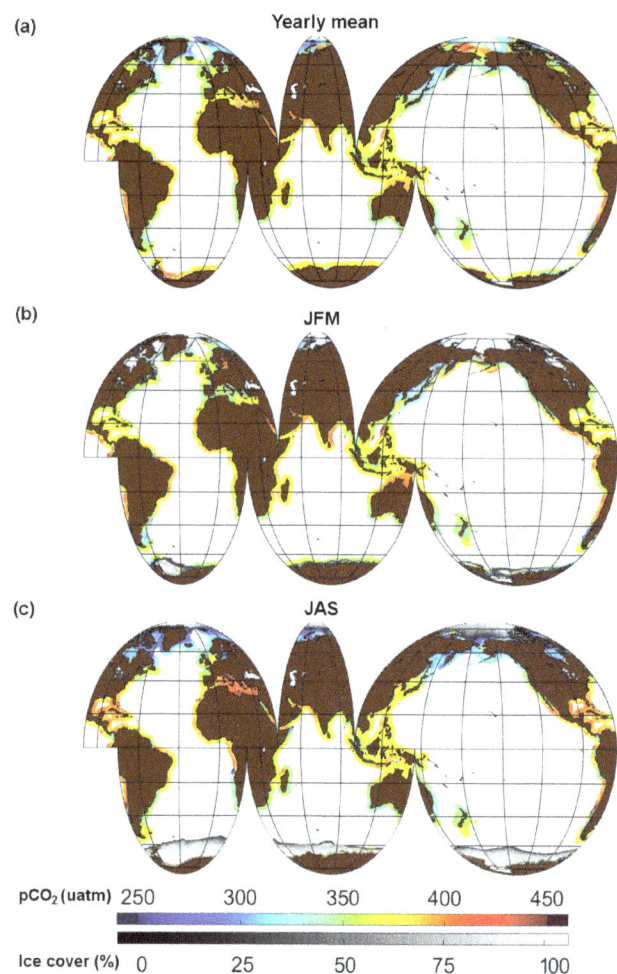

Figure 4. Climatological mean pCO_2 for **(a)** the long-term averaged pCO_2 (rainbow color scale) and sea-ice coverage (black–white color scale). The long-term average pCO_2 corresponds to roughly the nominal year 2006, as the average was formed over the full analysis period from 1998 to 2015: **(b)** the months of January, February and March; and **(c)** the months of July, August and September.

Bay while P9 (only $2.9 \times 10^6 \, \text{km}^2$) represents permanently deep and cold polar regions. P5 and P10 represent most of the polar shelves (P5 for the Antarctic and P10 for the Arctic) and are covered in sea ice at levels of 57 and 46 %, respectively. The regions experiencing most notable shifts in province allocation during the year include the northern polar regions and the temperate narrow shelves of the Atlantic and Pacific, particularly western Europe and eastern North America and East Asia (see Fig. B).

3.2 Performance of the coastal SOM-FFN

The mean climatological pCO_2 estimated by the coastal SOM-FFN for annually and seasonally averaged conditions are reported in Fig. 4. Before briefly analyzing the main spa-

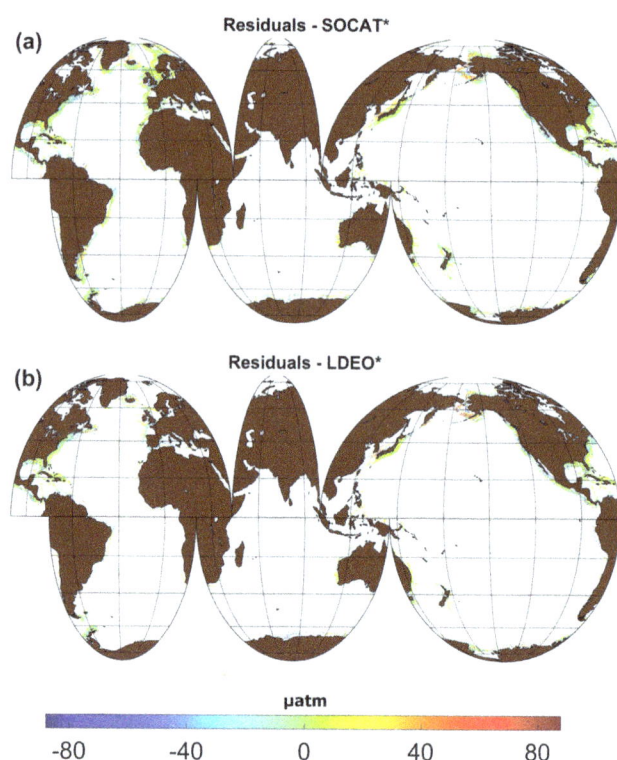

Figure 5. Mean residuals calculated as the difference between the SOM_FFM pCO_2 outputs and pCO_2 observations from SOCAT* **(a)** and LDEO* **(b)**.

tial and temporal variability of the pCO_2 fields (Sect. 3.3), we evaluate here the overall performance of our interpolation method globally and at the level of each province, including its ability to capture the seasonal cycle.

3.2.1 Comparison with training data (SOCAT*)

Within each province, the pCO_2 simulated by the coastal SOM-FFN are compared to the measurements extracted from SOCATv4 (Table 2). Globally, the average difference between observed and simulated pCO_2 is almost null (overall bias $= 0.0 \, \mu\text{atm}$) and the absolute bias is lower than 4 μatm in all 10 provinces. The average RMSE over all provinces of 32.9 μatm is comparable with those reported for other statistical reconstructions of coastal pCO_2 fields summarized by (Chen et al., 2016), although none of these studies were performed at global scale and many rely on different statistical approaches often using remote sensing data. This RMSE is about twice that achieved for the open ocean (Landschützer et al., 2014), reflecting the larger spatiotemporal variability in the coastal ocean and more complex processes governing that variability. Considering these complexities, achieving RMSE at the global scale in the same range as those reported for regional coastal studies is quite good.

Significant variations in both bias and RMSE can be observed between provinces (Table 2). P1 and P3 have the best

Table 2. List of the biogeochemical provinces, their geographic distribution and the environmental predictors used to calculate surface ocean pCO_2. SSS stands for sea-surface salinity, SST for sea-surface temperature, Bathy for bathymetry, Ice for sea-ice cover, Chl for chlorophyll concentration and Wind for wind speed.

Province	SSS	SST	Bathy	Ice	Chl	Wind
P1	X	X	X		X	X
P2	X	X	X		X	X
P3	X	X	X		X	X
P4	X	X	X		X	X
P5	X	X	X	X	X	X
P6	X	X	X	X	X	X
P7	X	X	X	X	X	X
P8	X	X	X	X		X
P9	X	X	X	X		X
P10	X	X	X	X		X

fit between simulated and observed pCO_2 with RMSE lower than 20 µatm. In five provinces that cover a cumulated surface area of 31.2×10^6 km^2 (P1, P2, P3, P6 and P9), RMSEs do not exceed 25 µatm. In P8, however, the maximum RMSE is found with a value of 46.8 µatm. Overall, the performance of the SOM-FFN deteriorates for provinces regularly covered by sea ice (P5, P8–10) in which data coverage is relatively low (RMSE > 34 µatm). This trend is consistent with the spatial distribution of the average residual errors between the pCO_2 field generated by the model and pCO_2 data extracted SOCAT* (Fig. 5a). The residuals are obtained by subtracting the observed values from model output in each grid cell for every month where observations are available. Thus, positive values correspond to cells where the simulated pCO_2 overestimates the field data, while negative values represent cells where the simulated pCO_2 underestimates the field data. The bulk of the residuals fall in the −20 to 20 µatm range in temperate and tropical regions, except for very shallow regions that are under the influence of a large river such as the Mississippi. There, the SOM-FFN often underestimates the observed pCO_2. There also exist coastal areas where the SOM-FFN underestimates the observed pCO_2 such as Nova Scotia, the southwestern coast of England and the shelves of California and Morocco. The complex hydrodynamics of those regions (some of them being characterized as upwelling regions) may explain the weaker performance of the SOM-FFN. At high latitudes, the performance of the model deteriorates somewhat. For example, Bering Sea contains cells with very high (> 50 µatm) and very low average residuals (< −50 µatm).

3.2.2 Evaluation with LDEO* data

The comparison of our results with the data from LDEO* yields a global bias of 0.0 µatm (calculated as the average difference between observed and SOM-FFN estimated pCO_2)

for the entire shelf domain. However, the spread is relatively large with an average RMSE of 39.2 µatm. This average RMSE is 19 % larger than the one obtained when comparing the SOM-FFN results with the SOCAT* dataset, which has been used to train the model. A province-based analysis reveals strong differences in the calculated RMSEs, ranging from 20 to 53 µatm (Table 2, LDEO*). A review of various statistical models used to generate continuous global ocean pCO_2 maps, including some using remote sensing data and algorithms, reports RMSE or uncertainties typically varying within the 10–35 µatm range (Chen et al., 2016) with outliers as high as 50 µatm in the Mississippi Delta (Lohrenz and Cai, 2006). This report also shows that open-ocean estimates generally yield RMSEs lower than 17 µatm, in agreement with Landschützer et al. (2014), whereas coastal estimates are associated with much higher uncertainties. This is likely because these coastal regions have complex biogeochemical dynamics and high frequency variability that cannot be fully captured with the current generation of data interpolation techniques using the limited available predictor data.

In our simulations, the province averaged biases are larger than those calculated with SOCAT* but their absolute value remains small and never exceed 3.9 µatm (P8). Provinces P1, P2, P3 and P6 have RMSE < 30 µatm, which compares with the most robust pCO_2 regional coastal estimates from the literature (Chen et al., 2016). Together, these four provinces account for 37 % of our domain. P4, P5 and P9 display RMSE of between 33 and 38 µatm for P4 and P9, respectively. Overall, these seven provinces, covering the entire tropical and temperate latitudinal bands and some subpolar regions, account for > 72 % of the shelf surface area and yield RMSEs of less than 38 µatm and absolute biases of less than 2.3 µatm. Provinces in the polar regions (P5, P7, P8 and P10) overall display larger deviations with respect to the LDEO* dataset, but the absolute value of their biases never exceeds 3.9 µatm. Their RMSEs all fall in the 51–53 µatm range. This suggests a significantly lower performance of the SOM-FFN in regions partly covered in sea ice. This can be attributed to the limited number of available data points and their very heterogeneous distribution in time and space, as well as to the very limited range of variation of some of the controlling variable such as temperature and salinity. The relatively good performance of the model in tropical region might be partly attributed to the relatively small seasonal variations in pCO_2 within these areas. The residuals calculated by subtracting the SOM-FFN results from LDEO* are very similar to those obtained by subtracting the SOM-FFN results from SOCAT* (Fig. 5b). The residual errors have a nearly Gaussian distribution for every biogeochemical province with the exception of province P8 (Fig. 6). In this case, the distribution not only has the highest spread but also is skewed toward high values.

In order to evaluate the contribution of the newly added predictors compared to the oceanic setup of the SOM-FFN (Landschützer et al., 2013), the model was also trained with-

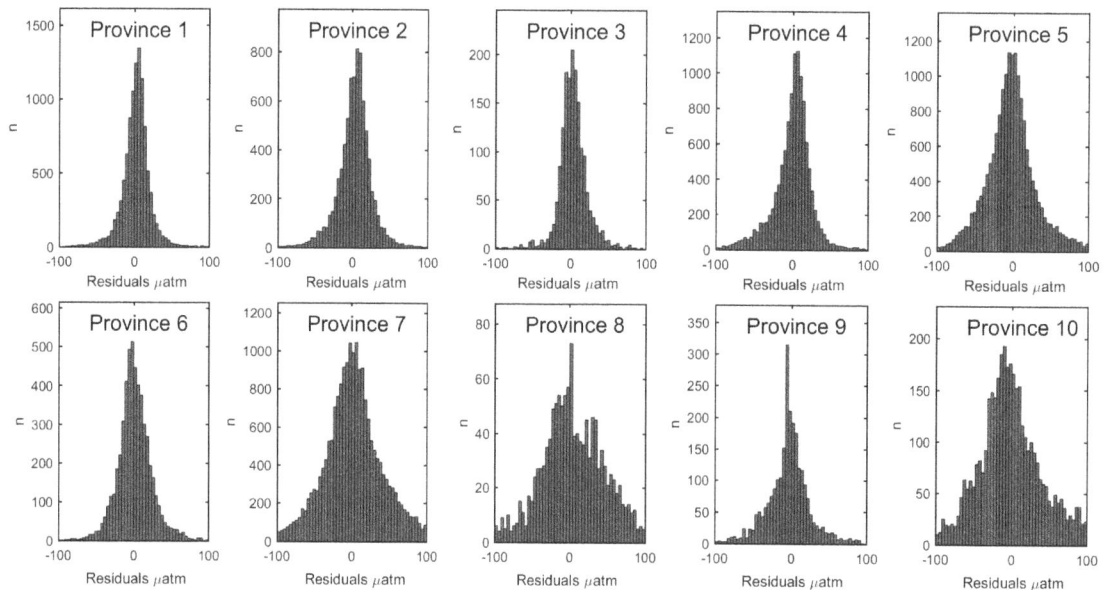

Figure 6. Histograms reporting the distribution of residuals between observed (LDEO*) and computed (SOM_FFN) pCO_2 in each biogeochemical province.

out wind speed and sea-ice cover. The RMSEs obtained with those simulations (Table 4) are significantly higher than those obtained using all predictors (Table 3). However, the overall bias remain small. The results of those simulations are presented in the table below and allow us to quantify the effect of addition of new predictors on the performance of the model. For instance, it can be noticed that the global RMSE increases significantly (from 39.2 to 48 µatm in the comparison with LDEO* when chlorophyll, sea ice and wind speed are not taken into account and from 39.2 to 45 µatm when only sea ice and wind speed are not taken into account). This deterioration of the performance of the model, however, does not evenly affect all provinces. Provinces located at high latitudes (i.e., P8, P9 and P10) perform significantly worse without the inclusion of wind speed and sea ice.

Finally, while the use of residuals and RMSE provides valid quantitative assessment of the model performance, it does not provide insights regarding its ability to reproduce the seasonal pCO_2 cycle. To address this issue, Fig. 7 displays observed mean monthly pCO_2 extracted from LDEO* and calculated by the coastal SOM-FFN for the 40 locations where the LDEO* database has the most data (> 40 month). The error bars associated with the observations reflect the interannual variability. Overall, the coastal SOM-FFN captures the timing of the seasonal pCO_2 cycle in most locations well with pCO_2 minima and maxima occurring at the same time in our results and in the uninterpolated LDEO* data. The pCO_2 maximum generally taking place in early summer is accurately captured by the coastal SOM-FFN. In terms of amplitudes in the pCO_2 signal, the coastal SOM-FFN and the LDEO* data reveal primarily how different the

seasonal pCO_2 cycle is from one region to the other, with very low amplitude (< 40 µatm) in some sub-tropical areas, amplitudes > 100 µatm at high northern and southern latitudes, and sometimes very sharp increases during summer, like off the coast of Japan. In most regions, the SOM-FFN-based reconstructions are able to capture these variations and predict seasonal amplitudes comparable to those observed in the data. However, in cells for which the difference between observed and simulated seasonal pCO_2 amplitude is larger than 20 %, the coastal SOM-FFN tends to systematically underestimate the amplitude of the seasonal pCO_2 cycle. This limitation of our model might result from the often short timescales associated with the continental influences in nearshore locations, which are not captured by the environmental predictors used in our calculation. It may also be the result of very short-term events that are aliased in our monthly average calculations.

3.2.3 Comparison with global SOM-FFN

The comparison of our coastal SOM-FFN results with those of Landschützer et al. (2016) for the overlapping grid cells (Table 3) reveals significant differences between both interpolated data products with a RMSE between 24 and 32 µatm for most provinces except P7, P9 and P10 (53, 55 and 37 µatm, respectively). These RMSE values are comparable but slightly lower than those obtained for the comparison with the LDEO* database, in line with those observed with the SOCAT* database. The differences (coastal SOM-FFN minus global SOM-FFN), however, are much larger than those observed between our results and the LDEO* database and highlight the current knowledge gap regarding the mean

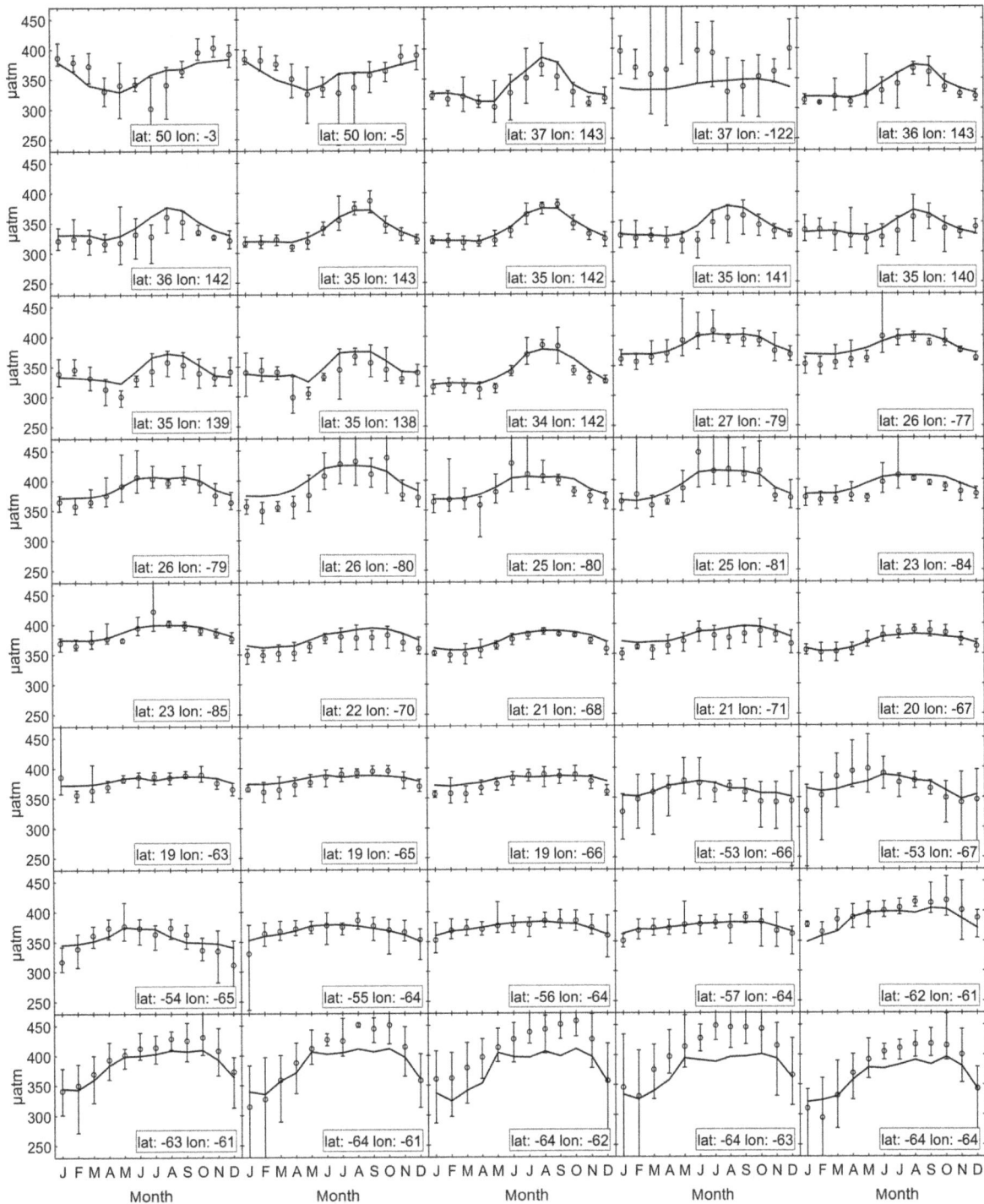

Figure 7. Climatological monthly mean $p\mathrm{CO_2}$ extracted from the LDEO* database (points) and generated by the artificial neural network (lines) for grid cells with more than 40 months of data. The error bars associated with the data represent the interannual variability, reported as the highest and lowest recorded values for a given month at a given location.

state and variability of the transition zone. They range from −17.9 to 11.7 µatm from one province to the next but only amount to −0.6 µatm when considering the cells from all provinces at once.

The overlapping cells used for the comparison with Landschützer et al. (2016) are mostly located over 100 km away from the coastline and therefore both the open ocean and our new shelf ocean dataset are constrained by fairly differ-

Table 3. Root mean square error between observed and calculated pCO_2 in the different biogeochemical provinces. The SOM-FFN results are compared to data extracted from the LDEO database (Takahashi et al., 2014) and the overlapping cells from the Landschützer et al. (2016) pCO_2 climatology.

Province	Surface area (km^2)	Ice cover (%)	SOCAT* bias (µatm)	RMSE (µatm)	Landschützer bias (µatm)	2016 RMSE (µatm)	LDEO bias (µatm)	RMSE (µatm)
P1	8.2×10^6	0	0.0	19.1	2.0	27.2	2.0	20.5
P2	10.9×10^6	0	0.2	24.7	9.3	24.2	1.3	27.2
P3	4.4×10^6	0	−0.3	16.1	2.2	37.9	2.3	22.7
P4	16.6×10^6	0	−0.2	31.2	8.0	21.1	−1.6	33.0
P5	7.5×10^6	57.1	0.0	34.2	11.5	30.9	−1.4	38.0
P6	4.8×10^6	0	0.0	24.3	6.8	18.1	1.3	27.9
P7	9.3×10^6	0.0	0.1	37.2	0.7	23.5	−0.2	52.5
P8	3.3×10^6	38.5	0.2	46.8	13.9	70.1	3.9	51.4
P9	2.9×10^6	54.3	−0.1	23.0	−5.2	42.5	−2.5	33.4
P10	9.0×10^6	45.8	0.0	35.7	−9.7	50.9	1.6	53.1
	76.9×10^6		0.0	32.9	3.9	34.7	0.0	39.2

Table 4. Biases and root mean square error (RMSE) between observed and calculated pCO_2 using only SST, SSS and bathymetry (STB) or SST, SSS, bathymetry and chlorophyll (STBC) as predictors.

Province	SOCAT* bias (µatm)		RMSE (µatm)		LDEO* bias (µatm)		RMSE (µatm)	
	STB	STBC	STB	STBC	STB	STBC	STB	STBC
P1	0.0	−0.2	20.8	21.0	2.4	2.0	21.7	21.5
P2	−0.1	0.1	26.9	27.8	0.5	0.8	29.0	29.6
P3	0.0	−0.5	22.7	21.3	3.0	2.3	27.1	26.8
P4	0.0	−0.2	33.0	33.0	−1.7	−2.3	33.8	33.8
P5	0.2	0.1	52.7	42.2	−1.7	−0.9	56.9	44.5
P6	0.0	0.1	26.8	26.5	−0.5	0.6	28.9	28.0
P7	0.4	0.3	44.3	44.1	1.2	0.3	59.3	58.8
P8	0.1	0.4	82.6	80.0	9.1	9.0	56.3	58.5
P9	0.1	0.9	34.7	36.5	−2.6	−2.8	39.8	41.8
P10	−0.3	0.7	49.8	49.5	−3.9	−3.0	76.5	75.4
Global	0.1	0.2	43.9	42.4	0.0	0.0	48.0	45.0

ent data because all the "shelf" cells from the open-ocean data product have a pCO_2 calculated by a model calibrated mostly for conditions representative of the open ocean. Overall, the occurrence of large residuals in the shallowest cells of our calculation domain in our results (Fig. 5) suggest that the very near-shore processes controlling the CO_2 dynamics likely are the most difficult to reproduce at the global scale. However, the added value of performing our simulations at the spatial resolution of 0.25° is exemplified by the ability of our model to capture the plumes of larger rivers such as the Amazon, where pCO_2 is significantly lower than that of the surrounding waters (Cooley et al., 2007; Ibanez et al., 2015).

3.3 Spatial and temporal variability of the coastal pCO_2

3.3.1 Spatial variability

Figure 4a presents the annual average pCO_2 estimated by the coastal SOM-FFN, representing the mean over 1998 through 2015 period (monthly climatological maps are shown in Fig. A in the Supplement). High annual mean values of pCO_2, close to or above atmospheric levels, are estimated around the Equator up to the tropics. This is consistent with previous studies that identified tropical and equatorial coastal regions as weak CO_2 sources for the atmosphere (Borges et al., 2005; Cai, 2011; Laruelle et al., 2010, 2014). A hotspot of very high pCO_2 emerges from our analysis around the Arabian Peninsula, extending into the eastern Mediterranean

Sea, as well as into the Red Sea and the Persian Gulf. These regions are poorly monitored and it remains difficult to assess if pCO_2 values in excess of 450 µatm are realistic or not, but the limited body of available literature suggests that very high pCO_2 are indeed observed in these regions (Ali, 2008; Omer, 2010). The very high temperature and salinity conditions observed in the Red Sea, in particular, reduce the CO_2 solubility and induce very high pCO_2 conditions. However, these predicted pCO_2 lie outside of the range used for the training of the SOM-FFN (typically 200–450 µatm) and should thus be considered with caution. Along the oceanic coast of the Arabian Peninsula, the SOM-FFN predicts pCO_2 ranging from 365 to 390 µatm all year round and thus does not capture the well-known increase in pCO_2 resulting from the monsoon-driven summer upwelling in the region (Sarma, 2003; Takahashi et al., 2009).

In both hemispheres, pCO_2 values in the 325 to 370 µatm range are generally reconstructed at temperate latitudes, i.e., up to 50° N and 50° S, respectively. The northern high latitudes generally have very low pCO_2 values, down to 300 µatm and below, a result that is consistent with the Arctic shelves contributing a large proportion (up to 60 %) of the global coastal carbon sink (Bates and Mathis, 2009; Cai, 2011; Laruelle et al., 2014). Several hotspots of pCO_2 with values as high as 450 µatm can be observed nevertheless north of 70° N, most notably along the eastern coast of Siberia in winter (see Fig. P in the Supplement), which displays a large zone characterized by $pCO_2 > 400$ µatm centered at the mouth of the Kolyma River. Such high pCO_2 values have been punctually observed in Arctic coastal waters (Anderson et al., 2009) and could result from the discharge of highly oversaturated riverine waters. But, overall, pCO_2 measurements over Siberian shelves are rare. Thus, our results should be considered with caution in this region because of the scarcity of data to train and validate the coastal SOM-FFN. It should also be noted that the vast majority of this high pCO_2 region is covered by sea ice (Fig. 4b, c) and, although the model estimates pCO_2 values over the entire domain, only ice-free (or partially ice-free) cells will contribute to the CO_2 exchange across the air–sea interface (Bates and Mathis, 2009; Laruelle et al., 2014).

3.3.2 Temporal variability

The reconstructed pCO_2 field is also subject to large seasonal variations (see Fig. P and A in the Supplement). To explore these variations further, Fig. 8 reports seasonal-mean latitudinal profiles of pCO_2 for continental shelves neighboring the eastern Pacific, Atlantic, Indian and western Pacific. The analysis excludes continental shelves at latitudes higher than 65° because a large fraction of these shelves are seasonally covered by sea ice. The latitudinal pCO_2 profiles reveal that, in most regions, highest and lowest pCO_2 values are observed during the warmest and coldest months, respectively. This trend is particularly pronounced at temperate latitudes

where the seasonal pCO_2 amplitude can reach 60 µatm and is exemplified by regions such as the western Mediterranean Sea or the eastern coast of America, which become supersaturated in CO_2 compared to the atmosphere during the summer months. However, there are a few other regions where the lowest pCO_2 is found in the summer, such as the Baltic Sea (Thomas and Schneider, 1999). Around the Equator, the magnitude of the seasonal variations in pCO_2 is limited and does not exceed 30 µatm.

Although the general latitudinal trend of the annual mean pCO_2 is similar across all continental shelves, significant differences in the seasonality can be observed across the largest ocean basins. In particular, most of the eastern Pacific shelves, except for latitudes north of 55° N, display limited seasonal change in pCO_2 (typically below 30 µatm), while the western Pacific shelves have seasonal pCO_2 amplitudes that can exceed 50 µatm in temperate regions and 100 µatm at high latitudes (above 55° N). Along the Atlantic shelves, the seasonal signal is more pronounced in the north compared to the south, in agreement with Laruelle et al. (2014). Overall, the northern Pacific (north of 55° N) displays the most pronounced seasonal change in pCO_2 with a difference of 80 µatm between summer and winter. In the Indian Ocean, the seasonal dynamics of pCO_2 is partly regulated by seasonal upwelling induced by the monsoon (Liu et al., 2010). In this basin north of the Equator, April, May and June are the months with the highest pCO_2 and the seasonal variations do not exceed 30 µatm. In contrast, the seasonal cycle is quite pronounced in the Indian Ocean south of the Equator (~ 50 µatm).

Latitudinal profiles of SST (Fig. 8b) are similar in all coastal oceans with minimal seasonal variations around the Equator and amplitudes as large as 20 °C at temperate latitudes. The comparison between the seasonal pCO_2 and SST profiles allows us to assess the contribution of temperature-induced changes in CO_2 solubility to the seasonal pCO_2 variations in the continental shelf waters. However, other factors such as seasonal upwelling and biological activity also strongly influence coastal pCO_2 and contribute to the complexity of the seasonal pCO_2 profiles. To quantify the effect of temperature on seasonal variations of pCO_2 the latter is normalized to the mean temperature at different latitudes in each oceanic basin (Fig. 8) using the formula proposed by Takahashi et al. (1993):

$$pCO_{2(SSTmean)} = pCO_{2,obs} \times \exp(0.0423 \times (T_{mean} - T_{obs})), \quad (1)$$

where $pCO_{2(SSTmean)}$ is the temperature normalized pCO_2, $pCO_{2,obs}$ is the observed pCO_2 at the observed temperature T_{obs} and T_{mean} is the yearly mean temperature at the considered location. In seawater, an increase in water temperature induces a decrease in gas solubility which leads to a higher water pCO_2. Thus, comparing $pCO_{2(SSTmean)}$ with observed pCO_2 monthly values provides a quantitative estimate of the influence of seasonal temperature change on the seasonality of pCO_2.

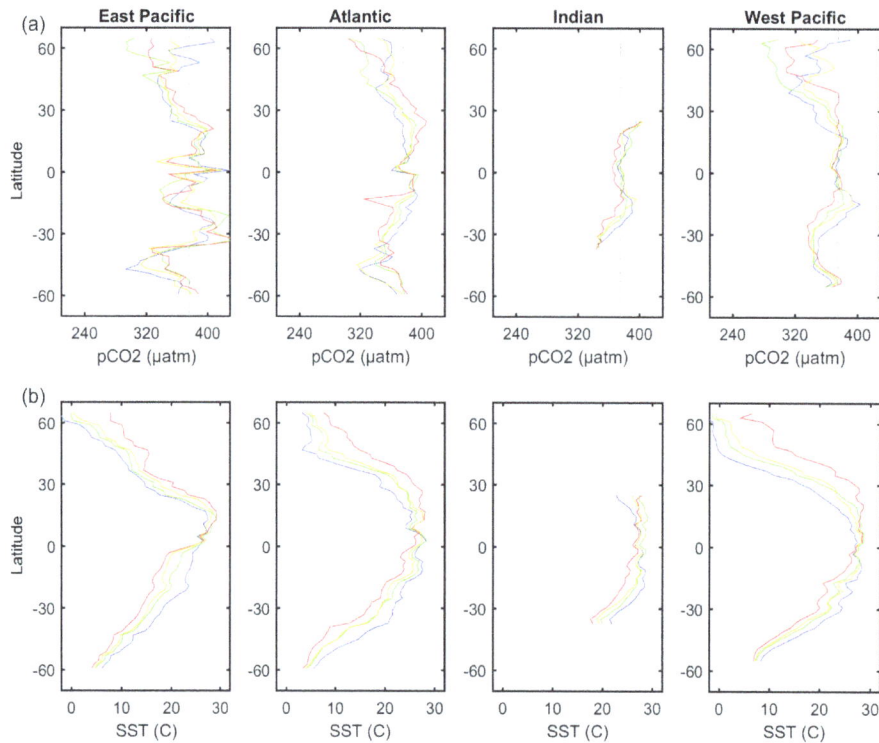

Figure 8. Seasonal-mean latitudinal profiles of pCO_2 **(a)** and SST **(b)** for the continental shelves surrounding four oceanic basins. Blue lines: averages over the months of January, February and March; green lines: averages over the months of April, May and June; red lines: averages over the months of July, August and September; yellow lines: averages over the months of October, November and December. The dashed line in the top panels represents the average atmospheric pCO_2 for year 2006.

For most latitudes and oceanic basins, pCO_2 is minimum in late winter or early spring, i.e., at the time when $pCO_{2(SSTmean)}$ has its maximum. pCO_2 also generally displays a maximum in summer, while $pCO_{2(SSTmean)}$ reaches its minimum then (Fig. 9). The amplitude of the changes in $pCO_{2(SSTmean)}$ is quite consistent across oceans and about 2 to 3 times larger than that of pCO_2. Between 45 and 60° N, the variations in $pCO_{2(SSTmean)}$ largely exceed 100 µatm (up to 220 µatm at 60° N in the western Pacific). In these regions, the magnitude of the seasonal temperature changes is also maximum and reaches 20 °C between winter and summer (Fig. 5). A seasonal signal in pCO_2 with a minimum in late winter or spring when $pCO_{2(SSTmean)}$ is maximal can also be identified. However, the magnitude of the seasonal variations in pCO_2 is significantly smaller than those of $pCO_{2(SSTmean)}$, suggesting that other processes such as biological uptake or transport/mixing partly offsets the temperature effect on solubility. In the subpolar western Pacific shelves (60° N), a second pronounced dip in pCO_2 following a weaker one in spring is observed in summer, which suggests the occurrence of a pronounced summer biological activity taking up large amounts of CO_2. This would also explain the sharp increase in pCO_2 in the following month, as a result of the degradation of organic matter synthesized during the summer bloom. Although this region is also the one sub-

jected to the strongest seasonal temperature, the amplitude of the seasonal $pCO_{2(SSTmean)}$ which reaches 220 µatm suggests that non thermal processes drive most of the seasonal pCO_2 variations in the regions. At 20° N, the amplitude of the changes in both pCO_2 and $pCO_{2(SSTmean)}$ are lower than at higher latitudes. pCO_2 varies by ~ 30 µatm between summer and winter in all oceanic basin while the seasonal variations in $pCO_{2(SSTmean)}$ are more pronounced in the Pacific (~ 60 µatm) than in the Atlantic or the Indian oceans. In the Southern Hemisphere, the seasonal variations in pCO_2 are not as pronounced as in the Northern Hemisphere, suggesting that the changes induced by the solubility pump are compensated by biological activities. At 10 and 30° S, the seasonal variations in pCO_2 rarely exceed 30 µatm in either basin with a minimum observed around August.

4 Summary

This study presents the first global high-resolution monthly pCO_2 maps for continental shelf waters at an unprecedented 0.25° spatial resolution. We show that when tailored for the specific conditions of shelf systems, the SOM-FFN method previously employed in the open ocean is capable of reproducing well-known and well-observed features of the pCO_2

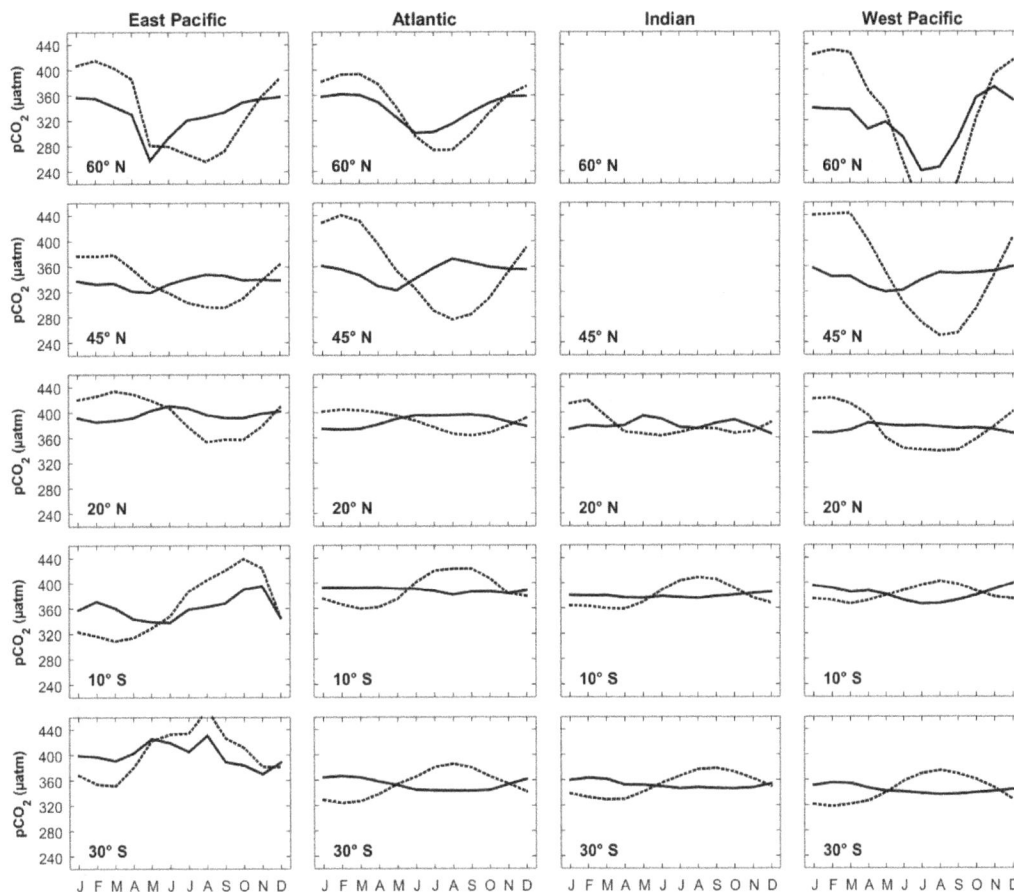

Figure 9. Seasonal cycle of observed (continuous lines) and temperature normalized pCO_2 ($pCO_{2(SSTmean)}$ dashed lines) at five different latitudes in four oceanic basins.

field in the coastal ocean. Our continuous shelf product allows us, for the first time, to analyze the dominant spatial patterns of pCO_2 across all ocean basins and their seasonality. The data product associated to this paper consists of a netcdf file containing the pCO_2 for ice-free cells at a 0.25° spatial resolution for each of the 216 month of the simulation period (from January 1998 to December 2015). Twelve maps representing mean pCO_2 fields calculated for each month over the simulation period are also provided. This data product can be combined with wind field products such as ERA-interim (Dee, 2010; Dee et al., 2011) or CCMP (Atlas et al., 2011) to compute spatially and temporally resolved air–sea CO_2 fluxes across the global shelf region, including the Arctic. Maps including pCO_2 for ice-covered cells are also available but should be treated with care because the dynamics of CO_2 fluxes through sea ice are still poorly understood and air–sea gas transfer velocities in partially sea-ice-covered areas cannot be predicted from classical wind speed relationships (Lovely et al., 2015)

Data availability. Version 4 of the SOCAT database (Bakker et al., 2016) can be downloaded from www.socat.info/upload/ SOCAT_v4.zip. The observation-based global monthly gridded sea-surface pCO_2 product is provided by Landschützer et al. (2015; https://doi.org/10.3334/CDIAC/OTG.SPCO2_ 1982_2011_ETH_SOM-FFN), was downloaded from http: //cdiac.ornl.gov/ftp/oceans/SPCO2_1982_2011_ETH_SOM_FFN and is now available at https://www.nodc.noaa.gov/ ocads/oceans/SPCO2_1982_2015_ETH_SOM_FFN.html. The LDEOv2015 database (Takahashi et al., 2015; https://doi.org/10.3334/CDIAC/OTG.NDP088(V2015)) was downloaded from http://cdiac.ornl.gov/oceans/LDEO_Underway_ Database/. The global atmospheric reanalysis ERA-interim datasets (Dee et al., 2011; http://doi.wiley.com/10.1002/qj.828) are accessible on the European Centre for Medium-Range Weather Forecasts (ECMWF) website. SST and SSS were extracted from the Met Office's EN4 dataset (Good et al., 2013; doi:10.1002/2013JC009067). The bathymetry used is the global ETOPO2 database (US Department of Commerce, 2006), which can be downloaded from http://www.ngdc.noaa.gov/mgg/fliers/06mgg01.html. The sea-ice concentrations are derived from the global 25 km resolution monthly data product compiled by the NSIDC (National Snow and Ice Cover Data; Cavalieri et al., 1996).

Competing interests. The authors declare that they have no conflict of interest.

Acknowledgements. Goulven G. Laruelle and Bruno Delille are a postdoctoral researcher and research associate, respectively, of FRS-FNRS. The Surface Ocean CO_2 Atlas (SOCAT) is an international effort, supported by the International Ocean Carbon Coordination Project (IOCCP), the Surface Ocean Lower Atmosphere Study (SOLAS) and the Integrated Marine Biogeochemistry and Ecosystem Research program (IMBER), in order to deliver a uniformly quality-controlled surface ocean CO_2 database. The many researchers and funding agencies responsible for the collection of data and quality control are thanked for their contributions to SOCAT. The research leading to these results has received funding from the European Union's Horizon 2020 research and innovation program under the Marie Skłodowska-Curie grant agreement no. 643052 744 (C-CASCADES project). NG acknowledges support by ETH Zürich. Peter Landschützer is supported by the Max Planck Society for the Advancement of Science.

Edited by: Jack Middelburg

References

Ali, E.: The Inorganic Carbon Cycle in the Red Sea, Master's thesis, University of Bergen, 2008.

Anderson, L. G., Jutterström, S., Hjalmarsson, S., Wåhlström, I., and Semiletov, I. P.: Out-gassing of CO_2 from Siberian Shelf seas by terrestrial organic matter decomposition, Geophys. Res. Lett., 36, L20601, https://doi.org/10.1029/2009GL040046, 2009.

Antonov, J. I., Seidov, D., Boyer, T. P., Locarnini, R. A., Mishonov, A. V., Garcia, H. E., Baranova, O. K., Zweng, M. M., and Johnson D. R.: in World Ocean Atlas 2009, Volume 2: Salinity, NOAA Atlas NESDIS, edited by: Levitus, S., US Gov. Print. Off., Washington, DC, Vol. 69, 2010.

Atlas, R., Hoffman, R. N., Ardizzone, J., Leidner, S. M., Jusem, J. C., Smith, D. K., and Gombos, D.: A cross-calibrated, multiplatform ocean surface wind velocity product for meteorological and oceanographic applications, Bull. Am. Meteorol. Soc., 92, 157–174, 2011.

Bakker, D. C. E., Pfeil, B., Smith, K., Hankin, S., Olsen, A., Alin, S. R., Cosca, C., Harasawa, S., Kozyr, A., Nojiri, Y., O'Brien, K. M., Schuster, U., Telszewski, M., Tilbrook, B., Wada, C., Akl, J., Barbero, L., Bates, N. R., Boutin, J., Bozec, Y., Cai, W.-J., Castle, R. D., Chavez, F. P., Chen, L., Chierici, M., Currie, K., De Baar, H. J. W., Evans, W., Feely, R. A., Fransson, A., Gao, Z., Hales, B., Hardman-Mountford, N. J., Hoppema, M., Huang, W.-J., Hunt, C. W., Huss, B., Ichikawa, T., Johannessen, T., Jones, E. M., Jones, S., Jutterstrom, S., Kitidis, V., Körtzinger, A., Landschützer, P., Lauvset, S. K., Lefèvre, N., Manke, A. B., Mathis, J. T., Merlivat, L., Metzl, N., Murata, A., Newberger, T., Omar, A. M., Ono, T., Park, G.-H., Paterson, K., Pierrot, D., Ríos, A. F., Sabine, C. L., Saito, S., Salisbury, J., Sarma, V. V. S. S., Schlitzer,

R., Sieger, R., Skjelvan, I., Steinhoff, T., Sullivan, K. F., Sun, H., Sutton, A. J., Suzuki, T., Sweeney, C., Takahashi, T., Tjiputra, J., Tsurushima, N., Van Heuven, S. M. A. C., Vandemark, D., Vlahos, P., Wallace, D. W. R., Wanninkhof, R., and Watson, A. J.: An update to the Surface Ocean CO_2 Atlas (SOCAT version 2), Earth Syst. Sci. Data, 6, 69–90, https://doi.org/10.5194/essd-6-69-2014, 2014.

Bakker, D. C. E. et al.: A multi-decade record of high-quality fCO_2 data in version 3 of the Surface Ocean CO_2 Atlas (SOCAT), Earth Syst. Sci. Data, 8, 383–413, https://doi.org/10.5194/essd-8-383-2016, 2016.

Bates, N. R. and Mathis, J. T.: The Arctic Ocean marine carbon cycle: Evaluation of air-sea CO_2 exchanges, ocean acidification impacts and potential feedbacks, Biogeosciences, 6, 2433–2459, https://doi.org/10.5194/bg-6-2433-2009, 2009.

Bates, N. R., Moran, S. B., Hansell, D. A., and Mathis, J. T.: An increasing CO_2 sink in the Arctic Ocean due to sea-ice loss, Geophys. Res. Lett., 33, L23609, https://doi.org/10.1029/2006GL027028, 2006.

Bauer, J. E., Cai, W.-J., Raymond, P. A., Bianchi, T. S., Hopkinson, C. S., and Regnier, P. A. G.: The changing carbon cycle of the coastal ocean, Nature, 504, 61–70, https://doi.org/10.1038/nature12857, 2013.

Borges, A. V., Delille, B., and Frankignoulle, M.: Budgeting sinks and sources of CO_2 in the coastal ocean: Diversity of ecosystems counts, Geophys. Res. Lett., 32, L14601, https://doi.org/10.1029/2005GL023053, 2005.

Bourgeois, T., Orr, J. C., Resplandy, L., Terhaar, J., Ethé, C., Gehlen, M., and Bopp, L.: Coastal-ocean uptake of anthropogenic carbon, Biogeosciences, 13, 4167–4185, https://doi.org/10.5194/bg-13-4167-2016, 2016.

Cai, W. J.: Estuarine and coastal ocean carbon paradox: CO_2 sinks or sites of terrestrial carbon incineration?, Annu. Rev. Mar. Sci., 3, 123–145, 2011.

Cavalieri, D. J., Parkinson, C. L., Gloersen, P., and Zwally, H.: Sea Ice Concentrations from Nimbus-7 SMMR and DMSP SSM/I-SSMIS Passive Microwave Data, years 1990–2011, NASA DAAC at the Natl. Snow and Ice Data Cent., Boulder, Colorado, 1996.

Chen, C. T. A. and Borges, A. V.: Reconciling opposing views on carbon cycling in the coastal ocean: continental shelves as sinks and near-shore ecosystems as sources of atmospheric CO_2, Deep-Sea Res. Pt. II, 56, 578–590, 2009.

Chen, C. T. A., Huang, T. H., Chen, Y. C., Bai, Y., He, X., and Kang, Y.: Air-sea exchanges of CO_2 in the world's coastal seas, Biogeosciences, 10, 6509–6544, https://doi.org/10.5194/bg-10-6509-2013, 2013.

Chen, S., Hu, C., Byrne, R. H., Robbins, L. L., and Yang, B.: Remote estimation of surface pCO_2 on the West Florida Shelf, Cont. Shelf Res., 128, 10–25, 2016.

Cooley, S. R., Coles, V. J., Subramaniam, A., and Yager, P. L.: Seasonal variations in the Amazon plume-related atmospheric carbon sink, Global Biogeochem. Cy., 21, GB3014, https://doi.org/10.1029/2006GB002831, 2007.

Crossland, C. J., Kremer, H. H., Lindeboom, H. J., Marshall Crossland, J. I., and LeTissier, M. D. A. (Eds.): Coastal Fluxes in the

Anthropocene, Global Change – The IGBP Series, Berlin, Heidelberg, Springer-Verlag, Germany, 232 pp., 2005.

Dee, D. P.: The ERA-Interim reanalysis: Configuration and performance of the data assimilation system, Q. J. R. Meteorol. Soc., 137, 553–597, 2010.

Dee, D. P., Uppala, S. M., Simmons, A. J., Berrisford, P., Poli, P., Kobayashi, S., Andrae, U., Balmaseda, M. A., Balsamo, G., Bauer, P., Bechtold, P., Beljaars, A. C. M., van de Berg, L., Bidlot, J., Bormann, N., Delsol, C., Dragani, R., Fuentes, M., Geer, A. J., Haimberger, L., Healy, S. B., Hersbach, H., Hòlm, E. V., Isaksen, L., Kallberg, P., Köhler, M., Matricardi, M., Mcnally, A. P., Monge-Sanz, B. M., Morcrette, J. J., Park, B. K., Peubey, C., de Rosnay, P., Tavolato, C., Thépaut, J. N., and Vitart, F.: The ERA-Interim reanalysis: Configuration and performance of the data assimilation system, Q. J. R. Meteorol. Soc., 137, 553–597, https://doi.org/10.1002/qj.828, 2011.

Doney, S. C.: The Growing Human Footprint on Coastal and Open-Ocean Biogeochemistry, Science, 328, 1210–1216, https://doi.org/10.1126/science.1185198, 2010.

Good, S. A., Martin, M. J., and Rayner, N. A.: EN4: quality controlled ocean temperature and salinity profiles and monthly objective analyses with uncertainty estimates, J. Geophys. Res.-Ocean., 118, 6704–6716, https://doi.org/10.1002/2013JC009067, 2013.

Gruber, N.: Ocean biogeochemistry: Carbon at the coastal interface, Nature, 517, 148–149, https://doi.org/10.1038/nature14082, 2015.

Grimm, R., Notz, D., Glud, R. N., Rysgaard, S., and Six, K. D.: Assessment of the sea-ice carbon pump: Insight from a three-dimensional ocean-sea-ice-biogeochemical model (MPIOM/HAMOCC), Elementa, Science of the Anthropocene, 4, 000136, https://doi.org/10.12952/journal.elementa.000136, 2016.

Ibánhez, J. S. P., Diverrès, D., Araujo, M., and Lefèvre, N.: Seasonal and interannual variability of sea-air CO_2 fluxes in the tropical Atlantic affected by the Amazon River plume, Global Biogeochem. Cy., 29, 1640–1655, https://doi.org/10.1002/2015GB005110, 2015.

Landschützer, P., Gruber, N., Bakker, D. C. E., Schuster, U., Nakaoka, S., Payne, M. R., Sasse, T., and Zeng, J.: A neural network-based estimate of the seasonal to inter-annual variability of the Atlantic Ocean carbon sink, Biogeosciences, 10, 7793–7815, https://doi.org/10.5194/bg-10-7793-2013, 2013.

Landschützer, P., Gruber, N., Bakker, D. C. E., and Schuster, U.: Recent variability of the global ocean carbon sink, Global Biogeochem. Cy., 28, 927–949, https://doi.org/10.1002/2014GB004853, 2014.

Landschützer, P., Gruber, N., Haumann, F. A. Rödenbeck, C. Bakker, D. C. E. , van Heuven, S. Hoppema, M., Metzl, N., Sweeney, C., Takahashi, T., Tilbrook, B., and Wanninkhof, R.: The reinvigoration of the Southern Ocean carbon sink, Science, 349, 1221–1224, https://doi.org/10.1126/science.aab2620, 2015.

Landschützer, P., Gruber, N., and Bakker, D.C. E.: Decadal variations and trends of the global ocean carbon sink, Global Biogeochem. Cy., 30, 1396–1417, https://doi.org/10.1002/2015GB005359, 2016.

Laruelle, G. G., Dürr, H. H., Slomp, C. P., and Borges, A. V.: Evaluation of sinks and sources of CO_2 in the global coastal ocean using a spatially-explicit typology of estuaries and continental shelves, Geophys. Res. Lett., 37, L15607, https://doi.org/10.1029/2010gl043691, 2010.

Laruelle, G. G., Dürr, H. H., Lauerwald, R., Hartmann, J., Slomp, C. P., Goossens, N., and Regnier, P. A. G.: Global multi-scale segmentation of continental and coastal waters from the watersheds to the continental margins, Hydrol. Earth Syst. Sci., 17, 2029–2051, https://doi.org/10.5194/hess-17-2029-2013, 2013.

Laruelle, G. G., Lauerwald, R., Pfeil, B., and Regnier, P.: Regionalized global budget of the CO_2 exchange at the air-water interface in continental shelf seas, Global Biogeochem. Cy., 28, 1199–1214, https://doi.org/10.1002/2014GB004832, 2014.

Laruelle, G. G., Lauerwald, R., Rotschi, J., Raymond, P. A., Hartmann, J., and Regnier, P.: Seasonal response of air – water CO_2 exchange along the land – ocean aquatic continuum of the northeast North American coast, Biogeosciences, 12, 1447–1458, https://doi.org/10.5194/bg-12-1447-2015, 2015.

Liu, K.-K., Atkinson, L., Quinones, R., and Talaue-McManus, L. (Eds.): Carbon and Nutrient Fluxes in Continental Margins, Global Change – The IGBP Series, 3, Springer-Verlag Berlin Heidelberg, 2010.

Locarnini, R. A., Mishonov, A. V., Antonov, J. I., Boyer, T. P., Garcia, H. E., Baranova, O. K., Zweng, M. M., and Johnson D. R.: World Ocean Atlas 2009, Volume 1: Temperature, NOAA Atlas NESDIS, edited by: Levitus, S., US Gov. Print. Off., Washington, DC, Vol. 69, 2010.

Lohrenz, S. E. and Cai, W.-J.: Satellite ocean color assessment of air-sea fluxes of CO_2 in a river-dominated coastal margin, Geophys. Res. Lett., 33, L01601, https://doi.org/10.1029/2005GL023942, 2006.

Lovely, A., Loose, B., Schlosser, P., McGillis, W., Zappa C., Perovich, D., Brown, S., Morell, T., Hsueh, D., and Friedrich, R.: The Gas Transfer through Polar Sea ice experiment: Insights into the rates and pathways that determine geochemical fluxes, J. Geophys. Res.-Ocean., 120, 8177–8194, 2015.

Moreau, S., Vancoppenolle, M., Bopp, L., Aumont, O., Madec, G., Delille, B., Tison, J.-L., Barriat, P.-Y., and Goosse, H.: Assessment of the sea-ice carbon pump: Insights from a three-dimensional ocean-sea-ice-biogeochemical model (NEMO-LIM-PISCES), Elementa, 4, 000122, https://doi.org/10.12952/journal.elementa.000122, 2016.

Nakaoka, S., Telszewski, M., Nojiri, Y., Yasunaka, S., Miyazaki, C., Mukai, H., and Usui, N.: Estimating temporal and spatial variation of ocean surface pCO_2 in the North Pacific using a self-organizing map neural network technique, Biogeosciences, 10, 6093–6106, https://doi.org/10.5194/bg-10-6093-2013, 2013.

NASA Goddard Space Flight Center, Ocean Ecology Laboratory, Ocean Biology Processing Group: MODIS-Aqua chlorophyll Data (Dataset Release 2016); NASA Goddard Space Flight Center, Ocean Ecology Laboratory, Ocean Biology Processing Group, 2016.

Omer, W. M. M.: Ocean acidification in the Arabian Sea and the Red Sea. Master's thesis, University of Bergen, 2011.

Parmentier, F.-J. W., Christensen, T. R., Sørensen, L. L., Rysgaard, S., McGuire, A. D., Miller, P. A., and Walker, D. A.: The impact of lower sea-ice extent on Arctic greenhouse-gas exchange, Nature Climate Change, 3, 195–202, https://doi.org/10.1038/nclimate1784, 2013.

Pfeil, B., Olsen, A., Bakker, D. C. E., Hankin, S., Koyuk, H., Kozyr, A., Malczyk, J., Manke, A., Metzl, N., Sabine, C. L., Akl, J.,

Alin, S. R., Bates, N., Bellerby, R. G. J., Borges, A., Boutin, J., Brown, P. J., Cai, W.-J., Chavez, F. P., Chen, A., Cosca, C., Fassbender, A. J., Feely, R. A., González-Dávila, M., Goyet, C., Hales, B., Hardman-Mountford, N., Heinze, C., Hood, M., Hoppema, M., Hunt, C. W., Hydes, D., Ishii, M., Johannessen, T., Jones, S. D., Key, R. M., Körtzinger, A., Landschützer, P., Lauvset, S. K., Lefèvre, N., Lenton, A., Lourantou, A., Merlivat, L., Midorikawa, T., Mintrop, L., Miyazaki, C., Murata, A., Nakadate, A., Nakano, Y., Nakaoka, S., Nojiri, Y., Omar, A. M., Padin, X. A., Park, G.-H., Paterson, K., Perez, F. F., Pierrot, D., Poisson, A., Ríos, A. F., Santana-Casiano, J. M., Salisbury, J., Sarma, V. V. S. S., Schlitzer, R., Schneider, B., Schuster, U., Sieger, R., Skjelvan, I., Steinhoff, T., Suzuki, T., Takahashi, T., Tedesco, K., Telszewski, M., Thomas, H., Tilbrook, B., Tjiputra, J., Vandemark, D., Veness, T., Wanninkhof, R., Watson, A. J., Weiss, R., Wong, C. S., and Yoshikawa-Inoue, H.: A uniform, quality controlled Surface Ocean CO_2 Atlas (SOCAT), Earth Syst. Sci. Data, 5, 125–143, https://doi.org/10.5194/essd-5-125-2013, 2013.

Regnier, P., Friedlingstein, P., Ciais, P., Mackenzie, F. T., Gruber, N., Janssens, I. A., Laruelle, G. G., Lauerwald, R., Luyssaert, S., Andersson, A. J., Arndt, S., Arnosti, C., Borges, A. V., Dale, A. W., Gallego-Sala, A., Goddéris, Y., Goossens, N., Hartmann, J., Heinze, C., Ilyina, T., Joos, F., LaRowe, D. E., Leifeld, J., Meysman, F. J. R., Munhoven, G., Raymond, P. A., Spahni, R., Suntharalingam, P., and Thullner, M.: Anthropogenic perturbation of the carbon fluxes from land to ocean, Nat. Geosci., 6, 597–607, https://doi.org/10.1038/ngeo1830, 2013.

Rödenbeck, C., Bakker, D. C. E., Gruber, N., Iida, Y., Jacobson, A. R., Jones, S., Landschützer, P., Metzl, N., Nakaoka, S., Olsen, A., Park, G.-H., Peylin, P., Rodgers, K. B., Sasse, T. P., Schuster, U., Shutler, J. D., Valsala, V., Wanninkhof, R., and Zeng, J.: Data-based estimates of the ocean carbon sink variability – first results of the Surface Ocean pCO_2 Mapping intercomparison (SOCOM), Biogeosciences, 12, 7251–7278, https://doi.org/10.5194/bg-12-7251-2015, 2015.

Sabine, C. L., Hankin, S., Koyuk, H., Bakker, D. C. E., Pfeil, B., Olsen, A., Metzl, N., Kozyr, A., Fassbender, A., Manke, A., Malczyk, J., Akl, J., Alin, S. R., Bellerby, R. G. J., Borges, A., Boutin, J., Brown, P. J., Cai, W.-J., Chavez, F. P., Chen, A., Cosca, C., Feely, R. A., González-Dávila, M., Goyet, C., Hardman-Mountford, N., Heinze, C., Hoppema, M., Hunt, C. W., Hydes, D., Ishii, M., Johannessen, T., Key, R. M., Körtzinger, A., Landschützer, P., Lauvset, S. K., Lefèvre, N., Lenton, A., Lourantou, A., Merlivat, L., Midorikawa, T., Mintrop, L., Miyazaki, C., Murata, A., Nakadate, A., Nakano, Y., Nakaoka, S., Nojiri, Y., Omar, A. M., Padin, X. A., Park, G.-H., Paterson, K., Perez, F. F., Pierrot, D., Poisson, A., Ríos, A. F., Salisbury, J., Santana-Casiano, J. M., Sarma, V. V. S. S., Schlitzer, R., Schneider, B., Schuster, U., Sieger, R., Skjelvan, I., Steinhoff, T., Suzuki, T., Takahashi, T., Tedesco, K., Telszewski, M., Thomas, H., Tilbrook, B., Vandemark, D., Veness, T., Watson, A. J., Weiss, R., Wong, C. S., and Yoshikawa-Inoue, H.: Surface Ocean CO_2 Atlas (SOCAT) gridded data products, Earth Syst. Sci. Data, 5, 145–153, https://doi.org/10.5194/essd-5-145-2013, 2013.

Sasse, T. P., McNeil, B. I., and Abramowitz, G.: A new constraint on global air-sea CO_2 fluxes using bottle carbon data, Geophys. Res. Lett., 40, 1594–1599, https://doi.org/10.1002/grl.50342, 2013.

Sarma, V. V. S. S.: Monthly variability in surface pCO_2 and net air-sea CO_2 flux in the Arabian Sea, J. Geophys. Res., 108, 3255, https://doi.org/10.1029/2001JC001062, 2003.

Shadwick, E. H., Thomas, H., Comeau, A., Craig, S. E., Hunt, C. W., and Salisbury, J. E.: Air-Sea CO_2 fluxes on the Scotian Shelf: seasonal to multi-annual variability, Biogeosciences, 7, 3851–3867, https://doi.org/10.5194/bg-7-3851-2010, 2010.

Signorini, S. R., Mannino, A., Najjar Jr., R. G., Friedrichs, M. A. M., Cai, W.-J., Salisbury, J., Wang, Z. A., Thomas, H., and Shadwick, E.: Surface ocean pCO_2 seasonality and sea-air CO_2 flux estimates for the North American east coast, J. Geophys. Res.-Ocean., 118, 5439–5460, https://doi.org/10.1002/jgrc.20369, 2013.

Takahashi, T., Olafsson, J., Goddard, J. G., Chipman, D. W., and Sutherland, S. C.: Seasonal variation of CO_2 and nutrients in the high-latitude surface oceans: A comparative study, Global Biogeochem. Cy., 7, 843–878, 1993.

Takahashi, T., Sutherland, S., and Kozyr A.: Global ocean surface water partial pressure of CO_2 database: Measurements performed during 1957–2011 (Version 2011), ORNL/CDIAC-160, NDP-088(V2011), Carbon Dioxide Information Analysis Center, Oak Ridge Natl. Lab., US Department of Energy, Oak Ridge, Tennessee, 2012.

Takahashi, T., Sutherland, S. C., and Kozyr, A.: Global Ocean Surface Water Partial Pressure of CO_2 Database: Measurements Performed During 1957–2015 (Version 2015), ORNL/CDIAC-160, NDP-088(V2015), Carbon Dioxide Information Analysis Center, Oak Ridge National Laboratory, US Department of Energy, Oak Ridge, Tennessee, https://doi.org/10.3334/CDIAC/OTG.NDP088(V2015), 2016.

Thomas, H. and Schneider, B.: The seasonal cycle of carbon dioxide in Baltic Sea surface waters, J. Mar. Syst., 22, 53–67, 1999.

Turi, G., Lachkar, Z., and Gruber, N.: Spatiotemporal variability and drivers of pCO_2 and air–sea CO_2 fluxes in the California Current System: an eddy-resolving modeling study, Biogeosciences, 11, 671–690, https://doi.org/10.5194/bg-11-671-2014, 2014.

US Department of Commerce, National Oceanic and Atmospheric Administration, National Geophysical Data Center, 2-minute Gridded Global Relief Data (ETOPO2v2), http://www.ngdc.noaa.gov/mgg/fliers/06mgg01.html (last access: 26 December 2008), 2006.

Vancoppenolle M., Meiners, K. M., Michel, C., Bopp, L., Brabant, F., Carnat, G., Delille, B., Lannuzel, D., Madec, G., Moreau, S., Tison, J.-L., and van der Merwe, P.: Role of sea ice in global biogeochemical cycles: Emerging views and challenges, Quaternary Sci. Rev., 79, 207–230, https://doi.org/10.1016/j.quascirev.2013.04.011, 2013.

Wanninkhof, R., Park, G.-H., Takahashi, T., Sweeney, C., Feely, R., Nojiri, Y., Gruber, N., Doney, S. C., McKinley, G. A., Lenton, A., Le Quéré, C., Heinze, C., Schwinger, J., Graven, H., and Khatiwala, S.: Global ocean carbon uptake: magnitude, variability and trends, Biogeosciences, 10, 1983–2000, https://doi.org/10.5194/bg-10-1983-2013, 2013.

Walsh, J. J.: On the nature of continental shelves, Academic Press, San Diego, New York, Berkeley, Boston, London, Sydney, Tokyo, Toronto, 1988.

Wijesekera, H. W., Allen, J. S., and Newberger, P. A.: Modeling study of turbulent mixing over the continental shelf: Comparison of turbulent closure schemes, J. Geophys. Res., 108, 3103, https://doi.org/10.1029/2001JC001234, 2003

Yasunaka, S., Murata, A., Watanabe, E., Chierici, M., Fransson, A., van Heuven, S., Hoppema, M., Ishii, M., Johannessen, T., Kosugi, N., Lauvset, S. K., Mathis, J. T., Nishino, S., Omar, A. M., Olsen, A., Sasano, D., Takahashi, T., and Wanninkhof, R.: Mapping of the air–sea CO_2 flux in the Arctic Ocean and its adjacent seas: Basin-wide distribution and seasonal to interannual variability, Polar Science, 10, 323–334, https://doi.org/10.1016/j.polar.2016.03.006, 2016.

Zeng, J., Nojiri, Y., Landschützer, P., Telszewski, M., and Nakaoka, S.: A global surface ocean fCO_2 climatology based on a feedforward neural network, J. Atmos. Ocean Technol., 31, 1838–1849, 2014.

Fire-regime variability impacts forest carbon dynamics for centuries to millennia

Tara W. Hudiburg[1], Philip E. Higuera[2], and Jeffrey A. Hicke[3]

[1]Department of Forest, Rangeland, and Fire Sciences, University of Idaho, 875 Perimeter Dr.,
Moscow, ID 83844-1133, USA

[2]Department of Ecosystem and Conservation Sciences, University of Montana, 32 Campus Dr.,
Missoula, MT 59812, USA

[3]Department of Geography, University of Idaho, 875 Perimeter Dr.,
Moscow, ID 83844-3021, USA

Correspondence to: Tara W. Hudiburg (thudiburg@uidaho.edu) and Philip E. Higuera (philip.higuera@umontana.edu)

Abstract. Wildfire is a dominant disturbance agent in forest ecosystems, shaping important biogeochemical processes including net carbon (C) balance. Long-term monitoring and chronosequence studies highlight a resilience of biogeochemical properties to large, stand-replacing, high-severity fire events. In contrast, the consequences of repeated fires or temporal variability in a fire regime (e.g., the characteristic timing or severity of fire) are largely unknown, yet theory suggests that such variability could strongly influence forest C trajectories (i.e., future states or directions) for millennia. Here we combine a 4500-year paleoecological record of fire activity with ecosystem modeling to investigate how fire-regime variability impacts soil C and net ecosystem carbon balance. We found that C trajectories in a paleo-informed scenario differed significantly from an equilibrium scenario (with a constant fire return interval), largely due to variability in the timing and severity of past fires. Paleo-informed scenarios contained multi-century periods of positive and negative net ecosystem C balance, with magnitudes significantly larger than observed under the equilibrium scenario. Further, this variability created legacies in soil C trajectories that lasted for millennia. Our results imply that fire-regime variability is a major driver of C trajectories in stand-replacing fire regimes. Predicting carbon balance in these systems, therefore, will depend strongly on the ability of ecosystem models to represent a realistic range of fire-regime variability over the past several centuries to millennia.

1 Introduction

Wildfire is a pervasive disturbance agent in forest ecosystems, strongly shaping ecosystem structure and function, including vegetation composition, nutrient cycling, and energy flow. While the immediate impacts of disturbance can be dramatic, the longevity of these impacts is less clear. In ecosystems where disturbance is historically prevalent, vegetation and biogeochemical properties typically return to pre-disturbance conditions over years to decades (Dunnette et al., 2014; McLauchlan et al., 2014), motivating the concept of "biogeochemical resilience" (Smithwick, 2011). Characterizing biogeochemical resilience emphasizes understanding pool sizes and changes to inputs or outputs of key elements (McLauchlan et al., 2014; Smithwick, 2011). In the context of wildfire, biogeochemical resilience is determined by pool sizes (e.g., carbon, nitrogen) prior to a fire event, elemental losses and transformations that occur during and shortly after a fire event (e.g., from volatilization and erosion), and post-fire changes in elemental pools, which in turn are determined by the rate and composition of post-fire revegetation (McLauchlan et al., 2014; Schlesinger et al., 2015; Smithwick, 2011).

Changes in the characteristic frequency or severity of fire (i.e., the fire regime) are therefore predicted to lead to compounding and potentially long-lasting changes or shifts in biogeochemical states. For example, increased disturbance frequency can deplete key growth-limiting nutrients (Yelenik et al., 2013), potentially influencing ecosystem trajectories for decades to centuries (McLauchlan et al., 2014). Net ecosystem carbon balance (NECB; the balance between net forest carbon uptake and forest losses through fire emissions; Chapin et al., 2006) is also highly sensitive to disturbance (Hudiburg et al., 2011), and while NECB trends towards 0 under a uniform disturbance regime, shifting disturbance regimes may alter NECB over centuries to millennia (Goetz et al., 2012; Kelly et al., 2016). While these ideas have a strong conceptual basis and empirical support on decadal timescales, we have lacked the data needed to test them over longer timescales – and to consider their implications for future projections – until only recently.

Coupling paleo-observations (i.e., "paleo-informed") with ecosystem modeling provides an important tool for assessing the impacts of fire-regime variability on biogeochemical dynamics by combining the mechanistic representation of ecosystem processes with actual patterns of fire activity reconstructed from the past. For example, in Alaskan boreal forests paleo-informed ecosystem modeling highlights fire as the dominant control on C cycling over the past millennium, far outweighing the effects of climate variability (Kelly et al., 2016). Given the significant influence of fire, estimates of modern C states (initial conditions for modeling future C states) can be highly sensitive to assumptions about the past fire activity. Ecosystem models typically require a spin-up period to equilibrate C and N pools and can include a fixed disturbance interval (e.g., a constant fire return interval), resulting in ecosystem C and N trajectories that are in equilibrium with climate, ecosystem properties, and the disturbance regime. To initiate the model, C and N pools need to develop, as they start from bare soil with no vegetation; as vegetation grows the modeled soil pools increase, and it takes hundreds to thousands of simulation years during this spin-up period for the C and N pools to equilibrate. Following centuries of equilibrium, known disturbance events from the historical record are included, and the final results are used for initial conditions (baseline) for future scenarios. However, paleo-informed disturbance histories spanning many centuries can result in initial conditions that differ from equilibrium runs. In the boreal example, forests were a small net C source over the past several decades in paleo-informed simulations, whereas forests were a small net C sink when a constant fire return interval was assumed (Kelly et al., 2016). We would expect a similar sensitivity of C dynamics to fire in other stand-replacing fire regimes, although specific trajectories and impacts on modern states could vary widely, contingent on the specific history of fire activity.

Here, we pair a paleoecological record of vegetation and wildfire activity in a subalpine forest (Dunnette et al., 2014) with an ecosystem model to evaluate the sensitivity of forest ecosystem processes to fire-regime variability over a 4500-year period. Our paleoecological record reveals the timing and severity of past wildfire activity within a subalpine forest watershed that was consistently dominated by lodgepole pine (*Pinus contorta*). We use this record to drive fire disturbances

in an ecosystem model and test alternative hypotheses that help reveal the potential patterns and mechanisms causing past ecosystem change, focusing on a slowly varying carbon pool (soil C) and NECB. The resulting trends provide theoretical insight into how observed fire-regime variability can affect carbon trajectories from decadal to millennial scales. Through a series of paleo-informed and control modeling scenarios, we address two key questions about the biogeochemical impacts and legacies of wildfire activity: (1) how does centennial-to-millennial-scale variability in fire activity impact biogeochemical processes that regulate soil C and NECB, and (2) for how long does the legacy wildfire activity impact current biogeochemical states? In addition to testing the general hypothesis that forest carbon storage will differ between equilibrium and paleo-informed simulations, we also evaluate the impact of increasing or decreasing fire frequency, relative to that inferred from the paleo record.

2 Materials and methods

2.1 Model description

DayCent is the globally recognized daily timestep version of the biogeochemical model CENTURY, widely used to simulate the effects of climate and disturbance on ecosystem processes including forests worldwide (Bai and Houlton, 2009; Hartman et al., 2007; Savage et al., 2013). DayCent is a logical choice for our purposes, because it includes soil C pools that have long turnover times, spanning months to 4000 years, and thus can represent long-term ecosystem change. As used here, DayCent is aspatial, representing our ca. 30 ha study watershed as a single point. Detailed model documentation and publication lists can be found on the following website: http://www.nrel.colostate.edu/projects/daycent-downloads.html.

Required inputs for the model include vegetation cover, daily precipitation and temperature, soil texture, and disturbance histories. DayCent calculates potential plant growth as a function of water, light, and soil temperature and limits actual plant growth based on soil nutrient availability. The model includes three soil organic matter (SOM) pools (active, slow, and passive) with different decomposition rates, above and belowground litter pools, and a surface microbial pool associated with the decomposing surface litter. Plant material is split into structural and metabolic material as a function of the lignin-to-nitrogen ratio of the litter (more structural with higher lignin-to-nitrogen ratios). The active pool (microbial) has short turnover times (1–3 months), and the slow SOM pool (more resistant structural plant material) has turnover times ranging from 10 to 50 years depending on the climate. The passive pool includes physically and chemically stabilized SOM with turnover times ranging from 400 to 4000 years. For this study, DayCent was parameterized to model soil organic carbon dynamics to a depth of 30 cm.

Model outputs include soil C and N stocks, live and dead biomass, above- and belowground net primary productivity (NPP), heterotrophic respiration (R_h), fire emissions, and net ecosystem production (NEP, defined as the difference between NPP and heterotrophic respiration). We define NECB as the difference between NEP and fire emissions.

Disturbances in DayCent are prescribed and can be parameterized to reflect severity through associated impacts to the ecosystem (e.g., biomass killed, nitrogen lost, soil eroded). The fire model in DayCent is parameterized to include the combusted and/or mortality fraction of each carbon pool (live and dead wood, foliage, coarse and fine roots, etc.) that occurs with each fire event. Erosion is also scheduled as an event in DayCent and was prescribed to occur in the same year of the observed high-severity fire events. The erosion events are thus decoupled from precipitation in the model.

2.2 Study sites

We studied the biogeochemical consequences of fire-regime variability by informing the DayCent model with fire history data derived from sedimentary charcoal preserved in Chickaree Lake, Colorado (Dunnette et al., 2014). Chickaree Lake (40.334° N, 105.841° W; 2796 m above sea level) is a small, deep lake (ca. 1.5 ha surface area; 7.9 m depth) in a lodgepole-pine-dominated subalpine forest in Rocky Mountain National Park. The even-aged forest surrounding the lake regenerated after a high-severity (i.e., stand-replacing) fire in 1782 CE (common era) (Sibold et al., 2007). The fire regime in subalpine forests of Rocky Mountain National Park is characterized by infrequent, high-severity crown fires (ca. 100–300-year mean return intervals) associated with severe seasonal drought (Sibold et al., 2006). The mean monthly temperature is −8.5 °C in January and 14° C in July, and the average total annual precipitation is 483 mm (Western Regional Climate Center 1940–2013 observations from Grand Lake, CO).

Detailed methods for the collection and analysis of the Chickaree Lake sediment record are found in Dunnette et al. (2014). Briefly, the 4500-year record has an average sample resolution of 4 years and a chronology constrained by 13 ^{210}Pb dates spanning the upper 20 cm and 25 accelerator mass spectrometry ^{14}C dates for deeper sediments. Pollen analysis indicates that the site was continuously dominated by lodgepole pine for the duration of the record presented here, with successional changes following inferred fire events (Dunnette et al., 2014). The persistence of subalpine forest over the past 4500 years is also supported by nearby pollen records in Rocky Mountain National Park (Caffrey and Doerner, 2012; Higuera et al., 2014). Dunnette et al. (2014) used macroscopic charcoal and magnetic susceptibility (a soil-erosion proxy) from Chickaree Lake to infer the timing and severity of wildfires, identifying high-severity catchment fires (those with associated erosion) and lower-severity/extra-local fires (those without associated soil ero-

sion). Thus, while all fire events were likely stand-replacing, the difference between these two fire types was the association with soil erosion. Here, we use the Chickaree Lake fire history record to inform the disturbance component of the DayCent ecosystem model by prescribing the timing and severity of past fire events within a simulated lodgepole-pine-dominated subalpine forest.

2.3 Model parameterization

DayCent submodels associated with tree physiological parameters, site characteristics, soil parameters, and disturbance events were modified using available site-specific observations (Dunnette et al., 2014; Sibold et al., 2007), values from the literature (Kashian et al., 2013; Turner et al., 2004), and publically available climate and soil databases. Climate data required for DayCent include daily minimum and maximum temperature and precipitation, which were obtained for a 30-year period from DAYMET (Thornton, 2012). For all model runs, the 30-year climate dataset was recycled for the duration of the run; thus, climate was functionally non-varying over the duration of the simulations (beyond the variability within the 30-year dataset). Soil texture and classification were identified using the United States National Resource Conservation Service (NRCS) SSURGO database (NRCS, 2010). Model input and parameterization files are available for download as supporting information files.

We defined two types of stand-replacing fire to distinguish between the two types of fires identified in the paleo record. The key difference between the two fire types simulated is the associated soil erosion. High-severity catchment fires from the paleo record were simulated by 95 % tree mortality and a soil-erosion event with ∼ 1 Mg ha^{-1} of soil loss from the watershed (Miller et al., 2011); we refer to these as high-severity fires with erosion. Lower-severity/extra-local fires from the paleo record were simulated by 95 % tree mortality with no associated soil-erosion event; we refer to these as high-severity fires without erosion. After parameterization, we evaluated modern modeled aboveground NPP, soil C, total ecosystem carbon, and disturbance C losses against observations of similar-aged lodgepole pine stands in the Central Rockies ecoregion (Hansen et al., 2015; Kashian et al., 2013; Turner et al., 2004).

2.4 Model experiments

We performed a series of modeling experiments to address our questions using the Chickaree Lake paleo-fire record, varied disturbance histories, and varied climate (Table 1). First, DayCent was spun-up and equilibrated to soil C and NPP levels characteristic of mature lodgepole pine stands in the region with a constant return interval of 145 years between high-severity fires with erosion, replicating the estimated fire rotation period (and mean fire return interval) for the broader study area (Sibold et al., 2007). This spin-up

Table 1. Model simulation scenarios, including climate, fire regime, duration, and summary description.

Scenario	Purpose	Climate[a]	Fire regime	Duration (yr)	Description
Spin-up	Spin-up C, N pools to equilibrium conditions	Ambient	Fixed 145-year return interval; high severity with erosion	2000	DayCent initialization run for NPP and C to reach equilibrium conditions
Equilibrium	Run with fixed fire interval	Ambient	Fixed 145-year return interval; high severity with erosion	4561	Equilibrium run extended from the spin-up run for the length of the paleo-fire record
Paleo-informed	Run with observed paleo-fire intervals and severity	Ambient	Paleo record; high severity with and without erosion	4561	A 4561-year simulation with fires matching the timing and severity from the paleo-fire record
Increased fire frequency	Run with paleo-fire intervals decreased by 25 %	Ambient	Modified paleo record; 90-year MFRI (mean fire return interval) with high severity with and without erosion	4561	A 4561-year simulation with the timing between fires in the paleo-informed scenario decreased by 25 %
Decreased fire frequency	Run with paleo-fire intervals increased by 25 %	Ambient	Modified paleo record; 155-year MFRI with high severity with and without erosion	4561	A 4561-year simulation with the timing between fires in the paleo-informed scenario increased by 25 %
$Paleo_{500} \ldots Paleo_{4000}$	Test influence of length of paleo record on modern states	Ambient	Paleo record; high severity with and without erosion	500–4000	Branches from the equilibrium scenario at varying points in time, in 500-year increments.[b]; all scenarios ends in 2010 CE

[a] Thirty-year recycled historical record (DayMet). [b] For example, the 500-year simulation starts in the year 1510 CE and runs until the end of 2009 CE.

period lasted for 2000 years, and it represents what would be done for model use, in the absence of the long-term fire history information from the paleo record. All experimental simulations were extended from this spin-up equilibrium simulation starting 4500 years before present (BP, where present is 1950 CE) and running through 2010 CE, for a total of 4561 simulation years. We defined our model simulation that would normally be used in the absence of paleo-informed disturbance histories (equilibrium scenario) as a continuation of the equilibrated spin-up with the same climate and fire regime, with only the last known fire event (1782 CE) explicitly simulated.

In addition to this equilibrium scenario, we implemented three additional scenarios that together helped illustrate the duration, magnitude, and relative importance of fire-induced changes to forest biogeochemistry. First, to test the impacts of variability in fire timing and severity on important biogeochemical states, we compared the equilibrium scenario to a paleo-informed scenario, which had a mean fire return interval of 120 years for all fires and 334 years for the high-severity fires with erosion. Climate was identical in each simulation (i.e., 30-year recycled modern climate), as we are not testing the influence of climate on the timing and severity of fire, but rather the influence of the known timing and severity of fires (from the charcoal record) versus a constant fire return interval.

Second, to identify the duration of a legacy effect from fire-regime variability, we constructed eight partially paleo-informed scenarios, which included increasingly longer periods of information from the paleo-fire record, spanning the past 500 to 4000 years, in 500-year increments that ended in 2010 CE ($Paleo_{500}$, $Paleo_{1000}$, ..., $Paleo_{4000}$; Fig. 1a). For example, the $Paleo_{500}$ scenario includes the most recent 500 years of fire history, while the $Paleo_{4000}$ scenario includes the most recent 4000 years of fire history.

Thirdly, to identify how a systematic shift in fire frequency would impact carbon balance, we created two additional scenarios with shortened and lengthened fire return intervals. Beginning with the observed paleo-fire record, we modified each interval between fires to be (a) shortened by 25 % (increased fire frequency) or (b) lengthened (decreased fire frequency) by 25 % (Fig. 1b). The corresponding mean fire return intervals of these two additional runs were 90 years for the increased fire frequency and 155 years for the decreased fire frequency scenarios.

Because fire events in DayCent are decoupled from climate, the climate data did not impact the timing or severity of fires in the simulations. We evaluated the results from each scenario in terms of the modern end points of soil C, soil N, and NECB as well as total cumulative changes in NECB over the entire record. We define cumulative NECB as a running total, such that the sum at any given year represents the inte-

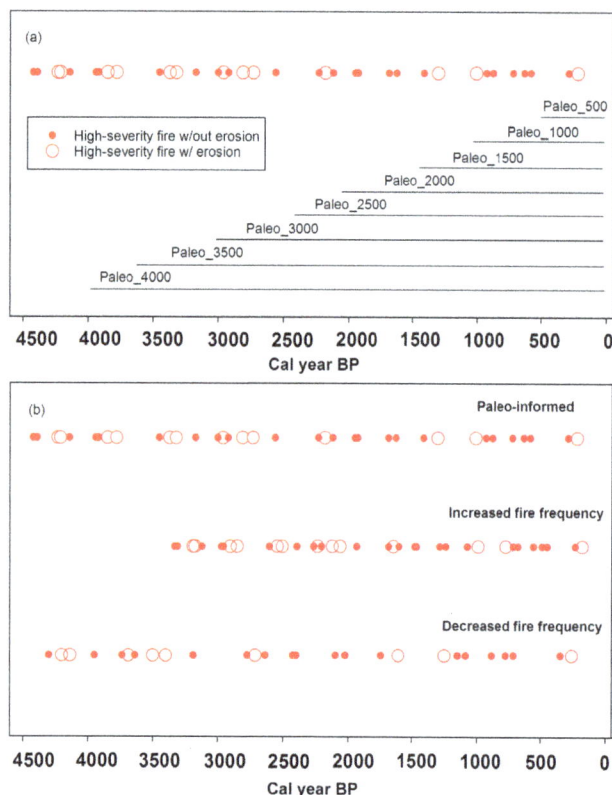

Figure 1. Paleo-informed fire history scenarios used to drive the DayCent model. **(a)** Fire history record form Chickaree Lake (red circles), with horizontal lines illustrating the duration of the record used in the incremental partially paleo-informed scenarios (Paleo_500…4000). **(b)** The same full Chickaree Lake fire history record used in the paleo-informed scenario **(a)**, with the two additional scenarios representing a 25 % increase and 25 % decrease in fire frequency.

grated impacts of past disturbance events. For example, when return intervals between disturbance events are shorter than C recovery times, cumulative NECB will remain negative. Finally, we considered uncertainty in our estimates based on the uncertainty in the reconstructed fire history record, our assumptions about soil erosion, and our use of recycled modern climate. While there is also uncertainty associated with modeled estimates of soil C, NECB, and other C fluxes presented, we are not attempting to provide estimates that are any more precise than measured modern states (e.g., STATSGO-derived soil C). Rather, we compare the variability in biogeochemical states arising from fire-regime variability to the uncertainties in the model that are revealed when evaluated against modern observations from the literature.

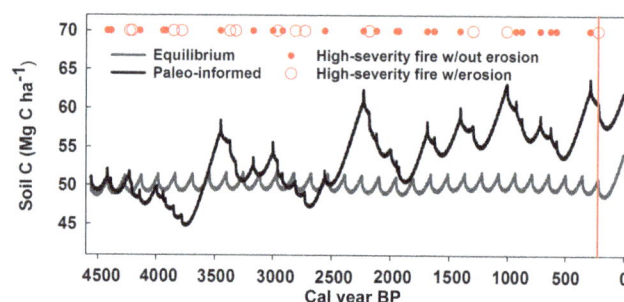

Figure 2. Model simulations of equilibrium (grey) and paleo-informed (black) total soil carbon (C) in $\mathrm{Mg\,C\,ha^{-1}}$. Each simulation branches from a 2000-year equilibrium spin-up starting at the same soil C baseline and runs for 4561 years (4500 BP to 2010 CE). The large open circles represent the years of the high-severity fires with erosion, and the small closed circles are high-severity fires without erosion used to drive the paleo-informed model run. A constant 145-year fire return interval was used for the equilibrium run. The vertical red line indicates the most recent stand-replacing fire (1782 CE), reconstructed from the tree-ring record (Sibold et al., 2007).

3   Results and discussion

3.1   Model parameterization and evaluation

We compared our model results with reported values from ecological studies in the region that examined some aspect of the carbon balance in the similar-aged subalpine forests in order to evaluate our model estimates. We found few reported observations (e.g., for C, N pools, NPP) for old (> 200 years) lodgepole pine stands in the Rocky Mountains in the literature. Therefore, we also compare our results with results for the same genus (*Pinus*) and with the soil C content reported by the NRCS as part of the national soil survey. Our modeled estimates of modern soil C (to 30 cm) of 54 and 62 $\mathrm{Mg\,C\,ha^{-1}}$, for the equilibrium and paleo-informed scenario, respectively (Fig. 2), compare well with the NRCS-derived estimates (STATSGO2, NRCS, 2010) of $66 \pm 16\,\mathrm{Mg\,C\,ha^{-1}}$ for the Chickaree Lake region and with measurements of current soil C (to 30 cm) ranging from 51 to 73 $\mathrm{Mg\,C\,ha^{-1}}$ in similarly aged (> 200 years) Rocky Mountain *Pinus* stands (Bradford et al., 2008). Modeled estimates of aboveground NPP were also in agreement with observations, averaging 156 and 172 $\mathrm{g\,C\,m^{-2}}$ for the equilibrium and paleo-informed simulations, respectively, compared to estimates from the northern or central Rockies ranging from 100 to 200 $\mathrm{g\,C\,m^{-2}}$ (Hansen et al., 2015). Finally, fire emissions from our modeled estimates range from 20 to 30 % loss of aboveground C, broadly in agreement with other studies (Campbell et al., 2007; Smithwick et al., 2009).

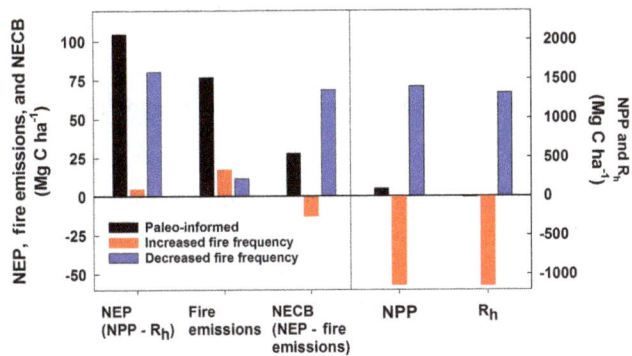

Figure 3. Accumulated anomalies in fluxes relative to the equilibrium scenario, in $Mg\,C\,ha^{-1}$, summed over the entire 4561-year simulation period. NEP, fire emissions, and NECB (left y axis) and NPP and R_h (right y axis) for the paleo-informed (black), increased fire frequency (red; 155-year mean FRI), and decreased fire frequency (blue; 90-year mean FRI) scenarios. Negative (positive) numbers indicate a decrease (increase) in total carbon flux compared to the equilibrium scenario.

3.2 Fire-regime variability impacts soil C and NECB

When DayCent was driven with the paleo-informed fire history, soil C accumulation was $8\,Mg\,ha^{-1}$ more at the end of the simulation than in the equilibrium scenario (Fig. 2). Total NEP summed over the 4561-year period was also higher in the paleo-informed scenario ($1276\,Mg\,C\,ha^{-1}$) compared with the equilibrium scenario ($1171\,Mg\,C\,ha^{-1}$), directly reflecting NPP rates that were higher than heterotrophic respiration (Fig. 3, black bars). In the paleo-informed scenario, cumulative emissions due to combustion losses (i.e., fire emissions) were lower than NEP over the entire record, resulting in a cumulative NECB of $27\,Mg\,C\,ha^{-1}$ more than the equilibrium scenario (Fig. 3; black bars).

The paleo-informed scenario showed substantial variability in soil C (Fig. 2) and NECB (Fig. 4) trajectories and higher total accumulations relative to the equilibrium scenario. In fact, the range of variability in soil C over the paleo-informed simulation, from ca. 45 to $65\,Mg\,C\,ha^{-1}$, nearly spanned the range of observations of current soil C (to 30 cm) in similarly aged (> 200 years) Rocky Mountain *Pinus* stands (Bradford et al., 2008). For the first ~ 2000 years of the paleo-informed scenario, long-term mean soil C was similar to baseline levels of soil C in the equilibrium scenario (Fig. 2), averaging around $54\,Mg\,C\,ha^{-1}$, though with substantial variability on centennial timescales. Following this period, the soil C trajectory increased distinctly in the paleo-informed scenario during a 500-year period with only one high-severity fire without erosion (ca. 2500 cal yr BP). Despite a return to a mean fire return interval closer to the equilibrium scenario, soil C persisted at this elevated level for the following 2000 years (ca. 2000 cal yr BP to present), resulting in $8\,Mg\,C\,ha^{-1}$ (15 %) more than the equilibrium

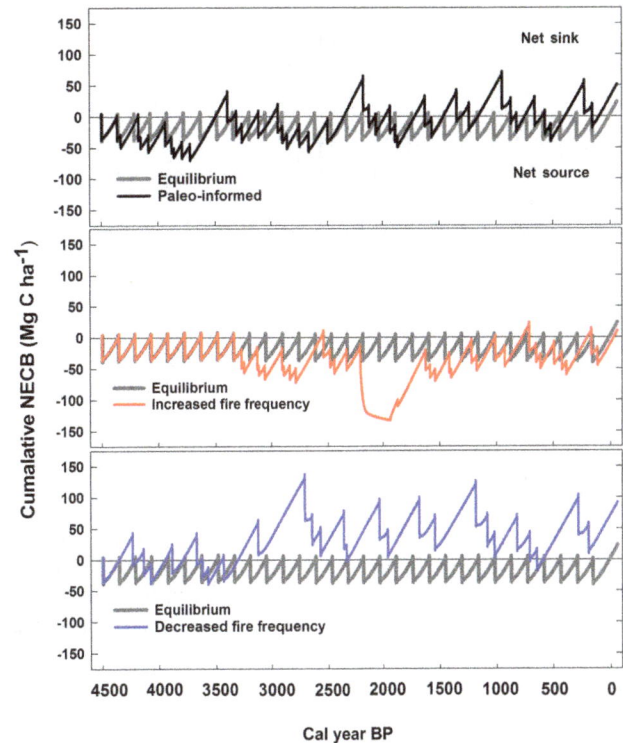

Figure 4. Trends in cumulative NECB over time for the paleo-informed, increased fire frequency, and decreased fire frequency scenarios compared to equilibrium over the last 4561 years. Positive numbers indicate a cumulative net sink, while negative numbers indicate a cumulative net source.

scenario at the end of the simulation (2010 CE). A similar trend was observed for NECB (Fig. 4), where the paleo-informed scenario maintained a lower NECB in the first half of the record compared to the second half. In the latter half of the record, NECB was more consistently positive, ultimately storing more ecosystem C than the equilibrium scenario. The dynamism in NECB over time is consistent with the findings of Kelly et al. (2016). Together, this work and ours highlights the value of examining the ecosystem impacts of past fire-regime variability, which may include disturbance-free or intensified disturbance periods that are not currently represented in or predicted by ecosystem models.

3.3 Impacts of fire-regime variability last for millennia

We compared the partially paleo-informed scenarios to the equilibrium scenario to determine the length of time necessary to arrive at the same inferences about soil C and NECB as in the full paleo-informed scenario. The 2010 CE end points for each partially informed scenario were compared to the 2010 CE end point for the equilibrium scenario. We found that disturbance-regime legacies lasted for millennia. The number of years needed to simulate the 2010 CE values was between 2000 and 2500 years (Fig. 5). Specifi-

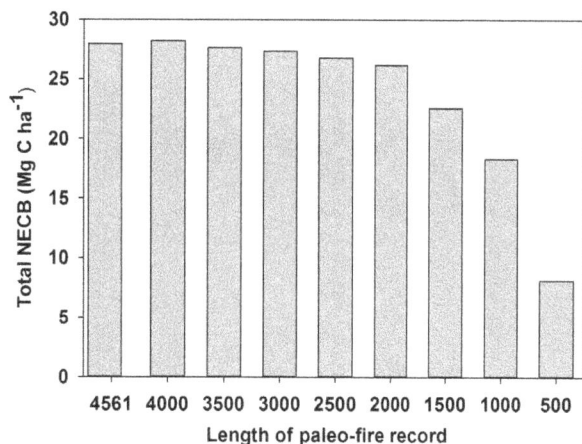

Figure 5. Total NECB (NPP – R_h – fire emissions) for the 4561-year simulated period and for each of the partially paleo-informed scenarios (Paleo_500, Paleo_1000, etc., in Fig. 1). Each partially paleo-informed scenario branches from the equilibrium scenario in the year indicated on the x axis. For example, the 500-year record only includes fires that occurred in the most recent 500 years of the paleo-fire record (1511–2010 CE).

cally, total NECB and soil C (end points that serve as initial conditions for future modeled states) were nearly the same when using 2500 to 4500 years of the paleo-fire record but differed by more than $1 \, \mathrm{Mg} \, \mathrm{C} \, \mathrm{ha}^{-1}$ when using only 500 to 2000 years of the paleo-fire record. We used the $1 \, \mathrm{Mg} \, \mathrm{C} \, \mathrm{ha}^{-1}$ as a significant threshold for changes in ecosystem C flux (total or soils) both because changes less than this indicate the ecosystem is stable and because it is a standard amount of annual C flux into or out of an ecosystem that is considered significant for carbon sequestration (mitigation) activities (Anderson-Teixeira et al., 2009).

Differences between the paleo-informed and equilibrium scenario can be interpreted in the context of other model parameters that are known to affect biogeochemical processes, including plant productivity and decomposition rates. Chief among these is growing season temperature, which strongly affects NPP and plant and microbial respiration in Day-Cent. In a simple sensitivity analysis where we repeated the equilibrium scenario with a uniform 2 °C warming during the growing season, we found that variability in the paleo-informed scenario was 1 order of magnitude greater than in the scenario with warming. Specifically, warming resulted in a small net decrease in soil C of $0.3 \, \mathrm{Mg} \, \mathrm{C} \, \mathrm{ha}^{-1}$ and a reduction in NECB by $0.2 \, \mathrm{Mg} \, \mathrm{C} \, \mathrm{ha}^{-1}$ relative to equilibrium scenario. Our results imply that C dynamics in lodgepole pine forests are far more sensitive to variability in the timing and severity of fire activity than to modeled changes to plant growth and decomposition introduced by climate warming alone. This inference is also consistent with findings from strand-replacing fire regimes in Alaskan boreal forests, where C dynamics over the past 1200 years were

more strongly shaped by fire activity than by climate variability (Kelly et al., 2016).

3.4 Implications for projecting future biogeochemical states

To evaluate the effects of changing fire regimes on our results, we varied the paleo-informed disturbance regimes by increasing and decreasing the frequency of events by 25 %. As expected, increased fire frequency (i.e., shorter return intervals) resulted in a cumulative loss of ecosystem C compared to equilibrium and paleo-informed scenarios, with NECB $13 \, \mathrm{Mg} \, \mathrm{C} \, \mathrm{ha}^{-1}$ lower compared to equilibrium over the entire simulation period (Fig. 3) and with periods of net carbon loss lasting nearly 800 years (Fig. 4; red line). The losses reflect large increases in fire emissions, without concurrent proportional increases in NEP (Fig. 3). In contrast, with decreased fire frequency (i.e., longer return intervals), NECB increases by $67 \, \mathrm{Mg} \, \mathrm{C} \, \mathrm{ha}^{-1}$ compared to equilibrium and by $40 \, \mathrm{Mg} \, \mathrm{C} \, \mathrm{ha}^{-1}$ compared to the original paleo-informed scenario. Again, this is primarily due to an unbalanced increase in NEP compared to fire emissions (Fig. 3).

While the differences in NECB (27 Mg C more) and soil C (8 Mg C more) between the paleo-informed and equilibrium scenarios are ultimately small for this single watershed, the impact of fire-regime variability will depend on the synchrony of events on the regional and sub-continental scales (Kelly et al., 2016). This is especially important when considering the trajectory of NECB compared to equilibrium simulations during the periods of the paleo record when fire frequency or severity were higher than in the past few centuries. Cumulative NECB was negative, serving as a net source of C to the atmosphere, for periods of up to 500 years in the paleo-informed scenario and up to 1000 years under scenarios with increased fire frequencies.

Given the strong correspondence between observed and simulated modern C stocks, we have high confidence that DayCent accurately simulated the key processes shaping biogeochemical properties in our study area. Important sources of uncertainty in our estimates of past carbon dynamics stem from uncertainty in the timing and severity of past fires. The fire history reconstruction has an estimated temporal precision of several decades ($\pm$10–20 years) (Dunnette et al., 2014), but because C dynamics unfold over centuries to millennia, this level of uncertainty has negligible effects on our inferences. A more important source of uncertainty is the potential for false positives or false negatives in the fire history reconstruction: failing to detect a fire that occurred in the past or identifying a fire that did not affect the Chickaree Lake watershed. While the Chickaree Lake record clearly identified the most recent high-severity fire in the watershed (Dunnette et al., 2014), we cannot quantify accuracy over the past four millennia. However, the range of variability in individual fire return intervals reconstructed at Chickaree Lake (20–330 years) is consistent with the range of intervals reconstructed from other lake-sediment records in Colorado subalpine forests (Calder et al., 2015; 75–885, 45–750, 30–645, 30–1035 years, Higuera et al., 2014), suggesting that the C dynamics highlighted here are not unique to this single fire history reconstruction.

In addition to fire timing, simulated C dynamics were also a function of variability in fire severity, which in this study reflects the degree of soil erosion associated with stand-replacing fire events. Watershed soil C losses were partially driven by the erosion events accompanying the high-severity catchment fires reconstructed in the paleo record. Because we have prescribed both fire and erosion, we cannot predict the range of soil C loss that may occur due to changes in precipitation regimes or if any erosion occurs with the lower-severity events; however, these results provide an estimate of expected changes in soil C for at least the higher-severity events. With expected changes to future precipitation regimes, including the intensification of rain events that could lead to increased erosion following fire (Larsen and MacDonald, 2007; Miller et al., 2011), ecosystem model development should include prognostic erosion to account for variability in this ecosystem process, especially on regional scales.

Finally, the most important limitation of our study is the fact that our modeling framework does not integrate realistic paleoclimate variability, nor does it represent the important coupling among climate, vegetation, and fire activity. Although paleoclimate proxies exist for nearby regions in Colorado, for example in the form of lake-level reconstructions and oxygen isotope records (Anderson 2011, 2012; Shuman et al., 2010), these records are far from the detailed climate information needed to drive DayCent. Thus, utilizing paleoclimate proxies to develop climate drivers for DayCent is an important next step but beyond the scope of this study. For example, it will involve developing methodologies to downscale paleoclimate proxies in space (to the elevation and location of Chickaree Lake), in time (to daily value), and to the specific metrics required by DayCent (e.g., from a relative moisture proxy to daily precipitation). While our simulated past carbon dynamics are limited by the lack of available paleoclimate data to drive DayCent, our temperature sensitivity analysis suggests that C dynamics are much more sensitive to the timing and severity of fire events than to even relatively large changes in climate (e.g., 2 °C warming). Further, because we have decoupled climate from fire by using prescribed fire events, the lack of a paleoclimate does not affect our conclusions about the impacts of fire-regime variability on C balance. While we used the paleo-informed modeling scenarios to test general hypotheses about the impacts of fire-regime variability on biogeochemical dynamics, future efforts to simulate the coupled climate–fire–ecosystem dynamics of the past clearly require independent paleoclimate drivers.

4 Summary and conclusions

Our simulations highlight fire-regime variability as a dominant driver of C dynamics in lodgepole pine forests, with periods of unusually high or low fire activity creating legacies lasting for centuries to millennia. Anticipating the impacts of future climate or disturbance-regime change on forest carbon balance, therefore, should be done in the context of past variability, with the duration dependent on the frequency and variability of relevant disturbance processes. In the case of stand-replacing wildfires this requires information spanning at least several centuries, and at Chickaree Lake this required several millennia, well beyond the length of both observational and tree-ring records. Many studies have reported ecosystem impacts or recovery times from individual fire events and then extrapolated to infer scenarios that would lead to C gain or loss (Dunnette et al., 2014; Kashian et al., 2013; Mack et al., 2011; Smithwick et al., 2009). In contrast, our paleo-informed scenario highlights the importance of variability in fire timing and severity over multiple fire events for carbon cycling dynamics, independent of complete shifts in a fire regime.

Our findings also have implications for ecosystem and Earth system model development, which are increasingly including prognostic fire components (Lasslop et al., 2014), primarily driven by climate and fuels. Some models are also representing post-fire C and N dynamics beyond the simple combustion of live and dead biomass or only the dead-wood pools (fuels). Development of these modules depends on observations of fire and climate interactions, fuel availability, and post-fire C and N dynamics. We suggest that this requires accurately accounting for the (often high) variability inherent in stand-replacing fire regimes, independent of or in response to climate variability. Our results indicate that even utilizing tree-ring records that span several centuries may not be sufficient to capture this variability. Further development of prognostic (predictive) fire processes in ecosystem models would benefit from the use of paleo-fire records to evaluate fire occurrence and severity, and if combined with paleoclimate data, model algorithms could be further improved to accurately reflect past variability.

The importance of fire-regime variability in determining ecosystem C dynamics implies that equilibrium scenarios are a poor assumption for conceptualizing and simulating fire regimes in ecosystem and Earth system models. Particularly on spatial scales larger than an individual site, such a simplification may result in C-balance projections that are grossly inaccurate. We demonstrate how variability in the timing and severity of disturbances can potentially have long-lasting and compounding impacts on biogeochemical states, such that modern (or future) states can reflect dynamics that have unfolded over centuries to millennia. For our modeling scenarios in lodgepole-pine dominated forests, the effects lasted approximately 2500 years. The duration of these legacies will depend on the ecosystem and the degree of variability in disturbance frequency and severity, relative to an equilibrium scenario. Ultimately, the implications of fire-regime variability for biogeochemical states will depend strongly on the synchrony of fire activity across spatial scales larger than a single watershed. If fire activity is synchronized on landscape to regional scales, as in the past (Calder et al., 2015; Marlon et al., 2012; Morgan et al., 2008) and as anticipated for the future (Westerling et al., 2011) in Rocky Mountain forests, we would expect to see similar centennial- to millennial-scale dynamics in biogeochemical states revealed here, which would have important implications for carbon cycling, including potential feedbacks to CO_2-induced warming.

Author contributions. TWH and PEH designed the study, analyzed the data, and prepared the paper with contributions from JAH.

Competing interests. The authors declare that they have no conflict of interest.

Acknowledgements. We thank K. McLauchlan and B. Shuman for valuable discussions on these topics. Tara W. Hudiburg was supported by the NSF Idaho EPSCoR Program and by the National Science Foundation under award number IIA-1301792. Philip E. Higuera was supported by the National Science Foundation under award numbers IIA-0966472 and EF-1241846, and JAH was supported by the Agriculture and Food Research Initiative of the USDA National Institute of Food and Agriculture (Grant 2013-67003-20652) and the National Science Foundation under award number DMS-1520873. The authors declare no competing financial conflicts of interests or other affiliations with conflicts of interest with respect to the results of the paper.

Edited by: Kirsten Thonicke

References

Anderson, L.: Holocene record of precipitation seasonality from lake calcite ^{18}O in the central Rocky Mountains, USA, Geology, 39, 211–214, 2011.

Anderson, L.: Rocky Mountain hydroclimate: Holocene variability and the role of insolation, ENSO, and the North American Monsoon, Glob. Planet. Change, 92–93, 198–208, 2012.

Anderson-Teixeira, K. J., Davis, S. C., Masters, M. D., and Delucia, E. H.: Changes in soil organic carbon under biofuel crops, Global Change Biology Bioenergy, 1, 75–96, 2009.

Bai, E. and Houlton, B. Z.: Coupled isotopic and process-based modeling of gaseous nitrogen losses from tropical rain forests, Global Biogeochem. Cy., 23, GB2011, https://doi.org/10.1029/2008gb003361, 2009.

Bradford, J. B., Birdsey, R. A., Joyce, L. A., and Ryan, M. G.: Tree age, disturbance history, and carbon stocks and fluxes in subalpine Rocky Mountain forests, Glob. Change Biol., 14, 2882–2897, https://doi.org/10.1111/j.1365-2486.2008.01686.x, 2008.

Caffrey, M. A. and Doerner, J. P.: A 7000-Year Record of Environmental Change, Bear Lake, Rocky Mountain National Park, USA, Phys. Geogr., 33, 438–456, 2012.

Calder, W. J., Parker, D., Stopka, C. J., Jiménez-Moreno, G., and Shuman, B. N.: Medieval warming initiated exceptionally large wildfire outbreaks in the Rocky Mountains, P. Natl. Acad. Sci. USA, 112, 13261–13266, 2015.

Campbell, J., Donato, D., Azuma, D., and Law, B.: Pyrogenic carbon emission from a large wildfire in Oregon, United States, J. Geophys. Res.-Biogeo., 112, G04014, https://doi.org/10.1029/2007JG000451, 2007.

Chapin, F., Woodwell, G., Randerson, J., Rastetter, E., Lovett, G., Baldocchi, D., Clark, D., Harmon, M., Schimel, D., Valentini, R., Wirth, C., Aber, J., Cole, J., Goulden, M., Harden, J., Heimann, M., Howarth, R., Matson, P., McGuire, A., Melillo, J., Mooney, H., Neff, J., Houghton, R., Pace, M., Ryan, M., Running, S., Sala, O., Schlesinger, W., and Schulze, E. D.: Reconciling carbon-cycle concepts, terminology, and methods, Ecosystems, 9, 1041–1050, 2006.

Dunnette, P. V., Higuera, P. E., McLauchlan, K. K., Derr, K. M., Briles, C. E., and Keefe, M. H.: Biogeochemical impacts of wildfires over four millennia in a Rocky Mountain subalpine watershed, New Phytol., 203, 900–912, 2014.

Goetz, S. J., Bond-Lamberty, B., Law, B. E., Hicke, J. A., Huang, C., Houghton, R. A., McNulty, S., O'Halloran, T., Harmon, M., Meddens, A. J. H., Pfeifer, E. M., Mildrexler, D., and Kasischke, E. S.: Observations and assessment of forest carbon dynamics following disturbance in North America, J. Geophys. Res.-Biogeo., 117, GO2022, https://doi.org/10.1029/2011jg001733, 2012.

Hansen, E. M., Amacher, M. C., Van Miegroet, H., Long, J. N., and Ryan, M. G.: Carbon Dynamics in Central US Rockies Lodgepole Pine Type after Mountain Pine Beetle Outbreaks, Forest Sci., 61, 665–679, 2015.

Hartman, M. D., Baron, J. S., and Ojima, D. S.: Application of a coupled ecosystem-chemical equilibrium model, DayCent-Chem, to stream and soil chemistry in a Rocky Mountain watershed, Ecol. Model., 200, 493–510, 2007.

Higuera, P. E., Briles, C. E., and Whitlock, C.: Fire-regime complacency and sensitivity to centennial-through millennial-scale climate change in Rocky Mountain subalpine forests, Colorado, USA, J. Ecol., 102, 1429–1441, 2014.

Hudiburg, T. W., Law, B. E., Wirth, C., and Luyssaert, S.: Regional carbon dioxide implications of forest bioenergy production, Nature Climate Change, 1, 419–423, https://doi.org/10.1038/nclimate1264, 2011.

Hudiburg, T. H., Higuera, P. E., and Hicke, J. H.: Data from: Fire-regime variability impacts forest carbon dynamics for centuries to millenia, Dryad Digital Repository, https://doi.org/10.5061/dryad.74b2c, 2017.

Kashian, D. M., Romme, W. H., Tinker, D. B., Turner, M. G., and Ryan, M. G.: Postfire changes in forest carbon storage over a 300-year chronosequence of Pinus contorta-dominated forests, Ecol. Monogr., 83, 49–66, 10.1890/11-1454.1, 2013.

Kelly, R., Genet, H., McGuire, A. D., and Hu, F. S.: Palaeodata-informed modelling of large carbon losses from recent burning of boreal forests, Nature Climate Change, 6, 79–82, 2016.

Larsen, I. J. and MacDonald, L. H.: Predicting postfire sediment yields at the hillslope scale: Testing RUSLE and Disturbed WEPP, Water Resour. Res., 43, W11412, https://doi.org/10.1029/2006WR005560, 2007.

Lasslop, G., Thonicke, K., and Kloster, S.: SPITFIRE within the MPI Earth system model: Model development and evaluation, Journal of Advances in Modeling Earth Systems, 6, 740–755, https://doi.org/10.1002/2013MS000284, 2014.

Mack, M. C., Bret-Harte, M. S., Hollingsworth, T. N., Jandt, R. R., Schuur, E. A. G., Shaver, G. R., and Verbyla, D. L.: Carbon loss from an unprecedented Arctic tundra wildfire, Nature, 475, 489–492, 2011.

Marlon, J. R., Bartlein, P. J., Gavin, D. G., Long, C. J., Anderson, R. S., Briles, C. E., Brown, K. J., Colombaroli, D., Hallett, D. J., and Power, M. J.: Long-term perspective on wildfires in the western USA, P. Natl. Acad. Sci. USA, 109, E535–E543, 2012.

McLauchlan, K. K., Higuera, P. E., Gavin, D. G., Perakis, S. S., Mack, M. C., Alexander, H., Battles, J., Biondi, F., Buma, B., and Colombaroli, D.: Reconstructing disturbances and their biogeochemical consequences over multiple timescales, BioScience, 64, 105–116, 2014.

Miller, M. E., MacDonald, L. H., Robichaud, P. R., and Elliot, W. J.: Predicting post-fire hillslope erosion in forest lands of the western United States, Int. J. Wildland Fire, 20, 982–999, 2011.

Morgan, P., Heyerdahl, E. K., and Gibson, C. E.: Multi-season climate synchronized forest fires throughout the 20th century, northern Rockies, USA, Ecology, 89, 717–728, 2008.

NRCS: Soil Survey Staff, Natural Resources Conservation Service, United States Department of Agriculture, available at: http://soildatamart.nrcs.usda.gov (last access: 1 November 2016), Soil Survey Geographic (SSURGO) Database for Eastern US, 2010.

Savage, K. E., Parton, W. J., Davidson, E. A., Trumbore, S. E., and Frey, S. D.: Long-term changes in forest carbon under temperature and nitrogen amendments in a temperate northern hardwood forest, Glob. Change Biol., 19, 2389–2400, 2013.

Schlesinger, W. H., Dietze, M. C., Jackson, R. B., Phillips, R. P., Rhoades, C. C., Rustad, L. E., and Vose, J. M.: Forest biogeochemistry in response to drought, Glob. Change Biol., 22, 2318–2328, 2015.

Shuman, B., Pribyl, P., Minckley, T. A., and Shinker, J.: Rapid hydrologic shifts and prolonged droughts in Rocky Mountain headwaters during the Holocene, Geophys. Res. Lett., 37, L06701, https://doi.org/10.1029/2009GL042196, 2010.

Sibold, J. S., Veblen, T. T., and Gonzalez, M. E.: Spatial and temporal variation in historic fire regimes in subalpine forests across the Colorado Front Range in Rocky Mountain National Park, Colorado, USA, J. Biogeogr., 33, 631–647, 2006.

Sibold, J. S., Veblen, T. T., Chipko, K., Lawson, L., Mathis, E., and Scott, J.: Influences of secondary disturbances on lodgepole pine stand development in Rocky Mountain National Park, Ecol. Appl., 17, 1638–1655, 2007.

Smithwick, E. A. H., Ryan, M. G., Kashian, D. M., Romme, W. H., Tinker, D. B., and Turner, M. G.: Modeling the effects of fire and climate change on carbon and nitrogen storage in lodgepole pine (Pinus contorta) stands, Glob. Change Biol., 15, 535–548, https://doi.org/10.1111/j.1365-2486.2008.01659.x, 2009.

Smithwick, E. A. H.: Pyrogeography and biogeochemical resilience, in: The Landscape Ecology of Fire, Springer, 143–163, 2011.

Thornton, P., Thornton, M. M., Mayer, B. W., Wilhelmi, N., Wei, Y., and Cook, R. B.: Daymet: Daily surface weather on a 1 km grid for North America, 1980–2008, available at: http://daymet.ornl.gov/ (last access: 15 September 2016), 20 September 2012 from Oak Ridge National Laboratory Distributed Active Archive Center, O. R., Tennessee, USA, https://doi.org/10.3334/ORNLDAAC/Daymet_V2, 2012.

Turner, M. G., Tinker, D. B., Romme, W. H., Kashian, D. M., and Litton, C. M.: Landscape patterns of sapling density, leaf area, and aboveground net primary production in postfire lodgepole pine forests, Yellowstone National Park (USA), Ecosystems, 7, 751–775, 2004.

Westerling, A. L., Turner, M. G., Smithwick, E. A. H., Romme, W. H., and Ryan, M. G.: Continued warming could transform Greater Yellowstone fire regimes by mid-21st century, P. Natl. Acad. Sci. USA, 108, 13165–13170, 2011.

Yelenik, S., Perakis, S., and Hibbs, D.: Regional constraints to biological nitrogen fixation in post-fire forest communities, Ecology, 94, 739–750, 2013.

9

Response of water use efficiency to summer drought in a boreal Scots pine forest in Finland

Yao Gao[1], Tiina Markkanen[1], Mika Aurela[1], Ivan Mammarella[2], Tea Thum[1], Aki Tsuruta[1], Huiyi Yang[3], and Tuula Aalto[1]

[1]Finnish Meteorological Institute, Helsinki, P.O. Box 503, 00101, Finland
[2]Department of Physics, University of Helsinki, Helsinki, P.O. Box 48, 00014, Finland
[3]Institute for Climate and Atmospheric Science, School of Earth and Environment, University of Leeds, Leeds, LS2 9JT, UK

Correspondence to: Yao Gao (yao.gao@fmi.fi)

Abstract. The influence of drought on plant functioning has received considerable attention in recent years, however our understanding of the response of carbon and water coupling to drought in terrestrial ecosystems still needs to be improved. A severe soil moisture drought occurred in southern Finland in the late summer of 2006. In this study, we investigated the response of water use efficiency to summer drought in a boreal Scots pine forest (*Pinus sylvestris*) on the daily time scale mainly using eddy covariance flux data from the Hyytiälä (southern Finland) flux site. In addition, simulation results from the JSBACH land surface model were evaluated against the observed results. Based on observed data, the ecosystem level water use efficiency (EWUE; the ratio of gross primary production, GPP, to evapotranspiration, ET) showed a decrease during the severe soil moisture drought, while the inherent water use efficiency (IWUE; a quantity defined as EWUE multiplied with mean daytime vapour pressure deficit, VPD) increased and the underlying water use efficiency (uWUE, a metric based on IWUE and a simple stomatal model, is the ratio of GPP multiplied with a square root of VPD to ET) was unchanged during the drought. The decrease in EWUE was due to the stronger decline in GPP than in ET. The increase in IWUE was because of the decreased stomatal conductance under increased VPD. The unchanged uWUE indicates that the trade-off between carbon assimilation and transpiration of the boreal Scots pine forest was not disturbed by this drought event at the site. The JSBACH simulation showed declines of both GPP and ET under the severe soil moisture drought, but to a smaller extent compared to the observed GPP and ET. Simulated GPP and ET led to a smaller decrease in EWUE but a larger increase in IWUE because of the severe soil moisture drought in comparison to observations. As in the observations, the simulated uWUE showed no changes in the drought event. The model deficiencies exist mainly due to the lack of the limiting effect of increased VPD on stomatal conductance during the low soil moisture condition. Our study provides a deeper understanding of the coupling of carbon and water cycles in the boreal Scots pine forest ecosystem and suggests possible improvements to land surface models, which play an important role in the prediction of biosphere–atmosphere feedbacks in the climate system.

1 Introduction

Terrestrial plants assimilate carbon dioxide (CO_2) through photosynthesis accompanied by a loss of water (H_2O) in transpiration. Both processes are strongly regulated by local environmental conditions and plant physiology (e.g. stomatal conductance; g_s). Plants protect themselves from excessive water losses (diffusion out of the leaf) under water-limited environments through a reduction of stomatal conductance, which in turn leads to less carbon uptake (diffusion of CO_2 into the leaf) and possibly subsequent physiological stress (McDowell et al., 2008; Will et al., 2013).

Soil water deficit can induce a reduction of transpiration (Bréda et al., 1993; Clenciala et al., 1998; Granier et al., 2007; Irvine et al., 1998), and it has been recognized as the main environmental factor limiting plant photosynthesis on the global scale (Nemani et al., 2003). Even though the occurrence of drought is low in northern Europe, the summer of 2006 in Finland was extremely dry and 24.4 % of the 603 forest health observation sites over entire Finland showed drought damage symptoms by visual examination, in comparison to 2–4 % damaged sites in a normal year (Muukkonen et al., 2015). According to the simulated regional soil moisture, the summer drought in 2006 in southern Finland was the most severe one over the past 30 years (1981–2010), and the spatial distribution of the drought damage has been found to be closely related to the plant available soil moisture (Gao et al., 2016).

Water use efficiency (WUE) is a critical metric that quantifies the trade-off between photosynthetic carbon assimilation and transpiration at the leaf level (Farquhar et al., 1982). WUE can be used to study ecosystem functioning which is in close connection to the global cycles of water, energy, and carbon (Keenan et al., 2013). With the use of the eddy covariance technique (EC) and associated data processing, i.e. the derivation of gross primary production (GPP) and evapotranspiration (ET) from measurements of CO_2 flux and latent heat flux, WUE can be calculated on the ecosystem scale as the ecosystem level water use efficiency (EWUE), which is the ratio of GPP to ET. EWUE is broadly adopted as a surrogate for the leaf level WUE in many studies, because more data are available at the ecosystem level than at the leaf level (Arneth et al., 2006; Law et al., 2002; Lloyd et al., 2002).

Reichstein et al. (2007) observed a small decrease in EWUE in the majority of the 11 studied EC sites during the 2003 summer heatwave in Europe. However, their findings are at odds with many models that describe the environmental controls on stomatal conductance, with increased EWUE predicted during drought periods (Schulze et al., 2005). Many of those models are based on the optimality theory by Cowan and Farquhar (1977) who proposed that plants are able to regulate stomatal conductance in order to maximize WUE. Granier et al. (2008) reported that EWUE increased linearly with soil water deficit duration and intensity at a young beech forest site in north-eastern France. Moreover, EWUE also increased substantially at two forest sites, but not at grassland sites, during the 2011 spring drought in Switzerland (Wolf et al., 2013). However, no differences in EWUE were shown between abundant- and low-rainfall years at a boreal Scots pine forest site in southeastern Finland, even though GPP was reduced during low-rainfall years with long-lasting drought periods (Ge et al., 2014). Therefore, the impact of drought on EWUE remains unclear. Beer et al. (2009) concluded that the impact of vapour pressure deficit (VPD) on canopy conductance disturbs responses of both GPP and ET to changing environmental conditions and proposed the ecosystem level inherent water use efficiency (IWUE), which is a quantity defined as EWUE multiplied with mean daytime VPD. IWUE has been found to increase during short-term moderate drought (Beer et al., 2009). Moreover, based on IWUE and an optimality-

theory-based (Cowan and Farquhar, 1977) stomatal model with the assumptions suggested by Farquhar et al. (1993) and Lloyd and Farquhar (1994), the underlying water use efficiency (uWUE) was introduced to exclude the nonlinear dependence of IWUE on VPD, and the linear relationship between GPP multiplied with a square root of VPD and ET was found on the half-hourly time scale by Zhou et al. (2014). Later on, the appropriateness of uWUE on the daily time scale was also demonstrated (Zhou et al., 2015).

Given the need to understand and project feedbacks between climate change and plant physiological responses, it is crucial to be able to realistically model the plant controls of stomatal conductance, and photosynthesis and transpiration responses under water stress (Berry et al., 2010; Knauer et al., 2015; Zhou et al., 2013). The various land ecosystem model simulations highlight the current uncertainty about plant physiology (water use) in response to drought in models (Huang et al., 2015; Jung et al., 2007).

The objectives of this study are (1) to understand the environmental controls on GPP and ET fluxes during a summer drought in boreal Scots pine (*Pinus sylvestris*) forests at a EC flux site in southern Finland; (2) to investigate the drought impact on WUE metrics, including EWUE, IWUE and uWUE; and (3) to evaluate how adequately the JSBACH land surface model captures plant responses to changes in environmental variables.

2 Data and methods

2.1 Study sites

The Hyytiälä flux site is located in southern Finland (61°51′ N, 24°17′ E; 180 m a.s.l.) at the SMEAR-II (Station for Measuring Ecosystem–Atmosphere Relations) field measurement station (Hari and Kulmala, 2005). The site is dominated by 55 year-old boreal Scots pine (*Pinus sylvestris*), which is homogeneous about 200 m in all directions from the site and extends to the north for about 1 km (Mammarella et al., 2007). The canopy height of trees is about 13–16 m and the mean all-sided leaf area index (LAI) is 6 m^2 m^{-2}. The soil at the site is Haplic podzol on glacial till (FAO-UNESCO, 1990). The 30 year (1961–1990) averaged annual mean air temperature is 2.9 °C and precipitation is 709 mm at the site (Vesala et al., 2005). Those details about the site are listed in Table 1. The ground vegetation consists mainly of blueberry (*Vaccinium myrtillus*), lingonberry (*Vaccinium vitis-idaea*), feather moss (*Pleurozium schreberi*) and other bryophytes (Kolari et al., 2009). We analysed the summer (June–August) from an 11-year period (1999–2009) according to data availability.

2.2 Flux measurement and data processing

Ecosystem carbon and water fluxes at the site were measured with the micrometeorological EC method. Turbulent fluxes

Table 1. Key characteristics relevant to this study from observation and the parameter settings in the JSBACH site level simulation at the Hyytiälä site.

						Observation				
Site	Location	Vegetation type	LAI $(m^2 m^{-2})$ (all-sided, annual)	Canopy height (m)	Measurement height (m)	Annual mean air temperature (°C) and precipitation (mm) (30 year averages)	Soil type	Analysed measurement depth of soil moisture (cm)	References
Hyytiälä	61°51′N, 24°17′E	Scots pine	6	13–16	23	2.9; 709	Mineral (Haplic podzol)	5–23; 23–60	Markkanen et al. (2001); Vesala et al. (2005)

						Settings in JSBACH				
Site	PFT	Maximum LAI $(m^2 m^{-2})$	Maximum electron transport rate (V_{max}) at 25 °C	Maximum carboxylation rate at 25 °C	Soil type	Analysed depth of soil moisture (cm)	Soil depth (m)	Root depth (m)
Hyytiälä	Evergreen needleleaf forest	16	37.5	71.3	Loamy sand	Average of layer-2 (6.5–31.9) and layer-3 (31.9–123.2)	5.416	1.265

were calculated as half-hourly averages following standard methodology (Aubinet et al., 2012) with EddyUH software (Mammarella et al., 2016). The vertical CO_2 flux was obtained as the covariance of high-frequency (10 Hz) observations of vertical wind speed and the CO_2 concentration (Baldocchi, 2003). The CO_2 flux was corrected for storage change to obtain net ecosystem CO_2 exchange (NEE), which was then partitioned into total ecosystem respiration (TER) and GPP according to Kolari et al. (2009). Data quality of 30 min values of NEE and latent heat flux (LE) was ensured by excluding records with low turbulent mixing (friction velocity below 0.25 m s^{-1}) as described in Markkanen et al. (2001), Mammarella et al. (2007), and Ilvesniemi et al. (2010). TER was modelled using an exponential equation with temperature at a depth of 2 cm in the soil organic layer as the explanatory factor. The value of GPP was then directly derived as residual from the measured NEE. When NEE was missing, GPP was gap-filled according to Kolari et al. (2009). LE was gap-filled using a linear regression against net radiation in a moving window of 5 days, and then ET was inferred from LE.

In addition to the EC measurements, a set of supporting meteorological variables were adopted as half-hourly averages; incoming shortwave radiation (R_s) and longwave radiation, air temperature (T_a), atmospheric humidity, and precipitation were used as meteorological forcing for the site level simulation. The soil moisture was monitored at 1 h intervals by the time domain reflectometry method (Tektronix 1502 C cable radar, Tektronix Inc., Redmond, USA). Three layers of mineral soil (0–5, 5–23, and 23–60 cm) were measured, as well as the organic layer on the top (−4 to 0 cm). In this study, soil moisture at the two lower levels of mineral soil (5–23 and 23–60 cm) at Hyytiälä was averaged over a day to represent daily soil moisture dynamics in the root zone at the

site. The reason to exclude layer 1 soil moisture is that it is too sensitive to temperature and precipitation variations.

The half-hourly data of GPP and ET, as well as meteorological variables were averaged over the selected time periods in a day. Prior to averaging, rainy days and a number of dry days after the rainy days were firstly excluded from the data. The number of excluded dry days was determined by the ratio of daily precipitation to potential evapotranspiration (PET). When precipitation was smaller than PET, no dry day after rainy day was excluded. When precipitation was equal or larger than twice that of PET, two dry days following the rainy day were excluded. Additionally, when precipitation was larger than PET but with the ratio less than 2, one dry day after the rainy day was excluded. PET was calculated using the Penman–Monteith equation and the "Evapotranspiration" package in R software was used (Guo et al., 2016). Second, in order to capture the daily time periods of effective photosynthesis, only half-hourly data with R_s larger than 100 W m^{-2} were selected. Finally, the half-hourly data of R_s, VPD, and T_a were also averaged over the selected time periods to get their daytime mean values respective to the GPP and ET data. The same data processing method was used for the simulation results.

2.3 JSBACH land surface model

JSBACH (Raddatz et al., 2007; Reick et al., 2013) is the land surface model of the Max Planck Institute for Meteorology Earth System Model (MPI–ESM) (Roeckner et al., 1996; Stevens et al., 2013). The land physics of JSBACH mainly follow those of the global atmosphere circulation model ECHAM5 (Roeckner et al., 2003), and the biogeochemical components are mostly taken from the biosphere model BETHY (Knorr, 2000). In JSBACH, land vegetation cover is described as plant functional types (PFTs) and a set of properties (e.g. maximum LAI and albedo) is attributed to each PFT with respect to the processes that are accounted for by JSBACH. The phenology model (Logistic Growth Phenology; LoGro-P) of JSBACH simulates the LAI dynamics to compute photosynthetic production (Böttcher et al., 2016). The models of Farquhar et al. (1980) and Collatz et al. (1992) are used for photosynthesis of C3 and C4 plants, respectively. A five-layer soil hydrology scheme was implemented in JSBACH by Hagemann and Stacke (2015). Gao et al. (2016) has demonstrated that JSBACH with its five-layer soil hydrology scheme is able to capture the soil moisture dynamics at sites and on the regional scale of Finland.

2.3.1 The stomatal conductance model in JSBACH

The current version of the stomatal conductance model in JSBACH considers the limitation from soil water availability on stomatal conductance (g_s), which further impacts on carbon assimilation and transpiration.

Firstly, the net assimilation rate (A_n; $mol\,m^{-2}\,s^{-1}$) and g_s ($mol\,m^{-2}\,s^{-1}$) are calculated without water limitation as the unstressed net assimilation rate ($A_{n,\,pot}$; $mol\,m^{-2}\,s^{-1}$) and the unstressed stomatal conductance ($g_{s,\,pot}$; $mol\,m^{-2}\,s^{-1}$). The $A_{n,\,pot}$ is calculated using the photosynthesis model in JSBACH, for which the intercellular CO_2 concentration under unstressed condition ($C_{i,\,pot}$; $mol\,mol^{-1}$) is needed. The $C_{i,\,pot}$ is prescribed using the atmospheric CO_2 concentration (C_a; $mol\,mol^{-1}$), where $C_{i,\,pot} = 0.87C_a$ for C3 plants and $C_{i,\,pot} = 0.67C_a$ for C4 plants (Knorr, 2000). After the $A_{n,\,pot}$ is determined, the $g_{s,\,pot}$ is derived using the following equation:

$$g_{s,\,pot} = \frac{1.6A_{n,\,pot}}{C_a - C_{i,\,pot}}. \tag{1}$$

Then, an empirical water stress factor, which is a function of volumetric soil moisture, is used to derive g_s ($mol\,m^{-2}\,s^{-1}$) from $g_{s,\,pot}$ as follows:

$$g_s = \beta g_{s,\,pot}. \tag{2}$$

where

$$\beta = \begin{cases} 1 & \theta \geq \theta_{crit} \\ \frac{\theta - \theta_{wilt}}{\theta_{crit} - \theta_{wilt}} & \theta_{wilt} < \theta < \theta_{crit} \\ 0 & \theta \leq \theta_{wilt} \end{cases}, \tag{3}$$

herein, θ ($m^3\,m^{-3}$) is the volumetric soil moisture, θ_{crit} ($m^3\,m^{-3}$) is the critical point, and θ_{wilt} ($m^3\,m^{-3}$) is the permanent wilting point.

Finally, the intercellular CO_2 concentration (C_i) and A_n are resolved using g_s. The canopy conductance (G_c; $mol\,m^{-2}\,s^{-1}$) and canopy-scale A_n are integrated over the leaf area. Unlike the BETHY approach (Knorr, 2000), the control of g_s in JSBACH does not include the influence of atmospheric humidity.

2.4 Site level simulation by JSBACH

For the site simulation, JSBACH was forced with the half-hourly local meteorological observations. Based on the site-specific information, PFT was assigned as evergreen needle-leaf forest and the soil type was set as loamy sand in JSBACH. The modelled LAI reached values close to the observed LAI when the parameter maximum LAI was set to $16\,m^2\,m^{-2}$. Also, the maximum carboxylation rate (Jmax) and maximum electron transport rate (V_{max}) at 25 °C were adjusted, for the simulated GPP to match the magnitude of the observed GPP. The V_{max} was set to be 37.5 and the Jmax was 71.3. The soil depth and root depth at the site were derived from maps for the regional JSBACH simulation presented in Gao et al. (2016) (see also Hagemann and Stacke, 2015). Those parameter settings in the JSBACH site level simulation for the site are listed in Table 1. Prior to the actual simulations, a 30 year spin-up run was conducted by cycling meteorological forcing that was used for the actual simulation to obtain equilibrium for soil water and soil heat balances.

2.5 Soil Moisture Index (SMI)

In this study, the soil moisture dynamics are represented by SMI (also referred to as Relative Extractable Water – REW), which has been demonstrated to represent summer drought in boreal forests in Finland (Gao et al., 2016). The SMI describes the ratio of plant available soil moisture to the maximum volume of water available to plants in the soil (Betts, 2004; Seneviratne et al., 2010):

$$SMI = (\theta - \theta_{WILT}) / (\theta_{FC} - \theta_{WILT}), \tag{4}$$

where θ is the volumetric soil moisture ($m^3\,H_2O\,m^{-3}$), θ_{FC} is the field capacity ($m^3\,H_2O\,m^{-3}$), and θ_{WILT} is the permanent wilting point ($m^3\,H_2O\,m^{-3}$). When θ exceeds θ_{FC}, soil water cannot be retained against gravitational drainage, while below θ_{WILT}, the soil water is strongly held by the soil matrix and cannot be extracted by plants (Hillel, 1998). In this study, soil moisture conditions were classified into five groups according to SMI values with an interval of 0.2: very dry, $0 \leq SMI < 0.2$; moderate dry, $0.2 \leq SMI < 0.4$; mid-range, $0.4 \leq SMI < 0.6$; moderate wet, $0.6 \leq SMI < 0.8$; and very wet, $0.8 \leq SMI < 1$.

From simulations, we used the average of the second layer (layer-2; 6.5–31.9 cm) and the third layer (layer-3; 31.9–123.2 cm) soil moisture together with model soil parameters to determine the simulated SMI for Hyytiälä, with the aim to correspond with the observed SMI that calculated with measured soil moisture at the two lower levels of mineral soil at the site. The layer 1 soil moisture was excluded in determining both simulated and observed SMIs because it is too sensitive to temperature and precipitation variations. For the observed SMI, the measured soil parameters derived based on water retention curves determined from soil samples taken at the site were adopted (i.e. volumetric soil moisture at saturation $(\theta_{SAT}) = 0.50\,m^3\,H_2O\,m^{-3}$, $\theta_{FC} = 0.30\,m^3\,H_2O\,m^{-3}$, and $\theta_{WILT} = 0.08\,m^3\,H_2O\,m^{-3}$). As θ_{FC} acts as a proxy for θ_{SAT} in the five-layer soil hydrology scheme in JSBACH (Hagemann and Stacke, 2015), θ_{SAT} was used instead of θ_{FC} for consistency when calculating SMI based on the observed soil moisture data.

2.6 Ecosystem water use efficiency (EWUE), inherent water use efficiency (IWUE), and underlying water use efficiency (uWUE)

The EWUE is calculated as,

$$EWUE = GPP/ET, \tag{5}$$

IWUE is defined as EWUE multiplied by daytime mean VPD in Beer et al. (2009),

$$IWUE = GPP \times VPD/ET, \tag{6}$$

uWUE is derived based on IWUE and an optimality-theory-based (Cowan and Farquhar, 1977) stomatal model with the assumptions suggested by Farquhar et al. (1993) and Lloyd and Farquhar (1994) in Zhou et al. (2014). The formulation of uWUE is,

$$uWUE = GPP \times VPD^{0.5}/ET \tag{7}$$

From EC data, EWUE and IWUE can only be calculated with ET, which, in addition to transpiration, contains evaporation of water intercepted by surfaces and soil evaporation. However, process-based ecosystem models do resolve evaporation and transpiration which together compose ET. Therefore, transpiration-based EWUE, IWUE, and uWUE can also be calculated using simulated transpiration instead of ET in those equations.

3 Results

3.1 Soil moisture drought at Hyytiälä in 2006

In the summer of 2006, a period with evidently lower SMI values (< 0.2) than in any other year during the 11-year time series was shown (Fig. 1a). According to the in situ observation, in the summer of 2006, there were 37 consecutive days (23 July–28 August) with SMI lower than 0.2, and 17 consecutive days (1–17 August) with SMI lower than 0.15. The observed SMI reached its minimum of 0.115 on 16 August 2006. The simulated SMI was generally smaller than the observed SMI in the summer of 2006, showing 42 consecutive days (17 July–27 August) with SMI lower than 0.2, and 33 consecutive days (26 July–27 August) with SMI lower than 0.15. The lowest SMI from simulation was 0.052 on 15 August. The simulated SMI agreed well with in situ observed SMI over the 11-year study period, with a correlation

coefficient of 0.63 and a root-mean-square error (RMSE) of 0.23. However, the simulated SMI showed a larger amplitude and a faster response to changes in climate conditions in comparison to the observed SMI. Nevertheless, a very good correlation coefficient of 0.97 between simulated and observed SMIs was found for the year 2006 (Fig. 1b), despite the simulated SMI being systematically lower than the observed SMI (RMSE = 0.12).

Concurrently with the low soil moisture, a high T_a anomaly was observed in August 2006 (Fig. 1c). In all the days in August 2006, the daily mean in situ T_a was higher than the 11-year averaged daily mean T_a. The monthly mean T_a in August 2006 ($18.1 \pm 1.9\,°C$) was $3.1\,°C$ higher than that of the 11-year average ($15.0 \pm 1.63\,°C$). Also, the daily mean VPD in August 2006 was higher than the 11-year averaged daily mean VPD in August in general (not shown), except on the days with precipitation. Especially, the mean value of the daily mean VPD in the period from 31 July to 16 August ($1.067 \pm 0.361\,kPa$) was substantially higher than the mean of the 11-year averaged daily mean VPD over this period ($0.582 \pm 0.200\,kPa$). The biggest difference between the daily mean VPD and the 11-year averaged daily mean VPD reached $1.054\,kPa$ on 5 August that was the day with highest T_a in August 2006. The daily mean R_s in the summer of 2006 was overall higher than the 11-year averaged daily mean R_s, with the monthly mean values by 15.4, 31.2, and 21.4 % higher in June, July, and August, respectively.

The precipitation events have a strong impact on the temporal pattern of SMI. The cumulative in situ precipitation of 34 mm in July 2006 was the lowest during the 11-year study period with the July average of 91 ± 31 mm. In contrast, the highest total precipitation in July was in 2007, reaching 146 mm. The cumulative precipitation of 48 mm in August 2006 was not as low as in July when compared to the 11-year average of 71 ± 43 mm. However, the lack of precipitation since the end of July led to the continuous drop of SMI till mid August 2006, followed by a small increase in soil moisture after a light precipitation event. The SMI increased to be above 0.2 in the end of August with a heavy precipitation event exceeding 25 mm in one day. Moreover, the precipitation in June 2006 was also less than the 11-year average (45 vs. 70 ± 24 mm) and temporally unevenly distributed, with only a small amount at the beginning of June and a large amount in the end of June. Therefore, there was a continuous decrease in soil moisture from the beginning of June and an abrupt increase in SMI of more than 0.1 at the end of June.

3.2 The relationship of GPP to ET categorized by environmental variables

In general, the daytime averaged GPP and ET from observations at Hyytiälä showed a non-linear relationship (Fig. 2a). When categorized according to environmental variables, there is a group of data under the very dry soil moisture

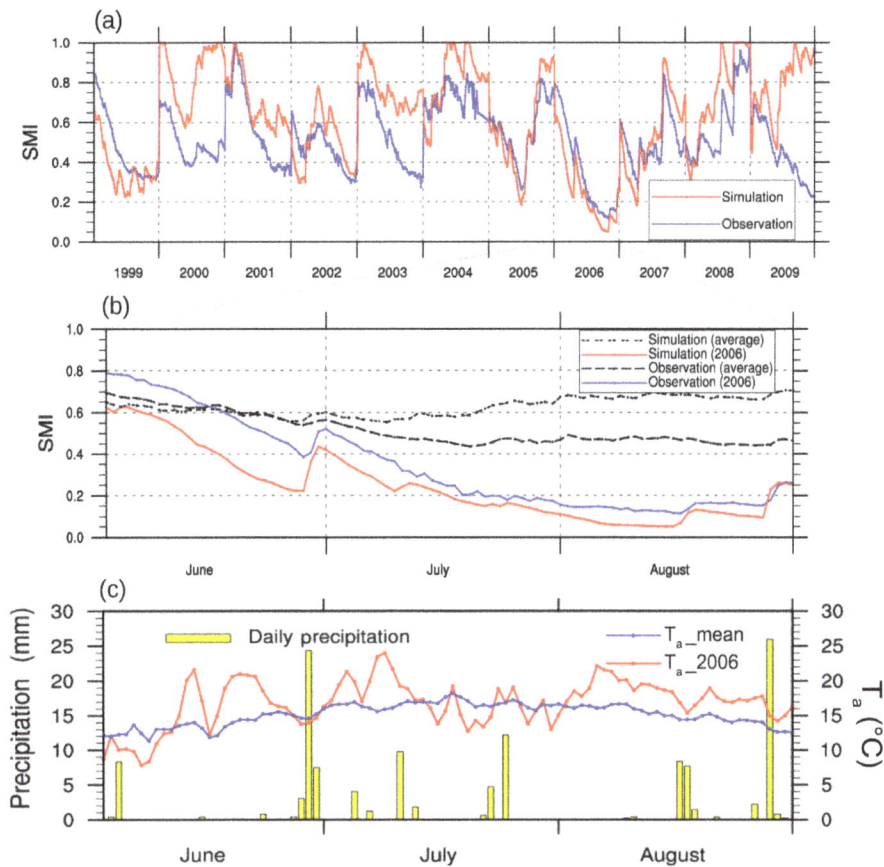

Figure 1. (a) Daily mean soil moisture index (SMI) at Hyytiälä from observation and the JSBACH simulation for the summer months (June, July, August) in the 11-year study period (from 1999 to 2009). **(b)** Daily mean SMI at Hyytiälä from observation and the JSBACH simulation for the summer months in 2006; the two black dashed lines represent the averaged daily SMI in the summer months over the 11-year study period from observation and the JSBACH simulation. **(c)** Daily mean air temperature (T_a) in the summer months of 2006 and the averaged daily mean T_a in the summer months over the 11-year study period at Hyytiälä from observation, meanwhile, the daily precipitation amount in 2006 is shown as the bar plot.

condition (encircled in grey in Fig. 2a) showing GPP values lower than other days. The ET values of this group are also located in the lower end, but just partly lower than ET values on other days. It is found that the days in this group are with SMI smaller than 0.15. Moreover, there are only two days with SMI values smaller than 0.15 that are not included in the encircled group due to their slightly higher GPP values. Most of the days in the group have high daytime mean T_a (18–24 °C), sufficient daytime mean R_s (mostly above 300 W m^{-2}), and relatively high daytime mean VPD (above 1 kPa).

The non-linear relationship between the daytime averaged GPP and ET was also found in the JSBACH simulated result (Fig. 2b). The decline of both GPP and ET during low SMI was captured by the model. However, under the very low soil moisture condition (SMI < 0.15) during the summer drought in 2006, the model simulated a much less reduction of GPP, while the ET decreased to be lower than the observation in a few days. The non-linear relationship between simulated

daytime averaged GPP and transpiration (Fig. S1 in Supplement) is similar to the relationship between simulated daytime averaged GPP and ET, which demonstrates that transpiration composes a large fraction of ET during daytime at the site, especially under soil water stress. Except the drought events, GPP and ET both increased with increasing R_s and VPD in the simulation, which was more evident than in the observational data.

3.3 Response of GPP and ET to environmental variables categorized by SMI

The dependence of GPP and ET on environmental variables was further investigated for different SMI ranges (Fig. 3). The exclusion of the night-time and the days affected by rain (see details in Sect. 2.2) also removed the small values of GPP and ET. Linear regressions were fitted between GPP (ET) and environmental variables for each soil moisture group to emphasize the deviating differences of dependence of GPP (ET) on environmental variables under different soil

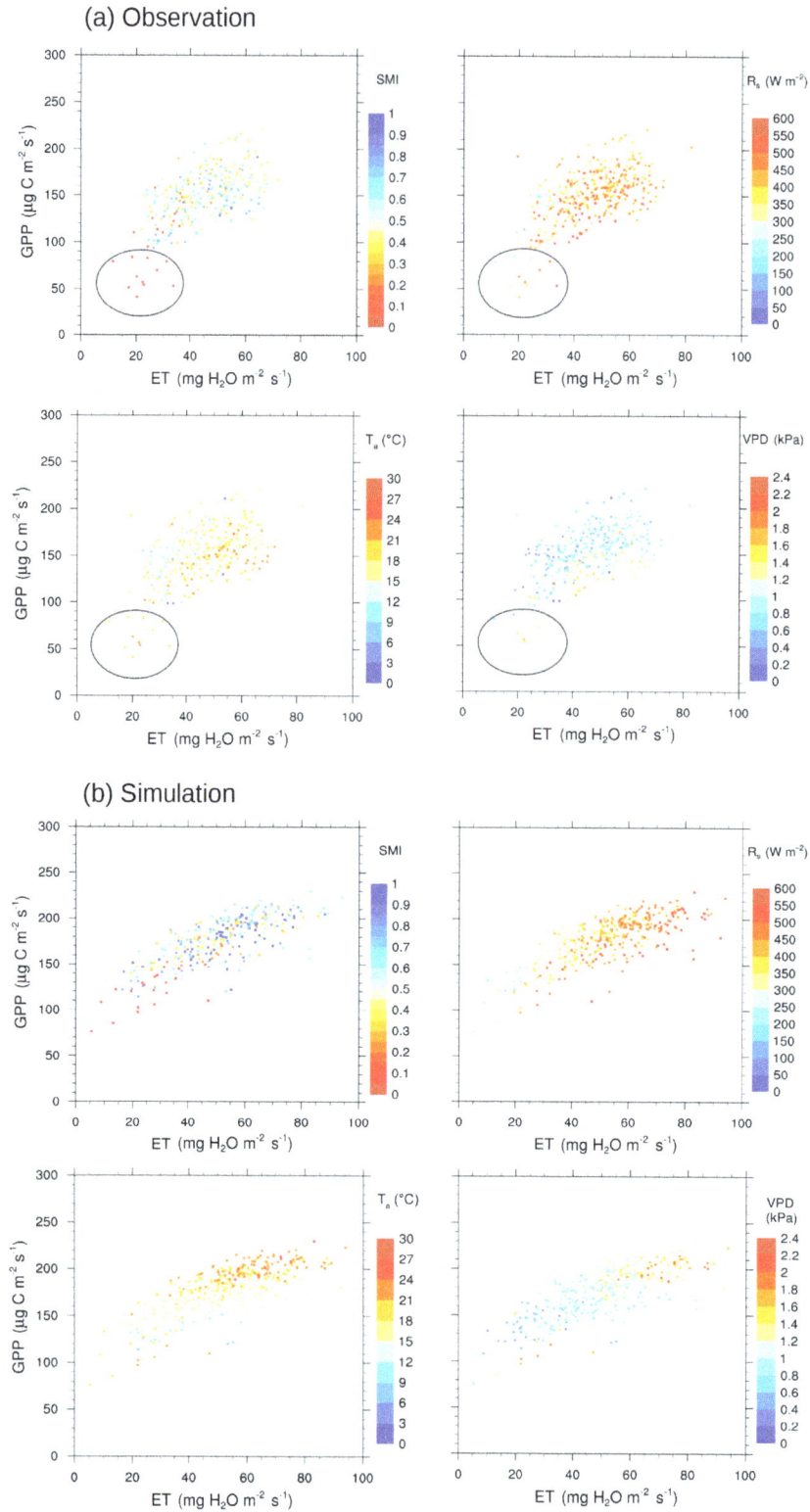

Figure 2. Relationship between the daytime averaged gross primary production (GPP in $\mu g\,C\,m^{-2}\,s^{-1}$) and evapotranspiration (ET in mg $H_2O\,m^{-2}\,s^{-1}$) at Hyytiälä in the summer months (June, July, August) of the 11-year study period (from 1999 to 2009) from **(a)** observation and **(b)** the JSBACH simulation. Data are categorized according to daily mean soil moisture index (SMI), daytime mean incoming shortwave radiation (R_s), daytime mean air temperature (T_a), and daytime mean vapour pressure deficit (VPD). In the observation, the group of data under the very dry soil moisture condition showing GPP values lower than other days is marked with a grey circle.

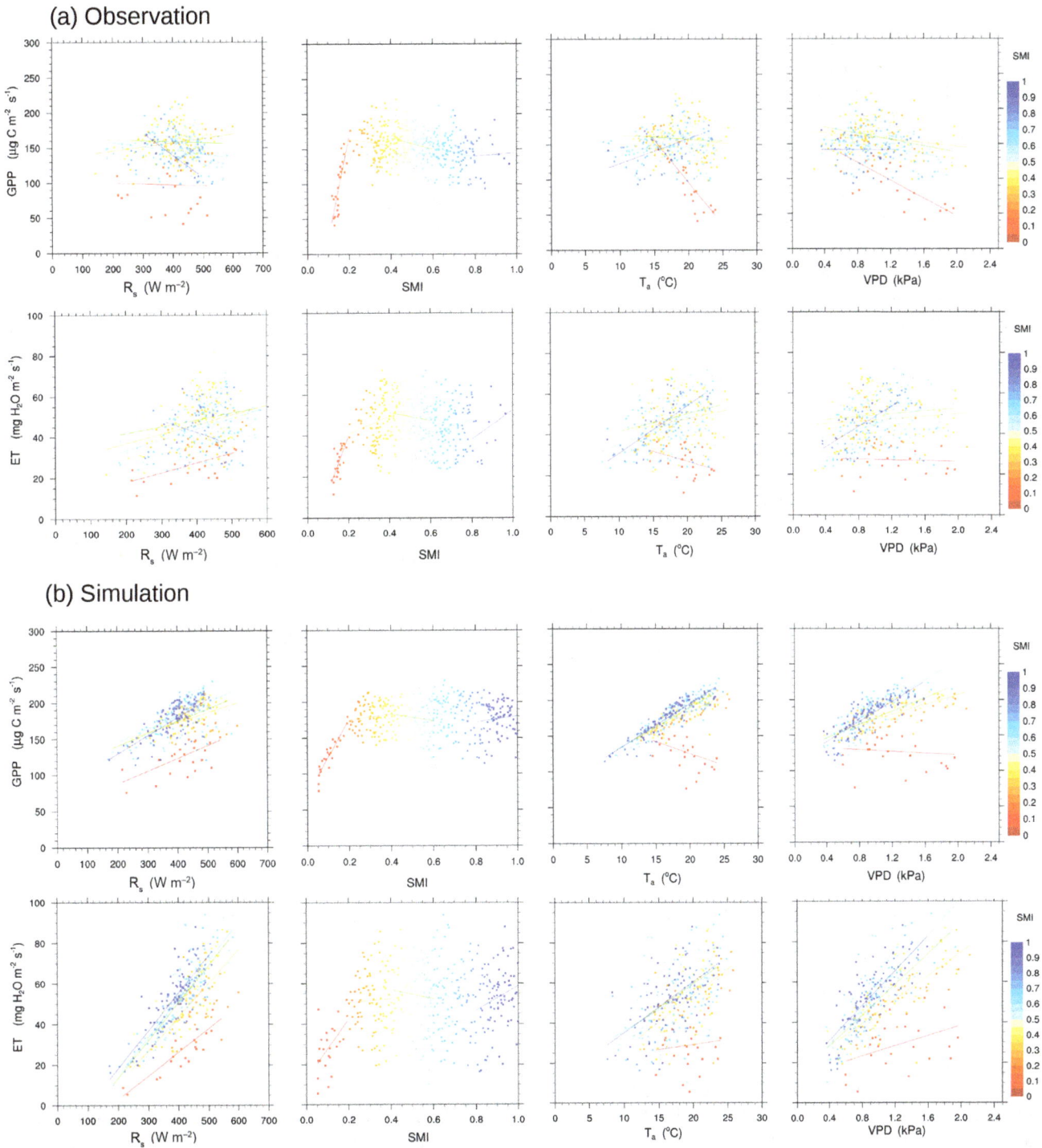

Figure 3. Response of daytime mean gross primary production (GPP in $\mu g\,C\,m^{-2}\,s^{-1}$) and evapotranspiration (ET in mg $H_2O\,m^{-2}\,s^{-1}$) to daytime mean incoming shortwave radiation (R_s), daytime mean air temperature (T_a), daytime mean vapour pressure deficit (VPD), and daily mean soil moisture index (SMI) at Hyytiälä, categorized by daily mean soil moisture index (SMI) in the summer months (June, July, August) of the 11-year study period (from 1999 to 2009) from **(a)** observation and **(b)** the JSBACH simulation. The regression lines are fitted for the five SMI groups (very dry, $0 \leq SMI < 0.2$; moderate dry, $0.2 \leq SMI < 0.4$; mid-range, $0.4 \leq SMI < 0.6$; moderate wet, $0.6 \leq SMI < 0.8$; and very wet, $0.8 \leq SMI < 1$).

moisture conditions. The regression parameters, correlation coefficient and statistical significance are summarized in Table S1 in the Supplement.

The very dry soil ($0 \leq$ SMI < 0.2) led to a response of observed daytime mean GPP and ET to daytime mean R_s, T_a, and VPD that deviated considerably from the responses of the daily mean SMI values greater than 0.2 (Fig. 3a). Under the very dry soil moisture condition, GPP decreased with the declining SMI with a high correlation of 0.79, whereas the other SMI groups showed a more scattered relationship between GPP and SMI. Unlike the other SMI groups, GPP was the most negatively correlated with T_a and VPD under the very dry soil moisture condition. Moreover, the group with SMI values less than 0.2 displayed lower GPP values (on average $97.6\,\mu g\,C\,m^{-2}\,s^{-1}$) than the other groups (on average $151\,\mu g\,C\,m^{-2}\,s^{-1}$). The response patterns of the observed ET to environmental variables were similar to those of GPP. As with GPP, the group under the very dry soil moisture condition deviated strongly from the other SMI groups. However, the decrease in ET under severe soil moisture drought was not as pronounced as in GPP.

For the simulated GPP and ET too, the group under the very dry soil moisture condition deviated from the other SMI groups, but not to the same extent as that in the observed GPP and ET. Under other soil moisture conditions (SMI > 0.2), the simulated GPP had stronger positive linear relationships with daytime mean R_s, T_a, and VPD than the observed GPP. Compared to the observed ET, some differences existed in the response of the simulated ET to environmental variables. First, the dependence of simulated ET on R_s tended to be more linear than the observed ET and R_s relationship. Second, unlike observed ET, the simulated ET increased concomitantly with VPD at high VPD. Nevertheless, simulated ET of the group under severe soil moisture drought deviated strongly from the other SMI groups, but to a lesser extent than observed ET.

3.4 Soil moisture drought impacts on EWUE, IWUE, and uWUE

From the observation, the decrease in GPP was much stronger than the decrease in ET during the soil moisture drought, which resulted in a largely decreased EWUE that reached the recorded minimum during the severe soil moisture drought (Figs. 2 and S2). In contrast to EWUE, IWUE increased from $3.25\,\mu g\,C\,kPa\,mg^{-1}\,H_2O$ (the mean value for the days with SMI equal or larger than 0.2) to $3.93\,\mu g\,C\,kPa\,mg^{-1}\,H_2O$ (the mean value for the days with SMI smaller than 0.2), and uWUE did not change under the severe soil moisture drought at Hyytiälä (Fig. 4a). The simulated EWUE decreased less and the simulated IWUE increased more (from 3.62 to $5.17\,\mu g\,C\,kPa\,mg^{-1}\,H_2O$) than the observation, which is mainly because of a smaller decrease in the simulated GPP than its observed counterpart during the soil moisture drought (Fig. 4b). The simulated uWUE remained insensitive to the severe soil moisture drought. In addition, the transpiration-based EWUE, IWUE, and uWUE (Fig. S3 in Supplement) showed similar results to those three metrics calculated with ET.

4 Discussion

4.1 Drought impacts on GPP and ET

Both GPP and ET were suppressed when there was the severe soil moisture drought in the summer of 2006 at Hyytiälä. In addition, the response of GPP and ET to the changes in environmental variables under severe water stress differed from those under other soil moisture conditions. The dominant reason is that low soil moisture leads to stomatal regulation of the plants, which limits plant carbon assimilation and transpiration. The decreased ET due to soil moisture drought may increase atmospheric VPD, which could in turn intensify stomatal closure (Eamus et al., 2013; Jarvis, 1976). Moreover, the GPP and ET were decoupled and EWUE decreased due to the soil moisture drought. Unlike EWUE, IWUE increased but uWUE showed no changes during the severe soil moisture drought at Hyytiälä. IWUE depends on the difference between ambient partial pressure of CO_2 (C_a) and a weighted average of inner leaf partial pressure of CO_2 (C_i) through the canopy within the tower footprint (Beer et al., 2009). It has been shown that the term $(1 - C_i/C_a)$ increases as VPD increases (Wong et al., 1979). Thus, the increase in IWUE during drought was a result of decreased stomatal conductance due to increased VPD. The uWUE was formulated to be more independent of a varying VPD than IWUE. According to Xie et al. (2016), both IWUE and uWUE at a flux site increased and reached their maximum values over the long-term during a severe drought in central and southern China in the summer of 2013. In this work, the unchanged uWUE during this drought event demonstrate that the tradeoff between carbon assimilation and transpiration of the boreal Scots pine forest was not disturbed by drought at the study site, even though the stomatal conductance decreased.

4.2 Differences between observations and site simulations

The model showed the limitations on GPP and ET under the very dry soil moisture condition ($0 \leq$ SMI < 0.2) at Hyytiälä. However, the discrepancies in response between observed and simulated GPP and ET to changing environmental variables were obvious. This is because the formulation for stomatal conductance in JSBACH does not include a response to air humidity, and therefore the stomatal conductance in JSBACH is insensitive to atmospheric VPD (Knauer et al., 2015). In Knauer et al. (2015), Ball–Berry model (Ball et al., 1987) has been found to be the best among a few stomatal conductance models in its response to atmospheric drought under non-limited soil moisture conditions.

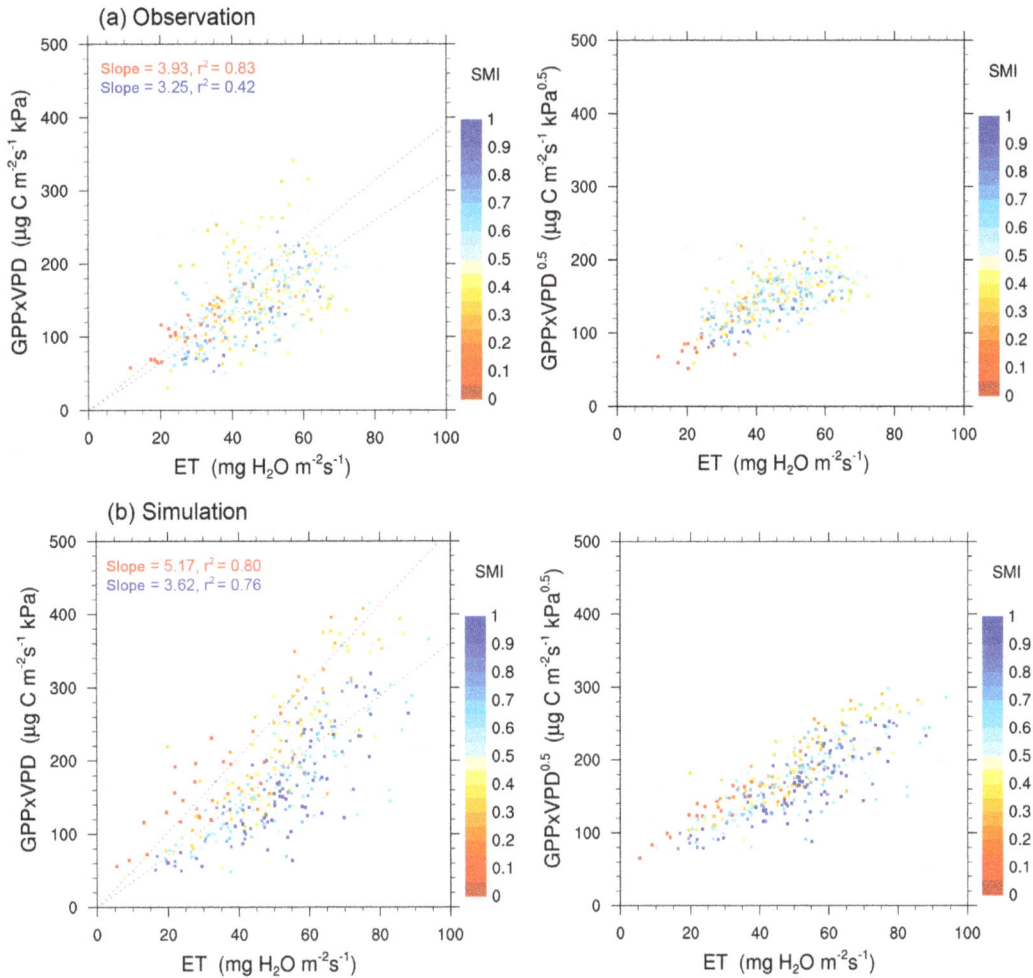

Figure 4. The dependence of the product of daytime mean gross primary production (GPP in $\mu g\,C\,m^{-2}\,s^{-1}$) and daytime mean vapour pressure deficit (VPD) on evapotranspiration (ET in mg $H_2O\ m^{-2}\,s^{-1}$) (i.e. GPP $\times$ VPD/ET, which represents the inherent water use efficiency, IWUE), and the dependence of the production of GPP and the square root of VPD on ET (i.e. GPP $\times$ VPD$^{0.5}$/ET, which represents the underlying water use efficiency, uWUE) in the summer months (June, July, August) of the 11-year study period (from 1999 to 2009) from **(a)** observation and **(b)** the JSBACH simulation. Data are categorized according to daily mean soil moisture index (SMI). The fitted lines for the dependence of the product of GPP and VPD on ET are for the data under SMI < 0.2 (red line) and the data under $0.2 \leq$ SMI < 1 (blue line); both fittings are statistically significant (p value < 0.05). No lines were fitted for the dependence of the production of GPP and the square root of VPD on ET, as the data under SMI < 0.2 and data under $0.2 \leq$ SMI < 1 are more converged in a line in comparison to the dependence of the product of GPP and VPD on ET.

In reality, low soil moisture and high T_a during drought are closely coupled with high atmospheric VPD. Our results indicate that the combined effects of soil moisture and atmospheric drought on stomatal conductance have to be taken into account. Moreover, model performance could be improved through the inclusion of non-stomatal limitations on plant photosynthesis, which have been considered to be important for the simulation of short-term plant responses to drought (Egea et al., 2011; Manzoni et al., 2011; Zhou et al., 2013). However, JSBACH is being continuously developed and the effect of soil water stress is to be accounted for according to Egea et al. (2011) for both stomatal and non-

stomatal processes, affecting both conductance and photosynthesis parameters.

Moreover, when comparing results from the EC data and simulations, it should be kept in mind that the EC method has its uncertainties. Due to the stochastic nature of the turbulent flow, there is always a random error component in the observations. In addition, imperfect spectral corrections and gap-filling procedures as well as calibration problems may be sources of systematic errors (Richardson et al., 2012; Wilson et al., 2002). The uncertainty of EC flux data is typically 20–30 % for annual carbon budget (Aubinet et al., 2012; Baldocchi, 2003). Nevertheless, the uncertainties of the GPP and ET estimated from EC measurements are likely to have negligible impacts on our findings of the three WUE metrics, as the same data with the same uncertainties were used.

5 Conclusions

In this study, the impact of the severe soil moisture drought in the summer of 2006 on the water use efficiency of a boreal Scots pine forest ecosystem at Hyytiälä flux site in southern Finland was investigated using both ground-based observations from a flux tower and the site-level simulation by the JSBACH land surface model. The SMI was used to indicate the soil moisture condition at the site. Finland is a high-latitude country and drought is uncommon. Nevertheless, the summer drought in 2006 caused severe forest damage in southern Finland (Muukkonen et al., 2015). The SMI calculated from regional soil moisture simulations over the past 30 years (1981–2010) indicated that such extreme drought affecting forest health was rare in Finland, and the summer drought in 2006 in southern Finland was the most severe one in the 30 year study period (Gao et al., 2016). According to climate scenarios, regardless of the anticipated increase in precipitation, a modest drying of soil is foreseen in northern Europe during the 21st century because of intensifying evapotranspiration (Ruosteenoja et al., 2017).

The impacts from the severe soil moisture drought on plant functioning at the site were clearly seen in the GPP and ET values. From both the observation and simulation results, the GPP and ET reached the recorded minimums during the drought event. The EWUE decreased, whereas the IWUE increased and the uWUE was unchanged during the severe soil moisture drought at the site. The EWUE is very sensitive to the daily changes of GPP and ET. The increase in IWUE during drought was due to the decreased stomatal conductance of plants under increased VPD. The unchanged uWUE indicates that the carbon assimilation and transpiration coupling of the boreal Scots pine forest was not disturbed by the drought event at this site, although the stomatal conductance of plants decreased.

The simulated response in plant functioning to the severe soil moisture drought predicted by JSBACH was weaker than those in the observed dataset, even though the strong limitation on GPP and ET through stomatal closure were seen at the very dry soil moisture condition ($0 \leq \text{SMI} < 0.2$) as in the observed data. The differences between the observed and the model results suggest that, in order to adequately simulate effects of drought on plant functioning, the combined effects of atmospheric and soil moisture drought on stomatal conductance have to be included in the stomatal conductance model in JSBACH. Moreover, inclusion of non-stomatal limitations on photosynthesis during drought, e.g. reduced mesophyll conductance or carboxylation capacity, may additionally improve the model results (Keenan et al., 2010).

This study gives a view of the response of water use efficiency to a summer drought event in a boreal Scots pine forest in Finland, and further suggests that improving our knowledge of ecosystem processes in land surface models are of great importance when estimating biosphere–atmosphere feedbacks of terrestrial ecosystems under climate change.

Competing interests. The authors declare that they have no conflict of interest.

Acknowledgements. We would like to thank the Academy of Finland Centre of Excellence (307331), Academy of Finland (266803), OPTICA (295874), CARB-ARC (285630), ICOS Finland (281255), ICOS-ERIC (281250), MONIMET (LIFE07 ENV/FIN/000133, LIFE12 ENV/FIN/000409), NCoE eSTICC (57001) and EU FP7 EMBRACE (282672) projects for support. The authors acknowledge MPI-MET and MPI-BGC for fruitful co-operation regarding use of the JSBACH model.

Edited by: Ivonne Trebs

References

Arneth, A., Veenendaal, E. M., Best, C., Timmermans, W., Kolle, O., Montagnani, L., and Shibistova, O.: Water use strategies and ecosystem-atmosphere exchange of CO_2 in two highly seasonal environments, Biogeosciences, 3, 421–437, https://doi.org/10.5194/bg-3-421-2006, 2006.

Aubinet, M., Vesala, T., and Papale, D.: Eddy Covariance: a Practical Guide to Measurement and Data Analysis, Springer, Dordrecht, 2012.

Aurela, M.: Carbon Dioxide Exchange in Subarctic Ecosystems Measured by a Micrometeorological Technique, Contributions, 51, Finnish Meteorological Institute, Helsinki, Finland, 132 pp., 2005.

Aurela, M., Lohila, A., Tuovinen, J.-P., Hatakka, J., Penttilä T., and Laurila, T.: Carbon dioxide and energy flux measurements in four northern-boreal ecosystems at Pallas, Boreal Environ. Res., 20, 455–473, 2015.

Baldocchi, D. D.: Assessing the eddy covariance technique for evaluating carbon dioxide exchange rates of ecosystems: past, present and future, Glob. Change Biol., 9, 479–492, 2003.

Ball, J. T., Woodrow, I. E., and Berry, J. A.: A model predicting stomatal conductance and its contribution to the control of photosynthesis under different environmental conditions, Progress in Photosynthesis Research, edited by: Biggins, J., Springer, Dordrecht, 221–224, 1987.

Beer, C., Ciais, P., Reichstein, M., Baldocchi, D., Law, B. E., Papale, D., Soussana, J. F., Ammann, C., Buchmann, N., Frank, D., Gianelle, D., Janssens, I. A., Knohl, A., Köstner, B., Moors, E., Roupsard, O., Verbeeck, H., Vesala, T., Williams, C. A., and Wohlfahrt, G.: Temporal and among-site variability of inherent water use efficiency at the ecosystem level, Global Biogeochem. Cy., 23, GB2018, https://doi.org/10.1029/2008GB003233, 2009.

Berry, J. A., Beerling, D. J., and Franks, P. J.: Stomata: key players in the earth system, past and present, Curr. Opin. Plant Biol., 13, 232–239, 2010.

Betts, A. K.: Understanding hydrometeorology using global models, B. Am. Meteorol. Soc., 85, 1673–1688, 2004.

Böttcher K., Markkanen, T., Thum, T., Aalto, T., Aurela, M., Reick, C. H., Kolari, P., Arslan, A. N., and Pulliainen. J.: Evaluating biosphere model estimates of the start of the vegetation active season in boreal forests by satellite observations, Remote Sens., 8, 580, https://doi.org/10.3390/rs8070580, 2016.

Bréda, N., Cochard, H., Dreyer, E., and Granier, A.: Water transfer in a mature oak stand (Quercus petraea): seasonal evolution and effects of a severe drought, Can. J. Forest Res., 23, 1136–1143, 1993.

Clenciala, E., Kucera, J., Ryan, M., G., and Lindroth, A.: Water flux in boreal forest during two hydrologically contrasting years; species specific regulation of canopy conductance and transpiration, Ann. For. Sci., 55, 47–61, 1998.

Collatz, G., Ribas-Carbo, M., and Berry, J.: Coupled photosynthesis-stomatal conductance model for leaves of C4 plants, Aust. J. Plant Physiol., 19, 519–538, https://doi.org/10.1071/PP9920519, 1992.

Cowan, I. R. and Farquhar, G. D.: Stomatal function in relation to leaf metabolism and environment, Sym. Soc. Exp. Biol., 31, 471–505, 1977.

Eamus, D., Boulain, N., Cleverly, J., and Breshears, D. D.: Global change-type drought-induced tree mortality: vapor pressure deficit is more important than temperature per se in causing decline in tree health, Ecol. Evol., 3, 2711–2729, 2013.

Egea, G., Verhoef, A., and Vidale, P. L.: Towards an improved and more flexible representation of water stress in coupled photosynthesis–stomatal conductance models, Agr. Forest Meteorol., 151, 1370–1384, 2011.

FAO-UNESCO: Soil Map of the World: Revised Legend, World Soil Resources Report 60, FAO, Rome, 1990.

Farquhar, G. D., Caemmerer, S., and Berry, J. A.: A biochemical model of photosynthetic CO_2 assimilation in leaves of C3 species, Planta, 149, 78–90, 1980.

Farquhar, G., O'Leary, M., and Berry, J.: On the relationship between carbon isotope discrimination and the intercellular carbon dioxide concentration in leaves, Funct. Plant Biol., 9, 121–137, 1982.

Farquhar, G. D., Lloyd, J., Taylor, J. A., Flanagan, L. B., Syvertsen, J. P., Hubick, K. T., Wong, S. C., and Ehleringer, J. R.: Vegetation effects on the isotope composition of oxygen in atmospheric CO_2, Nature, 363, 439–443, 1993.

Gao, Y., Markkanen, T., Thum, T., Aurela, M., Lohila, A., Mammarella, I., Kämäräinen, M., Hagemann, S., and Aalto, T.: Assessing various drought indicators in representing summer drought in boreal forests in Finland, Hydrol. Earth Syst. Sci., 20, 175–191, https://doi.org/10.5194/hess-20-175-2016, 2016.

Ge, Z.-M., Kellomaki, S., Zhou, X., and Peltola, H.: The role of climatic variability in controlling carbon and water budgets in a boreal Scots pine forest during ten growing seasons, Boreal Environ. Res., 19, 181–195, 2014.

Granier, A., Reichstein, M., Bréda, N., Janssens, I. A., Falge, E., Ciais, P., Grünwald, T., Aubinet, M., Berbigier, P., Bernhofer, C., Buchmann, N., Facini, O., Grassi, G., Heinesch, B., Ilvesniemi, H., Keronen, P., Knohl, A., Köstner, B., Lagergren, F., Lindroth, A., Longdoz, B., Loustau, D., Mateus, J., Montagnani, L., Nys, C., Moors, E., Papale, D., Peiffer, M., Pilegaard, K., Pita, G., Pumpanen, J., Rambal, S., Rebmann, C., Rodrigues, A., Seufert, G., Tenhunen, J., Vesala, T., and Wang, Q.: Evidence for soil water control on carbon and water dynamics in European forests during the extremely dry year: 2003, Agr. Forest Meteorol., 143, 123–145, 2007.

Guo, D., Westra, S., and Maier, H. R.: An R package for modelling actual, potential and reference evapotranspiration, Environ. Modell. Softw., 78, 216–224, 2016.

Granier, A., Bréda, N., Longdoz, B., Gross, P., and Ngao, J.: Ten years of fluxes and stand growth in a young beech forest at Hesse, North-eastern France, Ann. For. Sci., 65, 704, https://doi.org/10.1051/forest:2008052, 2008.

Hagemann, S. and Stacke, T.: Impact of the soil hydrology scheme on simulated soil moisture memory, Clim. Dynam., 44, 1731–1750, 2015.

Hari, P. and Kulmala, M.: Station for Measuring Ecosystem–Atmosphere Relations (SMEAR II), Boreal Environ. Res., 10, 315–322, 2005.

Hillel, D.: Environmental Soil Physics, Academic Press, San Diego, 1998.

Huang, M., Piao, S., Sun, Y., Ciais, P., Cheng, L., Mao, J., Poulter, B., Shi, X., Zeng, Z., and Wang, Y.: Change in terrestrial ecosystem water-use efficiency over the last three decades, Glob. Change Biol., 21, 2366–2378, 2015.

Ilvesniemi, H., Pumpanen, J., Duursma, R., Hari, P., Keronen, P., Kolari, P., Kulmala, M., Mammarella, I., Nikinmaa, E., Rannik, U., Pohja, T., Siivola, E., and Vesala, T.: Water balance of a boreal Scots pine forest, Boreal Environ. Res., 15, 375–396, 2010.

Irvine, J., Perks, M. P., Magnani, F., and Grace, J.: The response of Pinus sylvestris to drought: stomatal control of transpiration and hydraulic conductance, Tree Physiol., 18, 393–402, 1998.

Jarvis, P.: The interpretation of the variations in leaf water potential and stomatal conductance found in canopies in the field, Philos. T. Roy. Soc. B, 273, 593–610, 1976.

Jung, M., Le Maire, G., Zaehle, S., Luyssaert, S., Vetter, M., Churkina, G., Ciais, P., Viovy, N., and Reichstein, M.: Assessing the ability of three land ecosystem models to simulate gross carbon uptake of forests from boreal to Mediterranean climate in Europe, Biogeosciences, 4, 647–656, https://doi.org/10.5194/bg-4-647-2007, 2007.

Keenan, T., Sabate, S., and Gracia, C.: Soil water stress and coupled photosynthesis–conductance models: bridging the gap between conflicting reports on the relative roles of stomatal, mesophyll conductance and biochemical limitations to photosynthesis, Agr. Forest Meteorol., 150, 443–453, 2010.

Keenan, T. F., Hollinger, D. Y., Bohrer, G., Dragoni, D., Munger, J. W., Schmid, H. P., and Richardson, A. D.: Increase in forest water-use efficiency as atmospheric carbon dioxide concentrations rise, Nature, 499, 324–327, 2013.

Knauer, J., Werner, C., and Zaehle, S.: Evaluating stomatal models and their atmospheric drought response in a land surface scheme: a multibiome analysis, J. Geophys. Res.-Biogeosci., 120, 1894–1911, 2015.

Knorr, W.: Annual and interannual CO_2 exchanges of the terrestrial biosphere: process-based simulations and uncertainties, Global Ecol. Biogeogr., 9, 225–252, 2000.

Kolari, P., Kulmala, L., Pumpanen, J., Launiainen, S., Ilvesniemi, H., Hari, P., and Nikinmaa, E.: CO_2 exchange and component CO_2 fluxes of a boreal Scots pine forest, Boreal Environ. Res., 14, 761–783, 2009.

Law, B. E., Falge, E., Gu, L., Baldocchi, D. D., Bakwin, P., Berbigier, P., Davis, K., Dolman, A. J., Falk, M., Fuentes, J. D., Goldstein, A., Granier, A., Grelle, A., Hollinger, D., Janssens, I. A., Jarvis, P., Jensen, N. O., Katul, G., Mahli, Y., Matteucci, G., Meyers, T., Monson, R., Munger, W., Oechel, W., Olson, R., Pilegaard, K., Paw U, K. T., Thorgeirsson, H., Valentini, R., Verma, S., Vesala, T., Wilson, K., and Wofsy, S.: Environmental controls over carbon dioxide and water vapor exchange of terrestrial vegetation, Agr. Forest Meteorol., 113, 97–120, 2002.

Lloyd, J. and Farquhar, G. D.: ^{13}C discrimination during CO_2 assimilation by the terrestrial biosphere, Oecologia, 99, 201–215, https://doi.org/10.1007/BF00627732, 1994.

Lloyd, J. O. N., Shibistova, O., Zolotoukhine, D., Kolle, O., Arneth, A., Wirth, C., Styles, J. M., Tchebakova, N. M., and Schulze, E. D.: Seasonal and annual variations in the photosynthetic productivity and carbon balance of a central Siberian pine forest, Tellus B, 54, 590–610, 2002.

Mammarella, I., Kolari, P., Rinne, J., Keronen, P., Pumpanen, J., and Vesala, T.: Determining the contribution of vertical advection to the net ecosystem exchange at Hyytiälä forest, Finland, Tellus B, 59, 900–909, https://doi.org/10.1111/j.1600-0889.2007.00306.x, 2007.

Mammarella, I., Peltola, O., Nordbo, A., Järvi, L., and Rannik, Ü.: Quantifying the uncertainty of eddy covariance fluxes due to the use of different software packages and combinations of processing steps in two contrasting ecosystems, Atmos. Meas. Tech., 9, 4915–4933, https://doi.org/10.5194/amt-9-4915-2016, 2016.

Manzoni, S., Vico, G., Katul, G., Fay, P. A., Polley, W., Palmroth, S., and Porporato, A.: Optimizing stomatal conductance for maximum carbon gain under water stress: a meta-analysis across plant functional types and climates, Funct. Ecol., 25, 456–467, 2011.

Markkanen, T., Rannik, U., Keronen, P., Suni, T., and Vesala, T.: Eddy covariance fluxes over a boreal Scots pine forest, Boreal Environ. Res., 6, 65–78, 2001.

McDowell, N., Pockman, W. T., Allen, C. D., Breshears, D. D., Cobb, N., Kolb, T., Plaut, J., Sperry, J., West, A., Williams, D. G., and Yepez, E. A.: Mechanisms of plant survival and mortality during drought: why do some plants survive while others succumb to drought?, New Phytol., 178, 719–739, 2008.

Muukkonen, P., Nevalainen, S., Lindgren, M., and Peltoniemi, M.: Spatial occurrence of drought-associated damages in Finnish boreal forests: results from forest condition monitoring and GIS analysis, Boreal Environ. Res., 20, 172–180, 2015.

Nemani, R. R., Keeling, C. D., Hashimoto, H., Jolly, W. M., Piper, S. C., Tucker, C. J., Myneni, R. B., and Running, S. W.: Climate-driven increases in global terrestrial net primary production from 1982 to 1999, Science, 300, 1560–1563, 2003.

Raddatz, T. J., Reick, C. H., Knorr, W., Kattge, J., Roeckner, E., Schnur, R., Schnitzler, K. G., Wetzel, P., and Jungclaus, J.: Will the tropical land biosphere dominate the climate–carbon cycle feedback during the twenty-first century?, Clim. Dynam., 29, 565–574, 2007.

Reichstein, M., Ciais, P., Papale, D., Valentini, R., Running, S., Viovy, N., Cramer, W., Granier, A., OgÉE, J., Allard, V., Aubinet, M., Bernhofer, C., Buchmann, N., Carrara, A., GrÜNwald, T., Heimann, M., Heinesch, B., Knohl, A., Kutsch, W., Loustau, D., Manca, G., Matteucci, G., Miglietta, F., Ourcival, J. M., Pilegaard, K., Pumpanen, J., Rambal, S., Schaphoff, S., Seufert, G., Soussana, J. F., Sanz, M. J., Vesala, T., and Zhao, M.: Reduction of ecosystem productivity and respiration during the European summer 2003 climate anomaly: a joint flux tower, remote sensing and modelling analysis, Glob. Change Biol., 13, 634–651, 2007.

Reick, C. H., Raddatz, T., Brovkin, V., and Gayler, V.: Representation of natural and anthropogenic land cover change in MPI-ESM, J. Adv. Model. Earth Sy., 5, 459–482, 2013.

Richardson, A. D., Aubinet, M., Barr, A. G., Hollinger, D. Y., Ibrom, A., Lasslop, G., and Reichstein, M.: Uncertainty quantification, in: Eddy Covariance: a Practical Guide to Measurement and Data Analysis, edited by: Aubinet, M., Vesala, T., Papale, D., Springer Netherlands, Dordrecht, 2012.

Roeckner, E., Arpe, K., Bengtsson, L., Christoph, M., Claussen, M., Dümenil, L., Esch, M., Giogetta, M., Schlese, U., and Schultz-Weida, U.: The atmospheric general circulation model ECHAM4: Model description and simulation of the present-day climate, Max Planck Institute for Meterology, Hamburg, 1996.

Roeckner, E., Bäuml, G., Bonaventura, L., Brokopf, R., Esch, M., Giorgetta, M., Hagemann, S., Kirchner, I., Kornblueh, L., and Manzini, E.: The atmospheric general circulation model ECHAM5. PART I: Model description, MPI for Meteorology, Hamburg, Germany, 2003.

Ruosteenoja, K., Markkanen, T., Venäläinen, A., Räisänen, P., and Peltola, H.: Seasonal soil moisture and drought occurrence in Europe in CMIP5 projections for the 21st century, Clim. Dynam., https://doi.org/10.1007/s00382-017-3671-4, 2017.

Schulze, E.-D., Beck, E., and Müller-Hohenstein, K.: Plant Ecology, Springer-Verlag, Berlin Heidelberg, 2005.

Seneviratne, S. I., Corti, T., Davin, E. L., Hirschi, M., Jaeger, E. B., Lehner, I., Orlowsky, B., and Teuling, A. J.: Investigating soil moisture–climate interactions in a changing climate: a review, Earth-Sci. Rev., 99, 125–161, 2010.

Stevens, B., Giorgetta, M., Esch, M., Mauritsen, T., Crueger, T., Rast, S., Salzmann, M., Schmidt, H., Bader, J., Block, K., Brokopf, R., Fast, I., Kinne, S., Kornblueh, L., Lohmann, U., Pincus, R., Reichler, T., and Roeckner, E.: Atmospheric component of the MPI-M Earth System Model: ECHAM6, J. Adv. Model. Earth Sy., 5, 146–172, 2013.

Vesala, T., Suni, T., Rannik, Ü., Keronen, P., Markkanen, T., Sevanto, S., Grönholm, T., Smolander, S., Kulmala, M., Ilvesniemi, H., Ojansuu, R., Uotila, A., Levula, J., Mäkelä, A., Pumpanen, J., Kolari, P., Kulmala, L., Altimir, N., Berninger, F., Nikinmaa, E., and Hari, P.: Effect of thinning on surface fluxes in a boreal forest, Global Biogeochem. Cy., 19, GB2001, https://doi.org/10.1029/2004GB002316, 2005.

Will, R. E., Wilson, S. M., Zou, C. B., and Hennessey, T. C.: Increased vapor pressure deficit due to higher temperature leads to greater transpiration and faster mortality during drought for tree seedlings common to the forest–grassland ecotone, New Phytol., 200, 366–374, 2013.

Wilson, K., Goldstein, A., Falge, E., Aubinet, M., Baldocchi, D., Berbigier, P., Bernhofer, C., Ceulemans, R., Dolman, H., Field, C., Grelle, A., Ibrom, A., Law, B. E., Kowalski, A., Meyers, T., Moncrieff, J., Monson, R., Oechel, W., Tenhunen, J., Valentini, R., and Verma, S.: Energy balance closure at FLUXNET sites, Agr. Forest Meteorol., 113, 223–243, 2002.

Wolf, S., Eugster, W., Ammann, C., Häni, M., Zielis, S., Hiller, R., Stieger, J., Imer, D., Merbold, L., and Buchmann, N.: Contrasting response of grassland versus forest carbon and water fluxes to spring drought in Switzerland, Environ. Res. Lett., 8, 035007, 035007, https://doi.org/10.1088/1748-9326/8/3/035007, 2013.

Wong, S. C., Cowan, I. R., and Farquhar, G. D.: Stomatal conductance correlates with photosynthetic capacity, Nature, 282, 424–426, 1979.

Xie, Z., Wang, L., Jia, B., and Yuan, X.: Measuring and modeling the impact of a severe drought on terrestrial ecosystem CO_2 and water fluxes in a subtropical forest, J. Geophys. Res.-Biogeosci., 121, 2576–2587, https://doi.org/10.1002/2016JG003437, 2016.

Zhou, S., Duursma, R. A., Medlyn, B. E., Kelly, J. W. G., and Prentice, I. C.: How should we model plant responses to drought? An analysis of stomatal and non-stomatal responses to water stress, Agr. Forest Meteorol., 182–183, 204–214, 2013.

Zhou, S., Yu, B., Huang, Y., and Wang, G.: The effect of vapor pressure deficit on water use efficiency at the subdaily time scale, Geophys. Res. Lett., 41, 5005–5013, https://doi.org/10.1002/2014GL060741, 2014.

Zhou, S., Yu, B., Huang, Y., and Wang, G.: Daily underlying water use efficiency for AmeriFlux sites, J. Geophys. Res.-Biogeosci., 120, 887–902, https://doi.org/10.1002/2015JG002947, 2015.

10

Development of bacterial communities in biological soil crusts along a revegetation chronosequence in the Tengger Desert, northwest China

Lichao Liu[1]**, Yubing Liu**[1,2]**, Peng Zhang**[1]**, Guang Song**[1]**, Rong Hui**[1]**, Zengru Wang**[1]**, and Jin Wang**[1,2]

[1]Shapotou Desert Research & Experiment Station, Northwest Institute of Eco-Environment and Resources, Chinese Academy of Sciences, Lanzhou, 730000, China
[2]Key Laboratory of Stress Physiology and Ecology in Cold and Arid Regions of Gansu Province, Northwest Institute of Eco-Environment and Resources, Chinese Academy of Sciences, Lanzhou, 730000, China

Correspondence to: Yubing Liu (liuyb@lzb.ac.cn)

Abstract. Knowledge of structure and function of microbial communities in different successional stages of biological soil crusts (BSCs) is still scarce for desert areas. In this study, Illumina MiSeq sequencing was used to assess the compositional changes of bacterial communities in different ages of BSCs in the revegetation of Shapotou in the Tengger Desert. The most dominant phyla of bacterial communities shifted with the changed types of BSCs in the successional stages, from Firmicutes in mobile sand and physical crusts to Actinobacteria and Proteobacteria in BSCs, and the most dominant genera shifted from *Bacillus*, *Enterococcus* and *Lactococcus* to RB41_norank and JG34-KF-361_norank. Alpha diversity and quantitative real-time polymerase chain reaction (PCR) analysis indicated that bacterial richness and abundance reached their highest levels after 15 years of BSC development. Redundancy analysis showed that silt + clay content and total K were the prime determinants of the bacterial communities of BSCs. The results suggested that bacterial communities of BSCs recovered quickly with the improved soil physicochemical properties in the early stages of BSC succession. Changes in the bacterial community structure may be an important indicator in the biogeochemical cycling and nutrient storage in early successional stages of BSCs in desert ecosystems.

1 Introduction

Biological soil crusts (BSCs) are assemblages of cryptogamic species and microorganisms, such as cyanobacteria, green algae, diatoms, lichens, mosses, soil microbes and other related microorganisms that cement the surface soil particles through their hyphae, rhizines/rhizoids and secretions (Eldridge and Greene, 1994; Li, 2012; Pointing and Belnap, 2012; Weber et al., 2016). Due to their specialized structures and complicated assemblages of their members, BSCs constitute one of the most important landscapes and make up 40 % of the living cover of desert ecosystems, even exceeding 75 % in some special habitats (Belnap and Eldridge, 2003).

It is well known that BSCs play critical roles in the structure and function of semiarid and arid ecosystems (Eldridge and Greene, 1994; Li, 2012). They provide ecological services such as soil stabilization, reduction of wind and water erosion, and facilitation of higher plant colonization (Belnap, 2003; Belnap and Lange, 2001; Maier et al., 2014; Pointing and Belnap, 2012). BSCs are functionally important and variable, and may be a useful model system for diversity-function research. Their functional attributes are relatively well known, and estimation and manipulation of biodiversity in experiments are feasible, at least within some groups of BSC biota (Bowker et al., 2010). This relationship is more easily interpreted in artificially constructed BSCs. There are primary successional stages for BSCs in desert ecosystems: mobile sand, algal crust, lichen crust and moss crust (Lan et al., 2012a; Liu et al., 2006). The different successional stages of BSCs vary in their ecological function (Belnap, 2006; Bowker et al., 2006b; Li, 2012; Moquin et al., 2012).

During BSC succession, physical crusts in mobile sand contain the lowest carbon (C) and nitrogen (N) contents (Zhang et al., 2009). Algal crust is the earliest stage; it has a thin surface layer composed of eolian-borne materials and an organic layer formed by filamentous cyanobacteria associated with sand particles (Housman et al., 2006; Zhang, 2005; Zhang et al., 2009). Lichen and moss appear following stabilization of the algal filaments on the soil surface. The C and N fixation rates are increased in lichen crust (Evans and Lange, 2003; Lan et al., 2012b; Zhang et al., 2010), and there is higher photosynthesis, exopolysaccharide and nitrogenase activity in moss crust than in the early successional crusts (Housman et al., 2006; Lan et al., 2012b). In the BSC successional process, the microbial composition and community structure change greatly (Hu and Liu, 2003; Zhang et al., 2009). Crust succession is positively correlated with phospholipid fatty acid content and microbial biomass (Liu et al., 2013). The microbial biomass of soils is the greatest driving force in most terrestrial ecosystems, largely due to control of conversion rates and mineralization of organic matter (Albiach et al., 2000; Baldrian et al., 2010).

Bacteria present the highest proportion of the microbial biomass in BSCs (Bates et al., 2010; Green et al., 2008; Gundlapally and Garcia-Pichel, 2006; Maier et al., 2014; Wang et al., 2015) and thus have important roles in the BSC successional process. They can decompose organic material and release nutrients, mediating geochemical processes necessary for ecosystem functioning in the persistence of BSCs (Balser and Firestone, 2005). Species composition and community structure of bacteria change greatly during the successional process of BSCs (Gundlapally et al., 2006; Moquin et al., 2012; Zhang et al., 2016). Most research on prokaryotic diversity of BSCs has focused on cyanobacteria-dominated biocrusts in arid and semiarid regions (Abed et al., 2010; Garcia-Pichel et al., 2001; Nagy et al., 2005; Steven et al., 2013; Yeager et al., 2004). Recent studies of the bacterial community structure of bryophyte- or lichen-dominated crusts indicate that lichen-associated communities encompass a wide taxonomic diversity of bacteria (Bates et al., 2011; Cardinale et al., 2008; Maier et al., 2014). Heterotrophic bacteria may perform a variety of roles such as nutrient mobilization and N fixation and could be of considerable importance for the stability of lichen-dominated soil communities. However, there have been few studies on changes of bacterial diversity and their function in BSCs during the development process in desert zones, and these have only focused on the Sonoran (Nagy et al., 2005) and Gurbantunggut deserts (Zhang et al., 2016). What changes occur in bacterial community composition and in their potential roles in improving soil properties for different BSC successional stages? What is the significance of these changes to BSC succession in the recovery process of desert revegetation in temperate zones?

Bowker (2007) examined the role of BSCs in primary succession (vs. secondary succession) during a time when resources were available (e.g., light); however, they became less important once higher vegetation took over. In some environments of high abiotic stress (e.g., deserts), BSCs play a role in succession but remain a permanent component. Bowker's review and discussion is supported by work performed in southern Africa (Büdel et al., 2009) in which different successional BSCs are described. Büdel et al. (2009) also describe in detail crust types that were representative of successional stages. Castillo-Monroy et al. (2011) showed few BSC effects on ecosystem function could be ascribed to bacteria.

A recent study on crusts in the Tengger Desert, China, showed that bacterial diversity and richness were highest after 15 years, and at least 15 years might be needed for recovery of bacterial abundance of BSCs (Liu et al., 2017). To better understand these questions, we must analyze in detail the bacterial community composition of BSCs at all levels of classification and their corresponding function in the recovery process of BSCs. In the present study, bacterial community composition and potential function were analyzed in BSCs along a chronosequence of over 50-year-old revegetation. We investigated the following questions. What are the drivers of bacterial composition over time? What are the micro-processes that drive bacterial composition and function? Do bacteria drive changes in soil physicochemical properties which in turn have a direct influence on bacterial composition and function?

2   Materials and methods

2.1   Study site description

The study site is located in Shapotou, at the southeast fringe of the Tengger Desert, northwest China. The natural landscape is characterized by the reticulated chains of barchan dunes with a vegetation cover of less than 1 %. The mean annual precipitation is about 180 mm with large seasonal and interannual variation. The mean wind speed is $3.5 \, \mathrm{m \, s^{-1}}$, and on average $122 \, \mathrm{d \, yr^{-1}}$ days with dust events. The revegetation protection system for Bao–Lan railway in this area was established initially in 1956 and was expanded away from the railway in 1964, 1973, 1981 and 1987 through establishment of straw checkerboards and plantation of xerophilous shrubs. This unirrigated revegetation system has worked well in protecting the railroad line from sand burial and dust hazards for the past 60 years. Also, the experimental plots of less than 1 ha were established with the same plantation techniques by the Shapotou Desert Research and Experiment Station in 1987, 2000 and 2010 in the nearby sand dunes. These fixed-sand areas provide an ideal temporal succession sequence

Figure 1. Sand dune landscape before (MS, **a**) and after establishing sand-binding vegetation with physical crusts dominated by few algae, revegetated in 2010 (5YR, **b**); with BSCs dominated by algae and lichens, revegetated in 2000 (15YR, **c**); with BSCs dominated by lichens and few mosses, revegetated in 1987 (28YR, **d**); with BSCs dominated by few lichens and mosses, revegetated in 1981 (34YR, **e**); and with BSCs dominated by mosses, revegetated in 1964 (51YR, **f**). Five soil cores (3.5 cm diameter) with crust layers from four vertices of a square (20 m length) and a diagonal crossing point in each plot were sampled individually (as shown in **c**).

for studying the variation in environmental factors. As mentioned in the literature, the initial state of BSCs begins to form following the stabilization of sand dunes and develop with the colonization of cryptogam (Liu et al., 2006). BSCs can be divided into four types: physical, algae-dominated, lichen-dominated and moss-dominated crusts. In this study, we selected the whole BSC layer from the revegetation established in 1964, 1981, 1987, 2000 and 2010, and non-fixed mobile sand (MS) as the control (Fig. 1). BSCs were sampled in early November 2015 and named according to the fixed-sand time as 51YR (51 years of revegetation), 34YR, 28YR, 15YR, 5YR and MS, respectively. The main types of BSCs were algae-, lichen- and moss-dominated crusts from 15YR to 51YR.

2.2 BSC sampling

The detailed sampling method is shown in Fig. 1c, and BSCs were sampled individually using a sterile trowel. To decrease spatial heterogeneity, each BSC sample was taken from six individual plots (at least 20 m between two adjacent plots) from each revegetation time. Therefore, we obtained 30 BSC samples in total (5 cores × 6 individual plots), and these were mixed together to form one composite BSC sample. Triplicate composite samples for each revegetation time were collected, and the BSC samples were preserved in an ice box. Samples were then taken back to the laboratory, immediately sieved (by 1 mm) to remove stones and plant roots, homogenized thoroughly and stored at $-70\,°C$ for subsequent analyses.

2.3 DNA extraction and Illumina MiSeq sequencing

Microbial DNA was extracted from BSC samples using E.Z.N.A Soil DNA (Omega Bio-tek, Norcross, GA, USA) according to the manufacturer's protocols. The extracted DNA was diluted in TE buffer (10 mM Tris-HCl and 1 mM EDTA at pH 8.0) and stored at $-20\,^\circ$C until use. An aliquot of the extracted DNA from each sample was used as a template for amplification. The bacteria 16S ribosomal RNA (rRNA) gene was amplified by polymerase chain reaction (PCR) (95 $^\circ$C for 3 min; followed by 25 cycles at 95 $^\circ$C for 30 s, 55 $^\circ$C for 30 s and 72 $^\circ$C for 45 s; and a final extension at 72 $^\circ$C for 10 min) using primers 338F (5′-ACTCCTACGGGAGGCAGCA-3′) and 806R (5′-GGACTACHVGGGTWTCTAAT-3′). PCR analyses were performed in triplicate 20 µL mixture containing 2 µL of 5 × FastPfu Buffer, 2 µL of 2.5 mM dNTPs, 0.8 µL of each primer (5 µM), 0.2 µL of FastPfu Polymerase and 10 ng of template DNA. This was conducted according to Wang et al. (2015). Amplicons were extracted from 2 % agarose gels and purified using the AxyPrep DNA Gel Extraction Kit (Axygen Biosciences, Union City, CA, USA) according to the manufacturer's instructions and quantified using QuantiFluor™-ST (Promega Corporation, Madison, WI, USA).

Purified amplicons were pooled in equimolar and paired-end sequenced (2 × 300) on an Illumina MiSeq platform according to the standard protocols at Majorbio Bio-Pharm Technology Co. Ltd., Shanghai, China (http://www.majorbio.com). The raw reads were deposited in the NCBI Sequence Read Archive database (accession number: SRP091312).

2.4 Quantitative real-time PCR (qPCR)

qPCR was performed to determine the absolute 16S rRNA gene abundance. We used the primer sets of 515F (5′-GTGCCAGCMGCCGCGGTAA-3′) and 806R to quantify the total bacterial populations. The standard templates were made from 10-fold dilutions of linearized plasmids containing the gene fragment of interest that was cloned from amplified pure culture DNA. The 20 µL reaction mixtures contained 10 µL of 2 × SYBR mix (with ROX) (DBI Bioscience, Ludwigshafen, Germany), 0.4 µL each of 10 µM forward and reverse primers, 1 µL of total DNA template (1 ng µL^{-1}) and 8.2 µL of RNase-free ddH$_2$O. The reaction was conducted on a Stratagene Mx3000P real-time PCR system (Stratagene, Agilent Technologies Inc., Santa Clara, CA, USA) using the following program: 94 $^\circ$C for 3 min followed by 40 cycles of 94 $^\circ$C for 30 s, 58 $^\circ$C for 30 s and 72 $^\circ$C for 30 s, and then 72 $^\circ$C for 2 min. The detection signal was collected at 72 $^\circ$C for 30 s and analyzed. The melting curve was obtained to confirm that the amplified products were of the appropriate size. For each soil sample, the qPCRs were repeated six times.

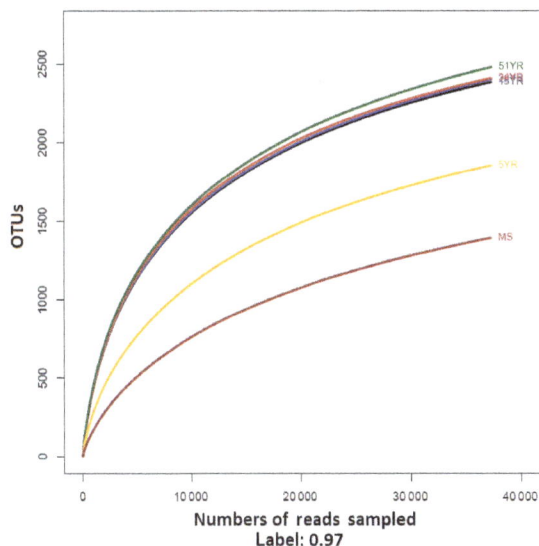

Figure 2. Rarefaction results of the 16S rDNA libraries based on 97 % similarity in different age of BSCs. MS, 5YR, 15YR, 28YR, 34YR and 51YR represent mobile sand, 5-, 15-, 28-, 34- and 51-year-old BSCs, respectively.

2.5 Processing of sequencing data

Raw FASTQ files were demultiplexed and quality-filtered using QIIME (version 1.17) with the following criteria: (i) the 300 bp reads were truncated at any site receiving an average quality score < 20 over a 50 bp sliding window, discarding the truncated reads shorter than 50 bp; (ii) there was exact barcode matching, with at most two mismatched nucleotides in primer matching, and reads containing ambiguous characters were removed; and (iii) only sequences that overlapped > 10 bp were assembled according to their overlap sequence. Reads that could not be assembled were discarded.

Operational taxonomic units (OTUs) were clustered with 97 % similarity cut-off using UPARSE (version 7.1 http://drive5.com/uparse/), and chimeric sequences were identified and removed using UCHIME. The taxonomy of each 16S rRNA gene sequence was analyzed by RDP Classifier (http://rdp.cme.msu.edu/) against the SILVA (SSU115) 16S rRNA database using a confidence threshold of 70 %. Hierarchical clustering analysis was performed using Cluster and visualized using TreeView, and other statistical analyses were performed with the Institute for Environmental Genomics (IEG) pipeline (http://ieg.ou.edu). The average data were calculated for BSCs of each revegetation before analyzing the unique and shared OTUs/genera. The figures were generated with OriginPro 9.1 and Excel 2013. Alpha-diversity analysis was used to reflect the richness and diversity of microbial communities. In order to investigate the overall differences in community composition among the samples, principal component analysis (PCA) was performed using unweighted UniFrac distance (Lozupone and Knight, 2005). Redundancy

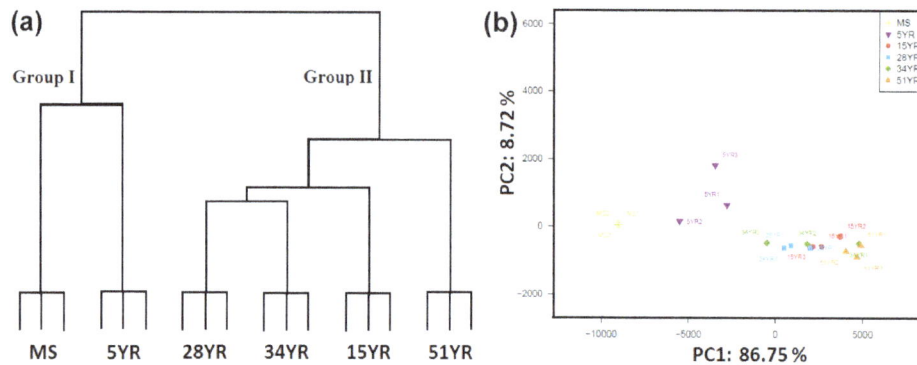

Figure 3. Hierarchical clustering analysis and PCA of bacterial communities in six different ages of BSCs at OTU level based on 97 % similarity (triplicate samples for each age). MS, 5YR, 15YR, 28YR, 34YR and 51YR represent mobile sand, 5-, 15-, 28-, 34- and 51-year-old BSCs, respectively.

analysis (RDA) was used to assess the relationship between bacterial compositions of BSCs and top soil physicochemical properties by permutation test analysis (Zhang et al., 2016). Phylogenetic analysis of the top abundance genus was aligned with closely related 16S rRNA gene sequences, previously selected according to initial BLAST analyses and downloaded from the NCBI website (http://www.ncbi.nlm.nih.gov), using CLUSTAL W (Gundlapally and Garcia-Pichel, 2006). Phylogenetic trees were constructed using the approximately maximum likelihood routine by FastTree (version 2.1.3 http://www.microbesonline.org/fasttree/).

3   Results

3.1   Overview of sequencing and bacterial diversity

Illumina MiSeq sequencing was used to assess the bacterial community composition and diversity of BSCs in successional stages for revegetation in Shapotou. In total, 18 libraries of bacterial 16S rRNA were constructed, and at least 37 332 effective sequences in each sample were obtained, with an average length of 437 bp. A total of 1197–2307 OTUs were generated using a threshold of 0.97 (Table S1 in the Supplement); 394 OTUs were shared and occupied a relatively high proportion among all samples (17.07–32.92 %) (Table S2), and these OTUs accounted for 41.96–84.88 % of the total sequences (Table S2). This indicated a high coherence of community among these soil crusts. Alpha-diversity analysis revealed the microbial richness and diversity. Rarefaction curves showed that the most bacterial OTUs were found in 51YR crust, whereas MS contained the fewest. The number of OTUs was almost the same from 15YR to 51YR (Fig. 2). Community richness estimation using ACE and Chao revealed a similar trend to that for community diversity, which was further supported by Shannon's indexes (Table S1). Hierarchical clustering analysis (Fig. 3a) and PCA (Fig. 3b) showed that the triplicate samples of each age of

BSCs were clustered, verifying that the sequencing results were reliable and the samples were reproducible.

3.2   Bacterial community composition at high taxonomic levels

In the bacterial community, a total of 28 phyla were retrieved at genetic distances of 3 % and clustered into four groups according to their relative abundance (Fig. 4). Of the total sequences, 4.48 % were not classified at the phylum level. The percentages of major phyla for each age of BSCs are shown in Fig. 5. The most abundant phylum shifted from Firmicutes (72.8 %) in MS and 5YR to Actinobacteria in BSCs (minimum of 27.4 % in 15YR and maximum of 30.7 % in 51YR). The following major phyla were at high abundance (> 10 % of total OTUs): Proteobacteria, Chloroflexi, Acidobacteria and Cyanobacteria. The low-abundance phyla (1 % < of total OTUs < 10 %) were Gemmatimonadetes, Bacteroidetes, Armatimonadetes, Verrucomicrobia and Deinococcus–Thermus. The percentages of Proteobacteria, Chloroflexi and Acidobacteria were nearly the same after 15 years of development of BSCs. Cyanobacteria, in addition to the high proportion for 15YR (16.13 %), also had a high proportion in 51YR (9.32 %). The other 17 phyla were all < 1 % of total OTUs and so were removed from further analysis.

At the class level (Table 1), 95.61 % of sequences were assigned, and there was considerable consistency in dominant classes among the crusts. Bacilli was the largest class in MS and 5YR with sequence percentages of 68.73 and 32.62 %, respectively; Actinobacteria was the predominant class from 15YR to 51YR. In addition to subdivisions of Proteobacteria, other major classes included Acidobacteria, Cyanobacteria, Chloroflexi, Clostridia, Cytophagia, Deinococci, Gemmatimonadetes, Ktedonobacteria, Sphingobacteria and Thermomicrobia. The percentages of high-abundance (> 10 % of total OTUs) and low-abundance (1 % < of total OTUs < 10 %) classes decreased from 98 % in

Table 1. Percentages of the major classes in each age of BSCs. MS, 5YR, 15YR, 28YR, 34YR and 51YR represent mobile sand, 5-, 15-, 28-, 34- and 51-year-old BSCs, respectively.

Dominant	MS	5YR	15YR	28YR	34YR	51YR
Bacilli	68.73281	32.6217	10.87003	18.88014	14.65767	2.809922
Actinobacteria	10.25572	17.22651	27.36705	28.34208	29.31533	30.65824
Alphaproteobacteria	4.058181	12.26026	19.93375	16.30594	18.98282	21.11772
Acidobacteria	1.404514	2.372406	11.75488	8.32619	7.703847	9.022644
Chloroflexia	0.886639	2.423301	4.006393	2.962606	3.367977	3.857281
Cyanobacteria	0.112504	16.13272	3.943891	2.275974	2.367049	9.32444
Clostridia	4.091218	1.661666	0.517876	1.017893	0.704489	0.15447
Cytophagia	0.265188	1.223258	0.93039	0.739312	1.022358	1.579521
Deinococci	0.048216	1.255402	0.342869	0.372335	0.249116	0.20715
Deltaproteobacteria	0.447337	0.740205	1.150934	0.993785	1.087539	1.255402
Gammaproteobacteria	5.715383	2.632237	1.011643	1.890246	1.417015	0.425908
Gemmatimonadetes	0.645559	2.400979	2.406336	2.646523	2.75992	2.40455
Ktedonobacteria	0.053573	0.113397	1.75542	1.121469	2.072395	1.657202
Sphingobacteriia	0.262509	0.666095	1.200043	0.897353	0.995571	0.889317
Thermomicrobia	0.449123	1.351834	3.24208	3.414408	3.008143	2.810815
Betaproteobacteria	0.572342	0.789314	0.939319	1.021465	1.073253	1.11254
Minor	0.018688	0.039555	0.080851	0.08194	0.081753	0.085887
Unclassified	0.000911	0.00142	0.005018	0.005822	0.009866	0.02084

MS to 89.29 % in 51YR, and minor and unclassified classes increased from 1.96 % in MS to 10.67 % in 51YR.

At the family level, there were 133 identified families (data not shown), with the most abundant families being Bacillaceae, Enterococcaceae and Streptococcaceae (Table S3). Other dominant families were Geodermatophilaceae, JG34-KF-161, JG34-KF-361, Methylobacteriaceae, Micromonosporaceae, Bradyrhizobiaceae and Enterobacteriaceae.

3.3   Characterization of major genera and species

A large proportion of sequences were not assigned to any genera. Even for genera with relative abundance > 1 % in any sample, unclassified sequences occupied a high proportion (4.87–8.59 %). Moreover, higher percentages of total sequences (from 13.51 % in MS to 37.28 % in 51YR) were found in low-abundance genera (< 1 % in any sample) (Table S4). A total of 460 genera were found in the crusts, of which 201 were shared by all BSC samples (data not shown). The major genera in each age of BSCs are summarized in Fig. 6. *Bacillus*, *Enterococcus* and *Lactococcus* were the primary genera; they represented 64.31 % of the total sequences in MS, and decreased to 30.20 % in 5YR and only 2.63 % in 51YR, indicating that these three genera were predominant in mobile sand or physical crusts. Enterobacteriaceae_unclassified and *Alkaliphilus* were low-abundance genera in MS. With the decrease in the three primary genera from MS to 51YR, a series of genera increased in BSCs compared with MS and 5YR, including RB41_norank, JG34-KF-361_norank, Acidimicrobiales_uncultured, JG34-KF-161_norank, JG30-

KF-CM45_norank, *Microvirga*, Actinobacteria_norank and *Rubrobacter* (relative abundance > 2 %).

The phylogenetic relationships of the 30 most abundant genera are shown in Fig. 7. They clustered into three groups at the phylum level: Actinobacteria formed one group and included 10 genera; another group was Firmicutes and Proteobacteria; and Cyanobacteria, Chloroflexi and Deinococcus–Thermus formed the third group. The genera *Bryobacter* and *Blastocatella* in phylum Acidobacteria were divided into two different groups.

Bacillus was the primary genus and represented 31 % of the sequences in MS (Table S4). An unclassified species in this genus reached nearly 30 % relative abundance in MS (Fig. 8). In the *Enterococcus* genus, another core component, there was also an unclassified species with high abundance. In the core species (Fig. 8), *Bacillus*_unclassified, *Enterococcus*_unclassified, *Lactococcus piscium*, Enterobacteriaceae_unclassified and *Alkaliphilus oremlandii* OhILAs were predominant and decreased from MS to 51YR; only *Acidimicrobiales*_unclassified increased, with the highest proportion in 51YR (2.62 %). The relative abundance of the primitive species in MS and physical crusts decreased in BSCs (from 15YR to 51YR) because of the increased numbers of species. There was little difference in numbers of genera and species among biocrusts (from 15YR to 51YR), only in sequence numbers.

3.4   Relationships between bacterial community structure and soil physicochemical properties

RDA (Fig. 9) and hierarchical clustering analysis (Fig. 3) were used to discern the correlations between bacterial com-

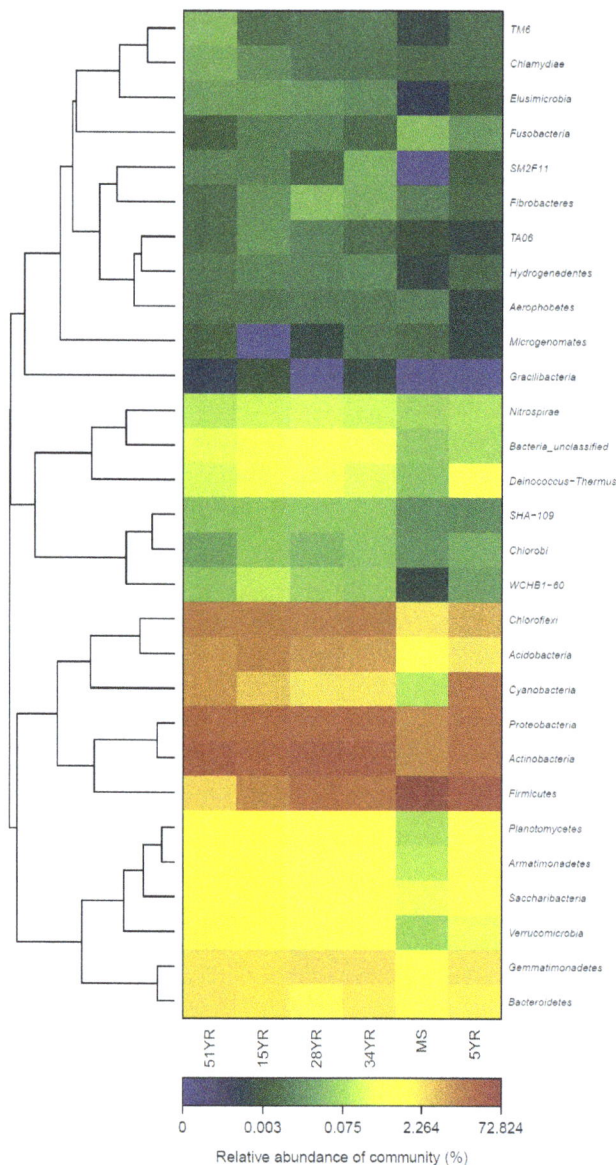

Figure 4. Heat map of bacterial communities in different ages of BSCs at phylum level. MS, 5YR, 15YR, 28YR, 34YR and 51YR represent mobile sand, 5-, 15-, 28-, 34- and 51-year-old BSCs, respectively.

munities and soil physicochemical properties. Taking into account the likely changes in the soil properties from samples with the same successional stages at the same experimental site, we selected soil biogeochemical data collected from 2005 in the RDA (data from Li et al., 2007a; Table S5). The BSC grouping patterns of bacterial communities at the phylum and genus levels were similar to the OTU level, with all divided into two groups. Group I contained two members, MS and 5YR, which dominated the physical crusts and algal crusts (Fig. 1a and b) and had the lowest diversities with Shannon indexes of 3.3 and 4.61, and Simpson indexes of

0.139 and 0.0531, respectively (Table S1). The remaining BSCs comprised the largest branch of group II, which dominated BSCs composed of algae, lichens or mosses (Fig. 1c–f) and had higher diversity with Shannon indexes > 6.0 (Table S1).

From Fig. 9, it can be inferred that BSC development was associated with soil physicochemical properties. The development of microbial community structure was positively correlated with the physicochemical index except for soil bulk density. The total variation in OTU data explained by the first four axes in the RDA (as constrained by the measured environmental variables) was 82.16 %, with the first axis explaining 75.27 % and the second axis explaining 4.42 %. Of all the environmental factors, silt + clay content and total K were most strongly related to axis 1, with highest correlated variable (silt + clay: −0.91; total K: −0.90). Therefore, silt + clay content and total K were closely related to bacterial community development of BSCs, shown by the positions of cluster groups along axis 1. Eight soil physicochemical variables were all significant as verified by the permutation test analysis ($P < 0.05$): pH; silt and clay content; organic C; total phosphorus (P), nitrogen (N) and potassium (K); electrical conductivity (EC); and water-holding capacity (WHC).

3.5 Quantification of bacterial abundance

The averaged bacterial abundance in MS was 1.12×10^6 copies (16S rRNA gene) per gram of soil (Table 2). Similar to the shift of bacterial richness, gene copies increased quickly in the initial 15 years of BSC development and reached the approximate highest level of 2.70×10^8 copies in 15YR. There were no significant differences among 28YR, 34YR and 51YR.

4   Discussion

On a landscape scale and in high-stress environments, the role of diversity hot spots of BSC microbes is crucial to establishing stability and regulating moisture and nutrient cycling (Bowker, 2007). Additionally, bacteria are the conduits between the larger BSC organisms and plants, facilitating micro-processes (Castillo-Monroy et al., 2011). Thus, bacteria are key contributors to the BSC primary succession process and no doubt also in terms of secondary succession.

4.1 Impact of BSC age on bacterial community composition

In the present study, we gained information concerning the diversity of bacterial communities in BSCs of different ages in restored vegetation in Shapotou in the Tengger Desert. The 16S rRNA gene-based amplicon survey revealed the dominance of Actinobacteria, Proteobacteria, Chloroflexi, Acidobacteria and Cyanobacteria in all BSCs, with Firmi-

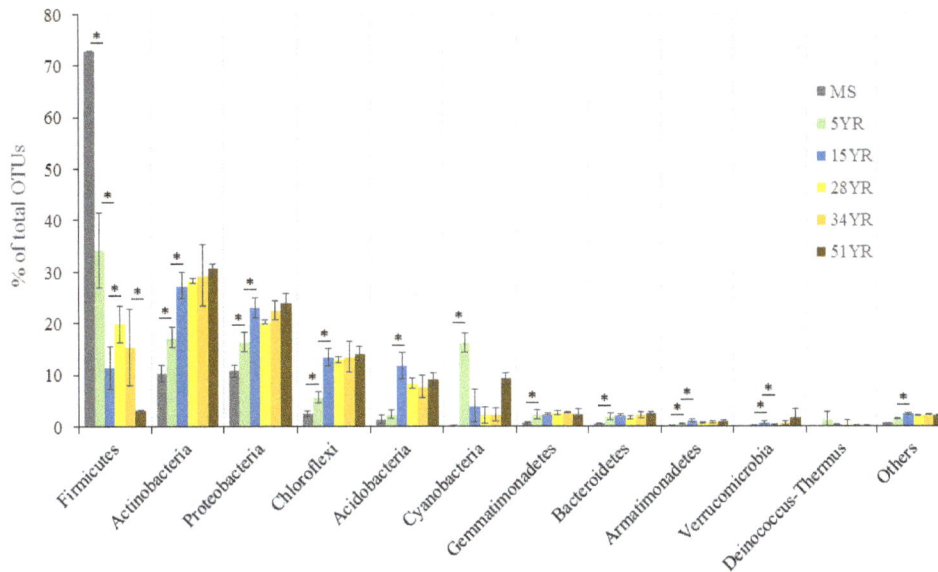

Figure 5. Abundant phyla (> 10 % of total OTUs) and low-abundance phyla (1 % < of total OTUs < 10 %) of bacteria distributed in different ages of BSCs. Data are defined at a 3 % OTU genetic distance. Data are presented as mean ± standard deviation; $n = 3$ per BSC sample. A paired t test (BSC samples) was used to assess the significance between adjacent ages of BSCs. * $P \leq 0.05$; ** $P \leq 0.001$. MS, 5YR, 15YR, 28YR, 34YR and 51YR represent mobile sand, 5-, 15-, 28-, 34- and 51-year-old BSCs, respectively.

Table 2. Absolute abundances of bacteria (copies of ribosomal genes per gram of soil) in BSCs quantified by qPCR (means ± standard deviation, $n = 6$). MS, 5YR, 15YR, 28YR, 34YR and 51YR represent mobile sand, 5-, 15-, 28-, 34- and 51-year-old BSCs, respectively.

Dominant	MS	5YR	15YR	28YR	34YR	51YR
Bacteria abundance	1.12×10^6 $\pm 4.19 \times 10^5$a	3.94×10^7 $\pm 2.21 \times 10^6$b	2.70×10^8 $\pm 1.91 \times 10^7$c	5.44×10^8 $\pm 4.23 \times 10^7$c	7.61×10^8 $\pm 8.5 \times 10^7$c	9.03×10^8 $\pm 2.55 \times 10^7$c

Means with different letters are significantly different ($P < 0.05$).

cutes dominating MS (72.8 %) and decreasing to 3.05 % in 51YR, and Actinobacteria increasing from 15YR (27.4 %) to 51YR (30.7 %). Due to different arid conditions, comparisons with other studies of BSCs should be viewed with caution. Cyanobacteria, Actinobacteria, Proteobacteria and Acidobacteria are ubiquitous in soils and sediments everywhere, in arid as well as wet landscapes (Fierer et al., 2012), and Proteobacteria are very common and diverse among all BSCs. We observed that Actinobacteria were the most abundant phylum in the developing (15YR, 28YR and 34YR) and relatively developed (51YR) BSCs, similar to BSCs from the Colorado Plateau and the Sonoran Desert, where Actinobacteria were dominant (Gundlapally and Garcia-Pichel, 2006; Nagy et al., 2005; Steven et al., 2013). Actinobacteria and Proteobacteria are usually predicted to be copiotrophic groups which increase in high-C environments (Fierer et al., 2007). These results differ from those reported in BSCs from Oman and the Gurbantunggut Desert (Abed et al., 2010; Moquin et al., 2012; Zhang et al., 2016), and even from BSCs of natural vegetation at the edge of the Tengger Desert (Wang et al., 2015), where Proteobacteria were the most abundant phy-

lum, followed by Cyanobacteria, Actinobacteria and Chloroflexi. Unexpectedly, Cyanobacteria had a high proportion in the developed BSCs, although they were prevalent in early successional stages of BSCs (5YR) and play crucial roles in initial crust development (Belnap and Lange, 2001). This is relatively similar to that in the natural habitat around the Tengger Desert, where Cyanobacteria (19.5 %) and Actinobacteria (19.4 %) were the most dominant phyla after Proteobacteria (25.0 %). Moreover, the results did not resemble those from arid Arizona soils (Dunbar et al., 1999) or the Gurbantunggut Desert (Zhang et al., 2016) due to the high proportion of Chloroflexi, an unexplained presence of thermophilic phyla (Gundlapally and Garcia-Pichel, 2006; Moquin et al., 2012; Nagy et al., 2005) that display good adaptation to drought conditions and the important roles in the development of BSCs in arid zones (Lacap et al., 2011; Wang et al., 2015).

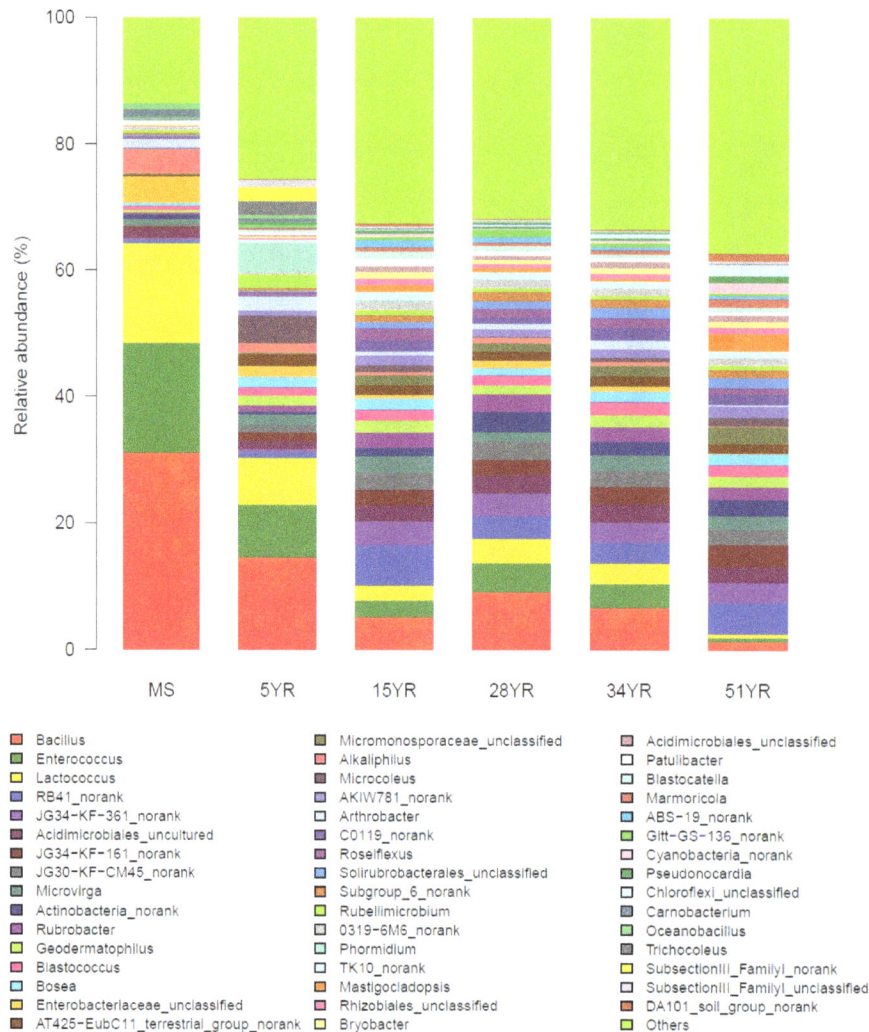

Figure 6. Bacterial community composition in six different ages of BSCs at the genus level. Data are defined at a 3 % OTU genetic distance. MS, 5YR, 15YR, 28YR, 34YR and 51YR represent mobile sand, 5-, 15-, 28-, 34- and 51-year-old BSCs, respectively.

4.2   Function of BSC bacteria

More recent information about BSC bacteria has been reported with the convenience of culture-independent sequencing methods, and studies of their function and classification in BSCs are increasingly detailed. The main function of these dominant bacteria involves the cycling and storage of C and N in desert ecosystems, which is vital to the functioning of arid land (Weber et al., 2016). Firmicutes are more frequently detected in below-biocrust soils (1–2 cm depth) (Elliott et al., 2014) and dominated in MS and 5YR, with the vast majority of abundant species being in Firmicutes in the Tengger Desert. Cyanobacteria are the main contributors to C and N fixation in soils during successional processes of BSCs (Belnap and Gardner, 1993). They are thought to serve as pioneers in the stabilization process of soils (Garcia-Pichel and Wojciechowski, 2009), of which the genus *Phormidium* is significantly more abundant in surface soils (0–1 cm

depth), and the genus *Microcoleus* is globally dominant as biocrust-forming microorganisms in most arid lands, and their production of polysaccharide sheaths aids in the formation of centimeter-long filament bundles (Belnap and Lange, 2003; Boyer et al., 2002; Garcia-Pichel et al., 2001; Pointing and Belnap, 2012). In addition to the filamentous bacteria of *Microcoleus* and *Phormidium*, *Mastigocladopsis* and *Trichocoleus* were also in the 30 most abundant genera of BSCs in Shapotou and mainly harvest energy from light. *Pseudonocardia*, a mycelial genus of Actinobacteria, was dominant and is likely important during BSC formation (Weber et al., 2016). Proteobacteria and Bacteroidetes can produce exopolysaccharides, so they could also play roles in soil stabilization and BSC formation (Gundlapally and Garcia-Pichel, 2006).

Owing to limited culture collections and curated sequence databases of BSC bacteria, most non-cyanobacterial se-

Phyla   **Genera**

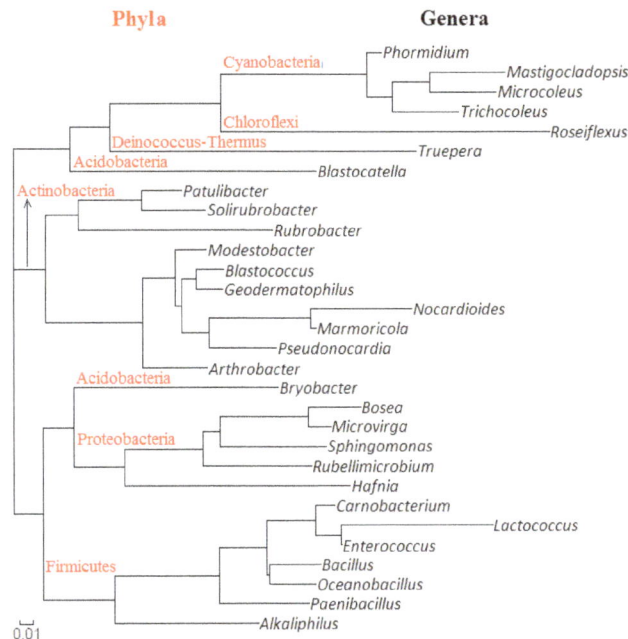

Figure 7. Phylogenetic relationship of the 30 most abundant genera in bacterial composition of BSCs.

quences from DNA-based bacterial surveys cannot be reliably named or taxonomically defined, especially in relatively abundant genera in Actinobacteria and Proteobacteria, such as *Bosea*, *Microvirga*, *Rubellimicrobium*, *Patulibacter*, *Solirubrobacter*, *Blastococcus* and *Arthrobacter* in the present study. Different compositions of bacterial communities play various roles in improving soil properties in different BSC successional stages, suggesting their positive potential function in soil biogeochemical cycle and ecosystem process. Further discovery and characterization of the functions of these dryland-adapted bacteria is a challenging area for future study.

4.3 Relationship between bacterial community shift and soil physicochemical properties

PCA and RDA showed that bacterial community compositions of MS and 5YR significantly differed from those of BSCs of more than 15 years in age and were positively correlated with soil physicochemical properties. Combined with the results of alpha-diversity analysis and qPCR, this means that the species richness and abundance reached their highest levels at 15 years of BSC development and then maintained similar levels thereafter. Similar trends were found in recovery of soil properties and processes after sand binding at five different-aged revegetated sites – proportions of silt and clay, and organic C increased with years since revegetation (Li et al., 2007a, b). The annual recovery rates of soil properties were greater at the initial revegetated sites (0–14 years) than at the old revegetated sites (43–50 years) (Li et al., 2007a).

These results suggest that bacterial communities of BSCs recovered quickly in the fastest recovery phase of soil properties (the initial 15 years), and the bacterial biomass increased with the improvement of soil texture and nutrients, especially silt, clay and total K content in the Tengger Desert. A significant positive correlation was found between silt and clay and the number of BSC types in southern Africa (Büdel et al., 2009), suggesting that fine grain size promotes BSC succession and their biomass content. This may be attributed to the diversity of BSCs, vegetation composition, soil temperature and soil moisture, because these are key factors regulating soil microbial composition and activity (Butenschoen et al., 2011; De Deyn et al., 2009; Sardans et al., 2008), soil nutrient uptake and release (Peterjohn et al., 1994; Rustad et al., 2001), especially in the BSCs of top soil. It would be good to understand more of the factors that together influenced the composition and function of BSC bacteria in long-term revegetation, including BSCs, plants, soil biochemical properties and climate conditions, and the microorganisms that in turn have the positive influence on soil improvement (Li et al., 2007b, 2010).

Many reports have interpreted correlations among soil properties and BSCs as an indicator that BSCs are drivers of soil fertility and development (Chamizo et al., 2012; Delgado-Baquerizo, 2013; Yu et al., 2014; Zhang et al., 2010); some have reported the opposite and suggest a direct influence of soil properties on BSC development (Belnap et al., 2014; Bowker et al., 2006a; Bowker and Belnap, 2008; Concostrina-Zubiri et al., 2013; Rivera-Aquilar et al., 2009; Root and McCune, 2012; Weber et al., 2016). These are important questions, and parsing out the interactions of BSCs and soil biogeochemical properties remains an important frontier in BSC research. However, further work to identify controlled experimental approaches is required because field correlations do not explain the directionality of causality over time.

4.4 The role of BSCs in succession

In temperate desert regions, BSCs are not well investigated regarding community structure and diversity. Furthermore, studies on succession are rare (Langhans et al., 2009). Most evidence indicates that BSC facilitate succession to later series, suggesting that assisted recovery of BSCs could speed up succession (Bowker, 2007). Because BSCs are ecosystem engineers in high-abiotic-stress systems, loss of BSCs may be synonymous with crossing degradation thresholds. Whether BSCs are deemed facilitative or inhibitory for later successional vegetation may depend on how exhaustively the interaction between plants and BSCs is investigated. In fixed-sand areas, BSCs may in some cases reduce infiltration (inhibitory effect) (Mitchell et al., 1998), but they also increase soil stability and serve as an N source for surviving and recolonizing trees (facilitative effects) (Tateno et al., 2003; Uchida et al., 2000). The BSC bacterial communities in the

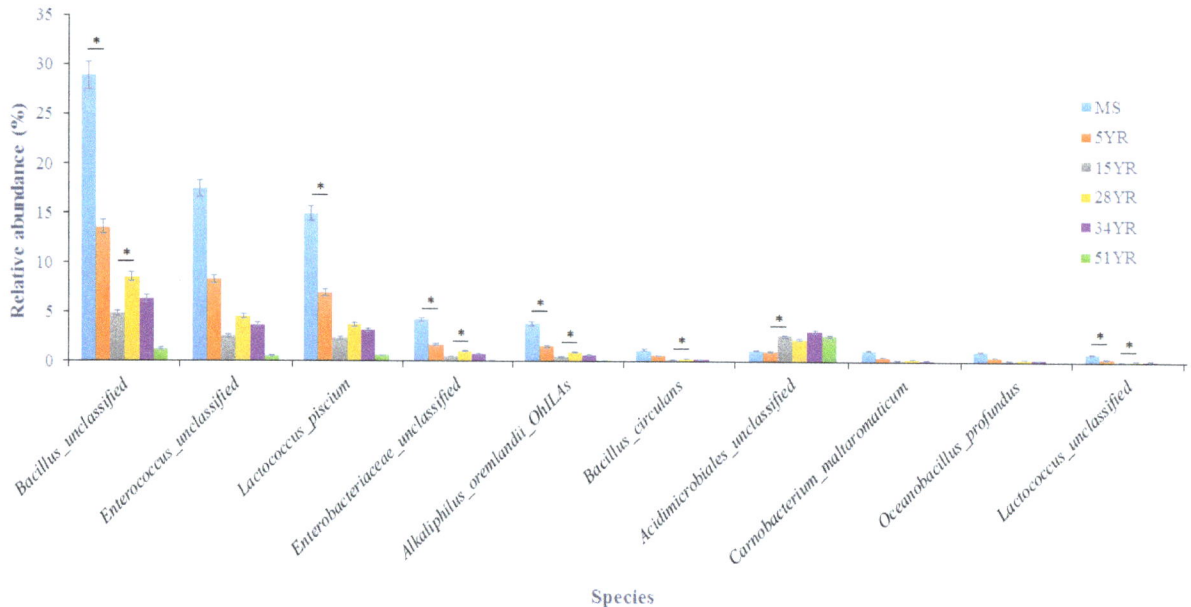

Figure 8. Abundant species (> 10 % of total OTUs) and low-abundance species (1 % < of total OTUs < 10 %) of bacteria distributed in different ages of BSCs. Data are defined at a 3 % OTU genetic distance. Data are presented as mean ± standard deviation; $n = 3$ per BSC samples. Paired t tests (BSC samples) were used to assess the significance between the adjacent ages of BSCs. * $P \leq 0.05$; ** $P \leq 0.001$. MS, 5YR, 15YR, 28YR, 34YR and 51YR represent mobile sand, 5-, 15-, 28-, 34- and 51-year-old BSCs, respectively.

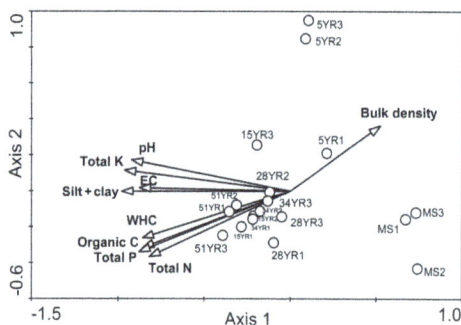

Figure 9. Redundancy analysis (RDA) of bacterial community structures in relation to soil physiochemical properties. Arrows indicate the direction and magnitude of soil physiochemical index associated with bacterial community structures. The lengths of arrows in the RDA plot correspond to the strength of the correlation between variables and community structure. Each circle represents the bacterial community structure for each sample.

successional stages may help establish stability and regulate nutrient and biogeochemical cycling. Castillo-Monroy et al. (2011) found that the BSC richness matrix had the greatest direct effect on the ecosystem function matrix. Despite this result, very few of the BSC effects on ecosystem function could be ascribed to changes within the bacterial community. This provides valuable insights concerning semiarid ecosystems where plant cover is spatially discontinuous and ecosystem function in plant interspaces is regulated largely by BSCs.

5 Conclusions

Illumina MiSeq sequencing showed that changes of BSC bacterial diversity and richness in BSC succession were consistent with the recovery phase of soil properties in vegetation succession of Shapotou in the Tengger Desert. The shift of bacterial community composition in BSCs at all levels of classification was related to their corresponding function in the BSC recovery process. BSC bacteria are crucial to establishing stability and nutrient cycling in desert ecosystem, and they are the conduits between the larger BSC organisms and plants facilitating micro-processes. These results have confirmed that bacteria are key contributors to the BSC succession process.

Author contributions. LL and YL designed the research. PZ, GS and RH collected samples from the field. YL and JW performed DNA extraction and quality detection. YL analyzed the high-throughput data and prepared the manuscript with consistent contributions from LL. ZW analyzed the soil biogeochemical data and made the RDA figure.

Competing interests. The authors declare that they have no conflict of interest.

Special issue statement. This article is part of the special issue "Biological soil crusts and their role in biogeochemical processes and cycling". It is not associated with a conference.

Acknowledgements. This work was financially supported by the Creative Research Group Program of the National Natural Science Foundation of China (grant no. 41621001) and the National Natural Science Foundation of China (grant nos. 41371100 and 41401112).

Edited by: Anita Antoninka

References

Abed, R. M. M., Kharusi, S. A., Schramm, A., and Robinson, M. D.: Bacterial diversity, pigments and nitrogen fixation of biological desert crusts from the Sultanate of Oman, FEMS Microbiol. Ecol., 72, 418–428, 2010.

Albiach, R., Canet, R., Pomares, F., and Ingelmo, F.: Microbial biomass content and enzymatic activities after the application of organic amendments to a horticultural soil, Bioresour. Technol., 75, 43–48, 2000.

Baldrian, P., Merhautova, V., Petrankova, M., and Cajthaml, T.: Distribution of microbial biomass and activity of extracellular enzymes in a hardwood forest soil reflect soil moisture content, Appl. Soil Ecol., 46, 177–182, 2010.

Balser, T. and Firestone, M.: Linking microbial community composition and soil processes in a California annual grassland and mixed-conifer forest, Biogeochemistry, 73, 395–415, 2005.

Bates, S. T., Nash, T. H., Sweat, K. G., and Garcia-Pichel, F.: Fungal communities of lichen-dominated biological soil crusts: Diversity, relative microbial biomass, and their relationship to disturbance and crust cover, J. Arid Environ., 74, 1192–1199, 2010.

Bates, S. T., Cropsey, G. W., Caporaso, J. G., and Knight, R.: Bacterial communities associated with the lichen symbiosis, Appl. Environ. Microbiol., 77, 1309–1314, 2011.

Belnap, J.: The world at your feet: desert biological soil crusts, Front Ecol. Environ., 1, 181–189, 2003.

Belnap, J.: The potential roles of biological soil crusts in dryland hydrologic cycles, Hydrol. Process., 20, 3159–3178, 2006.

Belnap, J. and Eldridge, D.: Disturbance and recovery of biological soil crusts, in: Biological soil crusts: structure, function, and management, ecological studies, edited by: Belnap, J. and Lange, O. L., Vol. 150, Springer, Berlin, 363–383, 2003.

Belnap, J. and Gardner, J. S.: Soil microstructure in soils of the Colorado Plateau: the role of the cyanobacterium Microcoleus vaginatus, Great Basin Nat., 53, 40–47, 1993.

Belnap, J. and Lange, O. L.: Biological Soil Crusts: Structure, Function, and Management, Springer-Verlag, Berlin, Germany, 2001.

Belnap, J., Miller, D. M., Bedford, D. R., and Phillips, S. L.: Pedological and geological relationships with soil lichen and moss distribution in the eastern Mojave Desert, CA, USA, J. Arid Environ., 106, 45–57, 2014.

Bowker, M. A.: Biological soil crust rehabilitation in theory and practice: an underexploited opportunity, Restor. Ecol., 15, 13–23, 2007.

Bowker, M. A. and Belnap J.: A simple classification of soil types as habitats of biological soil crusts on the Colorado Plateau, USA, J. Veg. Sci., 19, 831–840, 2008.

Bowker, M. A., Belnap, J., Davidson, D. W., and Goldstein, H.: Correlates of biological soil crust abundance across a continuum of spatial scales: support for a hierarchical conceptual model, J. Appl. Ecol., 43, 152–163, 2006a.

Bowker, M. A., Belnap, J., and Miller, M. E.: Spatial modeling of biological soil crusts to support rangeland assessment and monitoring, Rangeland Ecol. Manage, 59, 519–529, 2006b.

Bowker, M. A., Maestre, F. T., and Escolar, C.: Biological crusts as a model system for examining the biodiversity-ecosystem function relationship in soils, Soil Biol. Biochem., 42, 405–417, 2010.

Boyer, S. L., Johansen, J. R., Flechtner, V. R., and Howard, G. L.: Phylogeny and genetic variance in terrestrial Microcoleus (Cyanophyceae) species based on sequence analysis of the 16S rRNA gene and associated 16S–23S ITS region, J. Phycol., 38, 1222–1235, 2002.

Büdel, B., Darienko, T., Deutschewitz, K., Dojani, S., Friedl, T., Mohr, K. I., Salisch, M., Reisser, W., and Weber, B.: Southern African biological soil crusts are ubiquitous and highly diverse in drylands, being restricted by rainfall frequency, Microb. Ecol., 57, 229–247, 2009.

Butenschoen, O., Scheu, S., and Eisenhauer, N.: Interactive effects of warming, soil humidity and plant diversity on litter decomposition and microbial activity, Soil Biol. Biochem., 43, 1902–1907, 2011.

Cardinale, M., Castro Jr., J. V., Müller, H., Berg, G., and Grube, M.: In situ analysis of the bacterial community associated with the reindeer lichen Cladonia arbuscula reveals predominance of Alphaproteobacteria, FEMS Microbiol. Ecol., 66, 63–71, 2008.

Castillo-Monroy, A. P., Bowker, M. A., Maestre, F. T., Rodriguez-Echeverria, S., Martinez, I., Barraza-Zepeda C. E., and Escolar, C.: Relationships between biological soil crusts, bacterial diversity and abundance, and ecosystem functioning: Insights from a semi-arid Mediterranean environment, J. Veg. Sci., 22, 165–174, 2011.

Chamizo, S., Cantón, Y., Miralles, I., and Domingo, F.: Biological soil crust development affects physicochemical characteristics of soil surface in semiarid ecosystems, Soil Biol. Biochem., 49, 96–105, 2012.

Concostrina-Zubiri, L., Huber-Sannwald, E., Martínez, I., Flores, J. L., and Escudero, A.: Biological soil crusts greatly contribute to small-scale soil heterogeneity along a grazing gradient, Soil Biol. Biochem., 64, 28–36, 2013.

De Deyn, G. B., Quirk, H., Yi, Z., Oakley, S., Ostle, N. J., and Bardgett, R. D.: Vegetation composition promotes carbon and nitrogen storage in model grassland communities of contrasting soil fertility, J. Ecol., 97, 864–875, 2009.

Delgado-Baquerizo, M., Morillas, L., Maestre, F. T., and Gallardo, A.: Biocrusts control the nitrogen dynamics and microbial functional diversity of semi-arid soils in response to nutrient additions, Plant Soil, 372, 643–654, 2013.

Dunbar, J., Takala, S., Barns, S. M., Davis, J. A., and Kuske, C. R.: Levels of bacterial community diversity in four arid soils compared by cultivation and 16S rRNA gene cloning, Appl. Environ. Biol., 65, 1662–1669, 1999.

Eldridge, D. J. and Greene, R. S. B.: Microbiotic soil crusts – a review of their roles in soil and ecological processes in the rangelands of Australia, Aust. J. Soil Res., 32, 389–415, 1994.

Elliott, D. R., Thomas, A. D., Hoon, S. R., and Sen, R.: Niche partitioning of bacterial communities in biological crusts and soils under grasses, shrubs and trees in the Kalahari, Biodivers. Conserv., 23, 1709–1733, 2014.

Evans, R. D. and Lange, O. L.: Biological soil crusts and ecosystem nitrogen and carbon dynamics, in: Biological Soil Crusts: Structure, Function, and Management, edited by: Belnap, J. and

Lange, O. L., Springer, New York, 263–279, 2003.

Fierer, N., Bradford, M. A., and Jackson, R. B.: Toward an ecological classification of soil bacteria, Ecology, 88, 1354–1364, 2007.

Fierer, N., Leff, J. W., Adams, B. J., Nielsen, U. N., Bates, S. T., Lauber, C. L., Owens, S., Gilbert, J. A., Wall, D. H., and Caporaso, G. J.: Cross-biome metagenomic analyses of soil microbial communities and their functional attributes, P. Natl. Acad. Sci. USA, 109, 21390–21395, 2012.

Garcia-Pichel, F. and Wojciechowski, M. F.: The evolution of a capacity to build supra-cellular ropes enabled filamentous cyanobacteria to colonize highly erodible substrates, PLoS One, 4, e7801, https://doi.org/10.1371/journal.pone.0007801, 2009.

Garcia-Pichel, F., López-Cortés, A., and Nübel, U.: Phylogenetic and morphological diversity of cyanobacteria in soil desert crusts from the Colorado Plateau, Appl. Environ. Microb., 67, 1902–1910, 2001.

Green, L. E., Porras-Alfaro, A., and Sinsabaugh, R. L.: Translocation of nitrogen and carbon integrates biotic crust and grass production in desert grassland, J. Ecol., 96, 1076–1085, 2008.

Gundlapally, S. R. and Garcia-Pichel, F.: The community and phylogenetic diversity of biological soil crusts in the Colorado Plateau studied by molecular fingerprinting and intensive cultivation, Microb. Ecol., 52, 345–357, 2006.

Housman, D. C., Powers, H. H., Collins, A. D., and Belnap, J.: Carbon and nitrogen fixation differ between successional stages of biological soil crustsinthe Colorado Plateau and Chihuahuan Desert, J. Arid Environ., 66, 620–634, 2006.

Hu, C. X. and Liu, Y. D.: Primary succession of algal community structure in desert soil, Acta Bot. Sin., 45, 917–924, 2003.

Lacap, D. C., Warren-Rhodes, K. A., McKay, C. P., and Pointing, S. B.: Cyanobacteria and chloroflexi-dominated hypolithic colonization of quartz at the hyper-arid core of the Atacama Desert, Chile, Extremophiles, 15, 31–38, 2011.

Lan, S. B., Wu, L., Zhang, D. L., and Hu, C. X.: Successional stages of biological soil crusts and their microstructure variability in Shapotou region (China), Environ. Earth Sci., 65, 77–88, 2012a.

Lan, S. B., Wu, L., Zhang, D. L., and Hu, C. X.: Effects of drought and salt stresses on man-made cyanobacterial crusts, Eur. J. Soil Biol., 46, 381–386, 2012b.

Langhans, T. M., Storm, C., and Schwabe, A.: Community assembly of biological soil crusts of different successional stages in a temperate sand ecosystem, as assessed by direct determination and enrichment techniques, Microb. Ecol., 58, 394–407, 2009.

Li, X. R.: Eco-hydrology of biological soil crusts in desert regions of China, China Higher Education Press, 2012.

Li, X. R., He, M. Z., Duan, Z. H., Xiao, H. L., and Jia, X. H.: Recovery of topsoil physicochemical properties in revegetated sites in the sand-burial ecosystems of the Tengger Desert, northern China, Geomorphology, 88, 254–265, 2007a.

Li, X. R., Kong, D. S., Tan, H. J., and Wang, X. P.: Changes in soil and vegetation following stabilization of dunes in the southeastern fringe of the Tengger Desert, Plant Soil, 300, 221–231, 2007b.

Li, X. R., Tian, F., Jia, R. L., Zhang, Z. S., and Liu, L. C.: Do biological soil crusts determine vegetation changes in sandy deserts? Implications for managing artificial vegetation, Hydrol. Process., 24, 3621–3630, 2010.

Liu, L., Li, S., Duan, Z., Wang, T., Zhang, Z., and Li, X.: Effects of microbiotic crusts on dew deposition in the artificial vegetation area at Shapotou, northwest China, J. Hydrol., 328, 331–337, 2006.

Liu, L. C., Liu, Y. B., Hui, R., and Xie, M.: Recovery of microbial community structure of biological soil crusts in successional stages of Shapotou desert revegetation, northwest China, Soil Biol. Biochem., 107, 125–128, 2017.

Liu, Y., Li, X., Xing, Z., and Zhao, X.: Responses of soil microbial biomass and community composition to biological soil crusts in the revegetated areas of the Tengger Desert, Appl. Soil Ecol., 65, 52–59, 2013.

Lozupone, C. and Knight, R.: UniFrac: a new phylogenetic method for comparing microbial communities, Appl. Environ. Microbiol., 71, 8228–8235, 2005.

Maier, S., Schmidt, T. S. B., Zheng, L., Peer, T., Wagner, V., and Grube, M.: Analyses of dryland biological soil crusts highlight lichens as an important regulator of microbial communities, Biodivers Conserv., 23, 1735–1755, 2014.

Mitchell, D. J., Fullen, M. A., Trueman, I. C., and Fearnhough, W.: Sustainability of reclaimed desertified land in Ningxia, China, J. Arid. Environ., 39, 239–251, 1998.

Moquin, S. A., Garcia, J. R., Brantley, S. L., Takacs-Vesbach, C. D., and Shepherd, U. L.: Bacterial diversity of bryophyte-dominant biological soil crusts and associated mites, J. Arid Environ., 87, 110–117, 2012.

Nagy, M. L., Perez, A., and Garcia-Pichel, F.: The prokaryotic diversity of biological soil crusts in the Sonoran Desert (Organ Pipe Cactus National Monument, AZ), FEMS Microbiol. Ecol., 54, 233–245, 2005.

Peterjohn, W. T., Melillo, J. M., and Steudler, P. A.: Responses of trace gas fluxes and N availability to experimentally elevated soil temperature, Ecol. Appl., 4, 617–625, 1994.

Pointing, S. B. and Belnap, J.: Microbial colonization and controls in dryland systems, Nat. Rev. Microbiol., 10, 551–562, 2012.

Rivera-Aguilar, V., Godínez-Alvarez, H., Moreno-Torres, R., and Rodríguez-Zaragoza, S.: Soil physico-chemical properties affecting the distribution of biological soil crusts along an environmental transect at Zapotitlán drylands, Mexico, J. Arid Environ., 73, 1023–1028, 2009.

Rustad, L. E., Campbell, J. L., Marion, G. M., Norby, R. J., Mitchell, M. J., Hartley, A. E., Cornelissen, J. H. C., and Gurevitch, J.: A meta-analysis of the response of soil respiration, net nitrogen mineralization, and aboveground plant growth to experimental ecosystem warming, Oecologia, 126, 543–562, 2001.

Root, H. T. and McCune, B.: Regional patterns of biological soil crust lichen species composition related to vegetation, soils, and climate in Oregon, USA, J. Arid Environ., 79, 93–100, 2012.

Sardans, J., Penuelas, J., and Estiarte, M.: Changes in soil enzymes related to C and N cycle and in soil C and N content under prolonged warming and drought in a Mediterranean shrubland, Appl. Soil Ecol., 39, 223–235, 2008.

Steven, B., Gallegos-Graves, L. V., Belnap, J., and Kuske, C. R.: Dryland soil microbial communities display spatial biogeographic patterns associated with soil depth and soil parent material, FEMS Microbiol. Ecol., 86, 1–13, 2013.

Tateno, R., Katagiri, S., Kawaguchi, H., Nagayama, Y., Li, C., Sugimoto, A., and Koba, K.: Use of foliar ^{15}N and ^{13}C abundance to evaluate effects of microbiotic crust on nitrogen and water utilization in Pinus massoniana in deteriorated pine stands of south China, Ecol. Res., 18, 279–286, 2003.

Uchida, T., Ohte, N., Kimoto, A., Mizuyama, T., and Chnaghua, L.: Sediment yield on a devastated hill in southern China: effects of microbiotic crust on surface erosion process, Geomorphology, 32, 129–145, 2000.

Wang, J., Bao, J., Su, J., Li, X., Chen, G., and Ma, X.: Impact of inorganic nitrogen additions on microbes in biological soil crusts, Soil Biol. Biochem., 88, 303–313, 2015.

Weber, B., Büdel, B., and Belnap, J.: Biological Soil Crusts: An Organizing Principle in Drylands, Ecological studies 226, Springer International Publishing, Switzerland, 2016.

Yeager, C. M., Kornosky, J. L., Housman, D. C., Grote, E. E., Belnap, J., and Kuske, C. R.: Diazotrophic community structure and function in two successional stages of biological soil crusts from the Colorado Plateau and Chihuahuan Desert, Appl. Environ. Microbiol., 70, 973–983, 2004.

Yu, J., Glazer, N., and Steinberger, Y.: Carbon utilization, microbial biomass, and respiration in biological soil crusts in the Negev Desert, Biol. Fertil. Soils, 50, 285–293, 2014.

Zhang, B., Kong, W., Wu, N., and Zhang, Y.: Bacterial diversity and community along the succession of biological soil crusts in the Gurbantunggut Desert, Northern China, J. Basic Microb., 56, 670–679, 2016.

Zhang, B. C., Zhang, Y. M., Zhao, J. C., and Wu, N.: Microalgal species variation at different successional stages in biological soil crusts of the Gurbantunggut Desert, Northwestern China, Biol. Fert. Soils, 45, 539–547, 2009.

Zhang, Y. M.: The microstructure and formation of biological soil crust in their early developmental stage, Chinese Sci. Bull., 50, 117–121, 2005.

Zhang, Y. M., Wu, N., Zhang, B. C., and Zhang, J.: Species composition, distribution patterns and ecological functions of biological soil crusts in the Gurbantunggut Desert, J. Arid Land, 2, 180–189, 2010.

11

Amplification of global warming through pH dependence of DMS production simulated with a fully coupled Earth system model

Jörg Schwinger[1], **Jerry Tjiputra**[1], **Nadine Goris**[1], **Katharina D. Six**[2], **Alf Kirkevåg**[3], **Øyvind Seland**[3], **Christoph Heinze**[4,1], **and Tatiana Ilyina**[2]

[1]Uni Research Climate, Bjerknes Centre for Climate Research, Bergen, Norway
[2]Max Planck Institute for Meteorology, Hamburg, Germany
[3]Norwegian Meteorological Institute, Oslo, Norway
[4]Geophysical Institute, University of Bergen, Bjerknes Centre for Climate Research, Bergen, Norway

Correspondence to: Jörg Schwinger (jorg.schwinger@uni.no)

Abstract. We estimate the additional transient surface warming ΔT_s caused by a potential reduction of marine dimethyl sulfide (DMS) production due to ocean acidification under the high-emission scenario RCP8.5 until the year 2200. Since we use a fully coupled Earth system model, our results include a range of feedbacks, such as the response of marine DMS production to the additional changes in temperature and sea ice cover. Our results are broadly consistent with the findings of a previous study that employed an offline model set-up. Assuming a medium (strong) sensitivity of DMS production to pH, we find an additional transient global warming of 0.30 K (0.47 K) towards the end of the 22nd century when DMS emissions are reduced by 7.3 Tg S yr^{-1} or 31 % (11.5 Tg S yr^{-1} or 48 %). The main mechanism behind the additional warming is a reduction of cloud albedo, but a change in shortwave radiative fluxes under clear-sky conditions due to reduced sulfate aerosol load also contributes significantly. We find an approximately linear relationship between reduction of DMS emissions and changes in top of the atmosphere radiative fluxes as well as changes in surface temperature for the range of DMS emissions considered here. For example, global average T_s changes by -0.041 K per 1 Tg S yr^{-1} change in sea–air DMS fluxes. The additional warming in our model has a pronounced asymmetry between northern and southern high latitudes. It is largest over the Antarctic continent, where the additional temperature increase of 0.56 K (0.89 K) is almost twice the global average. We find that feedbacks are small on the global scale due to opposing regional contributions. The most pronounced feedback is found for the Southern Ocean, where we estimate that the additional climate change enhances sea–air DMS fluxes by about 9 % (15 %), which counteracts the reduction due to ocean acidification.

1 Introduction

Changes in emissions of marine dimethyl sulfide (DMS) have the potential to influence climate via a modification of aerosol and cloud properties. The implications of a DMS climate feedback were first described by Shaw (1983) and Charlson et al. (1987). The latter authors hypothesise that DMS production and emission, the number of cloud condensation nuclei (CCN), and the albedo of marine boundary layer clouds are interlinked in a negative feedback loop acting to stabilise the Earth's climate against external perturbations. This idea has become known as the CLAW hypothesis (after the initials of the authors, Charlson, Lovelock, Andreae, and Warren).

DMS is a by-product of marine primary production (e.g. Liss et al., 1994) and it is the main natural source of atmospheric sulfur (about 28 Tg S yr^{-1}; Lana et al., 2011). Once in the atmosphere, DMS is oxidised to SO_2 and methanesulfonic acid (MSA), and further to gaseous H_2SO_4, which rapidly condenses onto pre-existing aerosol particles or, in the absence of sufficient aerosol surface area, forms nucleation-mode sulfate particles (Carslaw et al., 2010). Hence, sulfate produced from DMS leads to both increased numbers of nucleation-mode sulfate and increased size and modified hygroscopicity of larger sulfate and other internally mixed particles. Korhonen et al. (2008) show that, although nucleation-mode sulfate particles do not directly contribute to CCN, the formation and subsequent growth of these particles in the free troposphere can be the main source of DMS-derived CCN in remote ocean areas. Sulfate from condensation can increase CCN numbers in areas where the pre-existing particles are too small or too hydrophobic, but may in other areas or situations contribute to a reduction of the CCN production, if pre-existing particles are already large enough to activate or more hygroscopic than sulfate.

The effect of anthropogenic climate change on DMS production and emission has been shown to be relatively small globally (Bopp et al., 2003; Gabric et al., 2004; Gunson et al., 2006; Vallina et al., 2007; Kloster et al., 2007), although, regionally, larger sensitivities have been reported (Bopp et al., 2003; Kloster et al., 2007). Most of these studies predict an increase in global surface DMS concentrations or sea–air fluxes of about 2 to 14 % in response to global warming, while Kloster et al. (2007) model a 10 % decrease under a future scenario. It has been further shown that the sensitivity of CCN numbers to changes in DMS emissions is relatively low (Woodhouse et al., 2010) such that there is currently no evidence for a significant CLAW-like feedback on the global scale.

More recently, results from mesocosm studies have indicated that DMS production in marine ecosystems might decrease with decreasing seawater pH (e.g. Archer et al., 2013; see also the discussion in the Supplement of Six et al., 2013, and references therein). These findings were confirmed by follow-up mesocosm studies in warmer ocean regions such as the coastal waters of Korea or in the sub-tropics off the Canaries (Park et al., 2014; S.D. Archer, personal communication, 2017).

Six et al. (2013) study the possible impacts of this phenomenon on the global radiation balance. Assuming a low, medium, and high sensitivity of DMS production to pH, they model the decrease in DMS sea–air fluxes due to warming and progressing ocean acidification for the A1B scenario from the Special Report on Emissions Scenarios (SRES). They find a global reduction of DMS emissions due to ocean acidification (in addition to reductions due to climate change) of 11 % in the medium-sensitivity run (19 % for the high-sensitivity). In a separate step, they feed the reduced marine DMS emissions into an atmospheric general circulation model with aerosol chemistry to derive an effective radiative forcing, and, finally, to estimate a range of additional equilibrium warming that results from this forcing (0.23–0.48 K for the medium pH sensitivity). Due to the offline design of the Six et al. (2013) study, some important aspects of the DMS sea–air flux reduction are not accessible – for example, the spatial pattern of the additional warming, as well as any feedback between atmospheric changes and DMS production itself.

The purpose of this study is to simulate the possible link between ocean acidification, sea–air DMS fluxes, and changes in surface climate in a fully coupled Earth system model that includes a prognostic marine DMS scheme as well as an aerosol module capable of modelling changes of the radiation balance due to altered atmospheric sulfur cycling. The marine biogeochemistry model employed in this study is very similar to the one used by Six et al. (2013), with only minor differences (see Sect. 2.1). The prognostic DMS scheme as well as the assumptions on the pH dependency of DMS production are the same as in Six et al. (2013). We simulate the transient climate change and its amplification by pH dependence of DMS production under the RCP8.5 scenario and its extension to the year 2200 (Meinshausen et al., 2011). This scenario and the simulation period up to 2200 is chosen to get a clear signal of transient change. For the same reason, we focus our analysis on the medium and strong sensitivity of DMS production to pH, without implying that the observational evidence is better for these than for a lower sensitivity.

We first present a brief description of our Earth system model and of the experimental set-up in Sect. 2. An evaluation of sea–air DMS fluxes and its drivers as well as the changes found in the baseline RCP8.5 simulation (that assumes no pH dependency of DMS production) is given in Sect. 3.1. We then present the results of two pH-sensitive simulations (medium and high sensitivity) and an analysis of feedbacks in Sect. 3.2 and 3.3, respectively. A summary of our results and our conclusions are provided in Sect. 4.

2 Model description and experimental set-up

The Norwegian Earth system model NorESM1-ME employed in this study is based on the Community Earth System Model (CESM1-BGC; Gent et al., 2011; Lindsay et al., 2014). It uses the same sea-ice and land models, but a different ocean component and a different aerosol module embedded in the atmosphere model. Here, we use the model version that participated in the Coupled Model Intercomparison Project Phase 5 (CMIP5; Taylor et al., 2012), and that has been described and evaluated in a series of papers (Bentsen et al., 2013; Iversen et al., 2013; Tjiputra et al., 2013; Kirkevåg et al., 2013). The only difference between the published model version and the model version employed for this study, is the coupling of prognostic sea–air DMS fluxes calculated by the ocean biogeochemistry component to the atmosphere model (as opposed to the use of prescribed climatological DMS emissions in the CMIP5 model version).

We do not give an extensive model description here, but refer the reader to the publications cited above. In the following sections, we provide a brief description of the marine biogeochemistry and aerosol modules with focus on the parameterisations relevant for this study.

2.1 The marine biogeochemistry model MICOM-HAMOCC

The physical ocean component of NorESM is the isopycnic MICOM (Miami Isopycnic Coordinate Ocean Model; Bleck et al., 1992), albeit with considerable modifications of numerics and physics as described in Bentsen et al. (2013). The HAMburg Ocean Carbon Cycle model (HAMOCC; Maier-Reimer, 1993; Maier-Reimer et al., 2005) has been implemented into MICOM by Assmann et al. (2010). The HAMOCC version used in NorESM1-ME is further described by Tjiputra et al. (2013) and Schwinger et al. (2016). We note that HAMOCC, since it was implemented into MICOM, has also been further developed by the biogeochemistry group at the Max Planck Institute in Hamburg. As a result, our model version is very similar but not identical to the one described by Ilyina et al. (2013). For example, the inorganic carbon chemistry has been updated for our model version and some technical modifications were necessary for the implementation into the isopycnic MICOM. Further, some parameters were differently tuned because of the different physical ocean models (see Schwinger et al., 2016, for details). None of these differences is expected to alter the findings and conclusions of this study.

HAMOCC simulates the marine biogeochemical cycles of carbon, phosphorous, nitrogen, silica, and iron. Biological production and export out of the euphotic zone is parameterised by a NPZD-type (nutrient–phytoplankton–zooplankton–detritus) ecosystem model extended to include dissolved organic carbon (Six and Maier-Reimer, 1996). The model has only one generic phytoplankton and one generic zooplankton type. Bacteria are not modelled explicitly; rather, processes related to bacterial activity (e.g. remineralisation of dissolved and particulate organic matter) are assumed to proceed at constant rates. A constant Redfield ratio of $1 : 16 : 122$ $(P : N : C)$ is used for the composition of organic matter. Primary production (PP) is limited by the least available macronutrient (phosphate, nitrate) or micronutrient (iron). Detritus is formed through grazing activity as well as phytoplankton and zooplankton mortality. Once formed, it sinks through the water column at a constant speed of $5\,\mathrm{m\,d^{-1}}$ and is remineralised at a constant rate. Although calcifying and silicifying plankton functional types are not modelled explicitly, the fraction of calcium carbonate and biogenic silica that is added to the pool of sinking shell material depends on the availability of silicic acid. Here, it is implicitly assumed that diatoms out-compete other phytoplankton species when the supply of silicic acid is ample. This is parameterised by assuming that a fraction $[\mathrm{Si}]/(K_{\mathrm{Si}}+[\mathrm{Si}])$ of detritus production contains opal shells while the remaining fraction contains calcareous shells, where $K_{\mathrm{Si}} = 1\,\mathrm{mmol\,Si\,m^{-3}}$ is the half-saturation constant for silicate uptake. We note that HAMOCC's ecosystem model has no dependency on inorganic carbon availability or pH – e.g. nei-

ther cell carbon quotas or stoichiometric ratios nor calcification rates are assumed to vary with changing pH.

The parameterisation of the DMS cycle in HAMOCC has been implemented and evaluated by Six and Maier-Reimer (2006) and Kloster et al. (2006). Processes included are DMS production and losses by bacterial consumption, photolysis, and sea–air DMS gas exchange. Observations indicate that when the cell membrane of phytoplankton is disrupted by senescence or due to viral attack and zooplankton grazing, the DMS precursor dimethylsulfoniopropionate (DMSP) is released into sea water and rapidly converted to DMS by bacterial and algal enzymes (Stefels et al., 2007). Therefore, the DMS production in our model is assumed to be a function of detritus production. It is further modified by the production of opal and $CaCO_3$ shell material – that is, calcite or opal producing organisms are assumed to have different sulfur-to-carbon ratios. This differentiation credits the fact that haptophytes have, in general, a higher DMSP to cell carbon ratio and, thus, have a higher contribution to DMS production (Keller et al., 1989). Bacterial consumption is a linear function of temperature and a monotypic saturation function of DMS concentration. Photolysis and sea–air gas exchange are linear functions of the seawater DMS concentration. The local photolysis rate depends on the intensity of incoming light, and the calculation of sea–air DMS fluxes follows the gas-exchange parameterisation of Wanninkhof (1992). For this study, we use the same set of tunable parameters as Kloster et al. (2006) and Six et al. (2013), except for the scaling factor for bacterial consumption, which is reduced by half to better reproduce the observed fluxes. We refer the reader to Kloster et al. (2006) and Six et al. (2013) for a detailed description of the DMS parameterisation used here.

The applied relationship between DMS production and seawater pH is motivated by a compilation of the results of several mesocosm studies (Six et al., 2013). In these mesocosm experiments, the temporal evolution of DMS concentration within a confined natural water volume is investigated under different levels of CO_2 partial pressure with corresponding seawater pH (further information on the method is given in Archer et al., 2013). As in Six et al. (2013) we assume a linear decrease in DMS production (P_{DMS}) with decreasing pH (Fig. 1 in Six et al., 2013), by multiplying P_{DMS} by a factor $F = 1 + (\mathrm{pH_{pi}} - \mathrm{pH})\gamma$. Here, pH is the modelled local pH value, and $\mathrm{pH_{pi}}$ is the local pH of the pre-industrial undisturbed ocean, relative to which a deviation of pH is measured. $\mathrm{pH_{pi}}$ is taken from a monthly climatology calculated from 10 years of a preindustrial control simulation. The constant γ defines the sensitivity of DMS production to pH, and is chosen following Six et al. (2013, see Sect. 2.3). We note that the reason for the observed decrease in DMS concentrations under low pH is still debated. Shifts in species composition seem to play a major role in general (Archer et al., 2013; Park et al., 2014), but also changes in the rate of DMSP-to-DMS conversion and the rate of loss through bacterial consumption might contribute to the observed pH

Table 1. Model experiments conducted for this study. All experiments have been run over the time period 1850–2200. The experiments BASE, SMED and SHIGH use prescribed CO_2 emissions (and other forcings) following the CMIP5 protocols for the historical simulation (1850–2005), the RCP8.5 scenario (2006–2100) and the RCP8.5 scenario extension (2101–2200).

Experiment name	CO_2 emissions	pH sensitivity
Control	none	$\gamma = 0$
BASE	historical/RCP8.5/RCP8.5-extension	$\gamma = 0$
SMED	historical/RCP8.5/RCP8.5-extension	$\gamma = 0.58$
SHIGH	historical/RCP8.5/RCP8.5-extension	$\gamma = 0.87$

dependency. All these processes cannot be resolved explicitly by our model, and, as a first approach, we assign a linear pH dependency to DMS production.

2.2 The aerosol module of CAM4-Oslo

As described in detail by Kirkevåg et al. (2013) and Seland et al. (2008), the aerosol life cycle scheme in the atmosphere component of NorESM1-ME (Community Atmosphere Model Version 4-Oslo, CAM4-Oslo) calculates and traces, for each constituent, aerosol mass mixing ratios which are tagged according to production mechanisms in clear and cloudy air. The processes treated in the model are gas phase and aqueous phase chemical production, gas-to-particle production (nucleation), condensation, and coagulation of small particles onto larger pre-existing particles. Primary particles are emitted as accumulation-mode sulfate; nucleation and accumulation mode black carbon (BC); Aitken-mode BC; internally mixed Aitken-mode organic matter (OM) and BC; Aitken-, accumulation-, and coarse-mode sea salt; and accumulation- and coarse-mode mineral dust.

Aerosol precursor gas-phase components accounted for are DMS and SO_2. DMS is produced in the surface ocean and emitted to the atmosphere as described above. It is depleted by oxidation to particulate methane sulfonic acid (MSA), thenceforth treated as primary ocean-biogenic OM, and by oxidation to SO_2. Oxidant fields (OH, O_3, and H_2O_2) for the sulfur chemistry are prescribed as detailed in Kirkevåg et al. (2013). Apart from being produced from DMS, SO_2 has important natural sources from volcanoes, and is otherwise emitted from combustion of fossil fuel and biomass containing sulfur. Gaseous sulfate (as H_2SO_4) produced in air by oxidation of SO_2 by OH is allowed to condense on pre-existing particles whenever sufficient particle surface area is available for condensation. Whatever gaseous sulfate is left after condensation (during a model time step) is assumed to form nucleation-mode sulfate by gas-to-particle production.

Internally mixed water from condensation of water vapour is treated separately through use of look-up tables (calculated offline) for optical parameters, accounting for the above-listed processes as well as hygroscopic swelling. Another

set of look-up tables is used to obtain dry size parameters (dry radius and standard deviation) of the aerosol population, which are used as input in the calculation of CCN activation following Abdul-Razzak and Ghan (2000). Aerosol components dissolved in cloud water are not kept as separate tracked variables but are either scavenged or distributed to accumulation-mode sulfate as well as accumulation and coarse-mode particles in internal mixtures.

2.3 Experimental set-up

We have run three sets of simulations (in addition to a control simulation with constant pre-industrial settings, Table 1) that assume different sensitivities of DMS production to ocean acidification. Each set consists of a historical simulation (1850 to 2005) followed by a RCP8.5 scenario simulation (2006 to 2100) and its extension to the year 2200. The first set of runs does not assume any sensitivity of DMS production to pH ($\gamma = 0$) and we refer to these simulations collectively as the "experiment BASE". Following Six et al. (2013), the second and third set of model runs assume a medium (experiment SMED, $\gamma = 0.58$) and a high (experiment SHIGH, $\gamma = 0.87$) sensitivity of DMS production to pH. We use the emission-driven configuration of NorESM, that is, atmospheric CO_2 concentrations evolve freely in response to anthropogenic emissions as well as ocean and land carbon sinks (the latter two being potentially different for each experiment).

Prior to these experiments, the NorESM1-ME was spun up for 900 years with a prescribed atmospheric CO_2 concentration fixed at 284.7 ppm. Subsequently, the model was further spun up for 550 years in its emission-driven configuration with prescribed zero emissions. Towards the end of the second spin-up period, the air–sea and air–land CO_2 fluxes closely balance one another and atmospheric CO_2 is stable at 284 ppm.

The control simulation is used to account for remaining model drift. Most importantly, there is a small decrease in primary production of about $0.57\,Pg\,C\,century^{-1}$, and a corresponding decrease in sea–air DMS fluxes of $0.51\,Tg\,S\,century^{-1}$, due to the fact that fluxes of nutrients to the sediment are not replenished by any mechanism in the model version employed for this study. Note that there is almost no drift ($< 0.01\,Pg\,C\,century^{-1}$) in CO_2 fluxes due to compensating effects in carbon and $CaCO_3$ export production. Since we wish to conserve the internal consistency of model fields, we do not correct for model drift. Rather, we express all results either relative to the control simulation or relative to the experiment BASE.

3 Results and discussion

To begin with, we examine changes of primary production and DMS sea–air fluxes in the BASE simulation relative to

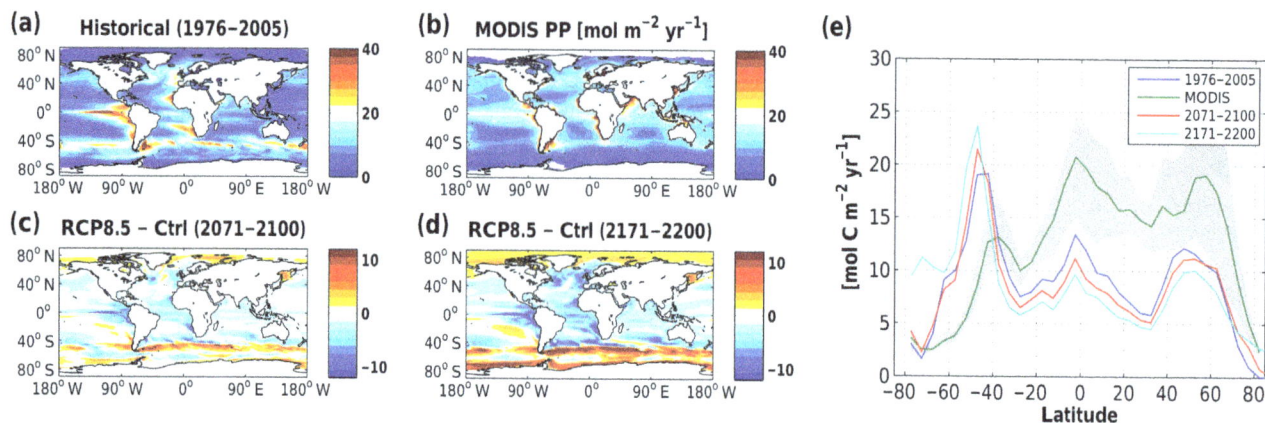

Figure 1. Vertically integrated annual primary production (PP, $mol\,C\,m^{-2}\,yr^{-1}$) **(a)** in the experiment BASE (historical simulation, average over 1976 to 2005), **(b)** mean of three satellite-based climatologies (derived from MODIS retrievals), **(c)** change found in the RCP8.5 scenario simulation relative to the control run towards the end of the 21st century (2071 to 2100), and **(d)** same as **(c)** but towards the end of the 22nd century (2171 to 2200). Panel **(e)** displays the zonal means of PP in the BASE historical simulation (dark blue) and for the RCP8.5 scenario (red, light blue). The green line with the grey shaded area represents the mean and range of the zonal means of the three satellite-based climatologies.

Table 2. Global mean atmospheric CO_2 concentration, surface temperature (T_s), marine primary production (PP), and sea–air DMS fluxes (F_{DMS}) in the control run and changes in experiment BASE relative to the control run.

		Control 2071–2100	BASE-Control 2071–2100	BASE-Control 2171–2022
CO_2	(ppm)	284	945–284	1951–284
T_s	(K)	285.8	4.2	8.1
PP	$(Pg\,C\,yr^{-1})$	44.2	−4.2	−5.3
F_{DMS}	$(Tg\,S\,yr^{-1})$	24.2	−0.97	0.12

the pre-industrial control run. Since, in our model, DMS production is tied to detritus production, which very closely follows PP, patterns of changes in PP are a useful indicator of changes in DMS production (with some modifications discussed below). In experiment BASE, these changes are caused by climate change alone (no influence of ocean acidification, since $\gamma = 0$). Additional changes relative to the BASE experiment that are caused by the pH sensitivity of DMS production in SMED and SHIGH will be analysed and discussed in Sect. 3.2 and 3.3.

3.1 Primary production and DMS emissions in experiment BASE

Results for the simulation BASE are summarised in Table 2. During the last 30 years of the 21st and 22nd century, atmospheric CO_2 concentrations reach 945 and 1951 ppm, respectively. These numbers are higher than the corresponding prescribed concentrations of RCP8.5, since our model shows a rather low carbon uptake by the land biosphere (Arora et al., 2013) due to the inclusion of nitrogen limitation of plant pro-

ductivity. The global average surface air temperature increase (ΔT_s) simulated by our model is 4.2 and 8.1 K for the time periods 2071 to 2100 and 2171 to 2200, respectively.

The spatial pattern of annual primary production in the control run and an estimate of PP derived from MODIS observations and three different processing algorithms (see Schwinger et al., 2016, for details) are shown in Fig. 1a and b. The spatial pattern of PP is well reproduced by the model, although PP is higher than the observation-based estimate in the equatorial Pacific upwelling region as well as south of 40° S. Lower than observed values are found in the subtropical gyres of the Pacific and the whole low-latitude western Pacific as well as in large parts of the Indian Ocean. The resulting annual globally integrated PP (44.2 Pg C in the control run) is lower than the satellite-based PP estimate (59.9 Pg C).

Under the RCP8.5 scenario, PP declines globally by 4.2 Pg C towards the end of the 21st century (Table 2, Fig. 1c) and by 5.3 Pg C towards the end of the 22nd century (Table 2, Fig. 1d). This decline is due to increased stratification (reduced mixed layer depth, not shown) and less nutrient supply almost everywhere north of 40° S and south of the seasonally ice-covered parts of the North Atlantic and North Pacific. One notable exception is the southeastern part of the Pacific, where increased mixing by surface winds outweighs the increased thermal stratification in our model, leading to increased nutrient supply and PP. Arctic regions that are seasonally or permanently ice covered in the control run experience an increase in PP due to reduction in both light and temperature limitations. In the Southern Ocean south of 40° S, we can identify three more or less distinct zones. In the southernmost part, along the coast of Antarctica, declining sea ice leads to an increase in PP. North of the control run sea

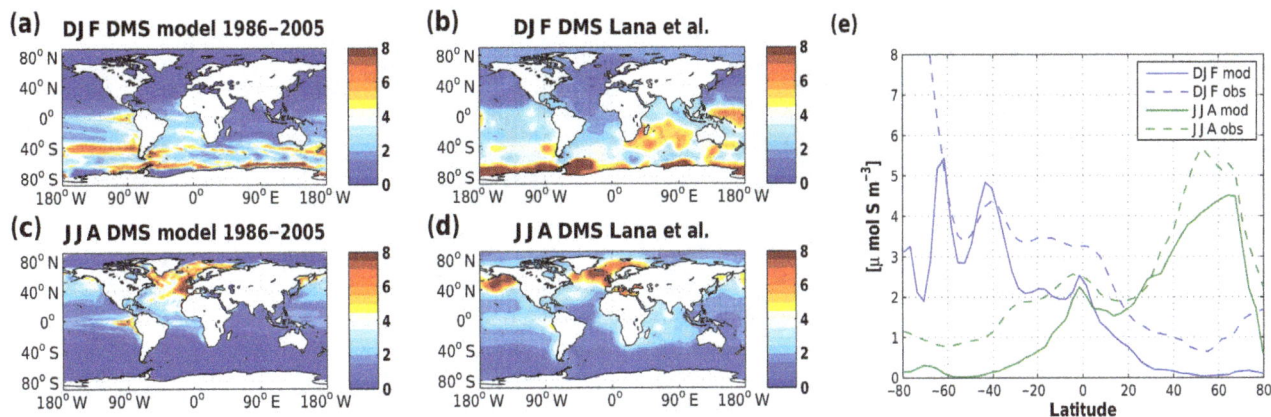

Figure 2. Surface DMS concentration (μmol S m^{-3}) during **(a, b)** boreal winter (DJF) and **(c, d)** boreal summer (JJA). Results for the BASE historical simulation averaged over 1986 to 2005 are shown in panels **(a)**, **(c)** and the observation-based climatology by Lana et al. (2011) is displayed in panels **(b)**, **(d)**. Panel **(e)** shows the zonal means of each field presented in panels **(a)**–**(d)**.

ice edge, we find a belt where PP partially decreases, while PP increases quite substantially in the northernmost part of the Southern Ocean. This pattern is caused by a southward shift of the circumpolar storm track (and hence cloud cover), which causes increased light limitation in the southern part of the Southern Ocean in our simulations. In the northern part of the Southern Ocean around 50° S increasing temperature and less light limitation sustain a substantially stronger PP under climate change.

We note that the pattern of PP changes found in our model is consistent with the results from Six et al. (2013) and with results from other state-of-the-art Earth system models (Steinacher et al., 2010; Bopp et al., 2013; Laufkötter et al., 2015). Although this is also true for the Southern Ocean, the increase in PP south of 40° S is, compared to other models, large relative to decreases elsewhere. This might indicate that the higher than observed PP south of 40° S also implies a relatively high sensitivity to climate change.

Modelled DMS surface concentrations for boreal winter (DJF) and summer (JJA) and the corresponding climatologies from Lana et al. (2011) are shown in Fig. 2. The too high and too low productivity in the eastern equatorial Pacific and the Indian Ocean, respectively, are visible in the surface DMS concentrations in these regions year round. The elevated DMS production around Antarctica in austral summer is well reproduced by our model, except for regions with too extensive summer sea ice cover (Weddell Sea and the eastern part of the Ross Sea). The zonal mean DJF DMS concentration south of 60° S is therefore significantly higher in the observation-based climatology than in our model (Fig. 2e). Due to too low PP in our model, DMS concentrations approach zero during winter poleward of 40° N and 40° S, while observations indicate winter values around 1 μmol m^{-3} in the zonal mean.

The change of DMS fluxes to the atmosphere in the RCP8.5 scenario simulation (Fig. 3) mainly follows the change in primary production. This is to be expected, since, in our model, detritus production (to which DMS production is tied) very closely follows PP on an annual timescale (about 20 % with little spatial variability). However, since the DMS production is modified by the fraction of opal and calcium carbonate contained in detritus, there are a few subtle differences. First, around Antarctica, the increase in DMS emissions towards the end of the 22nd century due to loss of sea ice is less than one would infer from PP increase alone. In this region, there is ample supply of silica and the increase in production is caused mainly by an increase in opal-producing organisms. Second, in latitudes between 60 and 40° S, opal production is limited by silica supply and therefore, the increase in calcium carbonate production is stronger than the increase in opal production. Hence, towards the end of the 22nd century, the sea–air DMS fluxes in this region increase more strongly than indicated by PP. These results are consistent with the model study by Bopp et al. (2003), who also find increased DMS production due to shifts from diatom to non-diatom species in the northern part of the Southern Ocean.

The global total marine DMS emissions to the atmosphere are 24.2 Tg S yr^{-1} in the control simulation, and average fluxes for the time period 2071 to 2100 decrease by 0.97 Tg S yr^{-1} in our BASE RCP8.5 simulation relative to the control run (Table 2). The increasing sea–air fluxes in the Southern Ocean and the Arctic Ocean outweigh the decreases elsewhere towards the end of the 22nd century, and we find DMS emissions that are globally slightly larger (0.12 Tg S yr^{-1} for 2171 to 2200) in RCP8.5 relative to the control simulation.

3.2 Experiments SMED and SHIGH

The average pH of the surface ocean declines from 8.16 in the control simulation to 7.72 (average over 2071 to 2100) and 7.43 (2171 to 2200) under the RCP8.5 scenario and its

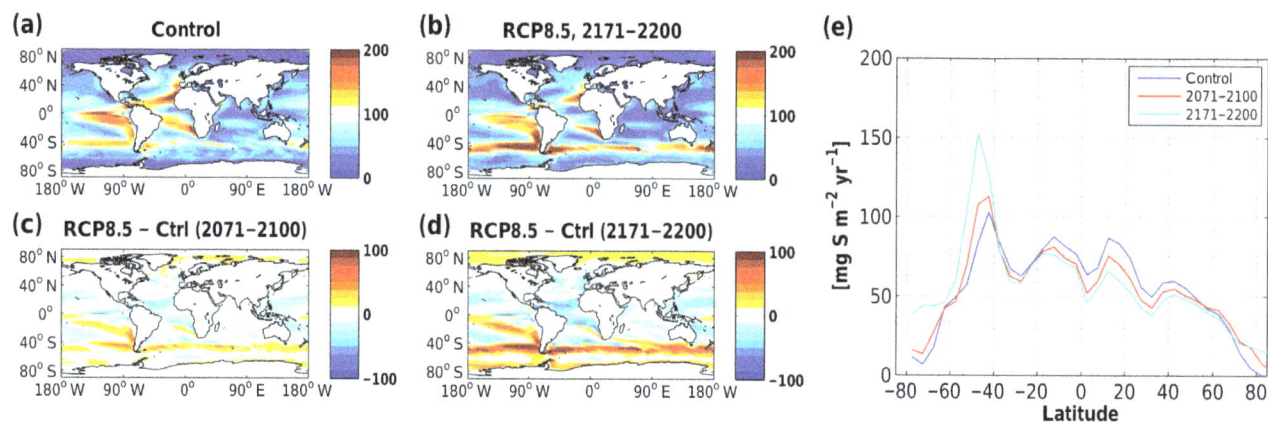

Figure 3. Modelled sea–air DMS fluxes (mg S m^{-2} yr^{-1}) in (**a**) the control run and (**b**) the RCP8.5 scenario simulation BASE (no sensitivity of DMS production to pH) towards the end of the 22nd century (2171 to 2200). Panels (**c**) and (**d**) display the change in sea–air DMS fluxes in the RCP8.5 scenario BASE relative to the control simulation towards the end of the 21st (2071 to 2100) and 22nd (2171 to 2200) century, respectively. Panel (**e**) shows the zonal mean sea–air DMS fluxes in the control and RCP8.5 simulations.

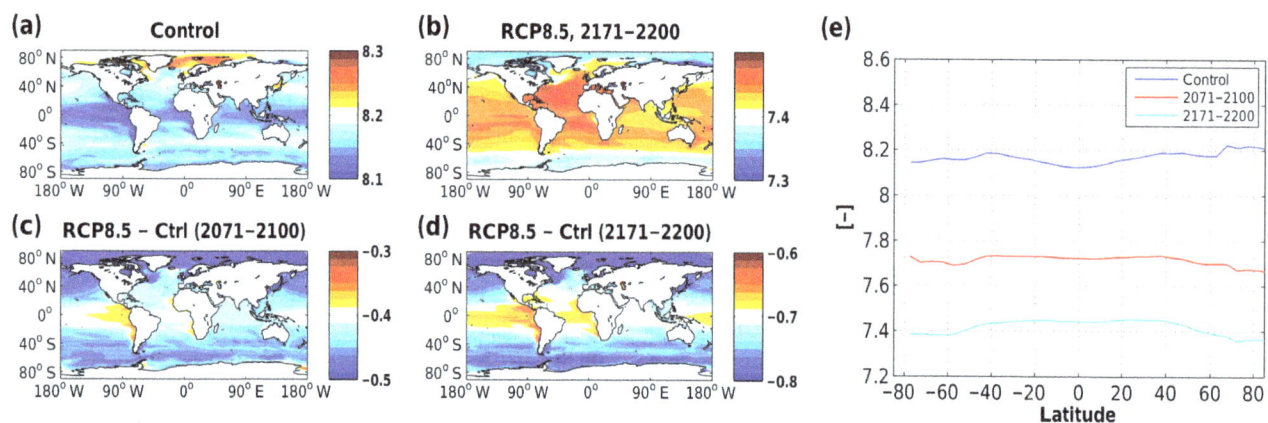

Figure 4. Surface ocean pH (**a**) in the control simulation, (**b**) in the RCP8.5 scenario simulation BASE (very similar for SMED and SHIGH) towards the end of the 22nd century (2171 to 2200), (**c**) change found in the RCP8.5 scenario simulation BASE relative to the control run towards the end of the 21st century (2071 to 2100), and (**d**) same as (**c**) but towards the end of the 22nd century (2171 to 2200). Panel (**e**) displays the zonal means of pH in the control run (dark blue) and for the RCP8.5 scenario BASE (red, light blue).

extension (Fig. 4). If we, following Six et al. (2013), introduce a sensitivity of DMS production to pH in our model as described above, global DMS emissions in 2071 to 2100 are reduced relative to experiment BASE by 17 (4 Tg S yr^{-1}) and 27 % (6.3 Tg S yr^{-1}) in the simulations SMED and SHIGH, respectively. Towards the end of the 22nd century (2171 to 2200), we find reductions of 31 (7.3 Tg S yr^{-1}) and 48 % (11.5 Tg S yr^{-1}). The most pronounced reduction in absolute sea–air DMS fluxes is found over the Southern Hemisphere (Fig. 5a–d), particularly between 40 and 50° S, where our model simulates the largest DMS emissions in the control run.

This drastic reduction of sulfur input to the atmosphere leads to a reduction of SO$_2$ and sulfate aerosol loads. Since sulfate particles usually have a longer atmospheric lifetime than DMS, particularly in the pristine atmosphere over re-

mote ocean regions, the pattern of changes in sulfate aerosol mass is regionally much smoother than the underlying sea–air DMS flux changes (Fig. 5e to h). There is a pronounced north–south gradient of sulfate load reduction, with the largest reductions found in the south, despite the fact that the decrease in pH is largest in the Arctic Ocean (Fig. 4). In the Arctic DMS sea–air fluxes remain relatively small in experiment BASE even towards the end of the 22nd century. Therefore, although the percentage reduction of fluxes (per m^2 of ocean area) in SMED and SHIGH relative to BASE are largest in the Arctic Ocean, the decrease in absolute fluxes (integrated over the whole region) is much smaller than in the Southern Ocean. As a result, the absolute change in SO$_4$ column burden is 3 times larger over the Southern Ocean than over the Arctic region.

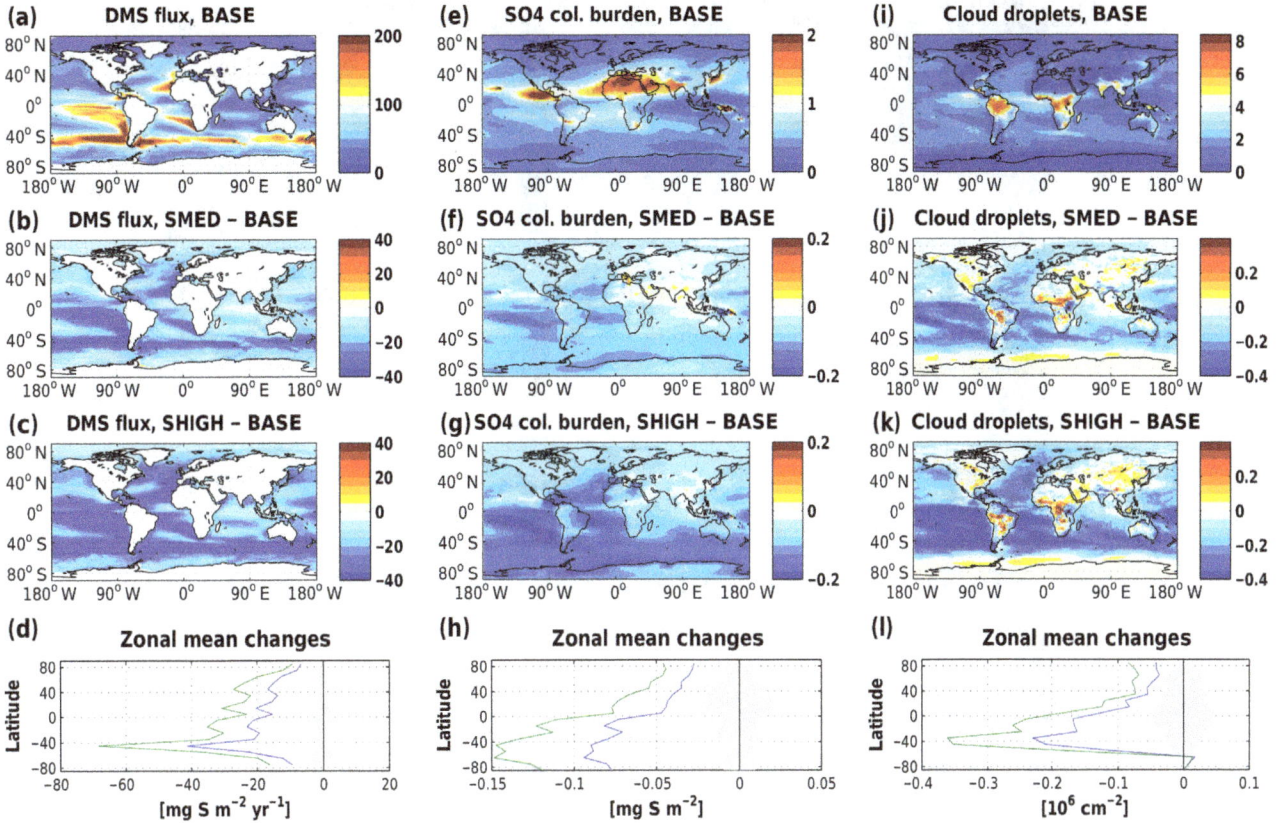

Figure 5. (a) Modelled sea–air DMS flux (mg S m^{-2} yr^{-1}) for scenario BASE (no sensitivity of DMS production to pH), and **(b, c)** changes of sea–air DMS fluxes relative to BASE in experiments SMED ($\gamma = 0.58$) and SHIGH ($\gamma = 0.87$), respectively. Panel **(d)** shows the zonal mean changes of sea–air DMS fluxes for SMED (blue) and SHIGH (green), and the grey shaded area displays a measure for natural variability (standard deviation over 100 years of the control run). Panels **(e)–(h)** and **(i)–(l)** show corresponding plots for atmospheric sulfate column burden (mg S m^{-2}) and vertically integrated cloud droplet number concentration (10^{6} cm^{-2}). All panels show results averaged over the period 2171 to 2200.

Close to strong natural (volcanic) and anthropogenic sulfur sources, we find a non-linear relationship between DMS emissions and sulfate, e.g. sulfate loads are slightly increased rather than reduced over the northern Indian Ocean and adjacent land regions. Owing to the reduced sulfate mass in aerosol particles and a reduced aerosol number concentration (Fig. 6a to c), we find a reduction in cloud droplet number concentrations (CDNC, Fig. 5i to l) over most ocean regions. Increases in CDNC are confined to land areas and to the Southern Ocean south of 60° S. Figure 6 demonstrates that the reduction of aerosol numbers and CDNC is found throughout the lower troposphere, being accompanied by an increase in the effective radius of liquid cloud droplets (R_{eff}).

Altered aerosol and cloud properties cause changes in the radiative balance of the atmosphere. Here, we investigate changes in the radiation balance at the top of our model (2.2 hPa). Changes in the shortwave net radiative flux (N_{s}) are the sum of changes in clear-sky net shortwave fluxes ($N_{\mathrm{s,c}}$) due to altered aerosol scattering and absorption (direct aerosol effect) and changes in the shortwave cloud radiative effect (CRE$_{\mathrm{s}}$), $\Delta N_{\mathrm{s}} = \Delta N_{\mathrm{s,c}} + \Delta$CRE$_{\mathrm{s}}$ (with all fluxes scaled to the total grid-cell area). The cloud radiative effect (CRE) is evaluated in the model by parallel calls to the radiation code with and without clouds. Differences in CRE between our transient simulations include contributions from changes in cloud albedo, cloud lifetime (first and second indirect effect), and from changes due to altered radiative heating by absorbing aerosols (semi-direct effect). Both $\Delta N_{\mathrm{s,c}}$ and ΔCRE$_{\mathrm{s}}$ also have contributions due to changes in surface albedo in our transient simulations (additional sea ice melt; see Sect. 3.3). The analysis of changes in the longwave radiation balance is complicated by the fact that the outgoing longwave fluxes are altered due to different climate conditions in our experiments. The longwave cloud radiative effect (CRE$_{\mathrm{l}}$) is available as model output, but our experiment design does not allow us to separate the direct aerosol effect on outgoing longwave radiation from other changes in the longwave band. However, we can assume that the longwave direct effect is generally negligible, since the diameter

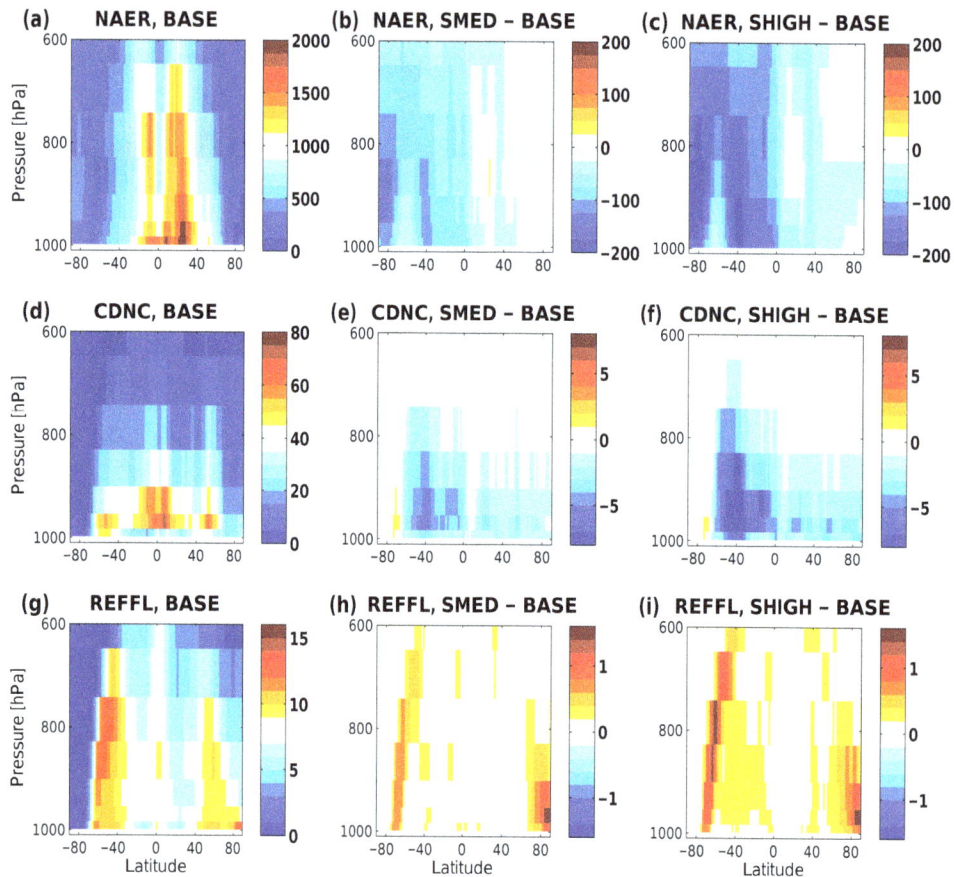

Figure 6. (a) Zonal mean aerosol number concentration (N_{aer}, cm^{-3}) averaged over the period 2171 to 2200 for the experiment BASE (no sensitivity of DMS production to pH), and **(b, c)** changes of N_{aer} in the experiments SMED and SHIGH (relative to BASE). Panels **(d)–(f)** and **(g)–(i)** display corresponding plots for cloud droplet number concentration (CDNC, cm^{-3}) and effective radius of liquid cloud droplets (R_{eff}, µm).

of DMS-derived aerosols is small compared to the longwave radiation's wavelength.

We find that changes in the radiative fluxes depend approximately linearly on the sea–air DMS flux anomaly, ΔF_{DMS}, in our model (Fig. 7). Although there is considerable interannual variability, linear trends are virtually identical whether we use data from the experiment SMED or SHIGH or use the data of both experiments combined. Clouds exert an average additional shortwave radiative effect of -0.055 W m^{-2} (Tg S yr^{-1})$^{-1}$. This effect is larger over the oceans than over land, where we also find a larger interannual variability (Fig. 7a to c). Changes in CRE$_l$ in our experiments are close to zero (Fig. 7d to f) and not discussed further. The reduced load of sulfate aerosols considerably alters $N_{s,c}$ by -0.047 W m^{-2} (Tg S yr^{-1})$^{-1}$, and again $\Delta N_{s,c}$ is significantly larger over the oceans than over land (Fig. 7g to i; see further discussion below). The global average transient change of the total net radiative flux at the top of our model (N_t) is -0.026 W m^{-2} (Tg S yr^{-1})$^{-1}$. This includes a contribution with a positive trend over land ($+0.037$ W m^{-2} (Tg S yr^{-1})$^{-1}$), and a stronger negative con-

tribution over the oceans (-0.055 W m^{-2} (Tg S yr^{-1})$^{-1}$; see Fig. 7j to l). The resulting negative radiative balance over land is due to the increased outgoing longwave radiation at higher surface temperatures, which outweighs the weak changes in CRE over land. Thus, part of the excess-energy gained through the reduced DMS load over the ocean is radiated back to space over land. The global surface temperature anomaly is -0.041 K (Tg S yr^{-1})$^{-1}$ in our experiments (Fig. 7m to o).

To set these results into perspective, we calculate the changes in radiative fluxes and in T_s towards the end of the 22nd century using the average ΔF_{DMS} over 2171 to 2200 for SMED (-7.3 Tg S yr^{-1}) and SHIGH (-11.5 Tg S yr^{-1}). We also break our results down into three broad latitude bands: northern high latitudes (NHL, north of 40° N), low latitudes (LL, 40° S to 40° N), and southern high latitudes (SHL, south of 40° S). For the southern high latitudes, we define an additional region SHLni, which excludes grid points that are covered by sea ice in the control run. The average changes in CRE$_s$ at the end of the 22nd century for experiment SMED are 0.40 W m^{-2} for the global domain,

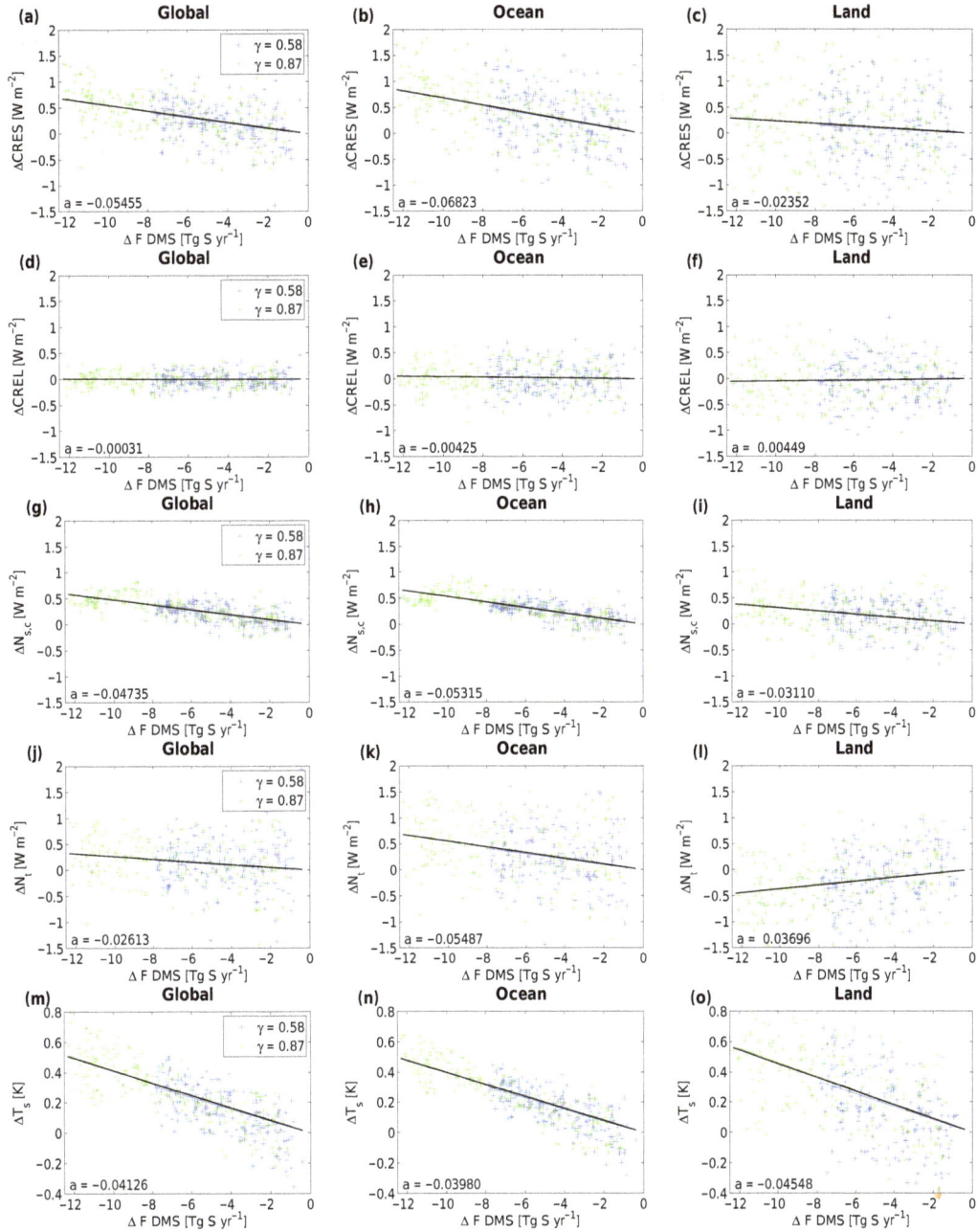

Figure 7. Scatter plots of anomalies of sea–air DMS fluxes (ΔF_{DMS}, $\mathrm{Tg\,S\,yr^{-1}}$) versus anomalies of **(a)**–**(c)** shortwave cloud radiative effect ($\Delta\mathrm{CRE_s}$, $\mathrm{W\,m^{-2}}$), **(d)**–**(f)** longwave cloud radiative effect ($\Delta\mathrm{CRE_l}$, $\mathrm{W\,m^{-2}}$), **(g)**–**(i)** clear-sky net shortwave radiative flux at the top of model ($\Delta N_{\mathrm{s,c}}$, $\mathrm{W\,m^{-2}}$), **(j)**–**(l)** total radiative flux at the top of model (ΔN_t, $\mathrm{W\,m^{-2}}$), and **(m)**–**(o)** near-surface temperature (ΔT_s, K) for the experiments SMED (blue) and SHIGH (green) relative to experiment BASE. The first column of plots displays global annual averages for each year in 2006 to 2200, while for the second and third column the annual mean values were calculated for ocean and land grid points, respectively. A linear fit to the data of each experiment (solid blue and green lines) as well as a linear fit to all data (solid black line) is shown in each panel. Note that in most cases the fit to individual experiments is indistinguishable from the fit to all data points and is hidden behind the black solid lines. The slope of the fit to all data is given in the lower left corner of each panel.

$0.50\,\mathrm{W\,m^{-2}}$ over the ocean, and $0.18\,\mathrm{W\,m^{-2}}$ over land. The corresponding values for the simulation SHIGH are 0.63, 0.78, and $0.28\,\mathrm{W\,m^{-2}}$ (Fig. 8a and b). The overall smaller cloud radiative effect over land is due to negative contribu-

tions from northern and southern high-latitude land areas. We note that the relatively small $\Delta\mathrm{CRE_s}$ for the Southern Ocean (despite the strongest reduction of DMS emissions) is due to the additional melt of sea ice around Antarctica in SMED

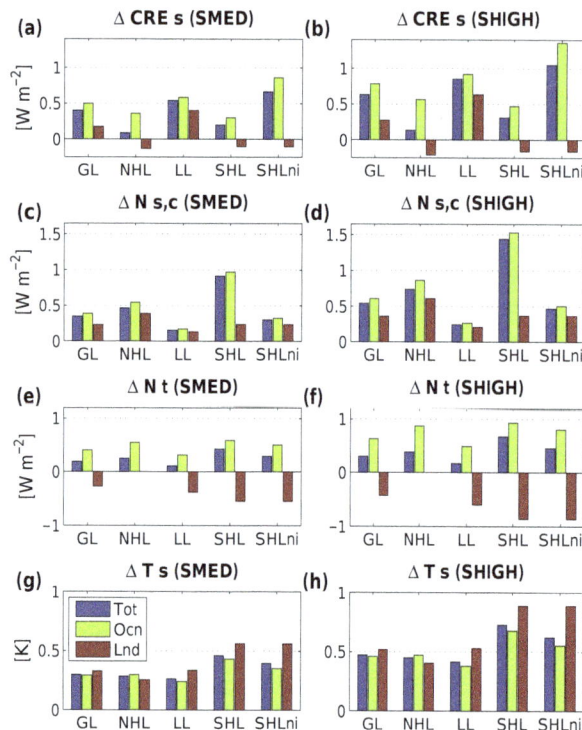

Figure 8. Changes of shortwave cloud radiative effect relative to experiment BASE (ΔCRE_s, W m^{-2}) for the experiment **(a)** SMED and **(b)** SHIGH towards the end of the 22nd century (2171 to 2200). The groups of bars indicate averages over different latitude bands (GL: global; NHL: northern high latitudes, north of 40° N; LL: low latitudes, 40° S to 40° N; SHL: southern high latitudes, south of 40° S). For the southern high latitudes there is an additional group of bars (SHLni) that indicates averages over the area that is sea-ice-free in the control run. Blue bars indicate the total area average, while green and brown bars indicate averages over ocean and land areas, respectively. **(c, d)** Same as panels **(a)** and **(b)**, but for changes of the top-of-model shortwave net flux under clear sky conditions ($\Delta N_{s,c}$). **(e, f)** Same as panels a and b, but for changes of the top-of-model total net radiative flux (ΔN_t). **(g, h)** Same as panels a and b, but for changes of surface temperature (ΔT_s).

and SHIGH. This feedback is further discussed in the next section. The effect of additional sea ice melt is also seen in changes of the shortwave clear-sky radiation balance, $\Delta N_{s,c}$, for the Southern Ocean (Fig. 8c and d). The very high values of about 1 W m^{-2} for SMED and about 1.5 W m^{-2} for SHIGH are due to less reflected incoming solar radiation where sea ice melts, but also due to the fact that modelled reductions in sulfate load are particularly high for the Southern Ocean.

The total transient changes in net radiative fluxes, ΔN_t, in our experiments (Fig. 8e and f) are smaller than the changes of shortwave fluxes, since they includes contributions of increased longwave radiation due to increased temperatures. We find values of 0.19, 0.40, and -0.27 W m^{-2} (SMED), and 0.30, 0.63, and -0.43 W m^{-2} (SHIGH) for the global

domain, over the ocean, and over land, respectively. Note that ΔN_t is much larger south of 40° S than north of 40° N (e.g. 0.68 versus 0.39 W m^{-2} in SHIGH) due to the much smaller land area and a stronger reduction of DMS emissions. Consequently, the additional surface warming ΔT_s shows a pronounced asymmetry between northern and southern high latitudes. South of 40° S, we find ΔT_s values as high as 0.46 (SMED) and 0.73 K (SHIGH) while the global average additional warming is 0.30 (SMED) and 0.47 K (SHIGH). The largest additional warming in our simulations is found over the Antarctic continent (0.56 SMED, and 0.89 K SHIGH), which is almost twice the global average.

The study of Six et al. (2013) uses a very similar version of the ocean biogeochemistry model HAMOCC and the same parameterisations of DMS production, pH sensitivity, and DMS emissions. However, compared to RCP8.5, the SRES A1B scenario used by these authors is much more moderate; the CO_2 concentration in this scenario reaches about 700 ppm in 2100.

If we, for comparison, consider the period 2046 to 2075, during which the average atmospheric CO_2 in our scenario simulations is 700 ppm, we find a reduction of DMS emissions of 2.9 (SMED) and 4.4 Tg S yr^{-1} (SHIGH), which is very similar to the values presented in Six et al. (3.2 and 5 Tg S yr^{-1}, respectively). Their estimate of the additional effective radiative forcing (0.4 W m^{-2} for SMED and 0.64 W m^{-2} for SHIGH in 2100) has been derived from 1-year simulations using a stand-alone atmospheric circulation model with aerosol chemistry, and does not include slow feedbacks. Therefore, the corresponding (transient) radiative imbalance of 0.08 and 0.12 W m^{-2} that we find in our SMED and SHIGH simulations for 2046 to 2075 is not comparable. If we calculate the expected equilibrium warming for NorESM using the Six et al. (2013) radiative forcing values with the equilibrium climate sensitivity for NorESM derived by Iversen et al. (2013, 2.87 K at 2×CO_2), we arrive at 0.31 (SMED) and 0.50 K (SHIGH). The additional transient surface warming of 0.12 and 0.18 K that we find for the period 2046 to 2075 in our simulations indicates that about 40 % of this equilibrium warming has been realised, which seems to be reasonable in view of the rapid changes in DMS fluxes ongoing in our simulations.

3.3 Feedbacks

3.3.1 Carbon cycle feedbacks

We ran our experiments with freely evolving atmospheric CO_2 – that is, atmospheric CO_2 concentrations are determined by anthropogenic emissions (including land use change emissions) and exchange with the land biosphere and the ocean. Hence, changes in climate, for example the additional surface warming due to reduced DMS emissions, can be amplified by carbon cycle feedbacks in our model simulations. This is indeed the case, but the effect is rather small.

By the end of the 22nd century, the ocean and land have released an additional amount of about 13 (26) Pg C in the experiments SMED (SHIGH) relative to BASE. This results in an atmospheric CO_2 concentration that is about 6 (12) ppm larger in SMED (SHIGH) than in BASE. Given that the atmospheric CO_2 concentration exceeds 2000 ppm at this point of time, this small feedback can be neglected.

3.3.2   Sea ice feedback

The additional warming in the experiments SMED and SHIGH leads to a reduced sea ice extent relative to BASE. We discuss the related feedbacks for the last 30 years of the 22nd century since the additional melt is largest for this period. At this time, Southern Hemisphere annual average sea ice extent has decreased from about 14.5×10^6 km^2 in the control simulation to 2.7×10^6 km^2 in experiment BASE. The additional decrease in SMED and SHIGH is 0.54×10^6 and 0.62×10^6 km^2. In contrast, in the Arctic Ocean, sea ice cover decreases from 11.5×10^6 to only 0.23×10^6 km^2 in simulation BASE. Hence, as the Northern Hemisphere is virtually sea-ice-free year-round towards the end of the 22nd century, we concentrate our analysis on the Southern Ocean.

The reduced sea ice area in SMED and SHIGH has a significant effect on the shortwave energy balance under clear-sky conditions (Fig. 8c and d). If we consider sea-ice-free regions only (SHLni), we find that $\Delta N_{s,c}$ is 0.65 (SMED) and 1.02 W m^{-2} (SHIGH) smaller than for the total ocean area south of 40° S (SHL). While this effect is straightforward to understand (more absorbed shortwave radiation due to less sea ice), the shortwave radiative effect due to clouds shows the opposite behaviour (Fig. 8a and b). We find increases in CRE$_s$ of 0.56 (SMED) and 0.89 W m^{-2} (SHIGH) if only sea-ice-free grid points are considered. This can be explained by the fact that clouds have only a weak shortwave effect over ice surfaces in our model (clouds reflects incident radiation marginally better than ice surfaces). If the sea ice is removed and replaced by open ocean, clouds have a much stronger negative radiative effect, since their reflectance increased relative to the underlying surface. The total effect of additional sea ice melt on the transient radiative imbalance over the ocean south of 40° S is small compared to the two opposing contributions (Fig. 8e and f). We find that N_t increases by about 0.08 (SMED) and 0.13 W m^{-2} (SHIGH) through the additional melt of sea ice.

3.3.3   DMS–climate feedbacks

Climate change caused by the pH sensitivity of DMS production also feeds back on the DMS production itself. Although there is no significant difference in globally integrated PP between the runs BASE, SMED, and SHIGH, we find compensating regional differences. The additional warming and increased incident shortwave radiation lead to an increase in PP south of 40° S in SMED and SHIGH compared to BASE, while, at low latitudes and north of 40° N, a reduction of PP in response to stronger stratification and less nutrient supply is found. We estimate the effect of these changes on DMS production by calculating P_{DMS}^*, which is the DMS production based on $CaCO_3$ and silica export fields from the simulation BASE, combined with temperature and pH fields (and the pH dependency) from the runs SMED or SHIGH. Hence, P_{DMS}^* is the DMS production that would arise in the simulations SMED and SHIGH if there was no climate change relative to the simulation BASE. This definition also includes differences in DMS production due to changes in sea ice cover. We then define the DMS–climate feedback on DMS production as $P_{DMS}^f = P_{DMS} - P_{DMS}^*$.

For our calculation of P_{DMS}^*, we use monthly averaged model output. This introduces a systematic bias, since there is a causal connection between pH and PP anomalies, and these anomalies occur on a sub-monthly timescale during bloom-periods (PP increases pH during strong phytoplankton blooms due to the draw-down of dissolved inorganic carbon). By reconstructing the known DMS production for SHIGH from monthly mean model output, we find that our calculations could underestimate the true value of P_{DMS}^* at high latitudes by up to 40 mg S m^{-2} yr^{-1} locally (with an average of about 10 mg S m^{-2} yr^{-1} south of 40° S, where the effect is largest). Nevertheless, the underestimation of P_{DMS}^* (or equivalently, the overestimation of P_{DMS}^f) is smaller than 3 % for the total production south of 40° S.

We find that the DMS–climate feedback P_{DMS}^f is generally small, except south of 40° S, where P_{DMS}^f locally exceeds 200 mg S m^{-2} yr^{-1} for simulation SHIGH (Fig. 9). This feedback is negative, that is, DMS production increases with climate change, which counteracts the decrease due to acidification. The feedback accounts for more than 15 % of the DMS production almost everywhere south of 40° S (Fig. 9d).

We estimate the DMS–climate feedback on DMS fluxes, F_{DMS}^f, by assuming that the fraction of the DMS production released as DMS to the atmosphere does not change between our simulations and the (hypothetical) simulation with the DMS–climate feedback excluded. By making this assumption, we neglect the fact that the fraction of DMS production released to the atmosphere generally increases with decreasing DMS production in our simulations (since the bacterial consumption decreases). The effect on the allocation of the DMS flux is small for the small differences between P_{DMS} and P_{DMS}^* discussed here. However, we might tend to overestimate the feedback F_{DMS}^f by this assumption.

Globally, the increase in DMS emissions due to the DMS–climate feedback in the Southern Ocean is nearly cancelled out by decreased fluxes at low latitudes and north of 40° N (Fig. 10). Towards the end of the 22nd century (2171 to 2200) the feedback south of 40° S accounts for 8.9 % (0.49 Tg S yr^{-1}) in SMED and 15.5 % (0.61 Tg S yr^{-1}) in SHIGH (see Table 3 for more details).

Table 3. Quantification of the DMS–climate feedback on sea–air DMS fluxes in our experiments (averages over the period 2171 to 2200). F_{DMS} denotes DMS fluxes including all feedbacks, while F_{DMS}^f gives the estimated fraction of these fluxes that arise due to feedbacks (see Sect. 3.3.3).

		Global	40–90° N	40° S–40° N	90–40° S
SMED					
F_{DMS}	(Tg S yr^{-1})	16.56	1.04	10.31	5.21
F_{DMS}^f	(Tg S yr^{-1})	0.10	−0.03	−0.27	0.46
Fraction	(%)	0.6	−3.4	−2.6	8.9
SHIGH					
F_{DMS}	(Tg S yr^{-1})	12.35	0.70	7.76	3.89
F_{DMS}^f	(Tg S yr^{-1})	0.30	−0.03	−0.21	0.61
Fraction	(%)	2.4	−3.9	−2.7	15.5

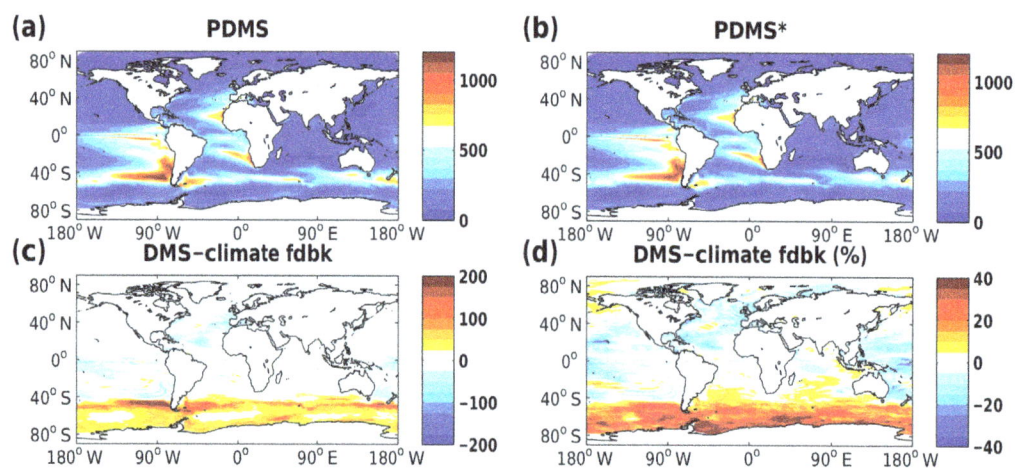

Figure 9. Averages over 2171 to 2200 of **(a)** DMS production for the experiment SHIGH, **(b)** an estimate of the DMS production that would occur in SHIGH without DMS–climate feedback (P_{DMS}^*), **(c)** the corresponding estimate of the DMS–climate feedback P_{DMS}^f in absolute values, and **(d)** the DMS–climate feedback expressed in percent of the total DMS production shown in panel **(a)**. Units in **(a)**–**(c)** are mg S m^{-2} yr^{-1}.

As mentioned in Sect. 3.1, the sensitivity of PP (and hence detritus and DMS production) to climate change in the Southern Ocean seems to be high in our model compared to other models. Therefore, and because of the issues discussed above (use of monthly mean output fields, assumption of constant sea–air flux fraction), the negative DMS–climate feedback found in our simulations south of 40° S might represent an upper estimate.

4   Summary and conclusions

We have simulated the impact of a possible reduction of marine DMS emissions due to ocean acidification in a fully coupled Earth system model. The atmospheric model employs a state-of-the-art aerosol module that reacts to changes in atmospheric DMS concentrations by changing the radiative balance of the Earth due to direct and indirect aerosol effects. All processes included in the ocean model's DMS scheme (DMS production as well as loss through bacterial consumption, photolysis, and sea–air fluxes) are affected by climate change due to their explicit dependence on environmental factors. Hence, we simulate the interplay between DMS production and emissions, and the radiation balance of the Earth in a fully interactive fashion. Our set of model experiments consists of RCP8.5 scenario simulations that assume no (BASE), medium (SMED), and high (SHIGH) sensitivity of DMS production to seawater pH.

We find a linear relationship between reduction of DMS sea–air fluxes and changes in radiative fluxes, as well as changes in surface temperature. Our study is consistent with the results of a previous offline study and confirms the order of magnitude of the additional warming found. The global average transient warming in our simulation BASE (8.1 K towards the end of the 22nd century) is amplified by 3.7 % (5.8 %) in experiment SMED (SHIGH) through the pH de-

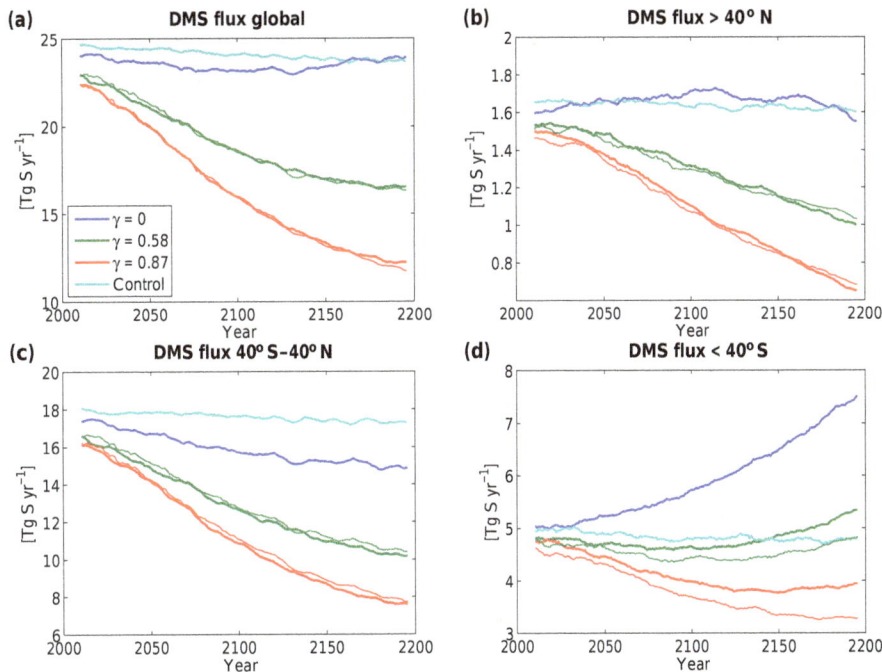

Figure 10. Time series of DMS fluxes (smoothed by a 10-year running mean filter) for the RCP8.5 scenario and its extension to 2200 for the experiments BASE (dark blue lines), SMED (green) and SHIGH (red). The control run fluxes are indicated by the light blue lines. Thin green and red lines give an estimate of the fluxes that would occur without DMS–climate feedback (F_{DMS}^*).

pendency of DMS fluxes. However, we find that the additional surface warming is not spatially homogeneous, but has a strong north–south gradient with much stronger surface warming in the Southern Hemisphere. The additional transient warming over the Antarctic continent of 0.56 K (0.89 K) is almost twice the global average of 0.30 K (0.47 K).

Since our model simulates large reductions in sea–air DMS fluxes (up to 48 % in experiment SHIGH towards the end of the 22nd century), a considerable negative feedback loop should, according to the CLAW hypothesis, counteract the original reduction. We indeed find a CLAW-like feedback in the Southern Ocean where the additional warming stimulates additional primary and DMS production. We estimate that south of 40° S DMS emissions would be further reduced by up to 15 % in a model run excluding this feedback. North of 40° S, however, the additional warming tends to additionally weaken (or to only slightly enhance) DMS fluxes, such that, at the global scale, there is no significant CLAW-like feedback in our simulations.

We have chosen a high-emission scenario for our model experiments, and we assume a medium to high sensitivity of DMS production to pH. Therefore, our results likely represent an upper limit of additional transient warming that might arise by this mechanism due to ocean acidification. If humankind decides to follow a low-emission pathway, and consequently, the global ocean experiences much less severe acidification than projected in this study, the global additional temperature change might appear less significant.

However, our study highlights that the additional warming due to a reduced DMS production could reach almost twice the global average over the Antarctic region, which might have implications for the atmospheric CO_2 level that is allowable to avoid a destabilisation of Antarctic ice shelves.

We finally note that the parameterisation for the pH dependence of DMS production is solely based on empirical evidence. A better mechanistic understanding of the processes leading to changes of DMS production on the plankton community level is needed to better assess the climatic implications that ocean acidification might have through altering the global sulfur cycle.

Competing interests. The authors declare that they have no conflict of interest.

Special issue statement. This article is part of the special issue "Progress in quantifying ocean biogeochemistry – in honour of Ernst Maier-Reimer". It is not associated with a conference.

Acknowledgements. We thank David Archer and an anonymous reviewer for their helpful and constructive comments, which improved this paper. Jörg Schwinger, Jerry Tjiputra, Nadine Goris, Alf Kirkevåg, and Øyvind Seland were supported by the Research Council of Norway through project EVA (229771). Supercomputer time and storage resources were provided by the Norwegian metacenter for computational science (NOTUR, project nn2345k), and the Norwegian Storage Infrastructure (NorStore, project ns2345k). This work was supported by the Bjerknes Centre for Climate Research. We gratefully acknowledge the CESM project, which is supported by the National Science Foundation and the Office of Science (BER) of the US Department of Energy.

Edited by: Matthias Hofmann

References

Abdul-Razzak, H. and Ghan, S. J.: A parameterization of aerosol activation: 2. Multiple aerosol types, J. Geophys. Res., 105, 6837–6844, https://doi.org/10.1029/1999JD901161, 2000.

Archer, S. D., Kimmance, S. A., Stephens, J. A., Hopkins, F. E., Bellerby, R. G. J., Schulz, K. G., Piontek, J., and Engel, A.: Contrasting responses of DMS and DMSP to ocean acidification in Arctic waters, Biogeosciences, 10, 1893–1908, https://doi.org/10.5194/bg-10-1893-2013, 2013.

Arora, V., Boer, G., Friedlingstein, P., Eby, M., Jones, C., Christian, J., Bonan, G., Bopp, L., Brovkin, V., Cadule, P., Hajima, T., Ilyina, T., Lindsay, K., Tjiputra, J., and Wu, T.: Carbon-concentration and carbon-climate feedbacks in CMIP5 Earth system models, J. Climate, 26, 5289–5314, https://doi.org/10.1175/JCLI-D-12-00494.1, 2013.

Assmann, K. M., Bentsen, M., Segschneider, J., and Heinze, C.: An isopycnic ocean carbon cycle model, Geosci. Model Dev., 3, 143–167, https://doi.org/10.5194/gmd-3-143-2010, 2010.

Bentsen, M., Bethke, I., Debernard, J. B., Iversen, T., Kirkevåg, A., Seland, Ø., Drange, H., Roelandt, C., Seierstad, I. A., Hoose, C., and Kristjánsson, J. E.: The Norwegian Earth System Model, NorESM1-M – Part 1: Description and basic evaluation of the physical climate, Geosci. Model Dev., 6, 687–720, https://doi.org/10.5194/gmd-6-687-2013, 2013.

Bleck, R., Rooth, C., Hu, D., and Smith, L.: Salinity-driven thermocline transients in a wind- and thermohaline-forced isopycnic coordinate model of the North Atlantic, J. Phys. Oceanogr., 22, 1486–1505, 1992.

Bopp, L., Aumont, O., Belviso, S., and Monfray, P.: Potential impact of climate change on marine dimethyl sulfide emissions, Tellus B, 55, 11–22, https://doi.org/10.1034/j.1600-0889.2003.042.x, 2003.

Bopp, L., Resplandy, L., Orr, J. C., Doney, S. C., Dunne, J. P., Gehlen, M., Halloran, P., Heinze, C., Ilyina, T., Séférian, R., Tjiputra, J., and Vichi, M.: Multiple stressors of ocean ecosystems in the 21st century: projections with CMIP5 models, Biogeosciences, 10, 6225–6245, https://doi.org/10.5194/bg-10-6225-2013, 2013.

Carslaw, K. S., Boucher, O., Spracklen, D. V., Mann, G. W., Rae, J. G. L., Woodward, S., and Kulmala, M.: A review of natural aerosol interactions and feedbacks within the Earth system, Atmos. Chem. Phys., 10, 1701–1737, https://doi.org/10.5194/acp-10-1701-2010, 2010.

Charlson, R. J., Lovelock, J. E., Andreae, M. O., and Warren, S. G.: Oceanic phytoplankton, atmospheric sulphur, cloud albedo and climate, Nature, 326, 655–661, 1987.

Gabric, A. J., Simó, R., Cropp, R. A., Hirst, A. C., and Dachs, J.: Modeling estimates of the global emission of dimethylsulfide under enhanced greenhouse conditions, Global Biogeochem. Cy., 18, GB2014, https://doi.org/10.1029/2003GB002183, 2004.

Gent, P., Danabasoglu, G., Donner, L., Holland, M., Hunke, E., Jayne, S., Lawrence, D., Neale, R., Rasch, P., Vertenstein, M., Worley, P., Yang, Z., and Zhang, M.: The Community Climate System Model version 4, J. Climate, 24, 4973–4991, https://doi.org/10.1175/2011JCLI4083.1, 2011.

Gunson, J. R., Spall, S. A., Anderson, T. R., Jones, A., Totterdell, I. J., and Woodage, M. J.: Climate sensitivity to ocean dimethylsulphide emissions, Geophys. Res. Lett., 33, L07701, https://doi.org/10.1029/2005GL024982, 2006.

Ilyina, T., Six, K. D., Segschneider, J., Maier-Reimer, E., Li, H., and Núñez Riboni, I.: Global ocean biogeochemistry model HAMOCC: Model architecture and performance as component of the MPI-Earth System Model in different CMIP5 experimental realizations, J. Adv. Model. Earth Syst., 5, 287–315, https://doi.org/10.1029/2012MS000178, 2013.

Iversen, T., Bentsen, M., Bethke, I., Debernard, J. B., Kirkevåg, A., Seland, Ø., Drange, H., Kristjansson, J. E., Medhaug, I., Sand, M., and Seierstad, I. A.: The Norwegian Earth System Model, NorESM1-M – Part 2: Climate response and scenario projections, Geosci. Model Dev., 6, 389–415, https://doi.org/10.5194/gmd-6-389-2013, 2013.

Keller, M., Bellows, W., and Guillard, R.: Dimethyl sulfide production in marine phytoplankton, in: Biogenic sulfur in the enviroment, edited by: Saltzman, E. and Cooper, W., ACS-Symposium series, New Orleans, Louisiana, American Chemical Society, 167–181, 1989.

Kirkevåg, A., Iversen, T., Seland, Ø., Hoose, C., Kristjánsson, J., Struthers, H., Ekman, A., Ghan, S., Griesfeller, J., Nilsson, E., and Schulz, M.: Aerosol-climate interactions in the Norwegian Earth System Model – NorESM1-M, Geosci. Model Dev., 6, 207–244, https://doi.org/10.5194/gmd-6-207-2013, 2013.

Kloster, S., Feichter, J., Maier-Reimer, E., Six, K. D., Stier, P., and Wetzel, P.: DMS cycle in the marine ocean-atmosphere system — a global model study, Biogeosciences, 3, 29–51, https://doi.org/10.5194/bg-3-29-2006, 2006.

Kloster, S., Six, K. D., Feichter, J., Maier-Reimer, E., Roeckner, E., Wetzel, P., Stier, P., and Esch, M.: Response of dimethylsulfide (DMS) in the ocean and atmosphere to global warming, J. Geophys. Res., 112, G03005, https://doi.org/10.1029/2006JG000224, 2007.

Korhonen, H., Carslaw, K. S., Spracklen, D. V., Mann, G. W., and Woodhouse, M. T.: Influence of oceanic dimethyl sulfide emissions on cloud condensation nuclei concentrations and seasonality over the remote Southern Hemisphere oceans: A global model study, J. Geophys. Res., 113, D15204, https://doi.org/10.1029/2007JD009718, 2008.

Lana, A., Bell, T. G., Simó, R., Vallina, S. M., Ballabrera-Poy, J., Kettle, A. J., Dachs, J., Bopp, L., Saltzman, E. S., Stefels, J., Johnson, J. E., and Liss, P. S.: An updated climatology of surface dimethylsulfide concentrations and emission fluxes in the global ocean, Global Biogeochem. Cy., 25, GB1004, https://doi.org/10.1029/2010GB003850, 2011.

Laufkötter, C., Vogt, M., Gruber, N., Aita-Noguchi, M., Aumont, O., Bopp, L., Buitenhuis, E., Doney, S. C., Dunne, J., Hashioka, T., Hauck, J., Hirata, T., John, J., Le Quéré, C., Lima, I. D., Nakano, H., Séférian, R., Totterdell, I., Vichi, M., and Völker, C.: Drivers and uncertainties of future global marine primary production in marine ecosystem models, Biogeosciences, 12, 6955–6984, https://doi.org/10.5194/bg-12-6955-2015, 2015.

Lindsay, K., Bonan, G. B., Doney, S. C., Hoffman, F. M., Lawrence, D. M., Long, M. C., Mahowald, N. M., Moore, J. K., Randerson, J. T., and Thornton, P. E.: Preindustrial-Control and Twentieth-Century Carbon Cycle Experiments with the Earth System Model CESM1(BGC), J. Climate, 27, 8981–9005, https://doi.org/10.1175/JCLI-D-12-00565.1, 2014.

Liss, P. S., Malin, G., Turner, S. M., and Holligan, P. M.: Dimethyl sulphide and Phaeocystis: A review, J. Marine Syst., 5, 41–53, https://doi.org/10.1016/0924-7963(94)90015-9, 1994.

Maier-Reimer, E.: Geochemical cycles in an ocean general circulation model. Preindustrial tracer distributions, Global Biogeochem. Cy., 7, 645–677, 1993.

Maier-Reimer, E., Kriest, I., Segschneider, J., and Wetzel, P.: The Hamburg Oceanic Carbon Cycle Circulation model HAMOCC5.1 – Technical Description Release 1.1, Reports on Earth System Science 14, Max Planck Institute for Meteorology, Hamburg, Germany, 2005.

Meinshausen, M., Smith, S. J., Calvin, K., Daniel, J. S., Kainuma, M. L. T., Lamarque, J.-F., Matsumoto, K., Montzka, S. A., Raper, S. C. B., Riahi, K., Thomson, A., Velders, G. J. M., and van Vuuren, D. P. P.: The RCP greenhouse gas concentrations and their extensions from 1765 to 2300, Climate Change, 109, 213–241, https://doi.org/10.1007/s10584-011-0156-z, 2011.

Park, K.-T., Lee, K., Shin, K., Yang, E. J., Hyun, B., Kim, J.-M., Noh, J. H., Kim, M., Kong, B., Choi, D. H., Choi, S.-J., Jang, P.-G., and Jeong, H. J.: Direct Linkage between Dimethyl Sulfide Production and Microzooplankton Grazing, Resulting from Prey Composition Change under High Partial Pressure of Carbon Dioxide Conditions, Environ. Sci. Technol., 48, 4750–4756, https://doi.org/10.1021/es403351h, 2014.

Schwinger, J., Goris, N., Tjiputra, J., Kriest, I., Bentsen, M., Bethke, I., Ilicak, M., Assmann, K., and Heinze, C.: Evaluation of NorESM-OC (versions 1 and 1.2), the ocean carbon-cycle stand-alone configuration of the Norwegian Earth System Model (NorESM1), Geosci. Model Dev., 9, 2589–2622, https://doi.org/10.5194/gmd-9-2589-2016, 2016.

Seland, Ø., Iversen, T., Kirkevåg, A., and Storelvmo, T.: Aerosol-climate interactions in the CAM-Oslo atmospheric GCM and investigation of associated basic shortcomings, Tellus A, 60, 459–491, https://doi.org/10.1111/j.1600-0870.2008.00318.x, 2008.

Shaw, G. E.: Bio-controlled thermostasis involving the sulfur cycle, Climate Change, 5, 297–303, 1983.

Six, K. D. and Maier-Reimer, E.: Effects of plankton dynamics on seasonal carbon fluxes in an ocean general circulation model, Global Biogeochem. Cy., 10, 559–583, 1996.

Six, K. D. and Maier-Reimer, E.: What controls the oceanic dimethylsulfide (DMS) cycle? A modeling approach, Global Biogeochem. Cy., 20, GB4011, https://doi.org/10.1029/2005GB002674, 2006.

Six, K. D., Kloster, S., Ilyina, T., Archer, S. D., Zhang, K., and Maier-Reimer, E.: Global warming amplified by reduced sulphur fluxes as a result of ocean acidification, Nature Climate Change, 3, 975–978, https://doi.org/10.1038/NCLIMATE1981, 2013.

Stefels, J., Steinke, M., Turner, S., Malin, G., and Belviso, S.: Environmental constraints on the production and removal of the climatically active gas dimethylsulphide (DMS) and implications for ecosystem modelling, Biogeochemistry, 83, 245–275, https://doi.org/10.1007/s10533-007-9091-5, 2007.

Steinacher, M., Joos, F., Frölicher, T. L., Bopp, L., Cadule, P., Cocco, V., Doney, S. C., Gehlen, M., Lindsay, K., Moore, J. K., Schneider, B., and Segschneider, J.: Projected 21st century decrease in marine productivity: a multi-model analysis, Biogeosciences, 7, 979–1005, https://doi.org/10.5194/bg-7-979-2010, 2010.

Taylor, K. E., Stouffer, R. J., and Meehl, G. A.: An overview of CMIP5 and the experiment design, B. Am. Meteorol. Soc., 93, 485–498, https://doi.org/10.1175/BAMS-D-11-00094.1, 2012.

Tjiputra, J. F., Roelandt, C., Bentsen, M., Lawrence, D. M., Lorentzen, T., Schwinger, J., Seland, Ø., and Heinze, C.: Evaluation of the carbon cycle components in the Norwegian Earth System Model (NorESM), Geosci. Model Dev., 6, 301–325, https://doi.org/10.5194/gmd-6-301-2013, 2013.

Vallina, S. M., Simó, R., and Manizza, M.: Weak response of oceanic dimethylsulfide to upper mixing shoaling induce by global warming, P. Natl. Acad. Sci. USA, 104, 16004–16009, 2007.

Wanninkhof, R.: Relationship between wind speed and gas exchange over the ocean, J. Geophys. Res., 97, 7373–7382, 1992.

Woodhouse, M. T., Carslaw, K. S., Mann, G. W., Vallina, S. M., Vogt, M., Halloran, P. R., and Boucher, O.: Low sensitivity of cloud condensation nuclei to changes in the sea-air flux of dimethyl-sulphide, Atmos. Chem. Phys., 10, 7545–7559, https://doi.org/10.5194/acp-10-7545-2010, 2010.

12

Exchange of CO$_2$ in Arctic tundra: impacts of meteorological variations and biological disturbance

Efrén López-Blanco[1,2], **Magnus Lund**[1], **Mathew Williams**[2], **Mikkel P. Tamstorf**[1], **Andreas Westergaard-Nielsen**[3], **Jean-François Exbrayat**[2,4], **Birger U. Hansen**[3], **and Torben R. Christensen**[1,5]

[1]Department of Biosciences, Arctic Research Center, Aarhus University, Frederiksborgvej 399, 4000 Roskilde, Denmark
[2]School of GeoSciences, University of Edinburgh, Edinburgh, EH93FF, UK
[3]Center for Permafrost (CENPERM), Department of Geosciences and Natural Resource Management, University of Copenhagen, Oester Voldgade 10, 1350 Copenhagen, Denmark
[4]National Centre for Earth Observation, University of Edinburgh, Edinburgh, EH93FF, UK
[5]Department of Physical Geography and Ecosystem Science, Lund University, Sölvegatan 12, 223 62 Lund, Sweden

Correspondence to: Efrén López-Blanco (elb@bios.au.dk)

Abstract. An improvement in our process-based understanding of carbon (C) exchange in the Arctic and its climate sensitivity is critically needed for understanding the response of tundra ecosystems to a changing climate. In this context, we analysed the net ecosystem exchange (NEE) of CO$_2$ in West Greenland tundra (64° N) across eight snow-free periods in 8 consecutive years, and characterized the key processes of net ecosystem exchange and its two main modulating components: gross primary production (GPP) and ecosystem respiration (R_{eco}). Overall, the ecosystem acted as a consistent sink of CO$_2$, accumulating -30 g C m^{-2} on average (range of -17 to -41 g C m^{-2}) during the years 2008–2015, except 2011 (source of 41 g C m^{-2}), which was associated with a major pest outbreak. The results do not reveal a marked meteorological effect on the net CO$_2$ uptake despite the high interannual variability in the timing of snowmelt and the start and duration of the growing season. The ranges in annual GPP (-182 to -316 g C m^{-2}) and R_{eco} (144 to 279 g C m^{-2}) were > 5 fold larger than the range in NEE. Gross fluxes were also more variable (coefficients of variation are 3.6 and 4.1 % respectively) than for NEE (0.7 %). GPP and R_{eco} were sensitive to insolation and temperature, and there was a tendency towards larger GPP and R_{eco} during warmer and wetter years. The relative lack of sensitivity of NEE to meteorology was a result of the correlated response of GPP and R_{eco}. During the snow-free season of the anomalous year of 2011, a biological disturbance related to a larvae outbreak reduced

GPP more strongly than R_{eco}. With continued warming temperatures and longer growing seasons, tundra systems will increase rates of C cycling. However, shifts in sink strength will likely be triggered by factors such as biological disturbances, events that will challenge our forecasting of C states.

1 Introduction

Quantifying the climate sensitivity of carbon (C) stocks of the terrestrial biosphere is a major challenge for Earth system science (Williams et al., 2005). In the Arctic, organic soil C storage has the potential for very large C releases following thaw (Koven et al., 2011) that could create a positive feedback on climate change and accelerate the rate of global warming. Recent reviews have estimated the Arctic terrestrial C pool to be 1400–1850 Pg C, more than twice the size of the atmospheric C pool (Hugelius et al., 2014; McGuire et al., 2009; Tarnocai et al., 2009) and approximately 50 % of the global soil organic C pool (AMAP, 2011; McGuire et al., 2009). Further, Arctic ecosystems have experienced an intensified warming tendency, reaching almost twice the global average (ACIA, 2005; AMAP, 2011; Callaghan et al., 2012c; Serreze and Barry, 2011). The projected Arctic warming is also expected to be more pronounced in coming years (AMAP, 2011; Callaghan et al., 2012a; Christensen et al., 2007; Grøndahl et al., 2008; Meltofte et al., 2008) and temperature, precipitation and growing season length will likely increase in the Arctic (ACIA, 2005; Christensen et al., 2007, 2004; IPCC, 2007). Given this situation, an improvement in our process-based understanding of CO_2 exchanges in the Arctic and their climate sensitivity is critical (McGuire et al., 2009).

Measuring the interannual C exchange variability in the Arctic tundra is challenging due to extreme conditions and the patchy nature of the landscape linked to microtopography. Different eco-types are linked to different C exchange rates (Bubier et al., 2003). Synthesis studies have found a significant spatial variability in NEE (Lafleur et al., 2012; Mbufong et al., 2014) between different tundra sites (Lindroth et al., 2007; Lund et al., 2010) and also large temporal variability within sites (Aurela et al., 2004, 2007; Christensen et al., 2012; Grøndahl et al., 2008; Lafleur et al., 2012). Minor variations in the key process of photosynthesis (gross primary production, GPP) and ecosystem respiration (R_{eco}) may promote important changes in the sign and magnitude of the C balance (Arndal et al., 2009; Elberling et al., 2008; IPCC, 2007; Lund et al., 2010; Tagesson et al., 2012; Williams et al., 2000). With continued warming temperature and longer growing seasons, tundra systems will likely have enhanced GPP and R_{eco} rates, but long-term data with which to investigate and quantify these responses are rare. Further, the effects on net CO_2 sequestration are not known, and may be altered by long-term processes such as vegetation shifts and short-term disturbances like insect pest outbreaks, complicating the prognostic forecast of upcoming C states (Callaghan et al., 2012b; McGuire et al., 2012). Consequently, there is a need to understand how the C cycle behaves over timescales from days to years and the links to environmental drivers. There is a lack of reference sites in the Arctic from which full measurement-based data are available, documenting carbon fluxes at the terrestrial catchment scales. Here we investigate the functional responses of C exchange to environmental characteristics across eight snow-free periods in 8 consecutive years in West Greenland.

In recent decades, eddy covariance has become a fundamental method for carbon flux measurements at the landscape scale (Lasslop et al., 2012; Lund et al., 2012; Reichstein et al., 2005). Eddy covariance measurements of land–atmosphere fluxes or net ecosystem exchange (NEE), of CO_2 can be gap-filled and subsequently separated into the modulating components of GPP and R_{eco} using flux partitioning algorithms (Reichstein et al., 2005). These techniques are critical for providing a better understanding of the C uptake vs. C release behaviour (Lund et al., 2010), but they also allow for an examination of the environmental effects on ecological processes (Hanis et al., 2015). However, large gaps in the measured fluxes may introduce significant uncertainties in the C budget estimations. Moreover, GPP and R_{eco} estimates can be calculated in different ways. Some algorithms fit an instantaneous temperature–respiration curve to night-time data to calculate R_{eco} and estimate GPP (Lasslop

et al., 2012; Reichstein et al., 2005); others calculate R_{eco} from a light-response curve (Gilmanov et al., 2003; Lindroth et al., 2007; Lund et al., 2012; Mbufong et al., 2014; Runkle et al., 2013). Unfortunately, different interpretations of the flux gap filling and partitioning lead to different estimates of NEE, GPP and R_{eco} as well as undefined uncertainties.

The main objectives of this paper are (1) to explore the uncertainties in NEE gap filling and partitioning obtained from different approaches, (2) to determine how C uptake and C storage respond to the meteorological variability, and (3) to identify how the environmental forcing affects not only the interannual variability, but also the hourly, daily, weekly and monthly variability of NEE, GPP and R_{eco}. The intention of this paper is to elaborate on the information gathered in an existing catchment area under an extensive cross-disciplinary ecological monitoring programme in low Arctic West Greenland, established under the auspices of the Greenland Ecosystem Monitoring (GEM) (http://www.g-e-m.dk). Using a long-term (8-year) data set to explore uncertainties in NEE gap-filling and partitioning methods and to characterize the interannual variability of C exchange in relation to driving factors can provide our understanding of land–atmosphere CO_2 exchange in Arctic regions with a novel input. Our overarching hypothesis was that both GPP and R_{eco} would respond positively to warmer and longer growing seasons. However NEE response to warming would be more complex and variable (positive or negative) depending on subtle balances between plant and microbial climate sensitivity.

2 Materials and methods

2.1 Site description

Field measurements were conducted in the low Arctic Kobbefjord drainage basin in south-western Greenland ($64°07'$ N; $51°21'$ W) (Fig. 1a). The study area is located ~ 20 km SE of Nuuk, the Greenlandic capital. Kobbefjord has been subject to extensive environmental research activities (the Nuuk Ecological Research Operations) since 2007 (http://www.nuuk-basic.dk). The lowland site is located 500 m from the south-eastern shore of the bottom of Kangerluarsunnguaq Fjord (Kobbefjord), and 500 m from the western shore of the 0.7 km^2 lake called "Badesø" (Fig. 1b). Three glaciated mountains, all above 1000 m a.s.l., surround the site. The landscape consists of a fen area surrounded by heath, copse and bedrock. The current fen vegetation is dominated by *Scirpus cespitosus*, whereas the surroundings are dominated by heath species such as *Empetrum nigrum*, *Vaccinium uliginosum*, *Salix glauca* and copse species such as *S. glauca* and *Eriophorum angustifolium* (Bay et al., 2008). Kobbefjord belongs to the "arctic shrub tundra" (bioclimate zone E) according to The Circumpolar Arctic Vegetation Map (CAVM Team, 2003; Walker et al., 2005). This map is

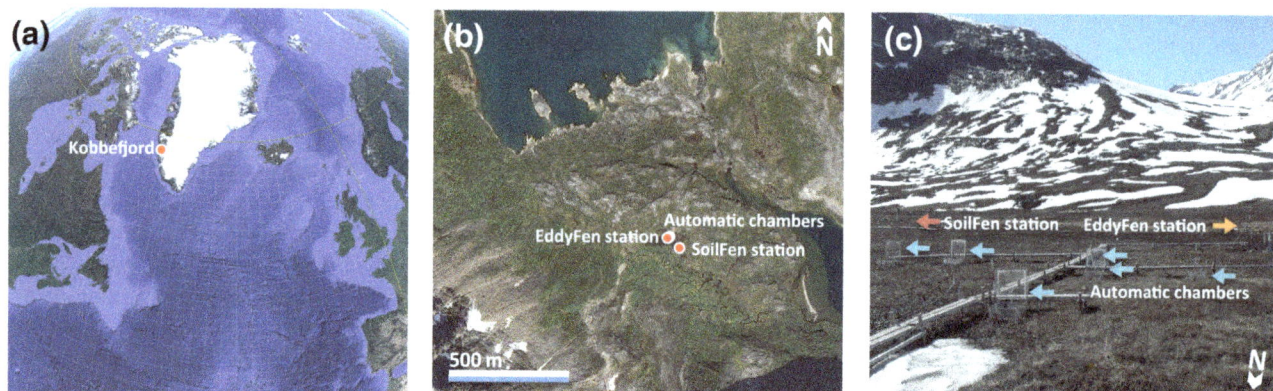

Figure 1. (a) Location of Kobbefjord in Greenland, 64°07′ N; 51°21′ W (source: Google Earth Pro). **(b)** Location of EddyFen station, automatic chambers and SoilFen station in Kobbefjord (source: Google Earth Pro, 16 July 2013). **(c)** Eddy covariance (orange arrow) from EddyFen station, six automatic chambers (light blue arrows) and SoilFen station (pale red arrow) (photo by Efrén López Blanco, 27 June 2015).

based on the summer warmth index (SWI), which is the sum of the monthly mean temperature above $0\,°C$ from May to September and the southernmost bioclimatic zone E has limits of 20–35. In 2010 and 2012, the weather conditions led the area to experience temperatures from warmer climatic zones (SWI ca. 36 and 35 respectively). For the 1961–1990 period, the mean annual air temperature was $-1.4\,°C$ and the annual precipitation was 750 mm (Cappelen, 2013). The sunlight hours between May and September range from 14 to 21 h. Outcalt's frost number (Nelson and Outcalt, 1987) indicates that discontinuous permafrost should be present, although no permafrost has been found. Nonetheless, thin lenses of ice may remain until late summer.

2.2 Measurements

We have used eddy covariance (EC) data on NEE, measured during the snow-free period from 2008 to 2015. Measurements typically started around the end of the snowmelt (ca. May–June) and extended until the freeze-in period (between September and October). Once the snow melts, the growing season (i.e. the part of the year when the weather conditions allow plant growth) has been reported as the most relevant period defining both spatial (Lund et al., 2010; Mbufong et al., 2014) and temporal (Aurela et al., 2004; Groendahl et al., 2007; Lund et al., 2012) CO_2 variability. The EC measurements were conducted at the EddyFen station (Fig. 1b and c), located in a wet lowland, 40 m a.s.l. The EC tower is equipped with a closed-path infrared CO_2 and H_2O gas analyser LI-7000 (LI-COR Inc, USA) and a 3-D sonic anemometer Gill R3-50 (Gill Instruments Ltd, UK). The anemometer was installed at a height of 2.2 m, while the air intake was attached 2.0 m above terrain on the steel stand. Adjacent to the EddyFen station, an independent system (Fig. 1b and c) measures round-the-clock net CO_2 fluxes using an automatic chamber (AC) method based on Goulden and Crill (1997).

The transparent chambers, each covering a known surface area of 60 cm by 60 cm, with a height of 30 cm, can be opened and closed by the computer in succession for 10 min every hour. When the chamber closes, a CO_2 analyser (SBA-4, PP Systems, UK) monitors both the CO_2 concentration by a close loop of tubing (further information about the set up can be found in Mastepanov et al. (2013). Nearly 20 m from the EddyFen station, the automated SoilFen (Fig. 1b and c) station provides environmental variables such as air and surface temperature (Vaisala HMP45C), soil temperature at different depths (Campbell scientific 10ST) and relative humidity (Vaisala HMP45C). Two kilometres from these stations, an automatic weather station provides complementary ancillary data such as short- and long-wave radiation (with a CNR1 instrument), photosynthetic active radiation (with a Kipp & Zonen PAR Lite instrument), precipitation (using an Ott Pluvio instrument) and snow depth (with a Campbell Scientific SR 50). The water table depth data were monitored using a piezometer located next to each of the six autochambers. Finally, a robust daily estimate of the timing of snowmelt was analysed at a pixel level from a time-lapse camera (HP e427) located at 500 m a.s.l. (Westergaard-Nielsen et al., 2013).

2.3 Data handling

2.3.1 Data collection and pre-processing

Data collection from the EddyFen station was performed using Edisol software (Moncrieff et al., 1997). Raw data files were processed using EdiRe software (version 1.5.0.32, R. Clement, University of Edinburgh) calculating the CO_2 fluxes on a half-hourly basis. The flux processing integrated despiking (Højstrup, 1993), 2-D rotation, time lag removal by covariance optimization, block averaging, frequency response correction (Moore, 1986) and Webb–Pearman–

Leuning correction (Webb et al., 1980). For more information, see Westergaard-Nielsen et al. (2013). Ancillary data (air temperature, soil temperature, incoming short-wave radiation, relative humidity, PAR and precipitation) were temporally resampled using R (R Development Core Team, 2015). Time-series-related packages such as *zoo* (Zeileis and Grothendieck, 2005), *xts* (Ryan and Ulrich, 2014) and *lubridate* (Grolemund and Wickham, 2011) were used to get the ancillary data aligned with the flux data on a half-hourly basis.

2.3.2 Generating robust and complete flux time series

Before the CO_2 flux time series were analysed, we applied three different processing techniques (u^* filtering, gap filling and partitioning) to (1) filter the NEE data for quality, (2) fill the NEE gaps and (3) separate NEE into GPP and R_{eco}. The identification of periods with insufficient turbulence conditions (indicated by low friction velocity u^*) is important for avoiding biases and uncertainties in EC fluxes. To control the data quality, the u^* thresholds were bootstrapped by identifying conditions with inadequate wind turbulence according to the method described in (Papale et al., 2006). We subsetted the data to similar environmental conditions, aside from friction velocity: 8 years and 7 temperature classes. Within each year/temperature subset the u^* threshold (5, 50 and 95 % of bootstrap) was estimated at 1000 samples per year. We used the subsequent gap filling and partitioning based on these different subsets to propagate the uncertainty of u^* threshold estimation across NEE, GPP and R_{eco}.

Our gap-filling method was similar to Falge et al. (2001), using the marginal distribution sampling (MDS) algorithm, re-adapted from Reichstein et al. (2005) in REddyProc (Reichstein et al., 2016). MDS takes into account similar meteorological data available with different window sizes (Moffat et al., 2007). Parallel to this approach, we also gap-filled the original EC NEE data with an independent AC NEE data set (2010–2013). AC data were collected simultaneously with EC data, and so we can used them as a cross check. The EC NEE was predicted from AC NEE based on linear regression models. The subsequent product was gap-filled using the MDS algorithm (REddyProc).

We separated NEE into its two main components (GPP and R_{eco}) using two approaches: (1) the REddyProc partitioning tool (Reichstein and Moffat, 2014) and (2) a light-response curve (LRC) approach (Lindroth et al., 2007; Lund et al., 2012). A brief description of each flux partitioning method is provided in the Supplement (Eq. S1). After the flux partitioning comparison, we used ReddyProc-based GPP and R_{eco} estimates on further analyses.

2.3.3 Flux uncertainties

In order to estimate the NEE gap-filling uncertainty, we assessed three different sources of uncertainty. First, we addressed the 95 % confidence interval of the EC prediction based on AC data. Second, we inferred the random uncertainty of filled half-hourly values from the spread of variables with otherwise very similar environmental conditions. REddyProc uses the gap filling to also estimate an observation uncertainty for the measured NEE, by temporarily introducing artificial gaps (T. Wutzler and M. Migliavacca (BGC-Jena), personal communication). Finally, we assessed the effect of uncertainty in the estimate of the u^* threshold. In the u^*-NEE relationship we want to exclude the probably false low fluxes (absolute NEE values) at low u^*. When choosing a lower u^* threshold, the associated lower flux will contribute to the gap filling and the annual sums. Therefore, there is a tendency of a lower absolute NEE associated with lower u^*. The difference between the 5 and 95 % of bootstrap provides a means of the uncertainties based on the u^* filters. We summed and propagated all these sources of uncertainties over time. The GPP and R_{eco} uncertainties include the bias from the one-to-one flux comparison obtained from each model. The micrometeorological sign convection used in this study present uptake fluxes (GPP) as negative, while the released fluxes (R_{eco}) are shown as positive.

2.4 Identifying environmental forcing

Snow- and phenology-related variables such as the end of the snowmelt period and the start, end and length of the growing season are important components that shape the Arctic CO_2 dynamics. In this study we defined the end of the snowmelt period as the day of year when more than 80 % of the surface of the fen was considered snow free; the threshold was chosen in agreement with suggestions previously reported in Hinkler et al. (2002) and Westergaard-Nielsen et al. (2015). For the start, end and length of the growing season (GS_{start}, GS_{end}, GS_{length}); the GS_{start} and the GS_{end} were defined as the first and last days on which the consecutive 3-day NEE average was negative (i.e. CO_2 uptake) and positive (i.e. CO_2 release) respectively (Aurela et al., 2004), while GS_{length} is the number of days between GS_{start} and GS_{end}).

A random forest machine-learning algorithm (Breiman, 2001; Pedregosa et al., 2011) was utilized in a data-mining exercise to identify how the environmental controls affect the variability of NEE, GPP and R_{eco}. Random forest calculates the relative importance of explanatory variables over the response variables. Here, we use photosynthetic active radiation (PAR), air temperature (T_{air}), precipitation (Prec) and vapour pressure deficit (VPD) to explain the response of C fluxes (NEE, GPP and R_{eco}) to climate variability. Each decision tree in the forest is trained on different random subset of the same training data set. The random forest is a classifier that groups explanatory variables and, in each fi-

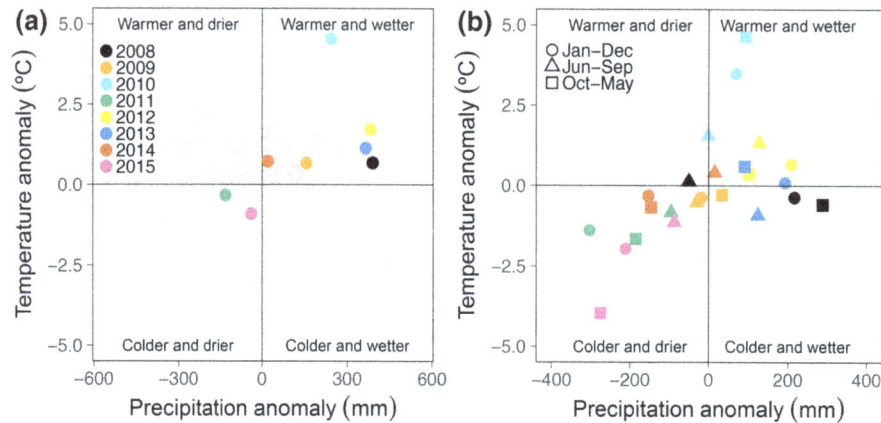

Figure 2. (a) Annual temperature (°C) and precipitation (mm) anomalies of the analysed years (2008–2015) compared to the 1866–2007 time series shown as empty circles (Cappelen, 2016), and **(b)** within the 2008–2015 period including annual (January to December), warm season (July to September) and cold season (October to May) averages.

nal cluster, a multiple linear regression is built to reproduce fluxes as function of driving factors. This approach has been used to extrapolate maps of biomass (Baccini et al., 2012; Exbrayat and Williams, 2015). This version of random forest sums the relative importance of each variable from 0 up to 100 %, which correspond to the fraction of decision in which a variable is involved to cluster the data. We applied random forests to assess the relative importance of PAR, T_{air}, Prec and VPD at different temporal scales (hourly, daily, weekly and monthly), aggregating them at the timescale indicated and lumping all the years together. (Table S1; Supplement). Moreover, we evaluated the diurnal, seasonal and annual pattern for each explanatory variable (data binned per hour; this is one random forest per hour of the day, day of the year and year respectively). To make sure that these results were not an artefact of the partitioning method that is based on a relationship between hourly R_{eco} and T_{air}, we performed the same analyses using daytime and night-time only hourly NEE as respective proxies for GPP and R_{eco}. Based on these results (Table S2) we concluded that the approach was robust for the Kobbefjord site.

3 Results

3.1 Interannual and seasonal variation of environmental and phenological variables

The annual mean temperature documented from Nuuk (−0.5 °C) and Kobbefjord (−0.4 °C) in the 2008–2015 period were generally warmer compared to the long time series between 1866 and 2007 (Cappelen, 2016; Fig. S1; Supplement), with an annual temperature average of −1.5 °C. The 2008–2015 period temperature also exhibited larger variability (coefficients of variation (CV) = 283.3 %) compared to the 1866–2007 period (CV = 79.3 %). The 2008–2015 mean annual temperature measured in Kobbefjord fluctuated be-

tween −1.7 °C in 2011 and 3.4 °C in 2010. Moreover, the mean annual precipitation documented from the nearby station of Nuuk (885 mm) and the one measured across the 8 year-study in Kobbefjord (862 mm) were both significantly higher than the 1931–2007 mean (689 mm), although less variable (CV = 30.8 and 24.5 %). Overall, 2008, 2009, 2010, 2012, 2013 and 2014 have shown warmer and wetter anomalies while 2011 and 2015 presented colder and drier anomalies compared to the long-term mean (Fig. 2a). Among the 8 study years (Fig. 2b), the temperature and precipitation anomalies in the warm season (June to September) ranged from about −1 °C (2011, 2013 and 2015) to +1.5 °C (2010) and −96 mm (2011) to about +125 mm (2012 and 2013). The cold season (October to May) anomalies have shown greater variability compared to the warm season, and 2010, 2012 and 2013 experienced warmer and wetter winters, while 2011 and 2015 were colder and drier.

The end of the snowmelt period and the growing season start and length presented high interannual variability (CVs were 9.5, 9.0 and 19.0 %). Kobbefjord became snow free on DOY 154 (3 June for non-leap years, SD = 15) on average. On average, the site switched from being a source of CO_2 to a sink (GS$_{start}$) on DOY 175 (24 June, SD = 20), and remained so (GS$_{end}$) until DOY 241 (29 July, SD = 8.4) (Table 1). The GS$_{start}$ and the GS$_{length}$ did not follow a consistent pattern among the analysed years, the growing season timing have fluctuated substantially. The high interannual variability of the GS$_{start}$ correlated with variations in temperature, end of snowmelt period and VPD ($p < 0.05$). The highest variability was observed during 2009–2012. The 2010's GS$_{length}$ was nearly twice as long as in 2011. Indeed, GS$_{start}$ in 2011 differs only by 26 days from the GS$_{end}$ in 2010.

Table 1. Summary of the phenology-related variables for the period 2008–2015.

	2008	2009	2010	2011	2012	2013	2014	2015
Maximum snow depth (m)	0.6	1.0	0.3	1.4	1.0	0.6	1.1	1.2
End of snowmelt period (DOY)	148	159	125	165	152	158	156	176
Beginning of growing season (DOY)	167	182	150	209	169	174	169	188
End of growing season (DOY)	230	249	235	256	247	237	–	246
Length of growing season (DOY)	63	67	85	47	78	63	–	58

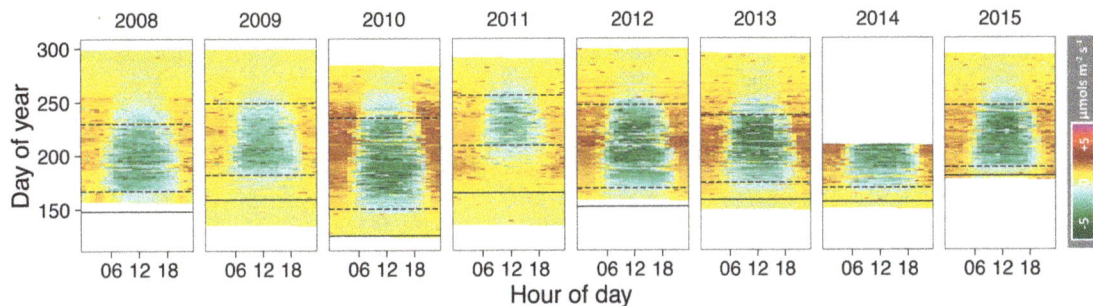

Figure 3. Time series of gap-filled NEE (2008–2015) based on autochamber data (2010–2013) and the MDS algorithm (from REddyProc). Green represents C uptake while the orange–dark-red denotes C release. The solid lines represent the end of the snowmelt period, while the area within the dashed lines represents the period between the start and the end of the growing season.

3.2 Data processing and quality

The NEE gap filling and subsequent partitioning obtained from different approaches exposed inconsistencies in performance and specific uncertainties in the seasonal C budget calculation. During the eight study snow-free periods, data gaps made up 46.5 % of the record from the EddyFen station due to unfavourable micro-meteorological conditions, instrument failures, maintenance and calibration (Jensen and Christensen, 2014) but also due to the rejection of low-quality flux measurements or too low u^*. In 2014 a major instrument failure forced the station to stop measurements in the middle of the season. In 2010 and 2012 there were two more interruptions in the measurements (data gaps of > 20 days), although the problems could be solved before the end of the season. Such prolonged gaps led to unreliable gap-filled NEE estimates. The REddyProc MDS algorithm tended to fill these large gaps with high peaks of respiration at noontime, coercing C uptake underestimation. For this reason, an independent AC NEE data set (2010–2013) was tested to gap-fill EC data (Figs. 3 and S2). The R^2 obtained from the EC-AC correlations was always > 0.70 (2010: $R^2 = 0.80$, $p < 0.001$; 2011: $R^2 = 0.72$, $p < 0.001$; 2012: $R^2 = 0.80$, $p < 0.001$; 2013: $R^2 = 0.84$, $p < 0.001$). By using AC data, the proportion of missing data was reduced to 28 %, and we found that the random uncertainty from the combination of AC and MDS algorithm decreased by 5 % on average. By using the u^* filtering and the AC data together with EC, there was an increase in ~ 6 % in terms of C sink strength. Moreover, the propagated uncertainty in

NEE never exceeded $\pm 1.8\,\mathrm{g\,C\,m^{-2}}$, mainly because the error related to u^* filtering was low. Further, we hypothesized that different flux partitioning approaches would lead to different estimates of GPP and R_{eco}. However, the results suggest a relatively good agreement (Fig. 4). There was a higher degree of agreement with regard to GPP ($R^2 = 0.83$) compared with R_{eco} ($R^2 = 0.30$). LRC tended to estimate 12 and 15 % larger GPP and R_{eco} respectively compared to REddyProc.

3.3 Interannual and seasonal variation of CO2 ecosystem fluxes

Overall, land–atmosphere CO_2 exchange measured for the snow-free periods of 2008–2015, omitting 2011, acted as a sink of CO_2, taking up $-30\,\mathrm{g\,C\,m^{-2}}$ on average (range -17 to $-41\,\mathrm{g\,C\,m^{-2}}$) (Fig. 5; Table 2). The cumulative NEE showed a characteristic pattern during the measurement period (Fig. 5), with an initial loss of carbon in early spring right after snowmelt (also observed in Fig. 3), followed by an intense C uptake as assimilation exceeded respiratory losses, triggered by increases in temperature, PAR and vegetation growth. This transition point matched the growing season start, when NEE switched from positive values (a net C source) to negative values (a net C sink). Eventually, the ecosystem turned again into a net C source, defining the growing season end. Even with high interannual variability in terms of the end of snowmelt time and growing season start/length (Table 1), the results do not show a marked meteorological effect on the NEE. The ranges in annual GPP (-182 to $-316\,\mathrm{g\,C\,m^{-2}}$) and R_{eco} (144–

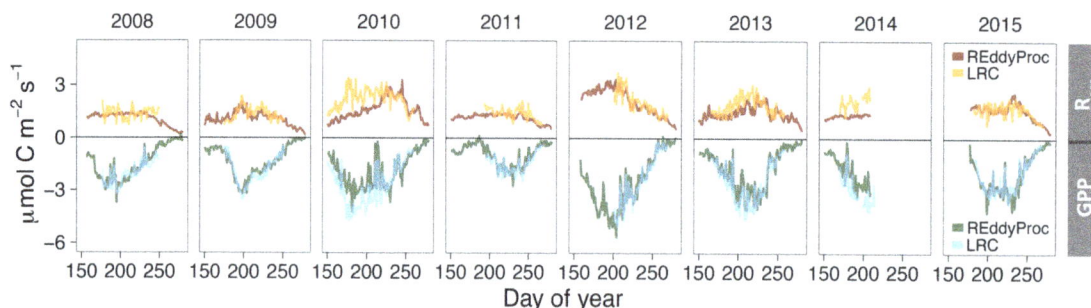

Figure 4. Time series of daily mean GPP (negative fluxes) and R_{eco} (positive fluxes) from 2008 to 2015 calculated by REddyProc (dark green and dark red) and LRC (orange and light blue).

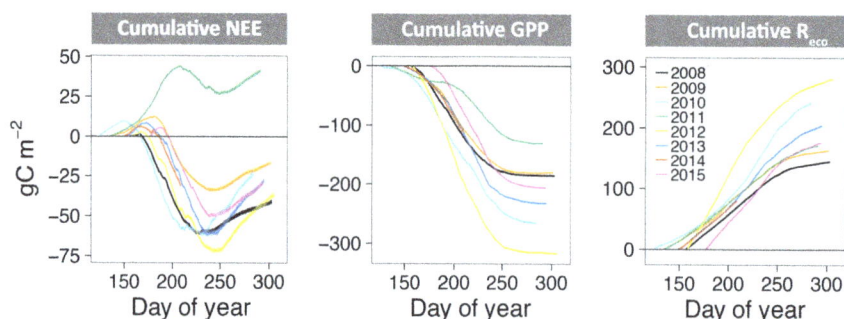

Figure 5. Cumulative NEE, GPP and R_{eco} from 2008 through 2015 including the u^* filtering and random errors.

$279 \, g \, C \, m^{-2}$) (Table 2) were > 5 fold larger and more variable (CVs are 3.6 and 4.1 % respectively) than for NEE (0.7 %). There was a tendency towards larger GPP and R_{eco} during warmer and wetter years (Fig. S3), but there were no warmer and drier years during the study period. The strongest growing season CO_2 uptake occurred in 2012 (NEE $= -74.2 \, g \, C \, m^{-2}$; $GS_{length} = 78$ days), followed by 2010 (NEE $= -70.0 \, g \, C \, m^{-2}$; $GS_{length} = 85$ days) (Tables 1 and 2). A lengthening of the growing season did not increase the net carbon uptake in this study. In other words, an earlier end of the snowmelt resulting in a longer growing season length did not lead to a stronger carbon sink.

The anomalous year, 2011, constituted a relatively strong source of CO_2 ($41 \, g \, C \, m^{-2}$) and was associated with a major pest outbreak, which reduced GPP more strongly than R_{eco}. Data on the larvae of the moth *Eurois occulta*, collected from pitfall traps in the surrounding *Salix*- and *Empetrum*-dominated plots, showed a strong peak at the beginning of the 2011 growing season (Lund et al., 2017), coinciding with high NEE and very low GPP (Fig. 4). In 2011 up to 2078 larvae were observed, while in other years only 14 (2008), 82 (2009), 186 (2010), 0 (2012) and 8 (2013) were observed. It is likely that the reduced primary production in the wetland area was a partial response to the *Eurois occulta* outbreak.

The daily aggregated NEE–GPP relationships displayed consistent linear correlation (2008–2015: $R^2 = 0.77$, $p < 0.001$) across the assessed years (Fig. 6a). The linear cor-

relations were weaker in 2010 and 2011. A hysteresis was detected in 2010 (i.e. long growing season with higher R_{eco} in autumn than in spring), while strong C releases were observed in 2011 across June and July. The relation between GPP and R_{eco}, which can be understood as the degree of coupling between inputs and outputs of C and therefore the degree of C sink strength, showed non-linear patterns (Fig. 6b). The curved behaviour is likely because GPP increased more than R_{eco} during early growing season, except for in 2011. Moreover, R_{eco} lagged behind GPP due to (1) the vegetation green-up in the first part of the growing season and (2) the higher respiration rates due to increased biomass in the second part. The years with clearer hysteresis coincide with the years with positive temperature anomalies (i.e. 2010, 2012 and 2013) of the 2008–2015 series. It is worth mentioning the different directions (clockwise or anticlockwise) in the hysteresis observed in these years between June, July and August. The data suggest that the clockwise 2012 hysteresis was due to greater gross C cycling (GPP and R_{eco}) in June and July favoured by warmer conditions, while in 2010 (anticlockwise hysteresis), the higher gross C fluxes were measured in August with warmer and wetter conditions (Fig. S4).

3.4 Environmental forcing

The varied importance of meteorological variables (such as PAR, T_{air}, VPD and precipitation) obtained from random forest at different temporal scales (hourly, daily, weekly and

Table 2. Summary of the measuring periods and the growing season CO_2 fluxes for the period 2008–2015.

	2008	2009	2010	2011	2012	2013	2014	2015
First measurement (DOY)	157	135	124	135	158	149	150	177
Last measurement (DOY)	303	304	282	287	305	295	209*	294
Missing data (%)	57.6	42.3	28.6	35.4	32.3	29.8	44.9*	40.0
NEE in measuring period (gCm^{-2})	−41.3	−16.9	−24.4	40.7	−37.0	−28.1	−28.7*	−31.5
	±1.4	±1.4	±1.9	±1.3	±1.8	±1.7	±1.1	±1.6
NEE in growing season (gCm^{-2})	−62.3	−45.9	−70.0	−16.2	−74.2	−69.7	−35.3[a]	−55.8
Maximum daily uptake (DOY)	195	205	182	230	204	220	192[a]	199
Maximum uptake ($\mu mols\ m^{-2}\ s^{-1}$)	−2.4	−1.7	−3.0	−1.4	−2.8	−2.5	−1.9[a]	−2.3
Estimated GPP (gCm^{-2})	−185.5	−181.8	−266.1	−130.6	−316.2	−230.7	−106.8*	−206.1
	±1.4	±1.4	±1.9	±1.3	±1.9	±1.7	±1.1	±1.6
Estimated R_{eco} (gCm^{-2})	144.2	164.9	241.6	171.3	279.2	202.6	78.1*	174.6
	±1.3	±1.3	±1.8	±1.2	±1.8	±1.7	±1.1	±1.5

Where applicable: ± sum of the autochamber, random and u^* filtering uncertainties. * incomplete growing season data set.

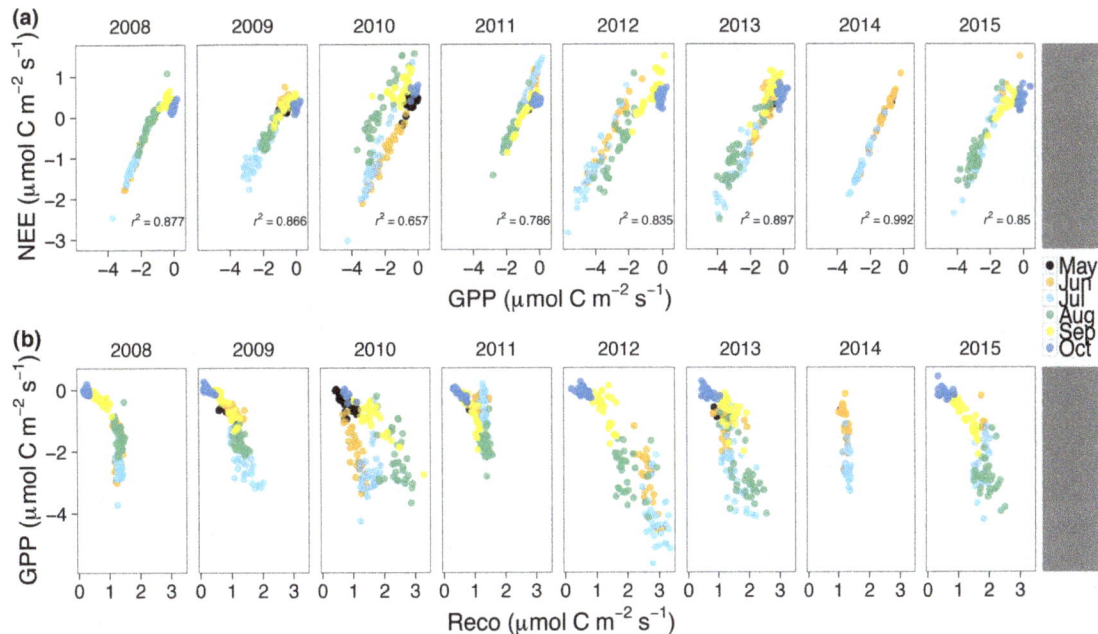

Figure 6. Interannual variability between (a) NEE–GPP and (b) GPP-R_{eco} relationships. The data were daily aggregated and coloured per month.

monthly) showed differences in behaviour depending on the time aggregation utilized (Fig. 7). PAR dominated NEE and GPP while T_{air} correlated the most with R_{eco} in hourly averages, whereas T_{air} became increasingly important at longer temporal aggregations for all the fluxes (Fig. 7). VPD and precipitation were not as important as the other variables while the use of water table depth in the analysis was discarded due to its very low impact on CO_2 fluxes. In general, NEE and GPP showed similar distributions of importance, reinforcing the linear relationships found between NEE and GPP (Fig. 6). The standard deviation of the variables' importance (across 1000 decision trees) tended to increase at coarser time aggregations.

Changes of environmental forcing (PAR, T_{air} and VPD) across diurnal, seasonal and annual timescales reveal patterns of functional responses to C fluxes. The diurnal cycle analyses on hourly data showed the changes in importance between day- and night-time (Fig. 8). NEE and GPP had two predominant variables (T_{air} and PAR) determining the variability at daytime. PAR was important at dawn (06:00 WGST) and dusk (20:00 WGST), while T_{air} was more important at other times. This performance indicates a threshold response to PAR, and a more continuous response to temperature. On the other hand, R_{eco} was mainly driven by T_{air} at both night-time and daytime. VPD and PAR had a negligible impact on R_{eco}. The seasonal pattern importance showed

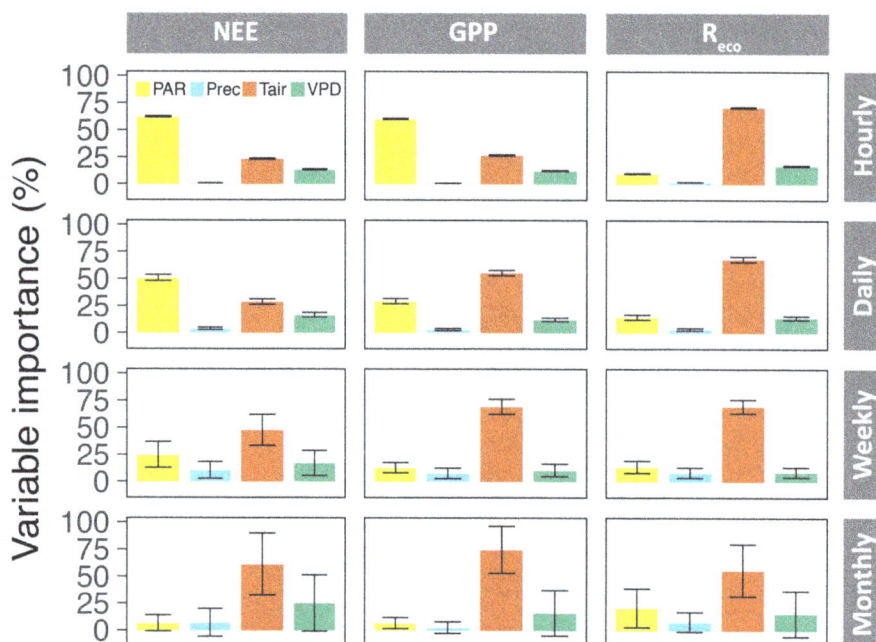

Figure 7. Importance of environmental variables PAR (yellow), T_{air} (orange), Prec (pink) and VPD (green), explaining variability in NEE, GPP and R_{eco} (partitioned by REddyproc) at different temporal aggregations (hourly, daily, weekly and monthly) when all the years were lumped together. Thick bars and error bars represent the mean ± standard deviation of the importance across 1000 decision trees.

PAR dominating NEE and GPP from early June to early October (Fig. 8), while T_{air} and VPD became more important before and after the snow-free conditions. In terms of CO_2 emission (R_{eco}) the pattern is less clear and noisier, although T_{air} appeared to be the most important variable. Finally, the annual pattern exposes a performance in line with previous results; i.e. PAR dominated NEE and GPP while R_{eco} was more sensitive to variations of T_{air}. Interestingly, the random forest analysis revealed a decrease in PAR's importance in 2011, the same year in which the sharp decrease in C sink strength was exposed.

4   Discussion

4.1   Data processing and quality

The NEE gap filling and subsequent partitioning into GPP and R_{eco} are needed to understand the CO_2 flux responses to the environmental forcing. However, these procedures expose unavoidable uncertainties in the seasonal C budget calculation (Table 2) and partial inconsistencies between approaches (Fig. 4). In this study, we used an MDS gap-filling technique, an enhancement to the standard look-up table. Both methods have shown a good overall performance compared to other procedures such as non-linear techniques or semi-parametric models but slightly inferior to artificial neural network (Moffat et al., 2007). However, the MDS gap filling alone introduced NEE estimates out of range across the

two extensive gaps in 2010 and 2012 (Fig. S2). Quantifying the uncertainty introduced by measurement gaps is complex (Falge et al., 2001; Moffat et al., 2007; Papale et al., 2006). One possibility would be a sensitivity analysis of time series with artificially introduced gaps (Dragomir et al., 2012; Pirk et al., 2017). However, the choice of gap length and position is difficult and would render uncertainty to the uncertainty assessment itself. Instead, we used the EC prediction based on independent autochamber (AC) measurements between 2010 and 2013. The agreement between EC and AC was always $R^2 > 0.72$ and $p < 0.001$, and the 95 % confidence intervals of the predictions were reported together with the resulting uncertainties (Table 2). Although the AC data itself incorporated a new source of uncertainty in the calculations, we consider this method to be less weak than an unreliable gap-filling estimate. We used the AC as platform with which to decrease the gap length and the total random uncertainty (Aurela et al., 2002) before the MDS algorithm was applied. AC was used together with MDS and was never used as an independent gap-filling procedure.

The NEE partitioning obtained from REddyProc and LRC suggests a relatively good agreement in model performance. The one-to-one comparison between different approaches found a better agreement with regard to GPP compared to R_{eco}. In this analysis, REddyProc produced smoother R_{eco} estimates compared to the noisier GPP estimates, whereas the LRC results were the other way around. This is mainly because measurement noise goes into GPP for the REddyProc method, and into R_{eco} for the LRC method. REd-

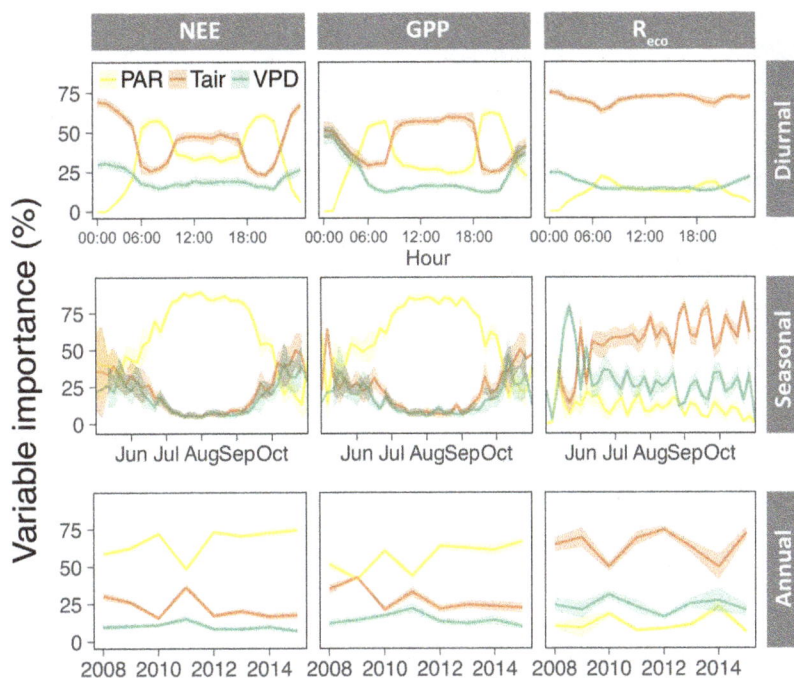

Figure 8. Diurnal, seasonal and annual importance of environmental variables PAR (yellow), T_{air} (orange), and VPD (green), explaining variability in NEE, GPP and R_{eco}. Thick lines and shading represent the mean $\pm$ standard deviation of the importance across 1000 decision trees.

dyProc retrieves positive GPP values, whereas the LRC method results in negative R_{eco} values. Both scenarios are not fully convincing, although it is not straightforward as to how they should be treated. Removing all positive GPP/negative R_{eco} values would risk removing only one side of the extremes. Besides night-time-based (REddyProc) and daytime-based (LRC) partitioning approaches, several implementations have been proposed to improve the algorithm's performance. Lasslop et al. (2010) has modified the hyperbolic LRC to account for the temperature sensitivity of respiration and the VPD limitation of photosynthesis. Further, Runkle et al. (2013) proposed a time-sensitive multi-bulk flux-partitioning model, where the NEE time series was analysed in 1-week increments as the combination of a temperature-dependent R_{eco} flux and a PAR-dependent flux (GPP). However, it remains uncertain as to under which circumstances each partitioning approach is more appropriate, especially in the boundaries between low- and high-Arctic due to the lack of dark night during polar days (when light is not a limiting factor for plant growth). Since there are few methods with an unclear precision, an evaluation study on the effect of using different partitioning approaches along latitudinal gradients would be very beneficial to assessing the suitability for each method.

4.2 Interannual and seasonal variation of CO_2 ecosystem fluxes

The balance between the two major gross fluxes in terrestrial ecosystems, photosynthetic inputs (GPP) and respiration outputs (R_{eco}), displayed larger temporal variability than did NEE. These results suggest that both GPP and R_{eco} were strongly coupled and sensitive to meteorological conditions such as insolation and temperature (Figs. 7 and 8). Interestingly, the tendency to warmer and wetter conditions led to greater rates of C cycling associated with larger GPP and R_{eco} (Fig. S3). This result does not entirely coincide with Peichl et al. (2014), even though they performed a similar analysis for a Swedish boreal fen. This finding points towards the complexity in the response of wetland ecosystems towards changing environmental conditions. The response is dependent on many things, such as hydrological settings, and these differ between sites. In this study, larger rates of C uptake (GPP) were linked to larger rates of C release (R_{eco}), with the exception of the anomalous year 2011. The relative insensitivity of NEE to meteorological conditions during the snow-free period could be the result of the correlated response of ranked cumulative GPP and R_{eco} (Fig. 5) (Richardson et al., 2007; Wohlfahrt et al., 2008). This site likely receives more precipitation than many other tundra ecosystems and has no permafrost; thus the NEE response to climate could be less variable. However, as Kobbefjord is located in a coastal area, it is not surprising that it receives high precipitation, and other ecosystems such as coastal blanket bogs often receive even more precipitation without a clear impact of drought effect on the NEE sensitivity (Lund et al., 2015). Furthermore, permafrost adds another layer of complexity to the C dynamics (Christensen et al., 2004; Koven et al., 2011; Schuur et al., 2015). Although some studies showed similarities of CO_2 fluxes in various northern wetland ecosystems with and without permafrost (Lund et al., 2015), permafrost has a strong influence on the hydrology of peatlands (Åkerman and Johansson, 2008), and therefore their topography and distribution of vegetation (Johansson et al., 2013). Especially in the context of climate warming permafrost thaw can cause large changes to the ecosystems. Further, this study agrees with Parmentier et al. (2011) and Lund et al. (2012), who suggested that a longer growing season does not necessarily increase the net carbon uptake. Here a more negative NEE indicated a stronger C sink (i.e.) in 2012 compared to 2010. Parmentier et al. (2011) hypothesized that this behaviour is due to site-specific differences, such as meteorology and soil structure, and that changes in the carbon cycle with longer growing seasons will not be uniform around the Arctic. Thus, the effects of climate change on the tundra C balance of are not straightforward to infer.

NEE measured at Kobbefjord from 2008 to 2015 indicates a consistent sink of CO_2 (within a range of -17 to $-41\,\mathrm{g\,C\,m^{-2}}$) with exception of the year 2011 ($+41\,\mathrm{g\,C\,m^{-2}}$) (Table 2). The year 2011, associated with a major pest outbreak, reduced GPP more strongly than R_{eco} (Fig. 5) and Kobbefjord turned into a strong C source within an episodic single growing season. The return to substantial cumulative CO_2 sink rates following the extreme year of 2011 shows the ability of the ecosystem to recover from the disturbance (Lund et al., 2017). Indeed, the ecosystem not only shifted back from being a C source to a C sink, but it also changed rapidly from one year to the next. Thus we found evidence in Kobbefjord of ecosystem resilience to the meteorological variability, similar to other cases described in other northern sites (Peichl et al., 2014; Zona et al., 2014). Only a few reference sites have reported similar decreases in net C uptake, but in no case as large as the one observed here. Zona et al. (2014) described an effect of delayed responses to an unusual warm summer in Alaska. Their results suggested that vascular plants, which have enhanced their physiological activity during the warmer summer, might have difficulties readapting to cooler, but not atypical, conditions, which have provoked a significant decrease in GPP and R_{eco} the following year. In their study, the ecosystem returned to be a fairly strong C sink after 2 years, suggesting strong ecosystem resilience. Moreover, Hanis et al., 2015 have reported comparable C sink–C source variations in a Canadian fen within the growing season due to changes in the water table depth. Drier and warmer than normal conditions have triggered an increase in C source strength. Finally, during an extensive outbreak of autumn and winter moths in a subarctic birch forest in Sweden, Heliasz et al. (2011) observed

a similar decrease in net sink of C (most likely due to weaker GPP) across the growing season. However, the C source strength (NEE = $40.7\,\mathrm{g\,C\,m^{-2}}$) found in 2011 at Kobbefjord was higher compared to these other cases. To our knowledge, such abrupt disturbance concerning C sink strength in Arctic tundra has not been previously reported, excluding severely burned landscapes (Rocha and Shaver, 2011).

A combination of different factors could have led to the sharp change in C balance observed between 2010 and 2011, both physical and biological. The year 2010 had the warmest mean annual temperature (3.4 °C compared to the -0.4 °C mean annual temperature for 2008–2015) and the warmest mean wintertime temperature (-2.7 °C compared to the -6.79 °C mean for 2008–2015) (Fig. 2a). These climatic conditions generated the thinnest (maximum daily snow depth of 0.3 m compared to an average of 0.9 m) (Table 1) and shortest-lasting snowpack. Consequently, 2010 had the longest growing season (85 days) and very high growing season C uptake ($-70\,\mathrm{g\,C\,m^{-2}}$). Increases in temperature can lead to high respiration rates during early winter (Commane et al., 2017; Zona et al., 2016) but also during the following summer (Helfter et al., 2015; Lund et al., 2012), which is related to soil temperature and snow dynamics. Further, in Kobbefjord the year 2011 had one of the lowest mean annual temperatures and mean wintertime temperatures (-1.7 and -6.1 °C respectively), which created the thickest (maximum daily snow depth of 1.4 m) and the longest-lasting snowpack, leading to the shortest growing season for the study period (only 47 days). According to Lund et al. (2012), soils will be insulated from low temperatures when below thick snowpack, which acts as a lid and prevents R_{eco} from being released to the atmosphere until the snowmelt period. Finally, an outbreak of larvae of the noctuid moth *Eurois occulta* occurred in 2011, overlapping the observed abrupt decrease in C sink strength. Although we cannot provide a quantification of change attributed to meteorological variations and biological disturbances, there is evidence showing that the moth outbreak could partially have decreased the C sink strength in Kobbefjord. In an undisturbed scenario, the meteorological conditions in 2015, colder and drier than the mean 2008–2015 period (Fig. 2) but similar to 2011, would have stimulated similar behaviours in terms of C fluxes. However, the cumulative fluxes in 2015 (Fig. 5) followed analogous patterns compared to other years. This evidence agrees with the literature (Callaghan et al., 2012b; Lund et al., 2017) on the fact that tundra systems can fluctuate in sink strength influenced by factors such as episodic disturbances or species shifts, events which are very difficult to predict.

4.3 Environmental forcing

Our data indicate that the importance of the main environmental controls (radiation and temperature) for C fluxes did vary across diurnal, seasonal and annual cycles, but also between time aggregations. The hourly variability of NEE and

GPP (Figs. 7 and 8) was mostly dependent on PAR because of the threshold nature on radiation control on GPP. Overall, the results indicate that environmental factors that can change rapidly such as PAR will have a high influence on short timescales (Stoy et al., 2014). The increased importance of PAR at 08:00 and 20:00 h WGST coincides with the sharp gradient in light at dawn and dusk (Fig. 8). The control of PAR on GPP is not a new finding itself, but the random forest approach helps to quantify its importance. There is no GPP at night, and therefore there will be a strong increase/decrease in GPP at dawn/dusk. The seasonal pattern also showed that radiation is the single main driver for NEE and GPP between early June and early October, supported by the longer daytime. Further, PAR appeared to be a limiting factor for annual NEE in 2011, increasing further the complexity around this anomalous year. These results agree with the literature (Groendahl et al., 2007; Stoy et al., 2014), suggesting that the uptake of CO_2 is partially controlled by radiation for the photosynthetic physiology at the leaf scale. Arctic plants are usually well adapted to environments with low light levels, reporting near-maximum rates ranging from 10 to 25 °C (Oechel and Billings, 1992; Shaver and Kummerow, 1992).

Photosynthesis is restricted by low temperature, so enzymatically driven processes such as carbon fixation are more sensitive to low temperature than the light-driven biophysical reactions (Chapin et al., 2011). In this paper the daily, weekly and monthly aggregated variability of C fluxes was primarily linked to T_{air}. Moreover, the random forest analyses revealed a strong diurnal pattern with a marked contribution of T_{air} to variations in NEE and GPP (both at night-time and between 08:00 and 18:00 h WGST). These results agree with Lindroth et al. (2007), who recognized T_{air} as the key driver of NEE seasonal trends in northern peatlands. However, in this analysis both NEE and GPP had similar responses to common environmental forcing, contrary to the results in Reichstein et al. (2007). In order to circumvent the potential circularity conflicts based on the use of partitioning products, we filtered daytime NEE (true GPP) and night-time NEE (true R_{eco}), obtaining very similar results (Table S2). Further, our data also suggest that R_{eco} is often dominated by air temperature. The patterns observed here are in agreement with findings on plant respiration dynamics (Heskel et al., 2016; Lloyd and Taylor, 1994; Tjoelker et al., 2001).

In this study, environmental drivers related to water availability such as VPD and precipitation were not found to be as influential as other assessed variables. We did not find significant relationships between CO_2 fluxes and the water table depth. Thus, there was no apparent water limitation on carbon dynamics during the 8-year period. However, the complex interactions based on changes in temperature and soil moisture particularly over full annual cycles and for sites with permafrost, should be further explored. Our results contrast with Strachan et al. (2015), who described water table depth as an important driver regulating the CO_2 balance, and

others, who found that CO_2 emissions increase during dry years due to increased decomposition rates and a reduction in GPP (Aurela et al., 2007; Lund et al., 2007; Oechel et al., 1993; Peichl et al., 2014), whereas other sites act as sinks during relatively wet years (Lafleur et al., 1997). The fen in Kobbefjord is probably quite resistant to droughts since it is fed with water from the surroundings.

5 Conclusions

We have analysed eight snow-free periods in 8 consecutive years in a West Greenland tundra (64° N) focusing on the net ecosystem exchange (NEE) of CO_2 and its photosynthetic inputs (GPP) and respiration outputs (R_{eco}). Here, the NEE gap filling exposed inherent uncertainties in the seasonal C budget calculation, but there were also inconsistencies between the flux partitioning approaches used. We find that Kobbefjord acted as a consistent sink of CO_2 during the years 2008–2015, except 2011, which was associated with a major pest outbreak. The results do not show a marked meteorological effect on the net C uptake. However, the relative insensitivity of NEE during the snow-free period was driven by the correlated, balancing responses of GPP and R_{eco}, both more variable than NEE and sensitive to temperature and insolation. In this paper we show a tendency towards larger GPP and R_{eco} during wetter and warmer years. The anomalous year 2011, affected by a biological disturbance, constituted a relatively strong source of CO_2 and reduced GPP more strongly than R_{eco}. A novel analysis assessing the changes of environmental forcing across diurnal, seasonal and annual timescales unmasked patterns of functional responses to C fluxes.

Despite the fact that we analysed an 8-year data set, the results do not provide a complete picture due to the lack of year-round data (Grøndahl et al., 2008). The snow season should be taken into account for a comprehensive understanding of complete C budget (Aurela et al., 2002; Commane et al., 2017; Zona et al., 2016) and the delayed effect of wintertime-based variables such as snow depth or snow cover on the C fluxes. Because some studies have suggested that GPP and R_{eco} increase with observed changes in climate and NEE trends remain unclear (Lund et al., 2012), it is challenging to produce strong evidence while the data remains scarce and fragmented. Hence, there is a need for increased efforts in monitoring Arctic ecosystem changes over the full annual cycle (Euskirchen et al., 2012; Grøndahl et al., 2008). Future work is also required with C flux modelling in order to explore process-based insights of C exchange balance in the Arctic tundra and the interactions of photosynthesis and R_{eco} with changes in C stocks.

Author contributions. EL-B, ML, MW, TRC and MPT designed the experiment. Data preparation and analysis were primarily performed by EL-B with contribution from ML (eddy covariance data processing, data quality control and LRC partitioning), MW and TRC (experimental set up), BUH (data gathering from Nuuk Ecological Research Operations, GeoBasis), AW-N (daily estimate of the timing of snowmelt) and J-FE (random forest approach). EL-B prepared the manuscript with active contributions from all co-authors.

Competing interests. The authors declare that they have no conflict of interest.

Acknowledgements. This work was supported in part by a scholarship from the Aarhus-Edinburgh Excellence in European Doctoral Education Project and by the eSTICC (eScience tools for investigating Climate Change in Northern High Latitudes) project, part of the Nordic Center of Excellence. The authors wish to thank the Nuuk Ecological Research Operations (nuuk-basic.dk) as well as GeoBasis programme, which is in charge of the eddy covariance and the autochamber systems. Both projects are being run under the Greenland Ecosystem Monitoring (GEM) programme funded by the Danish Environmental Protection Agency and the Danish Energy Agency.

Edited by: Xinming Wang

References

ACIA: Arctic Climate Impact Assessment, Cambridge University Press, New York, 1042 pp., 2005.

Åkerman, H. J. and Johansson, M.: Thawing permafrost and thicker active layers in sub-arctic Sweden, Permafrost Periglac., 19, 279–292, https://doi.org/10.1002/ppp.626, 2008.

AMAP: Snow, Water, Ice and Permafrost in the Arctic (SWIPA), Oslo, Norway, 2011.

Arndal, M. F., Illeris, L., Michelsen, A., Albert, K., Tamstorf, M., and Hansen, B. U.: Seasonal Variation in Gross Ecosystem Production, Plant Biomass, and Carbon and Nitrogen Pools in Five High Arctic Vegetation Types, Arct. Antarct. Alp. Res., 41, 164–173, https://doi.org/10.1657/1938-4246-41.2.164, 2009.

Aurela, M., Laurila, T., and Tuovinen, J.-P.: Annual CO_2 balance of a subarctic fen in northern Europe: importance of the wintertime efflux, J. Geophys. Res.-Atmos., 107, ACH 17-11–ACH 17-12, https://doi.org/10.1029/2002JD002055, 2002.

Aurela, M., Laurila, T., and Tuovinen, J.-P.: The timing of snow melt controls the annual CO_2 balance in a subarctic fen, Geophys. Res. Lett., 31, L16119, https://doi.org/10.1029/2004GL020315, 2004.

Aurela, M., Riutta, T., Laurila, T., Tuovinen, J.-P., Vesala, T., Tuittila, E.-S., Rinne, J., Haapanala, S., and Laine, J.: CO_2 exchange of a sedge fen in southern Finland – the impact of a drought period, Tellus B, 59, 826–837, https://doi.org/10.1111/j.1600-0889.2007.00309.x, 2007.

Baccini, A., Goetz, S. J., Walker, W. S., Laporte, N. T., Sun, M., Sulla-Menashe, D., Hackler, J., Beck, P. S. A., Dubayah, R., Friedl, M. A., Samanta, S., and Houghton, R. A.: Estimated carbon dioxide emissions from tropical deforestation improved by carbon-density maps, Nat. Clim. Change, 2, 182–185, 2012.

Bay, C., Aastrup, P., and Nymand, J.: The NERO line. A vegetation transect in Kobbefjord, West Greenland, National Environmental Research Institute, Aarhus University, Aarhus, 2008.

Breiman, L.: Random forests, Mach. Learn., 45, 5–32, https://doi.org/10.1023/a:1010933404324, 2001.

Bubier, L. J., Bhatia, G., Moore, R. T., Roulet, T. N., and Lafleur, M. P.: Spatial and temporal variability in growing-season net ecosystem carbon dioxide exchange at a large peatland in Ontario, Canada, Ecosystems, 6, 353–367, https://doi.org/10.1007/s10021-003-0125-0, 2003.

Callaghan, T. V., Johansson, M., Brown, R. D., Groisman, P. Y., Labba, N., Radionov, V., Barry, R. G., Bulygina, O. N., Essery, R. L. H., Frolov, D. M., Golubev, V. N., Grenfell, T. C., Petrushina, M. N., Razuvaev, V. N., Robinson, D. A., Romanov, P., Shindell, D., Shmakin, A. B., Sokratov, S. A., Warren, S., and Yang, D.: The changing face of arctic snow cover: a synthesis of observed and projected changes, AMBIO, 40, 17–31, https://doi.org/10.1007/s13280-011-0212-y, 2012a.

Callaghan, T. V., Johansson, M., Key, J., Prowse, T., Ananicheva, M., and Klepikov, A.: Feedbacks and interactions: from the arctic cryosphere to the climate system, AMBIO, 40, 75–86, https://doi.org/10.1007/s13280-011-0215-8, 2012b.

Callaghan, T. V., Johansson, M., Prowse, T. D., Olsen, M. S., and Reiersen, L.-O.: Arctic cryosphere: changes and Impacts, AMBIO, 40, 3–5, https://doi.org/10.1007/s13280-011-0210-0, 2012c.

Cappelen, J.: Greenland – DMI Historical Climate Data Collection 1873–2012, Danish Meteorological Institute, Copenhagen, 2013.

Cappelen, J.: Greenland – DMI Historical Climate Data Collection 1784–2015, Danish Meteorological Institute, Copenhagen, 2016.

CAVM Team: Circumpolar Arctic Vegetation Map, scale 1 : 7500000, in: Conservation of Arctic Flora and Fauna (CAFF) Map No. 1, US Fish and Wildlife Service, Anchorage, Alaska, 2003.

Chapin, F. S., Matson, P. A., and Vitousek, P. M.: Principles of Terrestrial Ecosystem Ecology, Springer, New York, 2011.

Christensen, J. H., Hewitson, B., Busuioc, A., Chen, A., Gao, X., Held, I., Jones, R., Kolli, R. K., Kwon, W.-T., Laprise, R., Magaña Rueda, V., Mearns, L., Menéndez, C. G., Räisänen, J., Rinke, A., Sarr, A., and Whetton, P.: Regional Climate Projections, in: Climate Change 2007: Climate Change 2007: The Physical Science Basis: Working Group I Contribution to the Fourth Assessment Report of the IPCC, edited by: Solomon, S., Qin, D., Manning, M., Chen, Z., Marquis, M., Averyt, K. B., Tignor, M., and Miller, H. L., Cambridge University Press, New York, 2007.

Christensen, T. R., Johansson, T., Åkerman, H. J., Mastepanov, M., Malmer, N., Friborg, T., Crill, P., and Svensson, B.: Thawing sub-arctic permafrost: effects on vegetation and methane emissions, Geophys. Res. Lett., 31, L04501, https://doi.org/10.1029/2003GL018680, 2004.

Christensen, T. R., Jackowicz-Korczyński, M., Aurela, M., Crill, P., Heliasz, M., Mastepanov, M., and Friborg, T.: Monitoring the multi-year carbon balance of a subarctic palsa mire with micrometeorological techniques, AMBIO, 41, 207–217, https://doi.org/10.1007/s13280-012-0302-5, 2012.

Commane, R., Lindaas, J., Benmergui, J., Luus, K. A., Chang, R. Y.-W., Daube, B. C., Euskirchen, E. S., Henderson, J. M., Karion, A., Miller, J. B., Miller, S. M., Parazoo, N. C.,

Randerson, J. T., Sweeney, C., Tans, P., Thoning, K., Ver-averbeke, S., Miller, C. E., and Wofsy, S. C.: Carbon dioxide sources from Alaska driven by increasing early winter respiration from Arctic tundra, P. Natl. Acad. Sci. USA, 114, 5361–5366, https://doi.org/10.1073/pnas.1618567114, 2017.

Dragomir, C. M., Klaassen, W., Voiculescu, M., Georgescu, L. P., and van der Laan, S.: Estimating annual CO_2 flux for Lutjewad Station using three different gap-filling techniques, Sci. World J., 2012, 10, https://doi.org/10.1100/2012/842893, 2012.

Elberling, B., Nordstrøm, C., Grøndahl, L., Søgaard, H., Friborg, T., Christensen, T. R., Ström, L., Marchand, F., and Nijs, I.: High-arctic soil CO_2 and CH_4 production controlled by temperature, water, freezing and snow, in: Advances in Ecological Research, Academic Press, 441–472, 2008.

Euskirchen, E. S., Bret-Harte, M. S., Scott, G. J., Edgar, C., and Shaver, G. R.: Seasonal patterns of carbon dioxide and water fluxes in three representative tundra ecosystems in northern Alaska, Ecosphere, 3, 1–19, https://doi.org/10.1890/ES11-00202.1, 2012.

Exbrayat, J.-F. and Williams, M.: Quantifying the net contribution of the historical Amazonian deforestation to climate change, Geophys. Res. Lett., 42, 2968–2976, https://doi.org/10.1002/2015GL063497, 2015.

Falge, E., Baldocchi, D., Olson, R., Anthoni, P., Aubinet, M., Bernhofer, C., Burba, G., Ceulemans, R., Clement, R., Dolman, H., Granier, A., Gross, P., Grünwald, T., Hollinger, D., Jensen, N.-O., Katul, G., Keronen, P., Kowalski, A., Lai, C. T., Law, B. E., Meyers, T., Moncrieff, J., Moors, E., Munger, J. W., Pilegaard, K., Rannik, Ü., Rebmann, C., Suyker, A., Tenhunen, J., Tu, K., Verma, S., Vesala, T., Wilson, K., and Wofsy, S.: Gap filling strategies for defensible annual sums of net ecosystem exchange, Agr. Forest Meteorol., 107, 43–69, https://doi.org/10.1016/S0168-1923(00)00225-2, 2001.

Gilmanov, T. G., Verma, S. B., Sims, P. L., Meyers, T. P., Bradford, J. A., Burba, G. G., and Suyker, A. E.: Gross primary production and light response parameters of four Southern Plains ecosystems estimated using long-term CO_2-flux tower measurements, Global Biogeochem. Cy., 17, https://doi.org/10.1029/2002GB002023, 2003.

Goulden, M. L. and Crill, P. M.: Automated measurements of CO_2 exchange at the moss surface of a black spruce forest, Tree Physiol., 17, 537–542, https://doi.org/10.1093/treephys/17.8-9.537, 1997.

Groendahl, L., Friborg, T., and Soegaard, H.: Temperature and snow-melt controls on interannual variability in carbon exchange in the high Arctic, Theor. Appl. Climatol., 88, 111–125, https://doi.org/10.1007/s00704-005-0228-y, 2007.

Grolemund, G. and Wickham, H.: Dates and Times Made Easy with lubridate, J. Stat. Softw., 40, 1–25, 2011.

Grøndahl, L., Friborg, T., Christensen, T. R., Ekberg, A., Elberling, B., Illeris, L., Nordstrøm, C., Rennermalm, Å., Sigsgaard, C., and Søgaard, H.: Spatial and inter-annual variability of trace gas fluxes in a heterogeneous high-arctic landscape, in: Advances in Ecological Research, edited by: Meltofte, H., Christensen, T. R., Elberling, B., Forchhammer, M. C., and Rasch, M., Academic Press, 473–498, 2008.

Hanis, K. L., Amiro, B. D., Tenuta, M., Papakyriakou, T., and Swystun, K. A.: Carbon exchange over four growing seasons for a subarctic sedge fen in northern Manitoba, Canada, Arctic Sci., 1, 27–44, https://doi.org/10.1139/as-2015-0003, 2015.

Helfter, C., Campbell, C., Dinsmore, K. J., Drewer, J., Coyle, M., Anderson, M., Skiba, U., Nemitz, E., Billett, M. F., and Sutton, M. A.: Drivers of long-term variability in CO_2 net ecosystem exchange in a temperate peatland, Biogeosciences, 12, 1799–1811, https://doi.org/10.5194/bg-12-1799-2015, 2015.

Heliasz, M., Johansson, T., Lindroth, A., Mölder, M., Mastepanov, M., Friborg, T., Callaghan, T. V., and Christensen, T. R.: Quantification of C uptake in subarctic birch forest after setback by an extreme insect outbreak, Geophys. Res. Lett., 38, L01704, https://doi.org/10.1029/2010GL044733, 2011.

Heskel, M. A., O'Sullivan, O. S., Reich, P. B., Tjoelker, M. G., Weerasinghe, L. K., Penillard, A., Egerton, J. J. G., Creek, D., Bloomfield, K. J., Xiang, J., Sinca, F., Stangl, Z. R., Martinez-de la Torre, A., Griffin, K. L., Huntingford, C., Hurry, V., Meir, P., Turnbull, M. H., and Atkin, O. K.: Convergence in the temperature response of leaf respiration across biomes and plant functional types, P. Natl. Acad. Sci. USA, 113, 3832–3837, https://doi.org/10.1073/pnas.1520282113, 2016.

Hinkler, J., Pedersen, S. B., Rasch, M., and Hansen, B. U.: Automatic snow cover monitoring at high temporal and spatial resolution, using images taken by a standard digital camera, Int. J. Remote Sens., 23, 4669–4682, https://doi.org/10.1080/01431160110113881, 2002.

Højstrup, J.: A statistical data screening procedure, Meas. Sci. Technol., 4, 153–157, 1993.

Hugelius, G., Strauss, J., Zubrzycki, S., Harden, J. W., Schuur, E. A. G., Ping, C.-L., Schirrmeister, L., Grosse, G., Michaelson, G. J., Koven, C. D., O'Donnell, J. A., Elberling, B., Mishra, U., Camill, P., Yu, Z., Palmtag, J., and Kuhry, P.: Estimated stocks of circumpolar permafrost carbon with quantified uncertainty ranges and identified data gaps, Biogeosciences, 11, 6573–6593, https://doi.org/10.5194/bg-11-6573-2014, 2014.

IPCC: Climate Change 2007: The Physical Science Basis: Working Group I Contribution to the Fourth Assessment Report of the IPCC, Cambridge University Press, New York, 2007.

Jensen, L. M. and Christensen, T. R.: Nuuk Ecological Research Operations, 7th Annual Report 2013, Aarhus University, DCE-Danish Centre for Environment and Energy, Roskilde, Denmark, 94, 2014.

Johansson, M., Callaghan, T. V., Bosiö, J., Åkerman, H. J., Jackowicz-Korczynski, M., and Christensen, T. R.: Rapid responses of permafrost and vegetation to experimentally increased snow cover in sub-arctic Sweden, Environ. Res. Lett., 8, 035025, https://doi.org/10.1088/1748-9326/8/3/035025, 2013.

Koven, C. D., Ringeval, B., Friedlingstein, P., Ciais, P., Cadule, P., Khvorostyanov, D., Krinner, G., and Tarnocai, C.: Permafrost carbon-climate feedbacks accelerate global warming, P. Natl. Acad. Sci. USA, 108, 14769–14774, https://doi.org/10.1073/pnas.1103910108, 2011.

Lafleur, P. M., McCaughey, J. H., Joiner, D. W., Bartlett, P. A., and Jelinski, D. E.: Seasonal trends in energy, water, and carbon dioxide fluxes at a northern boreal wetland, J. Geophys. Res.-Atmos., 102, 29009–29020, https://doi.org/10.1029/96JD03326, 1997.

Lafleur, P. M., Humphreys, E. R., St. Louis, V. L., Myklebust, M. C., Papakyriakou, T., Poissant, L., Barker, J. D., Pilote, M., and Swystun, K. A.: Variation in Peak Growing Season Net Ecosystem Production Across the Canadian Arctic, Environ. Sci. Technol., 46, 7971–7977, https://doi.org/10.1021/es300500m, 2012.

Lasslop, G., Reichstein, M., Papale, D., Richardson, A. D., Arneth, A., Barr, A., Stoy, P., and Wohlfahrt, G.: Separation of net ecosystem exchange into assimilation and respiration using a light response curve approach: critical issues and global evaluation, Glob. Change Biol., 16, 187–208, https://doi.org/10.1111/j.1365-2486.2009.02041.x, 2010.

Lasslop, G., Migliavacca, M., Bohrer, G., Reichstein, M., Bahn, M., Ibrom, A., Jacobs, C., Kolari, P., Papale, D., Vesala, T., Wohlfahrt, G., and Cescatti, A.: On the choice of the driving temperature for eddy-covariance carbon dioxide flux partitioning, Biogeosciences, 9, 5243–5259, https://doi.org/10.5194/bg-9-5243-2012, 2012.

Lindroth, A., Lund, M., Nilsson, M., Aurela, M., Christensen, T. R., Laurila, T., Rinne, J., Riutta, T., Sagerfors, J., Ström, L., Tuovinen, J.-P., and Vesala, T.: Environmental controls on the CO_2 exchange in north European mires, Tellus B, 59, 812–825, https://doi.org/10.1111/j.1600-0889.2007.00310.x, 2007.

Lloyd, J. and Taylor, J. A.: On the temperature dependence of soil respiration, Funct. Ecol., 8, 315–323, https://doi.org/10.2307/2389824, 1994.

Lund, M., Lindroth, A., Christensen, T. R., and Ström, L.: Annual CO_2 balance of a temperate bog, Tellus B, 59, 804–811, https://doi.org/10.1111/j.1600-0889.2007.00303.x, 2007.

Lund, M., Lafleur, P. M., Roulet, N. T., Lindroth, A., Christensen, T. R., Aurela, M., Chojnicki, B. H., Flanagan, L. B., Humphreys, E. R., Laurila, T., Oechel, W. C., Olejnik, J., Rinne, J., Schubert, P. E. R., and Nilsson, M. B.: Variability in exchange of CO_2 across 12 northern peatland and tundra sites, Glob. Change Biol., 16, 2436–2448, https://doi.org/10.1111/j.1365-2486.2009.02104.x, 2010.

Lund, M., Falk, J. M., Friborg, T., Mbufong, H. N., Sigsgaard, C., Soegaard, H., and Tamstorf, M. P.: Trends in CO_2 exchange in a high Arctic tundra heath, 2000–2010, J. Geophys. Res.-Biogeo., 117, G02001, https://doi.org/10.1029/2011JG001901, 2012.

Lund, M., Bjerke, J. W., Drake, B. G., Engelsen, O., Hansen, G. H., Parmentier, F. J. W., Powell, T. L., Silvennoinen, H., Sottocornola, M., Tømmervik, H., Weldon, S., and Rasse, D. P.: Low impact of dry conditions on the CO_2 exchange of a Northern-Norwegian blanket bog, Environ. Res. Lett., 10, 025004, https://doi.org/10.1088/1748-9326/10/2/025004, 2015.

Lund, M., Raundrup, K., Westergaard-Nielsen, A., López-Blanco, E., Nymand, J., and Aastrup, P.: Larval outbreaks in West Greenland: instant and subsequent effects on tundra ecosystem productivity and CO_2 exchange, AMBIO, 46, 26–38, https://doi.org/10.1007/s13280-016-0863-9, 2017.

Mastepanov, M., Sigsgaard, C., Tagesson, T., Ström, L., Tamstorf, M. P., Lund, M., and Christensen, T. R.: Revisiting factors controlling methane emissions from high-Arctic tundra, Biogeosciences, 10, 5139–5158, https://doi.org/10.5194/bg-10-5139-2013, 2013.

Mbufong, H. N., Lund, M., Aurela, M., Christensen, T. R., Eugster, W., Friborg, T., Hansen, B. U., Humphreys, E. R., Jackowicz-Korczynski, M., Kutzbach, L., Lafleur, P. M., Oechel, W. C., Parmentier, F. J. W., Rasse, D. P., Rocha, A. V., Sachs, T., van der Molen, M. K., and Tamstorf, M. P.: Assessing the spatial variability in peak season CO_2 exchange characteristics across the Arctic tundra using a light response curve parameterization, Biogeosciences, 11, 4897–4912, https://doi.org/10.5194/bg-11-4897-2014, 2014.

McGuire, A. D., Anderson, L. G., Christensen, T. R., Dallimore, S., Guo, L., Hayes, D. J., Heimann, M., Lorenson, T. D., Macdonald, R. W., and Roulet, N. T.: Sensitivity of the carbon cycle in the Arctic to climate change, Ecol. Monogr., 79, 523–555, https://doi.org/10.1890/08-2025.1, 2009.

McGuire, A. D., Christensen, T. R., Hayes, D., Heroult, A., Euskirchen, E., Kimball, J. S., Koven, C., Lafleur, P., Miller, P. A., Oechel, W., Peylin, P., Williams, M., and Yi, Y.: An assessment of the carbon balance of Arctic tundra: comparisons among observations, process models, and atmospheric inversions, Biogeosciences, 9, 3185–3204, https://doi.org/10.5194/bg-9-3185-2012, 2012.

Meltofte, H., Christensen, T. R., Elberling, B., Forchhammer, M. C., and Rasch, M. (Eds.): Introduction, in: Advances in Ecological Research, Academic Press, 1–12, 2008.

Moffat, A. M., Papale, D., Reichstein, M., Hollinger, D. Y., Richardson, A. D., Barr, A. G., Beckstein, C., Braswell, B. H., Churkina, G., Desai, A. R., Falge, E., Gove, J. H., Heimann, M., Hui, D., Jarvis, A. J., Kattge, J., Noormets, A., and Stauch, V. J.: Comprehensive comparison of gap-filling techniques for eddy covariance net carbon fluxes, Agr. Forest Meteorol., 147, 209–232, https://doi.org/10.1016/j.agrformet.2007.08.011, 2007.

Moncrieff, J. B., Massheder, J. M., de Bruin, H., Elbers, J., Friborg, T., Heusinkveld, B., Kabat, P., Scott, S., Soegaard, H., and Verhoef, A.: A system to measure surface fluxes of momentum, sensible heat, water vapour and carbon dioxide, J. Hydrol., 188–189, 589–611, https://doi.org/10.1016/S0022-1694(96)03194-0, 1997.

Moore, C. J.: Frequency response corrections for eddy correlation systems, Bound.-Layer Meteorol., 37, 17–35, https://doi.org/10.1007/BF00122754, 1986.

Nelson, F. E. and Outcalt, S. I.: A computational method for prediction and regionalization of permafrost, Arctic Alpine Res., 19, 279–288, https://doi.org/10.2307/1551363, 1987.

Oechel, W. C. and Billings, W.: Effects of global change on the carbon balance of arctic plants and ecosystems, in: Arctic Ecosystems in a Changing Climate: an Ecophysiological Perspective, Academic Press, San Diego, 139–168, 1992.

Oechel, W. C., Hastings, S. J., Vourlrtis, G., Jenkins, M., Riechers, G., and Grulke, N.: Recent change of Arctic tundra ecosystems from a net carbon dioxide sink to a source, Nature, 361, 520–523, 1993.

Papale, D., Reichstein, M., Aubinet, M., Canfora, E., Bernhofer, C., Kutsch, W., Longdoz, B., Rambal, S., Valentini, R., Vesala, T., and Yakir, D.: Towards a standardized processing of Net Ecosystem Exchange measured with eddy covariance technique: algorithms and uncertainty estimation, Biogeosciences, 3, 571–583, https://doi.org/10.5194/bg-3-571-2006, 2006.

Parmentier, F. J. W., van der Molen, M. K., van Huissteden, J., Karsanaev, S. A., Kononov, A. V., Suzdalov, D. A., Maximov, T. C., and Dolman, A. J.: Longer growing seasons do not increase net carbon uptake in the northeastern Siberian tundra, J. Geophys. Res.-Biogeo., 116, G04013, https://doi.org/10.1029/2011JG001653, 2011.

Pedregosa, F., Varoquaux, G., Gramfort, A., Michel, V., Thirion, B., Grisel, O., Blondel, M., Prettenhofer, P., Weiss, R., Dubourg, V., Vanderplas, J., Passos, A., Cournapeau, D., Brucher, M., Perrot, M., and Duchesnay, É.: Scikit-learn: machine learning in python, J. Mach. Learn. Res., 12, 2825–2830, 2011.

Peichl, M., Öquist, M., Löfvenius, M. O., Ilstedt, U., Sagerfors, J., Grelle, A., Lindroth, A., and B. Nilsson, M.: A 12-year record reveals pre-growing season temperature and water table level threshold effects on the net carbon dioxide exchange in a boreal fen, Environ. Res. Lett., 9, 055006, https://doi.org/10.1088/1748-9326/9/5/055006, 2014.

Pirk, N., Sievers, J., Mertes, J., Parmentier, F.-J. W., Mastepanov, M., and Christensen, T. R.: Spatial variability of CO_2 uptake in polygonal tundra: assessing low-frequency disturbances in eddy covariance flux estimates, Biogeosciences, 14, 3157–3169, https://doi.org/10.5194/bg-14-3157-2017, 2017.

R Development Core Team: R: A Language and Environment for Statistical Computing, R Foundation for Statistical Computing, Vienna, Austria, 2015.

Reichstein, M., Moffat, A. M., Wutzler, T., and Sickel, K.: REddyProc: Data processing and plotting utilities of (half-)hourly eddy-covariance measurements, R package version 0.8-2/r14, 2016.

Reichstein, M., Falge, E., Baldocchi, D., Papale, D., Aubinet, M., Berbigier, P., Bernhofer, C., Buchmann, N., Gilmanov, T., Granier, A., Grünwald, T., Havránková, K., Ilvesniemi, H., Janous, D., Knohl, A., Laurila, T., Lohila, A., Loustau, D., Matteucci, G., Meyers, T., Miglietta, F., Ourcival, J.-M., Pumpanen, J., Rambal, S., Rotenberg, E., Sanz, M., Tenhunen, J., Seufert, G., Vaccari, F., Vesala, T., Yakir, D., and Valentini, R.: On the separation of net ecosystem exchange into assimilation and ecosystem respiration: review and improved algorithm, Glob. Change Biol., 11, 1424–1439, https://doi.org/10.1111/j.1365-2486.2005.001002.x, 2005.

Reichstein, M., Papale, D., Valentini, R., Aubinet, M., Bernhofer, C., Knohl, A., Laurila, T., Lindroth, A., Moors, E., Pilegaard, K., and Seufert, G.: Determinants of terrestrial ecosystem carbon balance inferred from European eddy covariance flux sites, Geophys. Res. Lett., 34, L01402, https://doi.org/10.1029/2006GL027880, 2007.

Richardson, A. D., Hollinger, D. Y., Aber, J. D., Ollinger, S. V., and Braswell, B. H.: Environmental variation is directly responsible for short- but not long-term variation in forest-atmosphere carbon exchange, Glob. Change Biol., 13, 788–803, https://doi.org/10.1111/j.1365-2486.2007.01330.x, 2007.

Rocha, A. V. and Shaver, G. R.: Burn severity influences post-fire CO_2 exchange in arctic tundra, Ecol. Appl., 21, 477–489, https://doi.org/10.1890/10-0255.1, 2011.

Runkle, B. R. K., Sachs, T., Wille, C., Pfeiffer, E.-M., and Kutzbach, L.: Bulk partitioning the growing season net ecosystem exchange of CO_2 in Siberian tundra reveals the seasonality of its carbon sequestration strength, Biogeosciences, 10, 1337–1349, https://doi.org/10.5194/bg-10-1337-2013, 2013.

Ryan, J. A. and Ulrich, J. M.: xts: eXtensible Time Series. R package version 0.9-7, 2014.

Schuur, E. A. G., McGuire, A. D., Schadel, C., Grosse, G., Harden, J. W., Hayes, D. J., Hugelius, G., Koven, C. D., Kuhry, P., Lawrence, D. M., Natali, S. M., Olefeldt, D., Romanovsky, V. E., Schaefer, K., Turetsky, M. R., Treat, C. C., and Vonk, J. E.: Climate change and the permafrost carbon feedback, Nature, 520, 171–179, https://doi.org/10.1038/nature14338, 2015.

Serreze, M. C. and Barry, R. G.: Processes and impacts of Arctic amplification: a research synthesis, Global Planet. Change, 77, 85–96, https://doi.org/10.1016/j.gloplacha.2011.03.004, 2011.

Shaver, G. and Kummerow, J.: Phenology, resource allocation, and growth of arctic vascular plants, in: Arctic Ecosystems in a Changing Climate: An Ecophysiological Perspective, edited by: Svoboda, J., Academic Press, New York, 1992.

Stoy, P. C., Trowbridge, A. M., and Bauerle, W. L.: Controls on seasonal patterns of maximum ecosystem carbon uptake and canopy-scale photosynthetic light response: contributions from both temperature and photoperiod, Photosynth. Res., 119, 49–64, https://doi.org/10.1007/s11120-013-9799-0, 2014.

Strachan, I. B., Pelletier, L., and Bonneville, M.-C.: Inter-annual variability in water table depth controls net ecosystem carbon dioxide exchange in a boreal bog, Biogeochemistry, 127, 99–111, https://doi.org/10.1007/s10533-015-0170-8, 2015.

Tagesson, T., Mölder, M., Mastepanov, M., Sigsgaard, C., Tamstorf, M. P., Lund, M., Falk, J. M., Lindroth, A., Christensen, T. R., and Ström, L.: Land–atmosphere exchange of methane from soil thawing to soil freezing in a high-Arctic wet tundra ecosystem, Glob. Change Biol., 18, 1928–1940, https://doi.org/10.1111/j.1365-2486.2012.02647.x, 2012.

Tarnocai, C., Canadell, J. G., Schuur, E. A. G., Kuhry, P., Mazhitova, G., and Zimov, S.: Soil organic carbon pools in the northern circumpolar permafrost region, Global Biogeochem. Cy., 23, GB2023, https://doi.org/10.1029/2008GB003327, 2009.

Tjoelker, M. G., Oleksyn, J., and Reich, P. B.: Modelling respiration of vegetation: evidence for a general temperature-dependent Q10, Glob. Change Biol., 7, 223–230, https://doi.org/10.1046/j.1365-2486.2001.00397.x, 2001.

Walker, D. A., Raynolds, M. K., Daniëls, F. J. A., Einarsson, E., Elvebakk, A., Gould, W. A., Katenin, A. E., Kholod, S. S., Markon, C. J., Melnikov, E. S., Moskalenko, N. G., Talbot, S. S., Yurtsev, B. A., and the other members of the C.T.: The Circumpolar Arctic vegetation map, J. Veg. Sci., 16, 267–282, https://doi.org/10.1111/j.1654-1103.2005.tb02365.x, 2005.

Webb, E. K., Pearman, G. I., and Leuning, R.: Correction of flux measurements for density effects due to heat and water vapour transfer, Q. J. Roy. Meteor. Soc., 106, 85–100, https://doi.org/10.1002/qj.49710644707, 1980.

Westergaard-Nielsen, A., Lund, M., Hansen, B. U., and Tamstorf, M.: Camera derived vegetation greenness index as proxy for gross primary production in a low Arctic wetland area, ISPRS J. Photogramm., 86, 89–99, https://doi.org/10.1016/j.isprsjprs.2013.09.006, 2013.

Westergaard-Nielsen, A., Lund, M., Pedersen, S. H., Schmidt, N. M., Klosterman, S., Abermann, J., and Hansen, B. U.: Transitions in high-Arctic vegetation growth patterns and ecosystem productivity from 2000–2013 tracked with cameras, AMBIO, 46, Supplement 1, 39–52, https://doi.org/10.1007/s13280-016-0864-8, 2015.

Williams, M., Eugster, W., Rastetter, E. B., McFadden, J. P., and Chapin Iii, F. S.: The controls on net ecosystem productivity along an Arctic transect: a model comparison with flux measurements, Glob. Change Biol., 6, 116–126, https://doi.org/10.1046/j.1365-2486.2000.06016.x, 2000.

Williams, M., Schwarz, P. A., Law, B. E., Irvine, J., and Kurpius, M. R.: An improved analysis of forest carbon dynamics using data assimilation, Glob. Change Biol., 11, 89–105, https://doi.org/10.1111/j.1365-2486.2004.00891.x, 2005.

Wohlfahrt, G., Hammerle, A., Haslwanter, A., Bahn, M., Tappeiner, U., and Cernusca, A.: Seasonal and inter-annual variability of the net ecosystem CO_2 exchange of a temperate mountain grassland: effects of weather and management, J. Geophys. Res.-Atmos., 113, D08110, https://doi.org/10.1029/2007JD009286, 2008.

Zeileis, A. and Grothendieck, G.: zoo: S3 infrastructure for regular and irregular time series, J. Stat. Softw., 14, 1–27, 2005.

Zona, D., Lipson, D. A., Richards, J. H., Phoenix, G. K., Liljedahl, A. K., Ueyama, M., Sturtevant, C. S., and Oechel, W. C.: Delayed responses of an Arctic ecosystem to an extreme summer: impacts on net ecosystem exchange and vegetation functioning, Biogeosciences, 11, 5877–5888, https://doi.org/10.5194/bg-11-5877-2014, 2014.

Zona, D., Gioli, B., Commane, R., Lindaas, J., Wofsy, S. C., Miller, C. E., Dinardo, S. J., Dengel, S., Sweeney, C., Karion, A., Chang, R. Y.-W., Henderson, J. M., Murphy, P. C., Goodrich, J. P., Moreaux, V., Liljedahl, A., Watts, J. D., Kimball, J. S., Lipson, D. A., and Oechel, W. C.: Cold season emissions dominate the Arctic tundra methane budget, P. Natl. Acad. Sci. USA, 113, 40–45, https://doi.org/10.1073/pnas.1516017113, 2016.

13

Patterns and controls of inter-annual variability in the terrestrial carbon budget

Barbara Marcolla[1], **Christian Rödenbeck**[2], **and Alessandro Cescatti**[3]

[1]Fondazione Edmund Mach, IASMA Research and Innovation Centre, Sustainable Agro-ecosystems and Bioresources Department, 38010 San Michele all'Adige, Trento, Italy
[2]Max Planck Institute for Biogeochemistry, 07745 Jena, Germany
[3]European Commission, Joint Research Centre, Directorate for Sustainable Resources, 21027 Ispra (VA), Italy

Correspondence to: Alessandro Cescatti (alessandro.cescatti@ec.europa.eu)

Abstract. The terrestrial carbon fluxes show the largest variability among the components of the global carbon cycle and drive most of the temporal variations in the growth rate of atmospheric CO_2. Understanding the environmental controls and trends of the terrestrial carbon budget is therefore essential to predict the future trajectories of the CO_2 airborne fraction and atmospheric concentrations. In the present work, patterns and controls of the inter-annual variability (IAV) of carbon net ecosystem exchange (NEE) have been analysed using three different data streams: ecosystem-level observations from the FLUXNET database (La Thuile and 2015 releases), the MPI-MTE (model tree ensemble) bottom–up product resulting from the global upscaling of site-level fluxes, and the Jena CarboScope Inversion, a top–down estimate of surface fluxes obtained from observed CO_2 concentrations and an atmospheric transport model. Consistencies and discrepancies in the temporal and spatial patterns and in the climatic and physiological controls of IAV were investigated between the three data sources. Results show that the global average of IAV at FLUXNET sites, quantified as the standard deviation of annual NEE, peaks in arid ecosystems and amounts to $\sim 120\,gC\,m^{-2}\,y^{-1}$, almost 6 times more than the values calculated from the two global products (15 and $20\,gC\,m^{-2}\,y^{-1}$ for MPI-MTE and the Jena Inversion, respectively). Most of the temporal variability observed in the last three decades of the MPI-MTE and Jena Inversion products is due to yearly anomalies, whereas the temporal trends explain only about 15 and 20 % of the variability, respectively. Both at the site level and on a global scale, the IAV of NEE is driven by the gross primary productivity and in particular by the cumulative carbon flux during the months when land acts as a sink. Altogether these results offer a broad view on the magnitude, spatial patterns and environmental drivers of IAV from a variety of data sources that can be instrumental to improve our understanding of the terrestrial carbon budget and to validate the predictions of land surface models.

1 Introduction

Atmospheric CO_2 concentration has been constantly increasing since the Industrial Revolution, and has caused a corresponding rise of $0.85\,°C$ in the global air temperature from 1880 to 2012 (IPCC, 2013). Since the 1960s, terrestrial ecosystems have acted as a considerable sink of atmospheric CO_2, reabsorbing about one quarter of anthropogenic emissions (Friedlingstein et al., 2010; Le Quéré et al., 2014). The growth rate of atmospheric CO_2 concentration is characterized by a large inter-annual variability (IAV), which mostly results from the variability in the CO_2 net ecosystem exchange (NEE) on land (Bousquet et al., 2000; Le Quéré et al., 2009; Yuan et al., 2009). Multisite synthesis confirms that a large inter-annual variability in NEE is a common feature at all flux sites around the world (Baldocchi, 2008; Baldocchi et al., 2001). The reason why the IAV is so large is that NEE results from the small imbalance between two larger fluxes: the photosynthetic uptake of CO_2 (gross primary production, GPP) and the respiratory release of CO_2 (total ecosystem respiration, TER). As a consequence, even minor variation in either of the two fluxes can cause large variations in their difference.

It has been long debated whether it is GPP or TER that controls the spatial and temporal variability in NEE. Several studies have ascribed inter-annual variability in NEE to variability in either GPP (Ahlström et al., 2015; Janssens et al., 2001; Jung et al., 2011, 2017; Stoy et al., 2009; Urbanski et al., 2007) or TER (Morgenstern et al., 2004; Valentini et al., 2000) or both (Ma et al., 2007; Wohlfahrt et al., 2008b). GPP and TER show comparable ranges of IAV, typically larger in absolute terms than that observed for NEE due to the temporal correlation between the two gross fluxes (Richardson et al., 2007). Given that photosynthesis and respiration may respond differently to environmental drivers (Luyssaert et al., 2007; Polley et al., 2008), the interpretation of climate impacts on the variability in NEE requires an understanding of the relation between the IAV of NEE and that of GPP and TER (Polley et al., 2010).

The environmental factors driving the IAV of NEE (IAV_{NEE}) include climate, physiology, phenology, and natural and anthropogenic disturbances (Marcolla et al., 2011; Richardson et al., 2007; Shao et al., 2015). Understanding the spatio-temporal variability in NEE and its controlling mechanisms is essential to assess the vulnerability of the terrestrial carbon budget, to evaluate the land mitigation potentials and to quantify the ecosystem capacity to store carbon under future climatic conditions (Heimann and Reichstein, 2008). Besides, quantifying inter-annual variability in NEE is a prerequisite for detecting longer-term trends or step changes in flux magnitude in response to climatic or anthropogenic influences and identifying its drivers (Cox et al., 2000; Lombardozzi et al., 2014).

The temporal dynamic of NEE has been addressed in numerous studies, based on either "top–down" approaches, which primarily focus on aircraft atmospheric budgets (Leuning et al., 2004), tower-based boundary layer observations (Bakwin et al., 2004) and tracer transport inversion (Baker et al., 2006; Gurney et al., 2002; Rödenbeck et al., 2003), or on "bottom–up" methods that rely on data-driven gridded products derived from the upscaling of flux data (Jung et al., 2011, 2017; Papale et al., 2015; Papale and Valentini, 2003) or process-based biogeochemical models that simulate regional carbon budgets (Desai et al., 2007, 2008; Mahadevan et al., 2008).

Despite the broad literature on the subject, very few examples of IAV analysis based on multiple data streams are available in the literature (Desai et al., 2010; Pacala, 2001; Poulter et al., 2014). In the present study, patterns and controls of the inter-annual variability in NEE have been analysed using three different data streams: ecosystem-level data from the FLUXNET database, the MPI-MTE (model tree ensemble) bottom–up product resulting from the statistical upscaling of in situ flux data (le Maire et al., 2010) and the Jena CarboScope Inversion top–down product, which estimates land (and ocean) fluxes from atmospheric CO_2 concentration measurements and atmospheric transport modelling (Rödenbeck et al., 2003). In particular, this analysis aims to (i) assess the magnitude and the spatial pattern of the IAV of NEE (IAV_{NEE}), (ii) investigate the role of key climatic variables, like temperature and precipitation, in driving the spatial pattern of IAV, and (iii) identify the role of photosynthesis and respiration as sources of IAV_{NEE}. Finally, the consistencies and discrepancies among the different data products are analysed and critically evaluated.

2 Materials and methods

2.1 Datasets

Data on an ecosystem scale were retrieved from two releases of the FLUXNET dataset, namely La Thuile (http://fluxnet.fluxdata.org/data/la-thuile-dataset/) and 2015 (http://fluxnet.fluxdata.org/data/fluxnet2015-dataset/). These datasets contain half-hourly data of carbon dioxide, water vapour and energy fluxes that are harmonized, standardized and gap-filled. Time series of NEE and of the component fluxes GPP and TER, together with air temperature and precipitation, were used in the present analysis. Flux data have the advantage of representing direct observations of in situ IAV; however, at most sites the time series are still too short for a proper analysis of the temporal variability in NEE (Shao et al., 2015). For this reason only sites with a minimum of 5 years of observations and an open data distribution policy were selected. A subset of 89 sites satisfied the two criteria, among which there were 27 evergreen needle-leaf forests, 5 evergreen broadleaf forests, 12 deciduous broadleaf forests, 6 mixed forests, 12 grasslands, 8 croplands, 6 sites with closed and open shrublands, 7 wetlands, and 6 sites with savannahs and woody savannahs.

On a global scale, two sources of gridded data were used: a bottom–up data product, namely the MPI-MTE product (Jung et al., 2009), and, as a top–down product, the Jena CarboScope CO_2 Inversion (Rödenbeck et al., 2003). The MPI-MTE dataset is built with a machine learning technique to upscale in space and time the flux observations from the global network of eddy covariance sites (FLUXNET) integrated with climate and remote-sensing data for the time period 1982–2011 (Jung et al., 2009). Global maps for GPP and TER at $0.5°$ spatial resolution and monthly temporal resolution were used, while NEE fields were calculated as the difference between the gross fluxes. This product has become a reference dataset to evaluate process-oriented land models and remote-sensing estimates of primary productivity, despite the uneven distribution of eddy covariance sites on which it is trained. It integrates a large amount of in situ measurements and remote-sensing and meteorological observations using a machine learning technique and has been proved to reproduce spatial patterns and seasonal variability in fluxes well (Jung et al., 2009). On the other hand, the product has some shortcomings: for instance, the effects of land management, land use change and CO_2 fertilization are not represented. The MPI-MTE has been recognized as underestimating the inter-annual variability in carbon fluxes. These limits may be due to the missing representation of some key determinants of IAV like changes in soil and biomass pools, disturbances (e.g. fires), ecosystem age, management activities and land use history. Finally, the lag between external forcing and ecosystem response is not represented in the product (Jung et al., 2011).

To derive surface fluxes, the Jena CarboScope Inversion combines modelled atmospheric transport with high-precision measurements of atmospheric CO_2 concentrations. Atmospheric transport is simulated by a global three-dimensional transport model driven by meteorological data. For consistency with the MPI-MTE product, monthly averaged NEE land fluxes from the s81_v3.6 version of the product were used here, at a spatial resolution of $5° \times 3.75°$. The Jena Inversion is particularly suited for the analysis of temporal trends and variability since it is based on a temporally constant observation network (14 atmospheric stations for the version s81_v3.6). Weaknesses of the Jena Inversion product are linked (i) to the sparse density and biased spatial distribution of the sampling network, whose geometry affects the flux estimates in a systematic way, (ii) to data gaps, (iii) to incompleteness of the accounted fluxes, since the calculation is based on CO_2 data only, while atmospheric carbon comes also from CO and volatile organic compounds, and (iv) to potential systematic errors in the transport model (Baker et al., 2006).

As the inversion estimates the total land flux, calculated as the difference between the total surface flux and prescribed anthropogenic emissions, it also includes CO_2 emissions from fires in addition to NEE. For improving the consistency with the other two datasets (MPI-MTE and FLUXNET) that do not account for fire emissions, we subtracted fire emissions from the inversion estimates using a harmonized combination of the products RETRO (Schultz et al., 2008) for the period 1982–1996 and GFED4 (van der Werf et al., 2010) for the period 1997–2013. RETRO is a global gridded dataset (at $0.5°$ spatial resolution) for anthropogenic and vegetation fire emissions of several trace gases, covering the period from 1960 to 2000 with monthly time resolution. GFED4 combines satellite information on fire activity and vegetation productivity to estimate gridded monthly fire emissions at a spatial resolution of $0.25°$ since 1997. RETRO and GFED4 were harmonized using the overlapping years 1997–2000 to calculate calibration coefficients as the ratio of GFED4 to RETRO for latitudinal bands of $30°$. The RETRO time series was then multiplied by these coefficients and the resulting time series of fire emissions was finally subtracted from the land flux of the Jena Inversion. It is worth noting that the remaining flux from the inversion is the sum of land use change emissions and NEE while the MPI-MTE does not account for the land use change flux.

In order to analyse the role of climatic drivers in the inter-annual variability, global maps of temperature and precipitation were used. Gridded air temperatures were obtained from the Climatic Research Unit (CRU) at the University of East Anglia on a monthly timescale and $0.5° \times 0.5°$ spatial resolution, based on an archive of monthly mean temperatures provided by more than 4000 weather stations (Jones et al., 2012). Precipitation fields were obtained from the Global Precipitation Climatology Centre (GPCC) product at $0.5°$ and monthly time steps (Schneider et al., 2014). This product is based on a large dataset of monthly precipitation from more than 85 000 stations and is provided by NOAA ESRL Physical Sciences Division (PSD; Boulder, Colorado, USA). The MODIS MCD12C1 land cover product (Friedl and Brodley, 1997) was used to classify the land pixels and to calculate statistics by plant functional type. MCD12C1 provides the dominant land cover types at a spatial resolution of $0.05°$ using a supervised classification algorithm that is calibrated using a database of land cover training sites. Product resolutions were harmonized using the aggregate function of the raster R package (Hijmans and van Etten, 2014).

2.2 IAV analysis

The inter-annual variability in NEE was estimated as the standard deviation of annual NEE values generated by trend and residuals, computed on time windows of 12 months shifted with a monthly time step (Luyssaert et al., 2007; Shao et al., 2015; Yuan et al., 2009) and calculated with the same methodology for the three data streams used in the analysis. Average values of IAV for plant functional type (PFT) were determined using the PFT classification of FLUXNET sites and the MCD12C1 product (aggregated at the appropriate spatial resolution using the dominant PFT) for the MPI-MTE and the Jena Inversion. Map grid cells were also classified according to mean annual temperature and precipitation, and the mean value of IAV_{NEE} and normalized IAV_{NEE} were calculated for each climate bin. The IAV of both MPI-MTE and the Jena Inversion was normalized by the average GPP of the specific climate bin from the MPI-MTE.

For the two gridded products, which provide a 30-year-long time series (1982–2011), the IAV was partitioned into two components, namely the variance explained by the temporal trend and that due to annual anomalies (Ahlström et al., 2015). For this purpose a linear model was fitted to the time series at each pixel, and the determination coefficient of the regression was used to measure the fraction of variance explained by the trend, whereas its complement to 1 was the fraction of variance due to anomalies.

The spatial correlation between IAV and climatic drivers (air temperature and precipitation) was analysed on a global scale for the MPI-MTE by calculating the spatial correlation coefficient between the temporal standard deviation (IAV amplitude) of NEE and the average annual temperature or precipitation in moving spatial windows of $15° \times 11.5°$

(which means 31×21 pixels for MPI-MTE). The latitudinal averages of these correlation coefficients were calculated for latitudinal bands of $30°$. This analysis was not replicated on the Jena Inversion because on a fine scale the spatial variability in the fluxes in this product is mainly controlled by the prior estimates. In fact, the optimization algorithm of the inversion spatially allocates the fluxes proportionally to the prior; hence, grid cells with higher productivity will change more if compared to cells with a lower prior.

Finally, in order to identify which process between photosynthesis and respiration drives IAV_{NEE}, for FLUXNET and MPI-MTE linear regressions between NEE and GPP and NEE and TER were fitted for each site/pixel and the difference between the determination coefficients of the two linear regressions was computed. Since GPP and TER cannot be derived from inversion products, we performed a similar analysis using NEE of the carbon uptake period (CUP, sum of negative monthly NEE) and of the carbon release period (CRP, sum of positive monthly NEE) as proxies of GPP and TER for all the three data streams. Also in this case NEE was linearly regressed with NEE_{CUP}, and NEE_{CRP} to detect which of the two processes drives the variability in NEE. IAV_{NEE} and IAV controls were also analysed in a climatic space defined by mean annual temperature and precipitation. Finally, annual anomalies of the two global products used in the present analysis were compared with the estimates derived from the Global Carbon Project (GCP) (Le Quéré et al., 2016).

3   Results and discussion

3.1   IAV patterns

The spatial pattern of inter-annual variability for the three datasets is shown in Fig. 1. The IAV of NEE at individual FLUXNET sites ranges from 15 to $400\,gC\,m^{-2}\,y^{-1}$ with an average of $130\,gC\,m^{-2}\,y^{-1}$. On average the most northern sites show a lower temporal variability both in Europe and in North America (Fig. 1a). A global map of IAV_{NEE} is shown also for MPI-MTE (Fig. 1b) and the Jena Inversion (Fig. 1c) at the original spatial resolutions of the two products. The observed range of IAV is similar for the two gridded products and substantially lower than that observed at the site level, probably due to the spatial averaging of the land fluxes that dampens the temporal variability. The mean global value of IAV is in fact 15 and $20\,gC\,m^{-2}\,y^{-1}$ for MPI-MTE and the Jena Inversion, respectively, and hence about $1/6$ of the site-level IAV. The two gridded products confirm the decreasing trend of IAV toward northern latitudes observed at flux sites. A general decrease in IAV_{NEE} at higher latitude for both evergreen needleleaf forests (ENF) and deciduous broadleaf forests (DBF) was also observed by Yuan et al. (2009) although for none of the two PFTs were these trends significant.

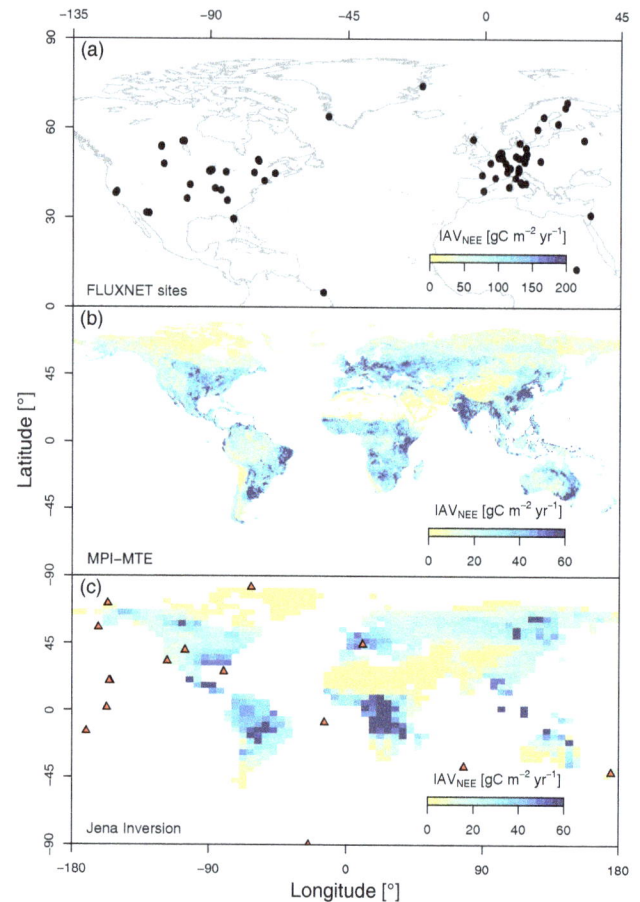

Figure 1. Spatial distribution of NEE standard deviation used as a measure of inter-annual variability (IAV_{NEE}). Results are shown for **(a)** FLUXNET sites with at least 5 years of observations, **(b)** the MPI-MTE NEE product and **(c)** the Jena Inversion product s81_v3.6; red triangles represent the CO_2 concentration measurement sites. Different scales were used for FLUXNET and the global product maps since the latter show a lower IAV.

In terms of IAV, the two global products show a reasonable qualitative correspondence for North America and Eurasia, whereas they disagree for South America, with MPI-MTE showing a minimum of IAV in the humid tropics, where the inversion product shows large variability. MPI-MTE in particular shows maximum values along the eastern coast of South America while the Jena Inversion shows an almost opposite pattern. A similar behaviour is also observed in Africa, where the top–down product shows a maximum in central Africa, while MPI-MTE shows a minimum in the Congo Basin and higher values in arid zones like the Sahel and southern Africa. These discrepancies could, on the one hand, be ascribed to the limits of the bottom–up approach in dealing with the low seasonality of the fraction of absorbed radiation (FaPAR) in evergreen broadleaf forests, given the relevance of this predictor in the MPI-MTE estimates. A second reason for the discrepancy could be the CO_2 emissions from land use change that are particular relevant in some tropical areas but are not accounted in the MPI-MTE estimates. On the other hand, the fine-scale estimates of the inversion are largely determined by the a priori weighting pattern, which has been chosen proportional to time-mean net primary production (NPP; from the LPJ model) as a vegetation proxy (Rödenbeck et al., 2003). As the atmospheric data can only constrain larger-scale patterns comparable to the distances between the stations, this means that IAV will be locally higher where mean NPP is high and vice versa.

As far as the Northern Hemisphere is concerned, a good correspondence is observed in western Eurasia, while some discrepancies are observed in other zones; for example MPI-MTE shows a large IAV in India, probably driven by the changes in FaPAR related to agricultural intensification, which does not emerge from the inversion product that has little observational constraint in this area. To summarize, the spatial pattern of IAV in the two products better agrees in the Northern Hemisphere for temperate and cold temperate zones, whereas for the Southern Hemisphere, and in particular for the humid evergreen forests, the two products show a poor match. In general, it has to be considered that both the MPI-MTE product and the Jena Inversion are driven by data from surface networks that are very limited in the tropics and the Southern Hemisphere, and, therefore, these observation-driven estimates are under-constrained in those areas. These results highlight that for achieving more robust and consistent estimates of the terrestrial carbon fluxes, it is of key importance to increase the availability of atmospheric and ecosystem flux observations in the tropical region, either by establishing new sites where the network is sparse or improving the sharing of data where the monitoring stations are available but not connected to global networks (e.g. flux stations in Amazonia).

The results presented in the maps of Fig. 1 are summarized in the climate space in Fig. 2. The left panels show that peak values of IAV are located in different climate regions for the two gridded products (temperate humid for MPI-MTE and tropical humid for the Jena Inversion). These results highlight that top–down and bottom–up estimates do not agree on the main sources of temporal variability in the terrestrial carbon budget and call for more investigation to pin down the reasons for these large discrepancies. Given that the IAV of NEE increases with the primary productivity at the FLUXNET sites (Fig. 3), in Fig. 2 (right column) we normalized the IAV of both MPI-MTE and the Jena Inversion by the average GPP of the specific climate bin from the MPI-MTE. Normalization using GPP (which is always positive) offers a more robust metric of relative IAV if compared to normalization with NEE (that spans 0). Figure 2 shows either the mean IAV (left column) or the ratio of the mean IAV and GPP (right column) in each climate bin, since this metric is less sensitive to outliers than the mean of ratios and gives more weight to points with larger fluxes. The normalized IAV shows a consistent pattern between the three differ-

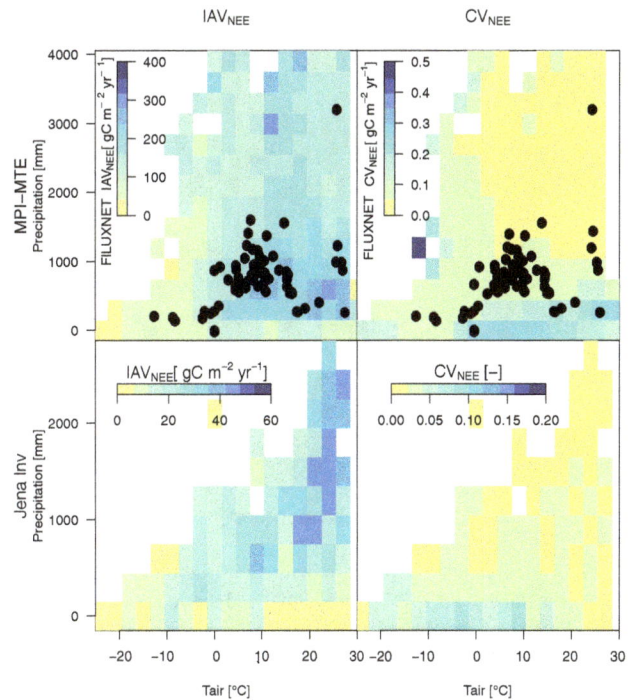

Figure 2. IAV_{NEE} (left panels) and normalized IAV_{NEE} (CV_{NEE}, right panels) plotted in a temperature–precipitation space for MPI-MTE (top panels) and the Jena Inversion (bottom panels). Dots represent FLUXNET site values. MPI-MTE and the Jena Inversion both refer to the scales of the bottom panels.

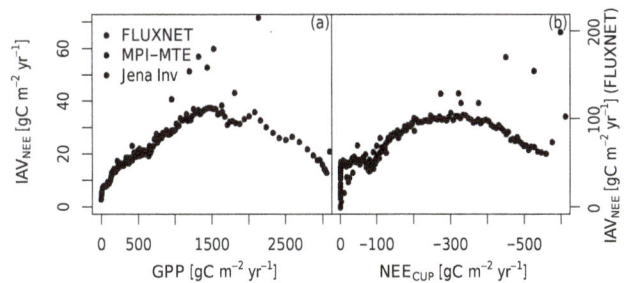

Figure 3. Dependency of NEE interannual variability on GPP and NEE_{CUP}. Results are shown for FLUXNET sites (green dots, different y scale on the right), for the MPI-MTE NEE (grey dots) and the Jena Inversion product (red dots).

ent data products, with a clear decreasing trend with increasing temperature and precipitation (i.e. increasing productivity). Ultimately arid regions seems to have the higher relative variation in land carbon fluxes, in accordance with previous findings (Ahlström et al., 2015). Interestingly, the two gridded products show slightly different climatic location for the peak in relative IAV, with MPI-MTE pointing to warm arid regions, whereas the Jena Inversion points to cold arid systems.

The dependency of IAV_{NEE} on GPP and on NEE_{CUP} is shown in Fig. 3 for the three datasets. Both for FLUXNET

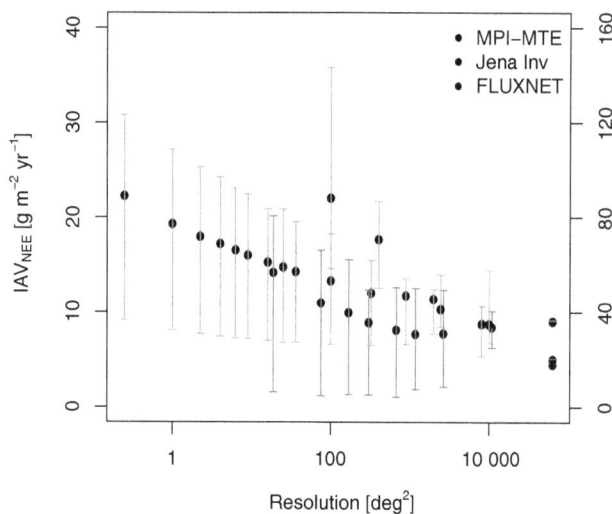

Figure 4. Dependence of IAV$_{NEE}$ on map resolution for FLUXNET sites (green dots), MPI-MTE (grey dots) and the Jena Inversion (red dots). Error bars represent the 25 and 75 % quantiles of the IAV in the aggregated sites/pixels.

and the Jena Inversion, IAV increases with both GPP and NEE$_{CUP}$. By contrast, the IAV$_{NEE}$ in the MPI-MTE dataset peaks at intermediate values of GPP and NEE$_{CUP}$, even if this trend is not evident in the FLUXNET data from which the MPI-MTE product is derived. As stressed previously, MPI-MTE seems to underestimate the temporal variability in evergreen tropical forests both in South America and Africa, where the highest values of GPP and NEE$_{CUP}$ are observed and where by contrast the inversion shows high values of IAV. We think that this mismatch is due to the prominent role that FaPAR has in the MTE approach. In fact, canopy greenness is particularly stable in the tropical humid forests, generating this unusual pattern of low relative IAV in regions of high productivity. These contrasting results for key regions like the Amazon and the Congo Basin confirm the large uncertainty of the IAV estimates in areas with limited observational constraints. In these regions, climate sensitivities derived from estimates of the inter-annual variability in the terrestrial carbon budget therefore have to be carefully interpreted (Fang et al., 2017).

The importance of the spatial scale of analysis for the IAV$_{NEE}$ has been explored both for FLUXNET sites and the global products (i.e. MPI-MTE and the Jena Inversion) (Fig. 4). The two global products show good agreement at the native Inversion resolution (5° × 3.75°) and at a global level, when only one global value is retrieved by spatially averaging all the pixels of the original maps. For the MPI-MTE product, the observed IAV decreases regularly at decreasing map resolution. By contrast, the Jena Inversion shows a rapid descent followed by a stabilization. FLUXNET sites and their aggregation at increasing distance also show a decreasing IAV with higher values compared to the global prod-

ucts. The slope of the lines in Fig. 4 reveals the degree of spatial compensation between anomalies (steeper slopes are generated by stronger compensation and therefore lower spatial coherence), which leads to a decrease in IAV$_{NEE}$ at the increase in the spatial extent of the observations. Of the three products, MPI-MTE shows the gentler slope and therefore the larger spatial coherence of the anomalies. This is possibly due to the missing representation of land disturbances in the MTE methodology, which may ultimately lead to an overestimation of the spatial coherence in the land CO_2 flux anomalies.

The fractions of IAV$_{NEE}$ generated either by temporal trends or by annual residuals are summarized in Fig. 5 for the two global gridded products. For MPI-MTE, more than 80 % of the IAV is explained by residuals at all latitudes. Only in limited zones like the Congo Basin and western Amazonia does MPI-MTE show a relative minimum in the importance of residuals, but this global product might underestimate the total variability in these zones (see Fig. 1b). Residuals explain the largest share (between 62 and 90 %, average 77 %) of the temporal variability also in the Jena Inversion, with a higher relevance of trends in the Southern Hemisphere. The inversion product shows several hotspots of trend-driven variability, like southern Africa, South America and northern Eurasia, which is indeed reported as an area of increasing productivity in the last decades (Forkel et al., 2016). In the interpretation of these results it is important to consider that MPI-MTE is generated by the statistical upscaling of FLUXNET data, using climate and FaPAR as predictors. This methodology relies on the assumption of a constant ecosystem response to climate drivers, and for this reason the product cannot reproduce the influence of some environmental factors (e.g. increasing CO_2 concentration or nitrogen deposition) that may alter these responses and that are not reflected in input variables like climate or FaPAR. By contrast, inversion products do not make any assumption on the climate dependence of ecosystem functioning but also includes emission from land management and land use change that may hide or emphasize the NEE trends. In summary, it is important to note that, despite the important climate trends, in the last 30 years the temporal variability in the land carbon balance has been driven by annual residuals, confirming the dominant role of climate variability in the terrestrial C budget (Le Quéré et al., 2014).

For the two gridded products the analysis of IAV (either in terms of absolute IAV$_{NEE}$ or normalized with NEE$_{CUP}$) was disaggregated by plant functional type (Fig. 6). The analysis in terms of absolute IAV$_{NEE}$ shows that savannahs and woody savannahs (WSAV-SAV) are the PFTs characterized by the larger IAV and variability within the PFT. This was found both for the MPI-MTE and the Jena Inversion product and confirms the results of a recent study (Ahlström et al., 2015) in which semi-arid ecosystems were found to account for the largest fraction (39 %) of the global IAV in net biome productivity. This variability was found to be signif-

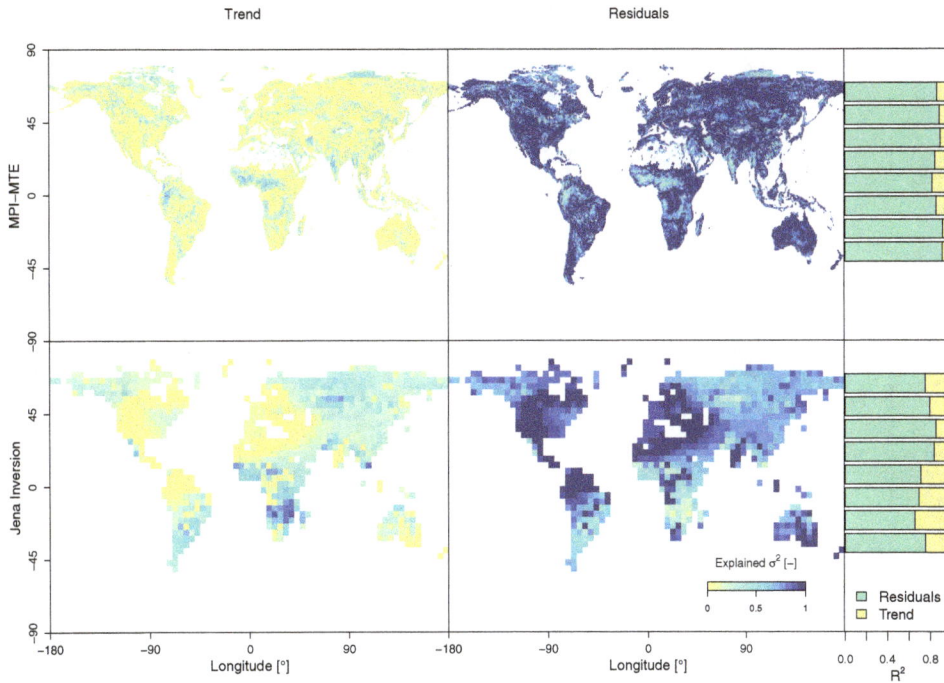

Figure 5. Maps of the fraction of NEE variance explained by temporal trends and anomalies for MPI-MTE NEE and the Jena Inversion; latitudinal band (15°) averages of the fractions are shown in the bar plots.

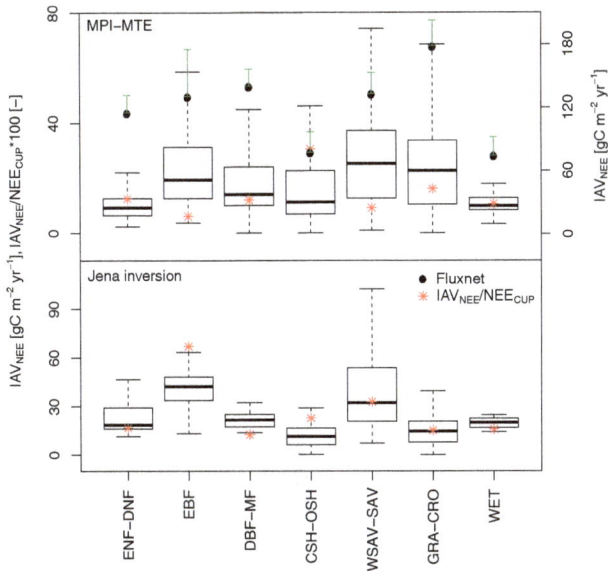

Figure 6. Boxplot of NEE inter-annual variability averaged in PFT classes for MPI-MTE NEE and the Jena Inversion; green dots represent observations at FLUXNET sites (different y scale on the right) plotted with their standard error. PFT classes are grouped as follows: evergreen needleleaf forests and deciduous needleleaf forests (ENF-DNF), evergreen broadleaf forests (EBF), deciduous broadleaf forests and mixed forests (DBF-MF), closed and open shrublands (CSH-OSH), woody savannahs and savannahs (WSAV-SAV), grasslands and croplands (GRA-CRO), and wetlands (WET).

icantly related to the length of the growing season (Ma et al., 2007) and is driven by the uncertainty in water supply in arid systems. In terms of normalized IAV the two gridded products show different behaviours, shrublands (CSH-OSH) being the most variable PFT for MPI-MTE, while the inversion data show a higher variability for evergreen broadleaf forests (EBF) and WSAV-SAV. As observed on a pixel scale in Fig. 1, even at PFT level the results obtained from FLUXNET sites show a higher variability than the gridded products. In general, at FLUXNET sites IAV is proportional to ecosystem productivity (Fig. 4) with the maximum values observed in EBF, deciduous broadleaf and mixed forests (DBF-MF), and grassland–croplands (GRA-CRO) and the minimum in wetlands (WET). The large value of IAV observed in GRA-CRO is presumably also affected by the potential large impact of management in these ecosystems that can either reduce (e.g. by irrigation) or increase the climate-induced variability (e.g. by changing crops or fertilization schemes). In general, the disaggregation of IAV_{NEE} by PFTs shows rather similar results between the two gridded products, both in terms of magnitude and distribution. The largest difference is observed in the evergreen broadleaf forests whose absolute and relative variability is much larger in the inversion, possibly as a result of the intensive disturbances that have occurred in these ecosystems over the last decades and that are not captured with the MTE methodology.

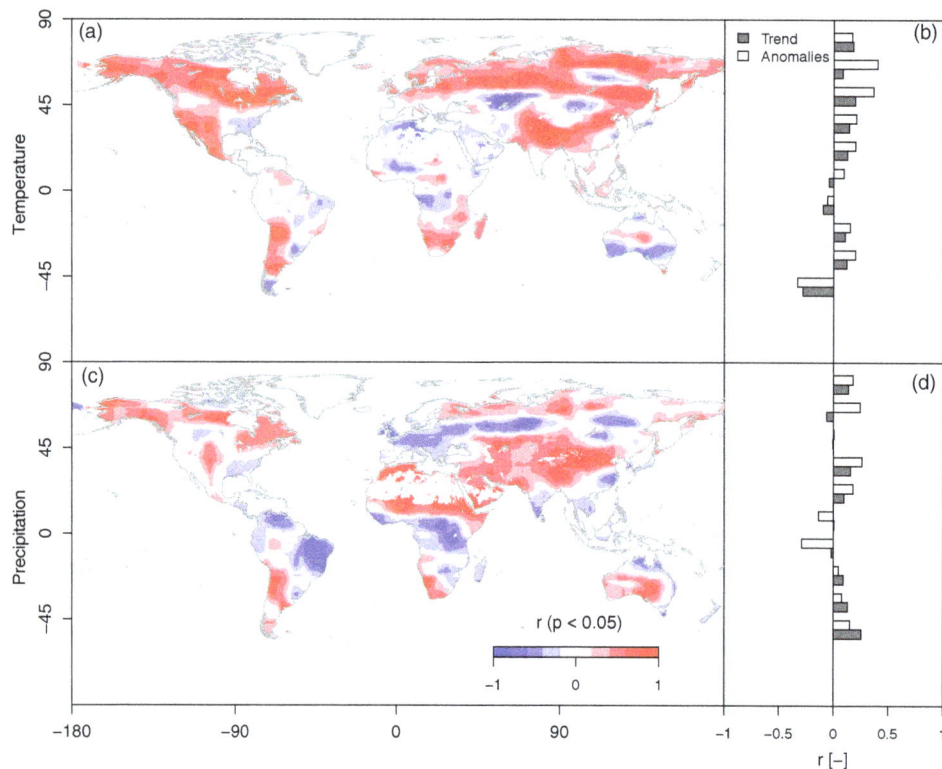

Figure 7. Climatic drivers of the spatial variability in NEE interannual variability. The panels **(a)** and **(c)** show maps of the spatial correlation coefficients (within moving spatial windows of $15° \times 11.5°$) of interannual NEE amplitude versus time-mean temperature and precipitation for the bottom–up product MPI-MTE. Pixels with non-significant correlation are left white. The bar plots in **(b)** and **(d)** show latitudinal averages of the correlation coefficients of NEE trend and anomalies versus temperature and precipitation.

3.2   Climate dependence of IAV

The climatic dependence of the spatial variability in IAV_{NEE} on a global scale for the MPI-MTE product (Fig. 7) shows a clear pattern, with positive correlations in temperature-limited areas at northern latitudes and a negative temperature dependence in water-limited zones (Braswell et al., 1997). These observations agree with Reichstein et al. (2007), who report that GPP shifts from soil water content to air temperature dependency at around $52°$ N. These opposite temperature dependences will probably lead to future contrasting changes in IAV. In fact, under a global warming scenario, the northern latitudes will be characterized by a larger sink (Zhao and Running, 2010) but also by a larger temporal variability, while arid zones like the Mediterranean Basin, central eastern Australia and sub-Saharan Africa will probably experience a reduction in IAV linked to large-scale droughts and consequent reduction in primary productivity (Ciais et al., 2005). Concerning precipitation, the MPI-MTE product shows more complex spatial patterns, with negative correlation in the humid tropics, temperate Europe and the southeast USA and positive correlation elsewhere. It is worth noting that this analysis does not account for the potential lag between climate events and ecosystem responses, in line with

the assumption adopted in the formulation of the MPI-MTE statistical model (Jung et al., 2011). Potential delays in the ecosystem responses to precipitation anomalies as observed from field studies (Doughty et al., 2015) may eventually explain the contrasting spatial pattern shown by this data product.

The climate dependencies of IAV are further separated between the variability due to trends and anomalies (Fig. 7b, d). The two components of IAV_{NEE} mostly show an agreement in the sign of the climatic controls, meaning that the environmental drivers have the same effects on trends and anomalies and therefore on the long and short timescales. This is a relevant finding because it supports the use of IAV to investigate medium-term climatic responses. In general, anomalies show a higher correlation than trends, probably due to the larger magnitude of the variance attributed to this component. In conclusion, the spatial patterns shown in the maps of Fig. 7 and the agreement between the two components of IAV shown in the bar plots indicate that the temperature controls of the IAV of NEE are in general the same as for the primary productivity (i.e. positive in colder biomes and negative in warmer regions), while the contrasting results observed for precipitation suggest that the role in the spatial and temporal variability played by water availability is unclear, probably because of the temporal correlation between precipitation and temperature anomalies, as shows by Jung et al. (2017). The analysis of the climate drivers of IAV was not performed for the Jena Inversion because for this product local variation in IAV is heavily driven by the prior estimates of NPP and therefore results have limited sensitivity to the atmospheric constraints.

3.3 Physiological drivers of IAV

An improvement in the mechanistic understanding of IAV_{NEE} can be achieved by partitioning the net flux into its two components: GPP and TER. Partitioned fluxes are available for FLUXNET sites and for derived products like MPI-MTE, while they cannot be derived from atmospheric inversions. For this latter product the fluxes during the CUP (NEE < 0) and during the CRP (NEE > 0) were used in this analysis as proxies of GPP and TER, respectively.

To investigate how good these proxies are, the ratios TER / GPP during CUP and GPP / TER during CRP were analysed at FLUXNET sites and for each pixel of the MPI-MTE product and averaged by PFT (Fig. 8a). As far as the MPI-MTE product is concerned, TER ranges from 55 to 78 % of GPP during the CUP, while GPP is 56 to 80 % of TER during the CRP; hence, on average about two-thirds of the signal comes from GPP (TER) in the CUP (CRP). These ratios show a certain variability among PFTs, with ENF having the larger imbalance between the two fluxes and the lowest TER / GPP ratio during CUP (due to the strong seasonality of GPP in this PFT), while the two fluxes are not so well partitioned for EBF (ratio ~ 0.8), which are characterized by a long growing season with consistently large fluxes of GPP and TER. The other PFTs show an average ratio value of ~ 0.65 both in CUP and CRP. In summary, it can be inferred that NEE during CUP is dominated by the GPP signal, while NEE during CRP is dominated by TER albeit to a smaller extent as is shown by the frequency distributions in Fig. 8bc calculated from the MPI-MTE product. The distribution of the TER / GPP ratio during the CUP is in fact narrower and peaks at a value of 0.7, while a broader distribution is observed for the GPP / TER ratio during the CRP. As expected, there is a larger spread in the composition of NEE during CRP across the world, and this is linked to the larger variability in the seasonality of GPP that may actually go to 0 in the dormancy season, while TER is always positive.

In order to identify which of the gross fluxes controls the variability in the net land flux, we assessed the fraction of variance (R^2) of NEE explained by GPP or TER (for MPI-MTE and FLUXNET) and CUP or CRP (for all products). Results given in Fig. 9 show the difference of the determination coefficients between the two regressions (NEE versus GPP or TER; NEE versus NEE_{CUP} or NEE_{CRP}) and are used to determine which component dominates the inter-annual variation in NEE. Blue zones in Fig. 9 are regions where IAV_{NEE} is driven by photosynthesis (GPP or CUP), the dif-

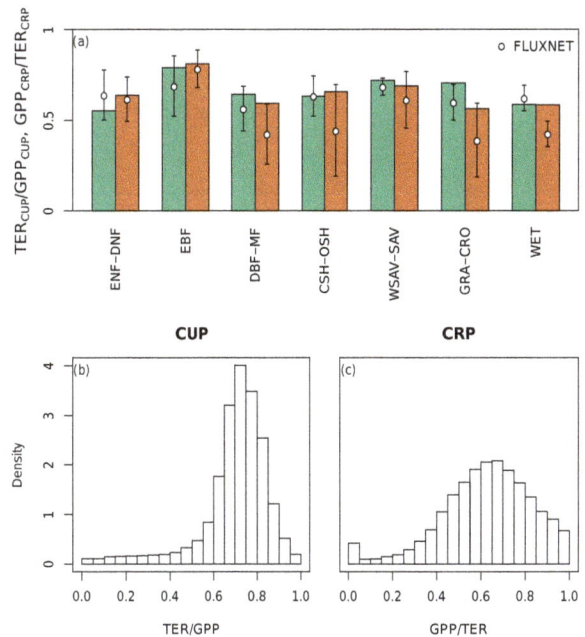

Figure 8. Bar plot of the TER / GPP ratio for the CUP (green bars, **a**) and of the GPP / TER ratio during the CRP (red bars, **a**); these values were calculated for the MPI-MTE product; dots refer to FLUXNET sites. Averages of yearly values are represented together with their standard deviation. The global frequency distributions of the ratios obtained from the MPI-MTE product are shown in the histograms (**b, c**) at the bottom.

ference $R^2_{GPP} - R^2_{TER}$ (or $R^2_{CUP} - R^2_{CRP}$) being positive, while in the red zones IAV_{NEE} is mainly controlled by respiration (TER or CRP). Figure 9a shows that, in most of the land area, the IAV_{NEE} is driven by GPP both at FLUXNET sites and for the MPI-MTE product. The same data products show an even clearer dominance of NEE_{CUP} over IAV (Fig. 9b). The Jena Inversion product shows that, although most of the globe is NEE_{CUP} driven, there are quite a few areas that are weakly CRP driven, like the eastern US, arid regions in Africa and the Amazon Basin, probably because these areas are estimated to be CO_2 sources in this inversion and therefore NEE is dominated by NEE_{CRP} (data not shown). When latitudinal profiles are considered, all the products show that GPP and NEE_{CUP} control the temporal variability in yearly NEE more than TER or NEE_{CRP} (le Maire et al., 2010). Results shown in the global maps of Fig. 9 are represented in the climatic space in Fig. 10. Map pixels were classified according to mean annual temperature and precipitation. For each climate bin the difference between the determination coefficients for NEE versus GPP and TER is shown. Across the whole climate space, the IAV retrieved from the MPI-MTE product is mostly controlled by CUP and GPP, although the difference in R^2 in the case of GPP and TER is low. The Jena Inversion by contrast shows climate areas where IAV is CRP driven, especially in intermediate to high temperature classes. Simi-

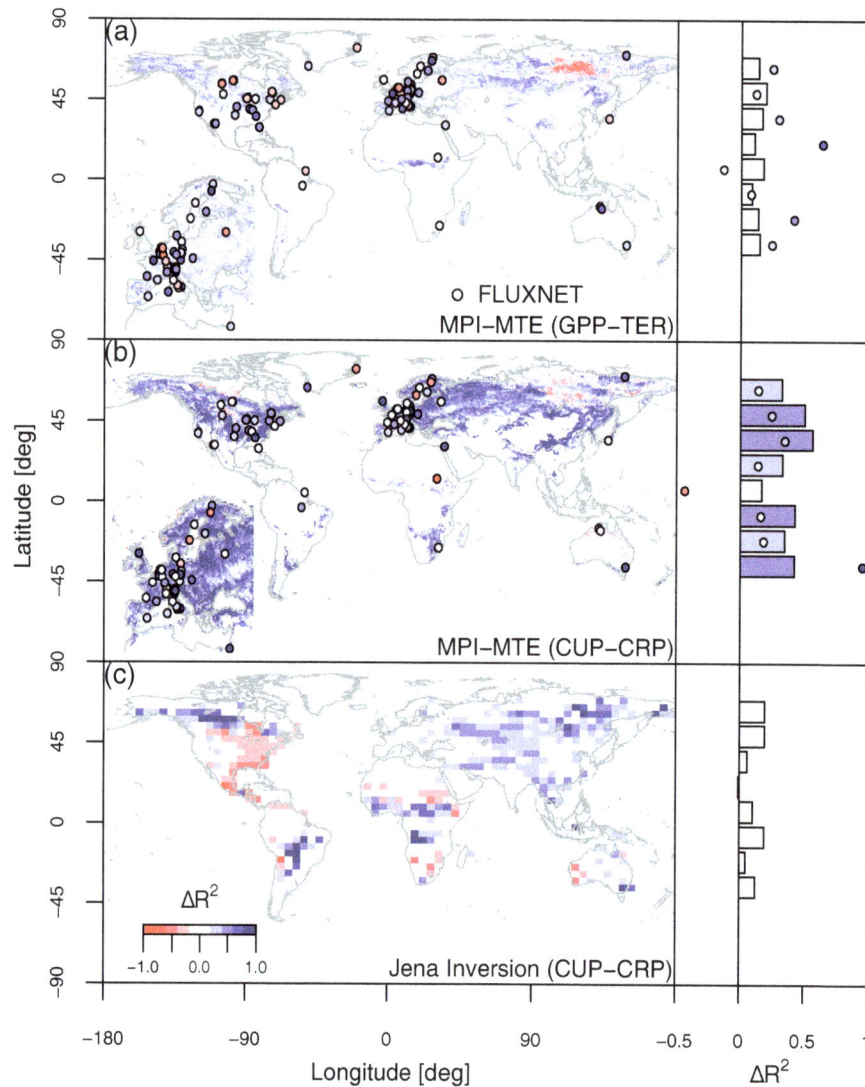

Figure 9. Control on IAV by GPP-TER and CUP-CRP, expressed as the difference of the determination coefficients ($R^2_{GPP} - R^2_{TER}$ for MPI-MTE and FLUXNET and $R^2_{CUP} - R^2_{CRP}$ for all products) for FLUXNET sites with at least 5 years of observations (dots), MPI-MTE and the Jena Inversion. Latitudinal averages are shown for latitudinal classes of 15°. Inset maps in (a) and (b) show an enlarged plot of Europe.

lar results have been reported across several PFTs by Yuan et al. (2009) and Ahlström et al. (2015) using FLUXNET site data and MTE products. A higher correlation of IAV with GPP rather than with TER in deciduous forests has also been reported by Barr et al. (2002) and Wu et al. (2012). These results suggest that ecosystem fluxes during the CUP, and in particular photosynthesis more than respiration, consistently control the inter-annual variability in NEE on all the spatial scales for both bottom–up and top–down data products (Janssens et al., 2001; Luyssaert et al., 2007; le Maire et al., 2010; Urbanski et al., 2007; Wohlfahrt et al., 2008a; Wu et al., 2012; Yuan et al., 2009). These results highlight that temporal variations in photosynthesis and in ecosystem CO_2 exchange during the carbon uptake period therefore drive the

short-term climate sensitivity of the global carbon cycle consistently across different regions and climates. The possibility of interpreting these short-term responses as long-term potential impacts of climate change is therefore to be disputed, given the limited role that respiration appears to play in modulating the rapid reactions of the terrestrial biosphere to environmental drivers.

Finally, in order to place our analysis in a broader context, global annual values of the gridded products used in the present analysis have been compared with the estimates of the Global Carbon Project (Fig. 11). On an annual timescale, the Jena Inversion shows excellent agreement with the GCP, and this is not surprising since GCP land fluxes are estimated as a residual term from the atmospheric CO_2 budget and are

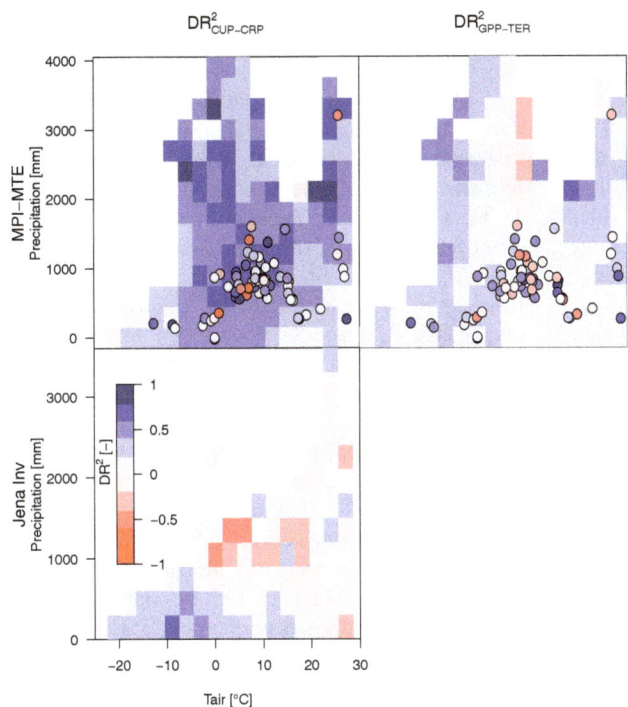

Figure 10. Control on IAV by GPP-TER and CUP-CRP, expressed as the difference of the determination coefficients plotted in a temperature–precipitation space. The two top panels refer to MPI-MTE, while the bottom panel refers to the Jena Inversion; dots refer to FLUXNET sites.

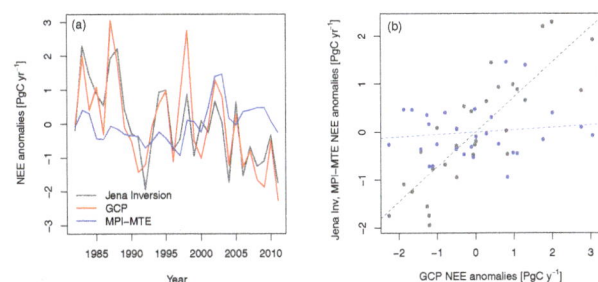

Figure 11. Comparison of the annual NEE anomalies between the Jena Inversion, MPI-MTE and the Global Carbon Project data. Time series of annual anomalies are shown in (a), while regressions of MPI-MTE and the Jena Inversion values are shown in (b) versus annual anomalies of the Global Carbon Project data.

therefore not completely independent of the Jena Inversion product. On the other hand, this analysis highlights how the MTE bottom–up approach is barely correlated ($R^2 = 0.015$, $p = 0.52$) with the top–down estimates, both in terms of trend and of anomalies. These discrepancies may be partially explained by the missing representation of land disturbances (land use change, land management) in the MTE product.

4 Conclusions

In conclusion, this study assessed the temporal variability in the terrestrial C budget with three different datasets to diagnose common patterns and emerging features. Some discrepancies between data products have emerged, in particular in the tropics, where a chronic deficiency of atmospheric and ecosystem observations severely limits the accuracy of large-scale assessments. On the other hand, several important global features have been identified and confirmed by the different products, like (i) the dominant role played by photosynthesis in the short-term variability in the land carbon budget, (ii) the high relative IAV in water-limited ecosystems, and (iii) the dependence of IAV on spatial scales and ecosystem productivity. Ultimately, the variability in the land fluxes observed in the recent decades proved to be extremely valuable to investigate the controlling mechanisms and the sensitivity and vulnerability of the terrestrial C balance to climate drivers.

Competing interests. The authors declare that they have no conflict of interest.

Acknowledgements. This study was supported by the JRC project AgForCC no. 442. The MCD12C1 product was retrieved from the online data pool, courtesy of the NASA EOSDIS Land Processes Distributed Active Archive Center (LP DAAC), USGS/Earth Resources Observation and Science (EROS) Center, Sioux Falls, South Dakota. This work used eddy covariance data acquired and shared by the FLUXNET community, including these networks: AmeriFlux, AfriFlux, AsiaFlux, CarboAfrica, CarboEuropeIP, CarboItaly, CarboMont, ChinaFlux, FLUXNET-Canada, GreenGrass, ICOS, KoFlux, LBA, NECC, OzFlux-TERN, TCOS-Siberia, and USCCC. The FLUXNET eddy covariance data processing and harmonization was carried out by the ICOS Ecosystem Thematic Center, AmeriFlux Management Project and Fluxdata project of FLUXNET, with the support of CDIAC, and the OzFlux, ChinaFlux and AsiaFlux offices.

Edited by: Alexey V. Eliseev

References

Ahlström, A., Raupach, M. R., Schurgers, G., Smith, B., Arneth, A., Jung, M., Reichstein, M., Canadell, J. G., Friedlingstein, P., Jain, A. K., Kato, E., Poulter, B., Sitch, S., Stocker, B. D., Viovy, N., Wang, Y. P., Wiltshire, A., Zaehle, S., and Zeng, N.: The dominant role of semi-arid ecosystems in the trend and variability of the land CO_2 sink, Science, 348, 895–899, https://doi.org/10.1126/science.aaa1668, 2015.

Baker, D. F., Law, R. M., Gurney, K. R., Rayner, P., Peylin, P., Denning, A. S., Bousquet, P., Bruhwiler, L., Chen, Y., Ciais, P., Fung, I. Y., Heimann, M., and Nin, E.: TransCom 3 inversion intercomparison?: Impact of transport model errors on the interannual variability of regional CO_2

fluxes, 1988–2003, Global Biogeochem. Cy., 20, GB1002, https://doi.org/10.1029/2004GB002439, 2006.

Bakwin, P., Davis, K., Yi, C., Wofsy, S. C., Munger, J. W., Haszpra, L., and Barcza, Z.: Regional carbon dioxide fluxes from mixing ratio data, Tellus B, 56, 301–311, https://doi.org/10.1111/j.1600-0889.2004.00111.x, 2004.

Baldocchi, D.: "Breathing" of the terrestrial biosphere: lessons learned from a global network of carbon dioxide flux measurement systems, Aust. J. Bot., 56, 1–26, https://doi.org/10.1071/BT07151, 2008.

Baldocchi, D., Falge, E., Gu, L., Olson, R., Hollinger, D., Running, S., Anthoni, P., Bernhofer, C., Davis, K., Evans, R., Fuentes, J., Goldstein, A., Katul, G., Law, B., Lee, X., Malhi, Y., Meyers, T., Munger, W., Oechel, W., Paw, K. T., Pilegaard, K., Schmid, H. P., Valentini, R., Verma, S., Vesala, T., Wilson, K., and Wofsy, S.: FLUXNET: A New Tool to Study the Temporal and Spatial Variability of Ecosystem–Scale Carbon Dioxide, Water Vapor, and Energy Flux Densities, B. Am. Meteorol. Soc., 82, 2415–2434, https://doi.org/10.1175/1520-0477(2001)082<2415:FANTTS>2.3.CO;2, 2001.

Barr, A. G., Griffis, T. J., Black, T. A., Lee, X., Staebler, R. M., Fuentes, J. D., Chen, Z., and Morgenstern, K.: Comparing the carbon budgets of boreal and temperate deciduous forest stands, Can. J. Forest Res., 32, 813–822, 2002.

Bousquet, P., Peylin, P., Ciais, P., Le Quéré, C., Friedlingstein, P., and Tans, P. P.: Regional Changes in Carbon Dioxide Fluxes of Land and Oceans Since 1980, Science, 290, 1342–1346, https://doi.org/10.1126/science.290.5495.1342, 2000.

Braswell, B., Schimel, D., Linder, E., and Moore, B.: The response of global terrestrial ecosystems to interannual temperature variability, Science, 278, 870–872, 1997.

Ciais, P., Reichstein, M., Viovy, N., Granier, A., Ogée, J., Allard, V., Aubinet, M., Buchmann, N., Bernhofer, C., Carrara, A., Chevallier, F., De Noblet, N., Friend, A. D., Friedlingstein, P., Grünwald, T., Heinesch, B., Keronen, P., Knohl, A., Krinner, G., Loustau, D., Manca, G., Matteucci, G., Miglietta, F., Ourcival, J. M., Papale, D., Pilegaard, K., Rambal, S., Seufert, G., Soussana, J. F., Sanz, M. J., Schulze, E. D., Vesala, T., and Valentini, R.: Europe-wide reduction in primary productivity caused by the heat and drought in 2003, Nature, 437, 529–533, https://doi.org/10.1038/nature03972, 2005.

Cox, P. M., Betts, R. A., Jones, C. D., Spall, S. A., and Totterdell, I. J.: Acceleration of global warming due to carbon-cycle feedbacks in a coupled climate model, Nature, 408, 184–187, https://doi.org/10.1038/35041539, 2000.

Desai, A. R., Moorcroft, P. R., Bolstad, P. V. and Davis, K. J.: Regional carbon fluxes from an observationally constrained dynamic ecosystem model: Impacts of disturbance, CO2 fertilization, and heterogeneous land cover, J. Geophys. Res., 112, G01017, https://doi.org/10.1029/2006JG000264, 2007.

Desai, A. R., Noormets, A., Bolstad, P. V., Chen, J., Cook, B. D., Davis, K. J., Euskirchen, E. S., Gough, C., Martin, J. G., Ricciuto, D. M., Schmid, H. P., Tang, J., and Wang, W.: Influence of vegetation and seasonal forcing on carbon dioxide fluxes across the Upper Midwest, USA: Implications for regional scaling, Agr. Forest Meteorol., 148, 288–308, https://doi.org/10.1016/j.agrformet.2007.08.001, 2008.

Desai, A. R., Helliker, B. R., Moorcroft, P. R., Andrews, A. E., and Berry, J. A.: Climatic controls of interan-

nual variability in regional carbon fluxes from top-down and bottom-up perspectives, J. Geophys. Res., 115, G02011, https://doi.org/10.1029/2009JG001122, 2010.

Doughty, C. E., Metcalfe, D. B., Girardin, C. A. J., Amézquita, F. F., Cabrera, D. G., Huasco, W. H., Silva-Espejo, J. E., Araujo-Murakami, A., da Costa, M. C., Rocha, W., Feldpausch, T. R., Mendoza, A. L. M., da Costa, A. C. L., Meir, P., Phillips, O. L., and Malhi, Y.: Drought impact on forest carbon dynamics and fluxes in Amazonia, Nature, 519, 78–82, https://doi.org/10.1038/nature14213, 2015.

Fang, Y., Michalak, A. M., Schwalm, C. R., Huntzinger, D. N., Berry, J. A., Ciais, P., Piao, S., Poulter, B., Fisher, J. B., Cook, R. B., Hayes, D., Huang, M., Ito, A., Jain, A., Lei, H., Lu, C., Mao, J., Parazoo, N. C., Peng, S., Ricciuto, D. M., Shi, X., Tao, B., Tian, H., Wang, W., Wei, Y., and Yang, J.: Global land carbon sink response to temperature and precipitation varies with ENSO phase, Environ. Res. Lett., 12, 064007, https://doi.org/10.1088/1748-9326/aa6e8e, 2017.

Forkel, M., Carvalhais, N., Rödenbeck, C., Keeling, R., Heimann, M., Thonicke, K., Zaehle, S., and Reichstein, M.: Enhanced seasonal CO_2 exchange caused by amplified plant productivity in northern ecosystems, Science, 351, 696–699, https://doi.org/10.1126/science.aac4971, 2016.

Friedl, M. A. and Brodley, C. E.: Decision tree classification of land cover from remotely sensed data, Remote Sens. Environ., 61, 399–409, https://doi.org/10.1016/S0034-4257(97)00049-7, 1997.

Friedlingstein, P., Houghton, R. A., Marland, G., Hackler, J. L., Boden, T. A., Conway, T. J., and Al, E.: Update on CO_2 emissions, Nat. Geosci., 3, 811–812, https://doi.org/10.1038/ngeo1022, 2010.

Gurney, K. R., Law, R. M., Denning, A. S., Rayner, P. J., Baker, D., Bousquet, P., and Bruhwiler, L.: Towards robust regional estimates of CO_2 sources and sinks using atmospheric transport models, Nature, 415, 626–630, 2002.

Heimann, M. and Reichstein, M.: Terrestrial ecosystem carbon dynamics and climate feedbacks, Nature, 451, 289–292, https://doi.org/10.1038/nature06591, 2008.

Hijmans, R. J. and van Etten, J.: raster: Geographic data analysis and modeling, R package version 2, R Foundation for Statistical Computing, Vienna, Austria, 2014.

IPCC: Climate Change 2013: The Physical Science Basis. Contribution of Working Group I to the Fifth Assessment Report of the Intergovernmental Panel on Climate Change, edited by: Stocker, T. F., Qin, D., Plattner, G.-K., Tignor, M., Allen, S. K., Boschung, J., Nauels, A., Xia, Y., Bex, V., and Midgley, P. M., Cambridge University Press, Cambridge, United Kingdom and New York, NY, USA, 1535 pp., https://doi.org/10.1017/CBO9781107415324, 2013.

Janssens, I. A., Lankreijer, H., Matteucci, G., Kowalski, A. S., Buchmann, N., Epron, D., Pilegaard, K., Kutsch, W., Longdoz, B., Grünwald, T., Montagnani, L., Dore, S., Rebmann, C., Moors, E. J., Grelle, A., Rannik, Ü., Morgenstern, K., Oltchev, S., Clement, R., Guomundsson, J., Minerbi, S., Berbigier, P., Ibrom, A., Moncrieff, J., Aubinet, M., Bernhofer, C., Jensen, N. O., Vesala, T., Granier, A., Schulze, E. D., Lindroth, A., Dolman, A. J., Jarvis, P. G., Ceulemans, R., and Valentini, R.: Productivity overshadows temperature in determining soil and ecosystem respiration across European forests, Glob. Change Biol., 7, 269–278, https://doi.org/10.1046/j.1365-2486.2001.00412.x, 2001.

Jones, P. D., Lister, D. H., Osborn, T. J., Harpham, C., Salmon, M., and Morice, C. P.: Hemispheric and large-scale land surface air temperature variations: an extensive revision and an update to 2010, J. Geophys. Res., 117, D05127, https://doi.org/10.1029/2011JD017139, 2012.

Jung, M., Reichstein, M., and Bondeau, A.: Towards global empirical upscaling of FLUXNET eddy covariance observations: validation of a model tree ensemble approach using a biosphere model, Biogeosciences, 6, 2001–2013, https://doi.org/10.5194/bg-6-2001-2009, 2009.

Jung, M., Reichstein, M., Margolis, H. A., Cescatti, A., Richardson, A. D., Arain, M. A., Arneth, A., Bernhofer, C., Bonal, D., Chen, J., Gianelle, D., Gobron, N., Kiely, G., Kutsch, W., Lasslop, G., Law, B. E., Lindroth, A., Merbold, L., Montagnani, L., Moors, E. J., Papale, D., Sottocornola, M., Vaccari, F., and Williams, C.: Global patterns of land-atmosphere fluxes of carbon dioxide, latent heat, and sensible heat derived from eddy covariance, satellite, and meteorological observations, J. Geophys. Res., 116, G00J07, https://doi.org/10.1029/2010JG001566, 2011.

Jung, M., Reichstein, M., Schwalm, C. R., Huntingford, C., and Sitch, S.: Compensatory water effects link yearly global land CO_2 sink changes to temperature, Nature, 541, 516–520, 2017.

le Maire, G., Delpierre, N., Jung, M., Ciais, P., Reichstein, M., Viovy, N., Granier, A., Ibrom, A., Kolari, P., Longdoz, B., Moors, E. J., Pilegaard, K., Rambal, S., Richardson, A. D., and Vesala, T.: Detecting the critical periods that underpin interannual fluctuations in the carbon balance of European forests, J. Geophys. Res., 115, G00H03, https://doi.org/10.1029/2009JG001244, 2010.

Le Quéré, C., Raupach, M. R., Canadell, J. G., Marland, G., Bopp, L., Ciais, P., Conway, T. J., Doney, S. C., Feely, R. A., Foster, P., Friedlingstein, P., Gurney, K., Houghton, R. A., House, J. I., Huntingford, C., Levy, P. E., Lomas, M. R., Majkut, J., Metzl, N., Ometto, J. P., Peters, G. P., Prentice, I. C., Randerson, J. T., Running, S. W., Sarmiento, J. L., Schuster, U., Sitch, S., Takahashi, T., Viovy, N., van der Werf, G. R., and Woodward, F. I.: Trends in the sources and sinks of carbon dioxide, Nat. Geosci., 2, 831–836, https://doi.org/10.1038/ngeo689, 2009.

Le Quéré, C., Peters, G. P., Andres, R. J., Andrew, R. M., Boden, T. A., Ciais, P., Friedlingstein, P., Houghton, R. A., Marland, G., Moriarty, R., Sitch, S., Tans, P., Arneth, A., Arvanitis, A., Bakker, D. C. E., Bopp, L., Canadell, J. G., Chini, L. P., Doney, S. C., Harper, A., Harris, I., House, J. I., Jain, A. K., Jones, S. D., Kato, E., Keeling, R. F., Klein Goldewijk, K., Körtzinger, A., Koven, C., Lefèvre, N., Maignan, F., Omar, A., Ono, T., Park, G.-H., Pfeil, B., Poulter, B., Raupach, M. R., Regnier, P., Rödenbeck, C., Saito, S., Schwinger, J., Segschneider, J., Stocker, B. D., Takahashi, T., Tilbrook, B., van Heuven, S., Viovy, N., Wanninkhof, R., Wiltshire, A., and Zaehle, S.: Global carbon budget 2013, Earth Syst. Sci. Data, 6, 235–263, https://doi.org/10.5194/essd-6-235-2014, 2014.

Le Quéré, C., Andrew, R. M., Canadell, J. G., Sitch, S., Korsbakken, J. I., Peters, G. P., Manning, A. C., Boden, T. A., Tans, P. P., Houghton, R. A., Keeling, R. F., Alin, S., Andrews, O. D., Anthoni, P., Barbero, L., Bopp, L., Chevallier, F., Chini, L. P., Ciais, P., Currie, K., Delire, C., Doney, S. C., Friedlingstein, P., Gkritzalis, T., Harris, I., Hauck, J., Haverd, V., Hoppema, M., Klein Goldewijk, K., Jain, A. K., Kato, E., Körtzinger, A., Landschützer, P., Lefèvre, N., Lenton, A., Lienert, S., Lombardozzi,

D., Melton, J. R., Metzl, N., Millero, F., Monteiro, P. M. S., Munro, D. R., Nabel, J. E. M. S., Nakaoka, S.-I., O'Brien, K., Olsen, A., Omar, A. M., Ono, T., Pierrot, D., Poulter, B., Rödenbeck, C., Salisbury, J., Schuster, U., Schwinger, J., Séférian, R., Skjelvan, I., Stocker, B. D., Sutton, A. J., Takahashi, T., Tian, H., Tilbrook, B., van der Laan-Luijkx, I. T., van der Werf, G. R., Viovy, N., Walker, A. P., Wiltshire, A. J., and Zaehle, S.: Global Carbon Budget 2016, Earth Syst. Sci. Data, 8, 605–649, https://doi.org/10.5194/essd-8-605-2016, 2016.

Leuning, R., Raupach, M. R., Coppin, P. A., Cleugh, H. A., Isaac, P., Denmead, O. T., Dunin, F. X., Zegelin, S., and Hacker, J.: Spatial and temporal variations in fluxes of energy, water vapour and carbon dioxide during OASIS 1994 and 1995, Bound.-Lay. Meteorol., 110, 3–38, https://doi.org/10.1023/A:1026028217081, 2004.

Lombardozzi, D., Bonan, G. B., and Nychka, D. W.: The emerging anthropogenic signal in land–atmosphere carbon-cycle coupling, Nature Climate Change, 4, 796–800, https://doi.org/10.1038/nclimate2323, 2014.

Luyssaert, S., Janssens, I. A., Sulkava, M., Papale, D., Dolman, A. J., Reichstein, M., Hollmén, J., Martin, J. G., Suni, T., Vesala, T., Loustau, D., Law, B. E., and Moors, E. J.: Photosynthesis drives anomalies in net carbon-exchange of pine forests at different latitudes, Glob. Change Biol., 13, 2110–2127, https://doi.org/10.1111/j.1365-2486.2007.01432.x, 2007.

Ma, S., Baldocchi, D. D., Xu, L., and Hehn, T.: Inter-annual variability in carbon dioxide exchange of an oak/grass savanna and open grassland in California, Agr. Forest Meteorol., 147, 157–171, https://doi.org/10.1016/j.agrformet.2007.07.008, 2007.

Mahadevan, P., Wofsy, S. C., Matross, D. M., Xiao, X., Dunn, A. L., Lin, J. C., Gerbig, C., Munger, J. W., Chow, V. Y., and Gottlieb, E. W.: A satellite-based biosphere parameterization for net ecosystem CO_2 exchange: Vegetation Photosynthesis and Respiration Model (VPRM), Global Biogeochem. Cy., 22, GB2005, https://doi.org/10.1029/2006GB002735, 2008.

Marcolla, B., Cescatti, A., Manca, G., Zorer, R., Cavagna, M., Fiora, A., Gianelle, D., Rodeghiero, M., Sottocornola, M., and Zampedri, R.: Climatic controls and ecosystem responses drive the inter-annual variability of the net ecosystem exchange of an alpine meadow, Agr. Forest Meteorol., 151, 1233–1243, https://doi.org/10.1016/j.agrformet.2011.04.015, 2011.

Morgenstern, K., Andrew Black, T., Humphreys, E. R., Griffis, T. J., Drewitt, G. B., Cai, T., Nesic, Z., Spittlehouse, D. L., and Livingston, N. J.: Sensitivity and uncertainty of the carbon balance of a Pacific Northwest Douglas-fir forest during an El Niño/La Niña cycle, Agr. Forest Meteorol., 123, 201–219, https://doi.org/10.1016/j.agrformet.2003.12.003, 2004.

Pacala, S. W.: Consistent Land- and Atmosphere-Based U.S. Carbon Sink Estimates, Science, 292, 2316–2320, https://doi.org/10.1126/science.1057320, 2001.

Papale, D. and Valentini, R.: A new assessment of European forests carbon exchanges by eddy fluxes and artificial neural network spatialization, Glob. Change Biol., 9, 525–535, https://doi.org/10.1046/j.1365-2486.2003.00609.x, 2003.

Papale, D., Black, T. A., Carvalhais, N., Cescatti, A., Chen, J., Jung, M., Kiely, G., Lasslop, G., Mahecha, M. D., Margolis, H., Merbold, L., Montagnani, L., Moors, E., Olesen, J. E., Reichstein, M., Tramontana, G., van Gorsel, E., Wohlfahrt, G., and Ráduly, B.: Effect of spatial sampling from European flux towers for estimating carbon and water fluxes with artificial neural networks, J. Geophys. Res.-Biogeo., 120, 1941–1957, https://doi.org/10.1002/2015JG002997, 2015.

Polley, H. W., Frank, A. B., Sanabria, J., and Phillips, R. L.: Interannual variability in carbon dioxide fluxes and flux–climate relationships on grazed and ungrazed northern mixed-grass prairie, Glob. Change Biol., 14, 1620–1632, https://doi.org/10.1111/j.1365-2486.2008.01599.x, 2008.

Polley, H. W., Emmerich, W., Bradford, J. A., Sims, P. L., Johnson, D. A., Saliendra, N. Z., Svejcar, T., Angell, R., Frank, A. B., Phillips, R. L., Snyder, K. A., and Morgan, J. A.: Physiological and environmental regulation of interannual variability in CO_2 exchange on rangelands in the western United States, Glob. Change. Biol., 16, 990–1002, https://doi.org/10.1111/j.1365-2486.2009.01966.x, 2010.

Poulter, B., Frank, D., Ciais, P., Myneni, R. B., Andela, N., Bi, J., Broquet, G., Canadell, J. G., Chevallier, F., Liu, Y. Y., Running, S. W., Sitch, S., and van der Werf, G. R.: Contribution of semiarid ecosystems to interannual variability of the global carbon cycle, Nature, 509, 600–603, https://doi.org/10.1038/nature13376, 2014.

Reichstein, M., Papale, D., Valentini, R., Aubinet, M., Bernhofer, C., Knohl, A., Laurila, T., Lindroth, A., Moors, E., Pilegaard, K., and Seufert, G.: Determinants of terrestrial ecosystem carbon balance inferred from European eddy covariance flux sites, Geophys. Res. Lett., 34, L01402, https://doi.org/10.1029/2006GL027880, 2007.

Richardson, A. D., Hollinger, D. Y., Aber, J. D., Ollinger, S. V., and Braswell, B. H.: Environmental variation is directly responsible for short- but not long-term variation in forest-atmosphere carbon exchange, Glob. Change Biol., 13, 788–803, https://doi.org/10.1111/j.1365-2486.2007.01330.x, 2007.

Rödenbeck, C., Houweling, S., Gloor, M., and Heimann, M.: CO_2 flux history 1982–2001 inferred from atmospheric data using a global inversion of atmospheric transport, Atmos. Chem. Phys., 3, 1919–1964, https://doi.org/10.5194/acp-3-1919-2003, 2003.

Schneider, U., Becker, A., Finger, P., Meyer-Christoffer, A., Ziese, M., and Rudolf, B.: GPCC's new land surface precipitation climatology based on quality-controlled in situ data and its role in quantifying the global water cycle, Theor. Appl. Climatol., 115, 15–40, https://doi.org/10.1007/s00704-013-0860-x, 2014.

Schultz, M. G., Heil, A., Hoelzemann, J. J., Spessa, A., Thonicke, K., Goldammer, J. G., Held, A. C., Pereira, J. M. C., and van Het Bolscher, M.: Global wildland fire emissions from 1960 to 2000, Global Biogeochem. Cy., 22, GB2002, https://doi.org/10.1029/2007GB003031, 2008.

Shao, J., Zhou, X., Luo, Y., Li, B., Aurela, M., Billesbach, D., Blanken, P. D., Bracho, R., Chen, J., Fischer, M., Fu, Y., Gu, L., Han, S., He, Y., Kolb, T., Li, Y., Nagy, Z., Niu, S., Oechel, W. C., Pinter, K., Shi, P., Suyker, A., Torn, M., Varlagin, A., Wang, H., Yan, J., Yu, G., and Zhang, J.: Biotic and climatic controls on interannual variability in carbon fluxes across terrestrial ecosystems, Agr. Forest. Meteorol., 205, 11–22, https://doi.org/10.1016/j.agrformet.2015.02.007, 2015.

Stoy, P. C., Richardson, A. D., Baldocchi, D. D., Katul, G. G., Stanovick, J., Mahecha, M. D., Reichstein, M., Detto, M., Law, B. E., Wohlfahrt, G., Arriga, N., Campos, J., McCaughey, J. H., Montagnani, L., Paw, U., K. T., Sevanto, S., and Williams, M.: Biosphere–atmosphere exchange of CO_2 in relation to climate: a cross-biome analysis across multiple time scales, Biogeosciences, 6, 2297–2312, https://doi.org/10.5194/bg-6-2297-2009, 2009.

Urbanski, S., Barford, C., Wofsy, S., Kucharik, C., Pyle, E., Budney, J., McKain, K., Fitzjarrald, D., Czikowsky, M., and Munger, J. W.: Factors controlling CO_2 exchange on timescales from hourly to decadal at Harvard Forest, J. Geophys. Res., 112, G02020, https://doi.org/10.1029/2006JG000293, 2007.

Valentini, R., Matteucci, G., Dolman, A. J., Schulze, E. D., Rebmann, C., Moors, E. J., Granier, A., Gross, P., Jensen, N. O., Pilegaard, K., Lindroth, A., Grelle, A., Bernhofer, C., Grünwald, T., Aubinet, M., Ceulemans, R., Kowalski, A. S., Vesala, T., Rannik, U., Berbigier, P., Loustau, D., Gudmundsson, J., Thorgeirsson, H., Ibrom, A., Morgenstern, K., and Clement, R.: Respiration as the main determinant of carbon balance in European forests, Nature, 404, 861–865, https://doi.org/10.1038/35009084, 2000.

van der Werf, G. R., Randerson, J. T., Giglio, L., Collatz, G. J., Mu, M., Kasibhatla, P. S., Morton, D. C., DeFries, R. S., Jin, Y., and van Leeuwen, T. T.: Global fire emissions and the contribution of deforestation, savanna, forest, agricultural, and peat fires (1997–2009), Atmos. Chem. Phys., 10, 11707–11735, https://doi.org/10.5194/acp-10-11707-2010, 2010.

Wohlfahrt, G., Hammerle, A., Haslwanter, A., Bahn, M., Tappeiner, U., and Cernusca, A.: Seasonal and inter-annual variability of the net ecosystem CO_2 exchange of a temperate mountain grassland: effects of climate and management, J. Geophys. Res., 113, D08110, https://doi.org/10.1029/2007jd009286, 2008a.

Wohlfahrt, G., Hammerle, A., Haslwanter, A., Bahn, M., Tappeiner, U., and Cernusca, A.: Seasonal and inter-annual variability of the net ecosystem CO_2 exchange of a temperate mountain grassland: Effects of weather and management, J. Geophys. Res., 113, D08110, https://doi.org/10.1029/2007JD009286, 2008b.

Wu, J., van der Linden, L., Lasslop, G., Carvalhais, N., Pilegaard, K., Beier, C., and Ibrom, A.: Effects of climate variability and functional changes on the interannual variation of the carbon balance in a temperate deciduous forest, Biogeosciences, 9, 13–28, https://doi.org/10.5194/bg-9-13-2012, 2012.

Yuan, W., Luo, Y., Richardson, A. D., Oren, R., Luyssaert, S., Janssens, I. A., Ceulemans, R., Zhou, X., Grunwald, T., Aubinet, M., Berhofer, C., Baldocchi, D. D., Chen, J., Dunn, A. L., Deforest, J. L., Dragoni, D., Goldstein, A. H., Moors, E., William Munger, J., Monson, R. K., Suyker, A. E., Starr, G., Scott, R. L., Tenhunen, J., Verma, S. B., Vesala, T., and Wofsy, S. C.: Latitudinal patterns of magnitude and interannual variability in net ecosystem exchange regulated by biological and environmental variables, Glob. Change Biol., 15, 2905–2920, https://doi.org/10.1111/j.1365-2486.2009.01870.x, 2009.

Zhao, M. and Running, S. W.: Drought-induced reduction in global terrestrial net primary production from 2000 through 2009, Science, 329, 940–943, https://doi.org/10.1126/science.1192666, 2010.

A global hotspot for dissolved organic carbon in hypermaritime watersheds of coastal British Columbia

Allison A. Oliver[1,2], **Suzanne E. Tank**[1,2], **Ian Giesbrecht**[2,7], **Maartje C. Korver**[2], **William C. Floyd**[3,4,2], **Paul Sanborn**[5,2], **Chuck Bulmer**[6], and **Ken P. Lertzman**[7,2]

[1]University of Alberta, Department of Biological Sciences, CW 405, Biological Sciences Bldg., University of Alberta, Edmonton, AB, T6G 2E9, Canada

[2]Hakai Institute, Tula Foundation, P.O. Box 309, Heriot Bay, BC, V0P 1H0, Canada

[3]Ministry of Forests, Lands and Natural Resource Operations, 2100 Labieux Rd, Nanaimo, BC, V9T 6E9, Canada

[4]Vancouver Island University, 900 Fifth Street, Nanaimo, BC, V9R 5S5, Canada

[5]Ecosystem Science and Management Program, University of Northern British Columbia, 3333 University Way, Prince George, BC, V2N 4Z9, Canada

[6]BC Ministry of Forests Lands and Natural Resource Operations, 3401 Reservoir Rd, Vernon, BC, V1B 2C7, Canada

[7]School of Resource and Environmental Management, Simon Fraser University, TASC 1 – Room 8405, 8888 University Drive, Burnaby, BC, V5A 1S6, Canada

Correspondence to: Allison A. Oliver (aaoliver@ualberta.ca)

Abstract. The perhumid region of the coastal temperate rainforest (CTR) of Pacific North America is one of the wettest places on Earth and contains numerous small catchments that discharge freshwater and high concentrations of dissolved organic carbon (DOC) directly to the coastal ocean. However, empirical data on the flux and composition of DOC exported from these watersheds are scarce. We established monitoring stations at the outlets of seven catchments on Calvert and Hecate islands, British Columbia, which represent the rain-dominated hypermaritime region of the perhumid CTR. Over several years, we measured stream discharge, stream water DOC concentration, and stream water dissolved organic-matter (DOM) composition. Discharge and DOC concentrations were used to calculate DOC fluxes and yields, and DOM composition was characterized using absorbance and fluorescence spectroscopy with parallel factor analysis (PARAFAC). The areal estimate of annual DOC yield in water year 2015 was $33.3\,\mathrm{Mg\,C\,km^{-2}\,yr^{-1}}$, with individual watersheds ranging from an average of 24.1 to $37.7\,\mathrm{Mg\,C\,km^{-2}\,yr^{-1}}$. This represents some of the highest DOC yields to be measured at the coastal margin. We observed seasonality in the quantity and composition of exports, with the majority of DOC export occurring during the extended wet period (September–April). Stream flow from catchments reacted quickly to rain inputs, resulting in rapid export of relatively fresh, highly terrestrial-like DOM. DOC concentration and measures of DOM composition were related to stream discharge and stream temperature and correlated with watershed attributes, including the extent of lakes and wetlands, and the thickness of organic and mineral soil horizons. Our discovery of high DOC yields from these small catchments in the CTR is especially compelling as they deliver relatively fresh, highly terrestrial organic matter directly to the coastal ocean. Hypermaritime landscapes are common on the British Columbia coast, suggesting that this coastal margin may play an important role in the regional processing of carbon and in linking terrestrial carbon to marine ecosystems.

1 Introduction

Freshwater aquatic ecosystems process and transport a significant amount of carbon (Cole et al., 2007; Aufdenkampe et al., 2011; Dai et al., 2012). Globally, riverine export is estimated to deliver around $0.9\,\mathrm{Pg\,C\,yr^{-1}}$ from land to the coastal ocean (Cole et al., 2007), with typically > 50 % quantified as dissolved organic carbon (DOC) (Meybeck, 1982; Ludwig et al., 1996; Alvarez-Cobelas et al., 2012; Mayorga et al., 2010). Rivers draining coastal watersheds serve as conduits of DOC from terrestrial and freshwater sources to marine environments (Mulholland and Watts, 1982; Bauer et al., 2013; McClelland et al., 2014) and can have important implications for coastal carbon cycling, biogeochemical interactions, ecosystem productivity, and food webs (Hopkinson et al., 1998; Tallis, 2009; Tank et al., 2012; Regnier et al., 2013). In addition, because the transfer of water and organic matter from watersheds to the coastal ocean represents an important pathway for carbon cycling and ecological subsidies between ecosystems, better understanding of these linkages is needed for constraining predictions of ecosystem productivity in response to perturbations such as climate change. In regions where empirical data are currently scarce, quantifying land-to-ocean DOC export is therefore a priority for improving the accuracy of watershed and coastal carbon models (Bauer et al., 2013).

While quantifying DOC flux within and across systems is required for understanding the magnitude of carbon exchange, the composition of DOC (as dissolved organic matter, or DOM) is also important for determining the ecological significance of carbon exported from coastal watersheds. The aquatic DOM pool is a complex mixture that reflects both source material and processing along the watershed terrestrial–aquatic continuum, and as a result it can show significant spatial and temporal variation (Hudson et al., 2007; Graeber et al., 2012; Wallin et al., 2015). Both DOC concentration and DOM composition can serve as indicators of watershed characteristics (Koehler et al., 2009), hydrologic flow paths (Johnson et al., 2011; Helton et al., 2015), and watershed biogeochemical processes (Emili and Price, 2013). DOM composition can also influence its role in downstream processing and ecological function, such as susceptibility to biological (Judd et al., 2006) and physiochemical interactions (Yamashita and Jaffé, 2008).

The coastal temperate rainforests (CTRs) of Pacific North America extend from the Gulf of Alaska through British Columbia to Northern California and span a wide range of precipitation and climate regimes. Within this rainforest region, the "perhumid" zone has cool summers and summer precipitation is common (> 10 % of annual precipitation) (Alaback, 1996) (Fig. 1). The perhumid CTR extends from southeast Alaska through the outer coast of central British Columbia and contains forests and soils that have accumulated large amounts of organic carbon above and below ground (Leighty et al., 2006; Gorham et al., 2012). Due to high amounts of precipitation and close proximity to the coast, this area represents a potential hotspot for the transport and metabolism of carbon across the land-to-ocean continuum, and quantifying these fluxes is pertinent for understanding global carbon cycling.

Within the large perhumid CTR, there is substantial spatial variation in climate and landscape characteristics that create uncertainty about carbon cycling and pattern. In Alaska, for example, riverine DOC concentrations vary with wetland cover (D'Amore et al., 2015a) and glacial cover (Fellman et al., 2014). Previous studies have shown that streams in southeast Alaska can contain high DOC concentrations (Fellman et al., 2009a; D'Amore et al., 2015a) and produce high DOC yields (D'Amore et al., 2015a, b; Stackpoole et al., 2016), but no known field estimates have been generated for the perhumid CTR of British Columbia, an area of approximately 97 824 km^2 (adapted from Wolf et al., 1995). Within the perhumid CTR of British Columbia, terrestrial ecologists have defined a large (29 935 km^2) "hypermaritime" subregion where rainfall dominates over snow, seasonality is moderated by the ocean, and wetlands are extensive (Pojar et al., 1991; area estimated using British Columbia Biogeoclimatic Ecosystem Classification Subzone/Variant mapping Version 10, 31 August 2016, available at https://catalogue.data.gov.bc. ca/dataset/f358a53b-ffde-4830-a325-a5a03ff672c3). Previous work in the hypermaritime CTR showed that DOC concentrations are high in small streams and tend to increase during rain events (Gibson et al., 2000; Fitzgerald et al., 2003; Emili and Price, 2013). Taken together, these conditions should be expected to generate high yields and fluxes of DOC from hypermaritime watersheds to the coastal ocean.

The objectives of this study were to provide the first field-based estimates of DOC exports from watersheds in the extensive hypermaritime region of British Columbia's perhumid CTR, to describe the temporal and spatial dynamics of exported DOC concentration and DOM composition, and to identify relationships between DOC concentration, DOM composition, and watershed characteristics.

2   Methods

2.1   Study sites

Study sites are located on northern Calvert Island and adjacent Hecate Island on the central coast of British Columbia, Canada (lat 51.650, long −128.035; Fig. 1). Average annual precipitation and air temperature at sea level from 1981 to 2010 was 3356 mm yr^{-1} and 8.4 °C (average annual min: 0.9 °C; average annual max: 17.9 °C) (available online at http://www.climatewna.com/; Wang et al., 2012), with precipitation dominated by rain and winter snowpack persisting only at higher elevations. Sites are located within the hypermaritime region of the CTR on the outer coast of British Columbia. Soils overlying the granodiorite bedrock (Roddick, 1996) are usually < 1 m thick and have formed in sandy colluvium and patchy morainal deposits, with limited areas of coarse glacial outwash. Chemical weathering and organic-matter accumulation in the cool, moist climate

Table 1. Watershed characteristics, discharge, DOC concentrations, and DOC yields for the seven study watersheds on Calvert and Hecate islands. Additional details on the methods used to determine watershed characteristics can be found in the Supplement.

Watershed	Area	Avg. slope	Lakes	Wetlands	Avg. depth organic soils	Avg. depth mineral soils	Total Q yield[1]	DOC[1,a]	Q-weighted avg. DOC[1]	DOC annual yield[b] WY2015[1]	DOC monthly yield[b] wet season[2]	DOC monthly yield[b] dry season[3]
	(km²)	(%)	(% Area)	(% Area)	(cm)	(cm)	(mm)	(mg L⁻¹)	(mg L⁻¹)	(Mg C km⁻²)	(Mg C km⁻²)	(Mg C km⁻²)
626	3.2	21.7	4.7	48.0	39.4 ± 24.3	30.8 ± 8.3	3673	11.0 ± 3.5	15.3	37.7 (31.9–44.2)	3.59 (3.05–4.18)	0.62 (0.49–0.77)
1015	3.3	34.2	9.1	23.8	39.5 ± 17.2	33.7 ± 8.6	3052	11.2 ± 1.6	12.9	24.7 (23.6–25.8)	2.56 (2.45–2.78)	0.27 (0.25–0.28)
819	4.8	30.1	0.3	50.2	37.9 ± 19.1	29.8 ± 5.7	3066	14.0 ± 3.5	19.3	35.7 (31.7–40.2)	3.80 (3.37–5.10)	0.57 (0.48–0.67)
844	5.7	32.5	0.3	35.2	35.4 ± 18.0	29.1 ± 6.4	4129	13.1 ± 3.6	15.9	43.6 (34.2–54.9)	4.24 (3.36–5.30)	0.54 (0.36–0.77)
708	7.8	28.5	7.5	46.3	36.2 ± 19.7	29.9 ± 6.0	3805	9.5 ± 2.4	10.9	24.1 (22.2–26.0)	2.67 (2.46–4.07)	0.38 (0.34–0.43)
693	9.3	30.2	4.4	42.8	35.4 ± 16.1	30.2 ± 6.4	5866	7.7 ± 2.5	8.4	29.7 (25.9–34.0)	3.19 (2.79–4.94)	0.41 (0.32–0.52)
703	12.8	40.3	1.9	24.3	37.3 ± 16.5	35.8 ± 13.4	6058	6.3 ± 2.6	9.0	37.0 (32.5–42.0)	3.48 (3.07–4.02)	0.64 (0.52–0.77)
All	46.9	32.7	3.7	37.1	37.4 ± 17.7	32.2 ± 9.2	4730	10.4 ± 3.8	11.1	33.3 (28.9–38.1)	3.35 (2.94–4.40)	0.50 (0.41–0.62)

[1] Calculated for water year 2015 (WY2015; 1 October 2014–30 September 2015). [2] Wet-period average monthly yield calculated from October to April and September, WY2015, and October to April, WY2016. [3] Dry-period average monthly yield calculated from May to August, WY2015. [a] Mean ± standard deviation. [b] Total ± 95 % confidence interval.

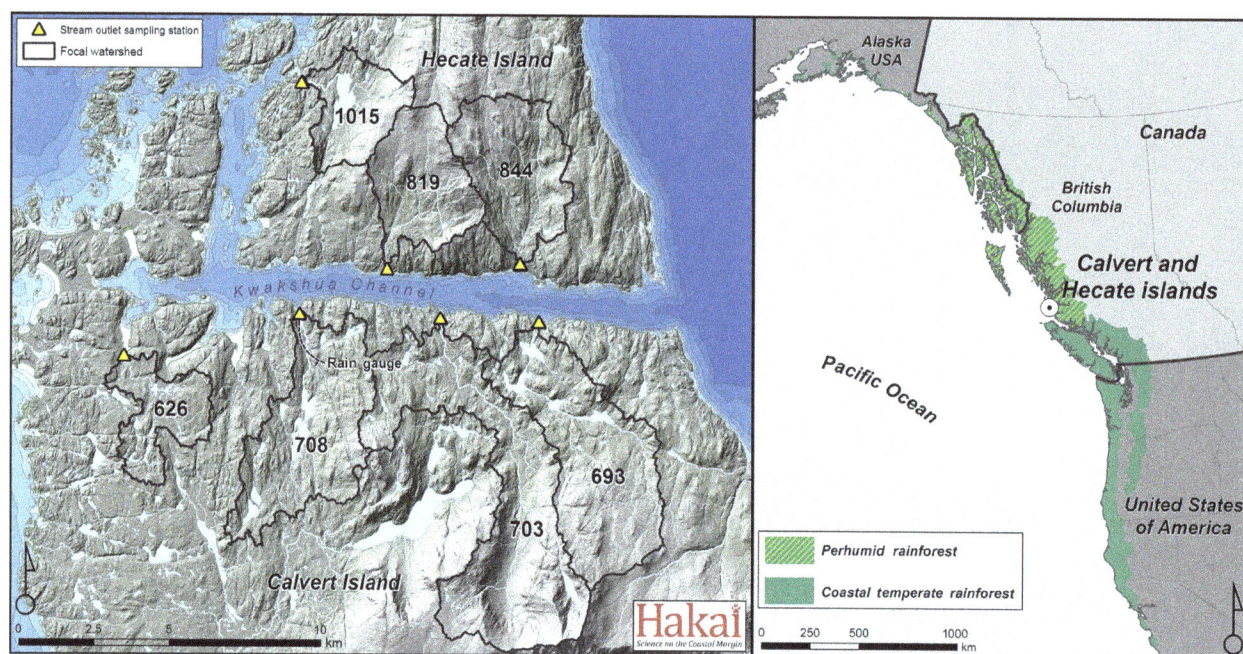

Figure 1. The location of Calvert Island, British Columbia, Canada, within the perhumid region of the coastal temperate rainforest (right) and the study area on Calvert and Hecate islands, including the seven study watersheds, corresponding stream outlet sampling stations, and location of the rain gauge (left). Characteristics of individual watersheds are described in Table 1.

have produced soils dominated by Podzols and Folic Histosols, with Hemists up to 2 m thick at depressional sites (IUSS Working Group WRB, 2015). The landscape is comprised of a mosaic of ecosystem types, including exposed bedrock, extensive wetlands, "bog forests", and woodlands, with organic-rich soils (Green, 2014; Thompson et al., 2016). Forest stands are generally short with open canopies reflecting the lower productivity of the hypermaritime forests compared to the rest of the perhumid CTR (Banner et al., 2005). Dominant trees are western red cedar, yellow cedar, shore pine, and western hemlock, with composition varying across topographic and edaphic gradients. Widespread understory plants include bryophytes, salal, deer fern, and tufted clubrush. Wetland plants are locally abundant including diverse *Sphagnum* mosses and sedges. Although the watersheds have no history of mining or industrial logging, archaeological evidence suggests that humans have occupied this landscape for at least 13 000 years (McLaren et al., 2014). This occupation has had a local effect on forest productivity near habitation sites (Trant et al., 2016) and on fire regimes (Hoffman et al., 2016). We selected seven watersheds with streams draining directly into the ocean (Fig. 1). These numbered watersheds (626, 693, 703, 708, 819 844, and 1015) range in size (3.2 to 12.8 km^2) and topography (maximum elevation 160 to 1012 m), are variably affected by lakes (0.3–9.1 % lake coverage), and – as is characteristic of the perhumid CTR – have a high degree of wetland coverage (24–50 %) (Table 1).

2.2 Soils and watershed characteristics

Watersheds and streams were delineated using a 3 m resolution digital elevation model (DEM) derived from airborne laser scanning (lidar) and flow accumulation analysis using geographic information systems (GISs) to summarize watershed characteristics for each watershed polygon and for all watersheds combined (Gonzalez Arriola et al., 2015; Table 1). Topographic measures were estimated from the DEM, and lake and wetland cover was estimated from the Province of British Columbia terrestrial ecosystem mapping (TEM) (Green, 2014), and soil material thickness was estimated from unpublished digital soil maps (Supplement S1). We recorded the thickness of organic soil material, the thickness of mineral soil material, and total soil depth to bedrock at a total of 353 field sites. Mineral soil horizons have ≤ 17 % organic C, while organic soil horizons have > 17 % organic C, per the Canadian System of Soil Classification (Soil Classification Working Group, 1998). In addition to field-sampled sites, 40 sites with exposed bedrock (0 cm soil depth) were located using aerial photography. Soil thicknesses were combined with a suite of topographic, vegetation, and remote-sensing (lidar and RapidEye satellite imagery) data for each sampling point and used to train a random forest model (randomForest package in R; Liaw and Wiener, 2002) that was used to predict soil depth values. Soil material thicknesses were then averaged for each watershed (Table 1). For additional details on field site selection and methods used for predictions of soil thickness, see Supplement S1.1.

2.3 Sample collection and analysis

From May 2013 to July 2016, we collected stream water grab samples from each watershed stream outlet every 2–3 weeks ($n_{total} = 402$), with less frequent sampling ($\sim$ monthly) during winter (Fig. 1). All samples were filtered in the field (Millipore Millex-HP Hydrophilic PES 0.45 µm) and kept in the dark, on ice, until analysis. DOC samples were filtered into 60 mL amber glass bottles and preserved with 7.5 M H_3PO_4. Fe samples were filtered into 125 mL HDPE bottles and preserved with 8 M HNO_3. DOC and Fe samples were analyzed at the BC Ministry of the Environment Technical Services Laboratory (Victoria, BC, Canada). DOC concentrations were determined on a total organic carbon (TOC) analyzer (Aurora 1030; OI Analytical) using wet chemical oxidation with persulfate followed by infrared detection of CO_2. Fe concentrations were determined on a dual-view inductively coupled plasma optical emission spectrometry (ICP-OES) spectrophotometer (Prodigy; Teledyne Leeman Labs) using a Seaspray pneumatic nebulizer.

In May 2014, we began collecting stream samples for stable isotopic composition of $\delta^{13}C$ in DOC ($\delta^{13}C$-DOC; $n = 173$) and the optical characterization of DOM using absorbance spectroscopy ($n = 259$). Beginning in January 2016, we also analyzed samples using fluorescence spectroscopy (see Sect. 2.6). Samples collected for $\delta^{13}C$-DOC were filtered into 40 mL EPA glass vials and preserved with H_3PO_4. $\delta^{13}C$-DOC samples were analyzed at GG Hatch Stable Isotope Laboratory (Ottawa, ON, Canada) using high-temperature combustion (TIC-TOC Combustion Analyser Model 1030; OI Analytical) coupled to a continuous-flow isotope ratio mass spectrometry (Finnigan Mat DeltaPlusXP; Thermo Fischer Scientific) (Lalonde et al., 2014). Samples analyzed for optical characterization using absorbance and fluorescence were filtered into 125 mL amber HDPE bottles and analyzed at the Hakai Institute (Calvert Island, BC, Canada) within 24 h of collection.

2.4 Hydrology: precipitation and stream discharge

We measured precipitation using a TB4-L tipping bucket rain gauge with a 0.2 mm resolution (Campbell Scientific Ltd.) located in watershed 708 (elevation: 16 m a.s.l.). The rain gauge was calibrated twice per year using a field calibration device, model 653 (HYQUEST Solutions Ltd).

We determined continuous stream discharge for each watershed by developing stage discharge rating curves at fixed hydrometric stations situated in close proximity to each stream outlet. Sites were located above tidewater influence and were selected based on favourable conditions (i.e., channel stability and stable hydraulic conditions) for the installation and operation of pressure transducers to measure stream stage. From August 2014 to May 2016 (21 months), we measured stage every 5 min using a pressure level sensor (OTT PLS-L, OTT Hydromet, Colorado, USA) pressure transducer (0–4 m range SDI-12) connected to a CR1000 (Campbell Scientific, Edmonton, Canada) data logger. Stream discharge was measured over various intervals using either the velocity area method (for flows < 0.5 m^3 s^{-1}; ISO Standard 9196, 1992; ISO Standard 748, 2007) or salt dilution (for flows > 0.5 m^3 s^{-1}; Moore, 2005). Rating curves were developed using the relationship between stream stage height and stream discharge (Supplement S2).

2.5 DOC flux

From 1 October 2014 to 30 April 2016, we estimated DOC flux for each watershed using measured DOC concentrations ($n = 224$) and continuous discharge recorded at 15 min intervals. The watersheds in this region respond rapidly to rain inputs and as a result DOC concentrations are highly variable. To address this variability, routine DOC concentration data (as described in Sect. 2.2) were supplemented with additional grab samples ($n = 21$) collected around the peak of the hydrograph during several high-flow events throughout the year. We performed watershed-specific estimates of DOC flux using the rloadest package (Lorenz et al., 2015) in R (version 3.2.5, R Core Team, 2016), which replicates functions developed in the US Geological Survey load-estimator program, LOADEST (Runkel et al., 2004). LOADEST is a multiple-regression adjusted maximum likelihood (ML) estimation model that calibrates a regression between measured constituent values and stream flow across seasons and time and then fits it to combinations of coefficients representing nine predetermined models of constituent flux. To account for potentially small sample size, the best model was selected using the second-order Akaike Information Criterion (AICc) (Akaike, 1981; Hurvich and Tsai, 1989). Input data were log-transformed to avoid bias and centered to reduce multicollinearity. For additional details on model selection, see Supplement Table S3.1.

2.6 Optical characterization of DOM

Prior to May 2014, absorbance measures of water samples ($n = 99$) were conducted on a Varian Cary-50 (Varian, Inc.) spectrophotometer at the BC Ministry of the Environment Technical Services Laboratory (Victoria, BC, Canada) to determine specific UV absorption at 254 nm (SUVA$_{254}$). After May 2014, we conducted optical characterization of DOM by absorbance and fluorescence spectroscopy at the Hakai Institute field station (Calvert Island, BC, Canada) using an Aqualog fluorometer (Horiba Scientific, Edison, New Jersey, USA). Strongly absorbing samples (absorbance units > 0.2 at 250 nm) were diluted prior to analysis to avoid excessive inner filter effects (Lakowicz, 1999). Samples were run in 1 cm quartz cells and scanned from 220 to 800 nm at 2 nm intervals to determine SUVA$_{254}$ as well as the spectral slope ratio (S_R). SUVA$_{254}$ has been shown to positively correlate with increasing molecular aromaticity associated with the fulvic acid fraction of DOM (Weishaar et al., 2003), and it is calculated by dividing the decadic absorption coefficient at 254 nm by DOC concentration (mg C L^{-1}). To account for potential Fe interference with absorbance values, we corrected SUVA$_{254}$ values by Fe concentration according the method described in Poulin et al. (2014). S_R has been shown to negatively correlate with molecular weight (Helms et al., 2008) and is calculated as the ratio of the spectral slope from 275 to 295 nm ($S_{275-295}$) to the spectral slope from 350 to 400 nm ($S_{350-400}$).

We measured excitation and emission spectra (as excitation emission matrices, EEMs) on samples every three weeks from January to July 2016 ($n = 63$). Samples were run in 1 cm quartz cells and scanned from excitation wavelengths of 230–550 nm at 5 nm increments and emission wavelengths of 210–620 nm at 2 nm increments. The Horiba Aqualog applied the appropriate instrument corrections for excitation and emission, inner filter effects, and Raman signal calibration. We calculated the fluorescence index and freshness index for each EEM. The fluorescence index is often used to indicate DOM source, where higher values are more indicative of microbially derived sources of DOM and lower values indicate more terrestrially derived sources (McKnight et al., 2001), and is calculated as the ratio of emission intensity at 450 to 500 nm, at an excitation of 370 nm. The freshness index is used to indicate the contribution of autochthonous or recently microbial-produced DOM, with higher values suggesting greater autochthony (i.e., microbial inputs), and is calculated as the ratio of emission intensity at 380 nm to the maximum emission intensity between 420 and 435 nm, at an excitation of 310 nm (Wilson and Xenopoulos, 2009).

To further characterize features of DOM composition, we performed parallel factor analysis (PARAFAC) using EEM data within the drEEM toolbox for Matlab (Mathworks, MA, USA) (Murphy et al., 2013). PARAFAC is a statistical technique used to decompose the complex mixture of the fluorescing DOM pool into quantifiable, individual components (Stedmon et al., 2003). We detected a total of six unique components and validated the model using core consistency and split-half analysis (Murphy et al., 2013; Stedmon and Bro, 2008). Components with similar spectra from previous studies were identified using the online fluorescence repository, OpenFluor (Murphy et al., 2014), and additional components with similar peaks were identified through literature review. Since the actual chemical structure of fluorophores is unknown, we used the concentration of each fluorophore as maximum fluorescence of excitation and emission in Raman units (F_{max}) to derive the percent contribution of each fluorophore component to total fluorescence. Relationships between PARAFAC components were also evaluated using Pearson correlation coefficients in the R package Hmisc (Harrell et al., 2016).

2.7 Evaluating relationships in DOC concentration and DOM composition with stream discharge and temperature

We used linear mixed-effects (LME) models to assess the relationship between DOC concentration or DOM composition (δ^{13}C-DOC, S_R, SUVA$_{254}$, fluorescence index, freshness index, PARAFAC components), stream discharge, and stream temperature. Analysis was performed in R using the nlme package (Pinheiro et al., 2016). Watershed was included as a random intercept to account for repeat measures on each watershed. For some parameters, a random slope of either discharge or temperature was also included based on data assessment and model selection. Model selection was performed using AIC to compare models fit using ML (Burnham and Anderson, 2002; Symonds and Moussalli, 2010). The final model was fit using restricted maximum likelihood (REML). Marginal R^2, which represents an approximation of the proportion of the variance explained by the fixed factors alone, and conditional R^2, which represents an approximation of the proportion of the variance explained by both the fixed and random factors, were calculated based on the methods described in Nakagawa and Schielzeth (2013) and Johnson (2014).

2.8 Redundancy analysis: relationships between DOC concentration, DOM composition, and watershed characteristics

We evaluated relationships between stream water DOC and watershed characteristics by relating DOC concentration and measures of DOM composition to catchment attributes using redundancy analysis (RDA; type 2 scaling) in the package rdaTest (Legendre and Durand, 2014) in R (version 3.2.2, R Core Team, 2015). To maximize the amount of information available, we performed RDA analysis on samples collected from January to July 2016 and therefore included all parameters of optical characterization (i.e., all PARAFAC components and spectral indices). We assessed the collinearity of DOM compositional variables using a variance inflation factor (VIF) criteria of > 10, which resulted in the removal of PARAFAC components C2, C3, and C5 prior to RDA analysis. Catchment attributes for each watershed included average slope, percent area of lakes, percent area of wetlands, average depth of mineral soil, and average depth of organic soil. Relationships between variables were linear, so no transformations were necessary and variables were standardized prior to analysis. To account for repeat monthly measures per watershed and potential temporal correlation associated with monthly sampling, we included sample month as a covariable ("partial-RDA"). To test whether the RDA axes significantly explained variation in the dataset, we compared permutations of residuals using ANOVA (9999 iterations; test.axes function of rdaTest).

3 Results

3.1 Hydrology

We present work for water year 2015 (WY2015; 1 October 2014–30 September 2015) and water year 2016 (WY2016; 1 October 2015–30 September 2016). Annual precipitation for both water years was lower than historical mean annual precipitation (WY2015: 2661 mm; WY2016: 2587 mm). It is worth noting that mean annual precipitation at our rain gauge location (2890 mm yr^{-1}; elevation: 16 m) is substantially lower than the average amount received at higher elevations, which from 1981 to 2010 was approximately 5027 mm yr^{-1} at an elevation of 1000 m within our study area. This area receives a very high amount of annual rainfall but also experiences seasonal variation, with an extended wet period from fall through spring and a much shorter, typically drier period during summer. In WY2015 and WY2016, 86–88 % of the annual precipitation on Calvert Island occurred during the 8 months of wetter and cooler weather between September and April ($\sim$ 75 % of the year), designated the "wet period" (WY2015 wet: 2388 mm, average air temp: 7.97 °C; WY2016 wet: 2235 mm, average air temp: 7.38 °C). The remaining annual precipitation occurred during the drier and warmer summer months of May–August, designated the "dry period" (WY2015 dry: 314 mm, average air temp: 13.4 °C; WY2016 dry: 352 mm, average air temp: 13.1 °C). Overall, although WY2015 was slightly wetter than WY2016, the two years were comparable in relative precipitation during the wet versus dry periods.

Stream discharge (Q) responded rapidly to rain events and as a result, closely tracked patterns in total precipitation (Fig. 2). Total Q for all watersheds was on average 22 % greater for the wet period of WY2015 (total Q: 223.02 × 10^6; range: 5.13 × 10^6–111.51 × 10^6 m^3) compared to the wet period of WY2016 (total Q: 182.89 × 10^6; range: 4.17 × 10^6–91.45 × 10^6 m^3). Stream discharge and stream temperature were significantly different for wet versus dry periods (Mann–Whitney tests, $p < 0.0001$).

3.2 Temporal and spatial patterns in DOC concentration, yield, and flux

Stream waters were high in DOC concentration relative to the global average for freshwater discharged directly to the ocean (average DOC for Calvert and Hecate islands: 10.4 mg L^{-1}, SD: 3.8; average global DOC: $\sim$ 6 mg L^{-1}) (Meybeck, 1982; Harrison et al., 2005) (Table 1; Fig. 3). Q-weighted average DOC concentrations were higher than average measured DOC concentrations (11.1 mg L^{-1}, Table 1), and also resulted in slightly different ranking of the watersheds for highest to lowest DOC concentration. Within watersheds, Q-weighted DOC concentrations ranged from a low of 8.4 mg L^{-1} (watershed 693) to a high of 19.3 mg L^{-1} (watershed 819), and concentrations were sig-

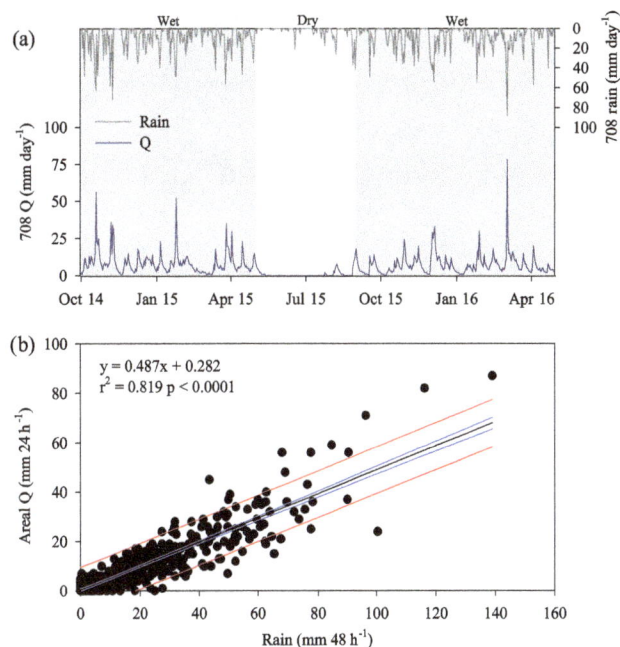

Figure 2. Hydrological patterns typical of watersheds located in the study area. **(a)** The hydrograph and precipitation record from Watershed 708 for the study period of 1 October 2015–30 April 2016. Grey shading indicates the wet period (1 September–30 April) and the unshaded region indicates the dry period (1 May–30 August). **(b)** Correlation of daily (24 h) areal runoff (discharge of all watersheds combined) to 48 h total rainfall recorded at watershed 708. For the period of study, comparisons of daily runoff to 48 h rainfall (runoff : rainfall mean: 0.92; SD ± 0.27) indicated rapid discharge response to rainfall.

nificantly different between watersheds (Kruskal–Wallis test, $p < 0.0001$). Seasonal variability tended to be higher in watersheds where DOC concentration was also high (watersheds 626, 819, and 844) and lower in watersheds with greater lake area (watersheds 1015 and 708) (Table 1; box plots, Fig. 3). On an annual basis, DOC concentrations generally decreased through the wet period and increased through the dry period, and concentrations were significantly lower during the wet period compared to the dry period (Mann–Whitney test, $p = 0.0123$). Results of our LME model (Table S6.1) indicate that DOC concentration was positively related to both discharge ($b = 0.613$, $p < 0.001$) and temperature ($b = 0.162$, $p = 0.011$) (model conditional $R^2 = 0.57$, marginal $R^2 = 0.09$).

Annual and monthly DOC yields are presented in Table 1. For the total period of available Q (1 October 2014–30 April 2016; 19 months), areal (all watersheds) DOC yield was 52.3 Mg C km^{-2} (95 % CI: 45.7 to 68.2 Mg C km^{-2}) and individual watershed yields ranged from 24.1 to 43.6 Mg C km^{-2}. For WY2015, areal annual DOC yield was 33.3 Mg C km^{-2} yr^{-1} (95 % CI: 28.9 to 38.1 Mg C km^{-2} yr^{-1}). Total monthly rainfall was strongly

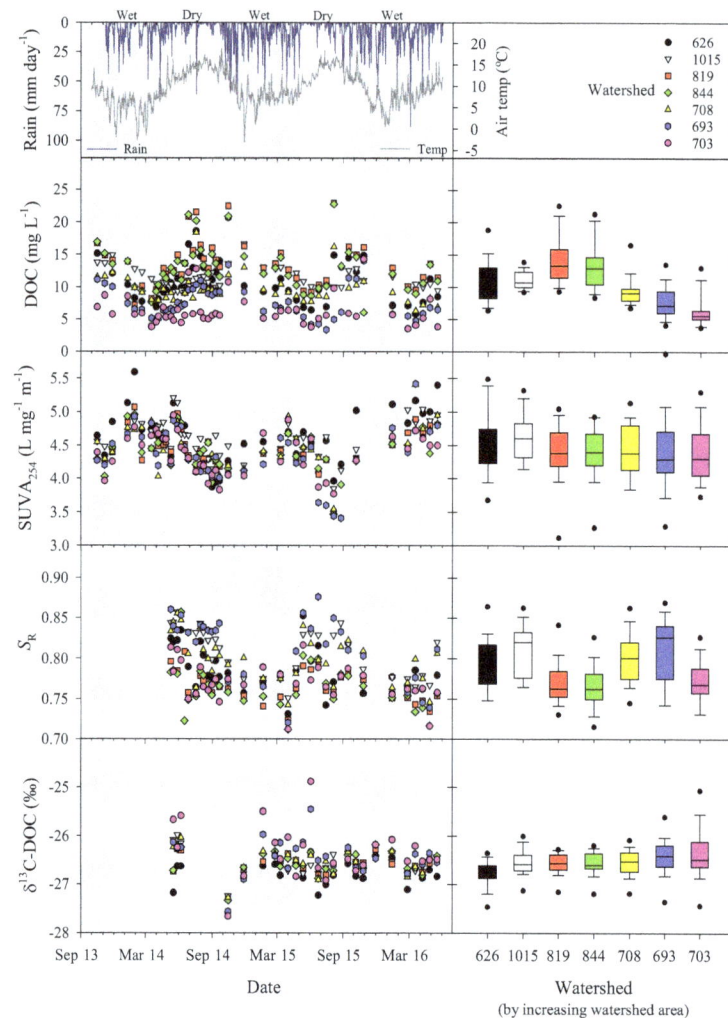

Figure 3. Seasonal (timelines, by date) and spatial (box plots, by watershed) patterns in DOC concentration and DOM composition for stream water collected at the outlets of the seven study watersheds on Calvert and Hecate islands. Boxes represent the 25th and 75th percentile, while whiskers represent the 5th and 95th percentile. Daily precipitation and annual temperature are shown in the top left panel. Grey shading indicates the wet period (1 September–30 April) and the unshaded region indicates the dry period of each water year.

correlated with monthly DOC yield (Fig. 4), and average monthly yield for the wet period (3.35 Mg C km^{-2} month^{-1}; 95 % CI: 2.94 to 4.40 Mg C km^{-2} month^{-1}) was significantly greater than average monthly yield during the dry period (0.50 Mg C km^{-2} month^{-1}; 95 % CI: 0.41 to 0.62 Mg C km^{-2} month^{-1}) (Mann–Whitney test, $p < 0.0001$).

Across our study watersheds, DOC flux generally increased with increasing watershed area (Fig. 5). In WY2015, total DOC flux for all watersheds included in our study was 1562 Mg C (95 % CI: 1355 to 1787 Mg C), and individual watershed flux ranged from ranged from 82 to 276 Mg C. DOC flux was significantly different in wet versus dry periods (Mann–Whitney test, $p < 0.0001$). Overall, 94 % of the export in WY2015 occurred during the wet period, and ex-

port for the wet period of WY2015 was lower than export for the wet period of WY2016 (Fig. 5).

3.3 Temporal and spatial patterns in DOM composition

The stable isotopic composition of dissolved organic carbon (δ^{13}C-DOC) was relatively tightly constrained over space and time (average δ^{13}C-DOC: -26.53‰, SD: 0.36; range: -27.67 to -24.89‰). Values of S_R were low compared to the range typically observed in surface waters (average $S_R = 0.78$, SD: 0.04; range: 0.71 to 0.89), and Fe-corrected SUVA$_{254}$ values were at the high end of the range compared to most surface waters (average SUVA$_{254}$ for Calvert and Hecate islands: 4.42 L mg^{-1} m^{-1}, SD: 0.46; range of SUVA$_{254}$ in surface waters: 1.0 to 5.0 L mg^{-1} m^{-1}) (Spencer et al., 2012). Values for both the fluorescence index (average fluorescence index: 1.36, SD: 0.04; range: 1.30 to

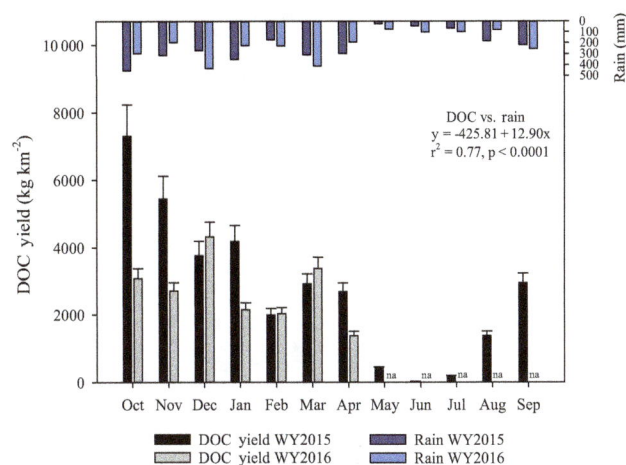

Figure 4. Monthly areal DOC yields and precipitation for water year 2015 (WY2015) and the wet period (1 October–30 April) of water year 2016 (WY2016). Error bars represent standard error. Total rain and DOC yield were significantly correlated ($r^2 = 0.77$), and months of higher rain produced higher DOC yields. In WY2015, the majority of DOC export ($\sim 94\%$ of annual flux) occurred during the wet period ($\sim 88\%$ of annual precipitation).

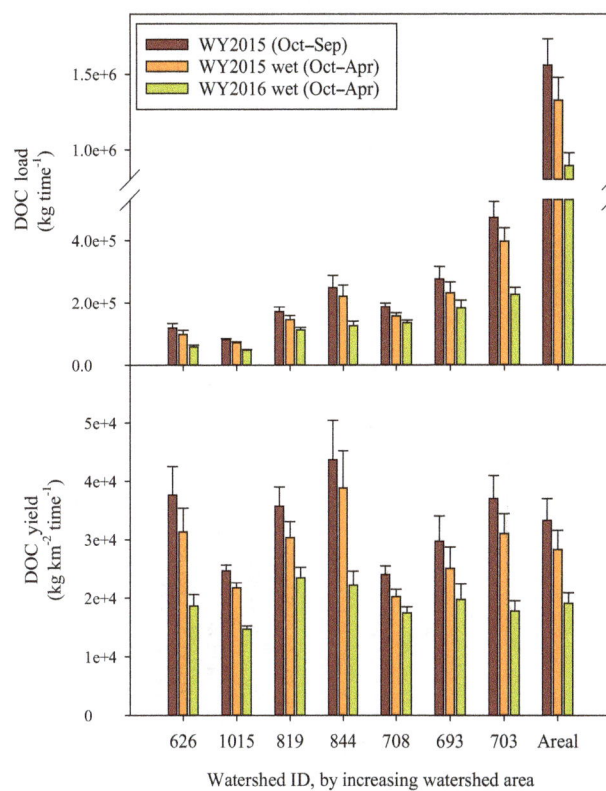

Figure 5. DOC fluxes and yields for the seven study watersheds and the total area of study ("areal", all watersheds combined) on Calvert and Hecate islands for water year 2015 (WY2015; 1 October–30 September), and 1 October–30 April of the wet period for water year 2015 (WY2015 wet) and water year 2016 (WY2016 wet). Because DOC yields were only available for September in WY2015, this month was excluded from the wet-period totals in order to make similar comparisons between years. Error bars represent standard error.

1.44) and freshness index (average freshness index: 0.46, SD: 0.02; range: 0.41 to 0.49) were relatively low compared to the typical range found in surface waters (Fellman et al., 2010; Hansen et al., 2016). Differences between watersheds were observed for δ^{13}C-DOC (Kruskal–Wallis test, $p = 0.0043$), S_R (Kruskal–Wallis test, $p = 0.0001$), fluorescence index (Kruskal–Wallis test, $p = 0.0030$), and freshness index (Kruskal–Wallis test, $p = 0.0099$), but watersheds did not differ in SUVA$_{254}$ (Kruskal–Wallis test, $p = 0.4837$).

We observed seasonal variability in δ^{13}C-DOC throughout the period of sampling (Fig. 3 and our LME model (Table S6.1) indicate that δ^{13}C-DOC declined with increasing discharge ($b = -0.049$, $p = 0.014$) and stream temperature ($b = -0.024$, $p < 0.001$) (model conditional $R^2 = 0.35$, marginal $R^2 = 0.10$). In contrast, although SUVA$_{254}$ appeared to exhibit a general seasonal trend of values increasing over the wet period and decreasing over the dry period, SUVA$_{254}$ was not significantly related to either discharge or stream temperature in the LME model results. S_R also appeared to fluctuate seasonally, with lower values during the wet season and higher values during the dry season. S_R was negatively related to discharge ($b = -0.026$; $p < 0.001$) and positively related to the interaction between discharge and stream temperature ($b = 0.0015$; $p < 0.001$) (model conditional $R^2 = 0.62$; marginal $R^2 = 0.28$). The freshness index was negatively related to stream temperature ($b = -0.003$; $p = 0.008$) (model conditional $R^2 = 0.59$; marginal $R^2 = 0.23$), while the fluorescence index was not significantly related to either discharge or stream temperature.

3.4 PARAFAC characterization of DOM

Six fluorescence components were identified through PARAFAC ("C1" through "C6") (Table 2). Additional details on PARAFAC model results are provided in Supplement Table S4.1 and Figs. S4.2 and S4.3. Of the six components, four were found to have close spectral matches in the Open-Fluor database (C1, C3, C5, C6; minimum similarity score > 0.95), while the remaining two (C2 and C4) were found to have similar peaks represented in the literature. The first four components (C1 through C4) are described as terrestrially derived, whereas components C5 and C6 are described as autochthonous or microbially derived (Table 2). In general, the rank order of each component's percent contribution to total fluorescence was maintained over time, with C1 comprising the majority of total fluorescence across all watersheds (Fig. 6).

Across watersheds, components fluctuated synchronously over time and variation between watersheds was relatively

Table 2. Spectral composition for the six fluorescence components identified using PARAFAC, including excitation (Ex.) and emission (Em.) peak values, percent composition across all samples, and likely structure and characteristics of the fluorescent component based on previous studies.

Component	Ex. (nm)	Em. (nm)	% composition*	Potential structure/ characteristics	Previous studies with comparable results
C1	315	436	34.1 ± 2.2 (31.1–39.3)	Humic-like, less processed terrestrial, high molecular weight, widespread but highest in wetland and forest environment	Garcia et al. (2015) (C1); Graeber et al. (2012) (C1); Walker et al. (2014) (C1); Yamashita et al. (2011) (C1); Cory and McKnight (2005) (C1)
C2	270/380	484	20.2 ± 1.9 (16.1–25.6)	Humic-like, resembles fulvic acid, widespread, high molecular weight terrestrial	Stedmon and Markager (2005) (C2); Stedmon et al. (2003) (C3); Cory and McKnight (2005) (C5)
C3	270	478	17.8 ± 1.8 (12.8–20.8)	Humic-like, highly processed terrestrial; suggested as refractory	Stedmon and Markager (2005) (C1); Yamashita et al. (2010) (C2)
C4	305/435	522	14.8 ± 2.6 (9.4–22.3)	Not commonly reported, similarities to fulvic-like, contributed from soils	Lochmuller and Saavedra (1986) (E)
C5	325	442	9.8 ± 3.5 (0.0–15.9)	Aquatic humic-like from terrestrial environments; autochthonous, microbial produced; may be photoproduced	Boehme and Coble (2000) (Peak C); Coble et al. (1998) (Peak C); Stedmon et al. (2003) (C3)
C6	285	338	3.4 ± 2.5 (0.0–9.3)	Amino acid-like/tryptophan-like. Freshly added from land, autochthonous. Rapidly photodegradable	Murphy et al. (2008) (C7); Shutova et al. (2003) (C4); Stedmon et al. (2007) (C7); Yamashita et al. (2003) (C5)

* Mean ± SD (min–max) from all samples.

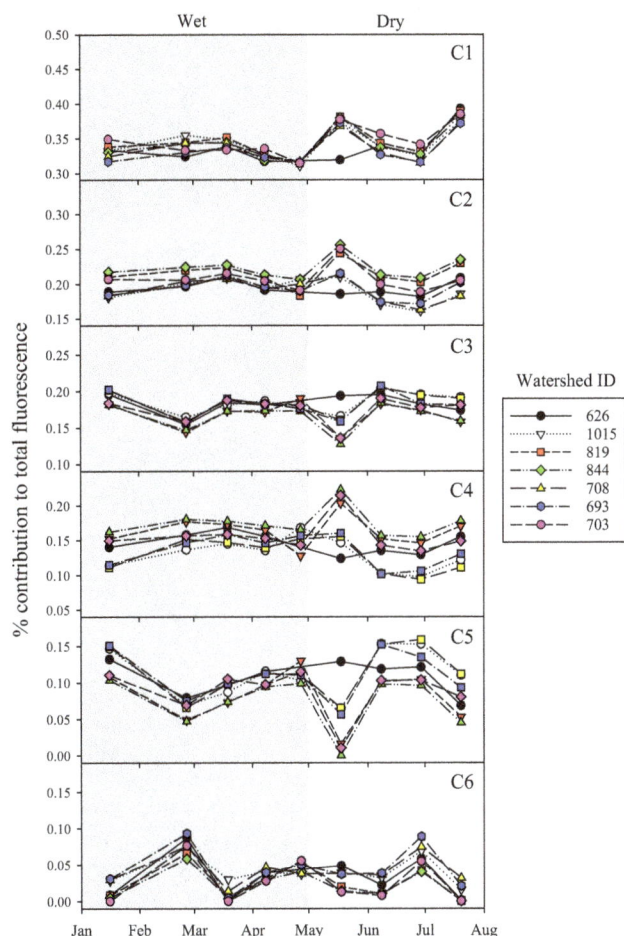

Figure 6. Percent contribution of the six components identified in parallel factor analysis (PARAFAC) for samples collected every three weeks from January–July 2016 from the seven study watersheds on Calvert and Hecate islands. The grey shading indicates the wet period and the unshaded region indicates the dry period. Note that while the y axis for each panel has a range of 20 %, the max and min for each y axis varies by panel.

low, although slightly more variation between watersheds was observed during the beginning of the dry period relative to other times of the year (Fig. 6). The percent contributions of components C1, C3, C5, and C6 to total fluorescence were not significantly different across watersheds (for all components Kruskal–Wallis test, $p > 0.05$); however, the percent composition of both C2 and C4 was different (Kruskal–Wallis test, $p = 0.0306$ and $p = 0.0307$, respectively) and higher for watersheds 819 and 844 relative to the other watersheds (Fig. S4.4).

PARAFAC components exhibited significant relationships with stream discharge and stream temperature, although predicted changes (beta, or b) in fluorescence components with discharge and/or stream temperature were small (Supplement Table S6.2). C3 increased with discharge ($b = 0.006$, $p = 0.003$), whereas C2, C4, and C5 decreased

with discharge (C2: $b = -0.005$, $p = 0.022$; C4: $b = -0.008$, $p = 0.002$; C5: $b = -0.008$, $p = 0.002$). C1, C4, and C6 increased with temperature (C1: $b = 0.001$, $p = 0.050$; C4: $b = 0.003$, $p < 0.001$; C6: $b = 0.005$, $p = 0.005$), while both C3 and C5 decreased with temperature (C3: $b = -0.003$, $p = 0.003$; C5: $b = -0.003$, $p = 0.027$). Conditional R^2 values for the models ranged from 0.28 to 0.69, while marginal R^2 ranged from 0.20 to 0.46. Overall, greater changes in component contribution to total fluorescence were observed with changes in discharge relative to changes in stream temperature.

3.5 Relationships between watershed characteristics, DOC concentrations, and DOM composition

Results of the partial RDA (type 2 scaling) were significant in explaining variability in DOM concentration and composition (semi-partial $R^2 = 0.33$, $F = 7.90$, $p < 0.0001$) (Fig. 7). Axes 1 through 3 were statistically significant at $p < 0.001$, and the relative contribution of each axis to the total explained variance was 47, 30, and 22 %, respectively. Additional details on the RDA test are provided in Figs. S5.1–S5.2 and Tables S5.3–S5.5. Axis 1 described a gradient of watershed coverage by water-inundated ecosystem types, ranging from more wetland coverage to more lake coverage. Total lake coverage (area) and mean mineral soil material thickness showed a strong positive contribution, and wetland coverage (area) showed a strong negative contribution to this axis. The freshness index, the fluorescence index, S_R, and the fluorescence component C6 were positively correlated with Axis 1, while component C4 showed a clear negative correlation. Axis 2 described a subtler gradient of soil material thickness ranging from greater mean organic soil material thickness to greater mean mineral soil material thickness. DOC concentration, δ^{13}C-DOC, SUVA$_{254}$, and fluorescence component C1 all showed a strong, positive correlation with Axis 2. Axis 3 described a gradient of watershed steepness, from lower gradient slopes with more wetland area and thicker organic soil material to steeper slopes with less developed organic horizons. Average slope contributed negatively to Axis 3 (see Table S5.5), followed by positive contributions from both wetland area and the thickness of organic soil material. δ^{13}C-DOC showed the most positive correlation with Axis 3, whereas fluorescence components C1 and C4 showed the most negative.

4 Discussion

4.1 DOC export from small catchments to the coastal ocean

In comparison to global models of DOC export (Mayorga et al., 2010) and DOC exports quantified for southeastern Alaska (D'Amore et al., 2015a, 2016; Stackpoole et al., 2017), our estimates of freshwater DOC yield from Calvert

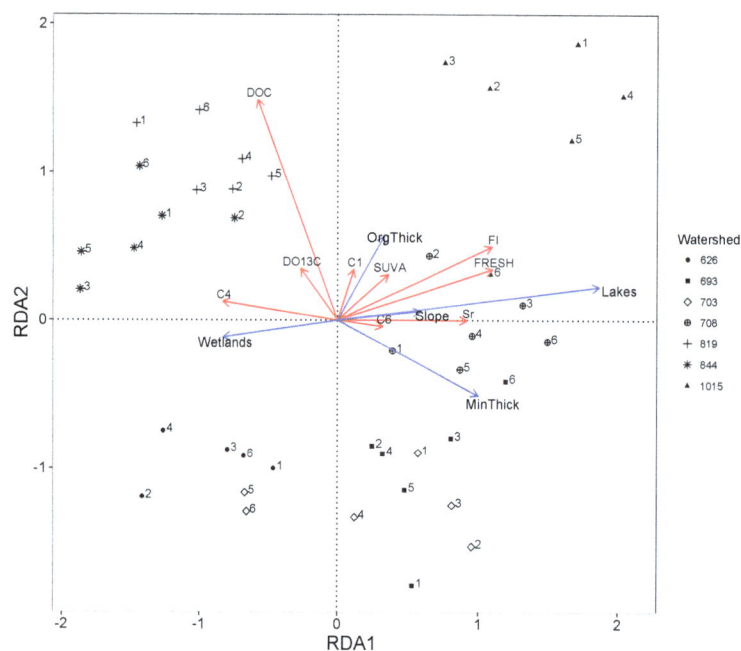

Figure 7. Results from the partial redundancy analysis (RDA; type 2 scaling) of DOC concentration and DOM composition versus watershed characteristics. Angles between vectors represent correlation; i.e., smaller angles indicate higher correlation. Symbols represent different watersheds, and numbers on symbols represent the sample month in 2016: 1 – January; 2 – February; 3 – March; 4 – early April; 5 – late April; and 6 – May.

and Hecate island watersheds are in the upper range predicted for the perhumid rainforest region. When compared to watersheds of similar size, DOC yields from Calvert and Hecate island watersheds are some of the highest observed (see reviews in Hope et al., 1994; Alvarez-Cobelas et al., 2012), including DOC yields from many tropical rivers, despite the fact that tropical rivers have been shown to export very high DOC (e.g., Autuna River, Venezuela, DOC yield: $56\,946\,kg\,C\,km^{-2}\,yr^{-1}$; Castillo et al., 2004) and are often regarded as having disproportionately high carbon export compared to temperate and Arctic rivers (Aitkenhead and McDowell, 2000; Borges et al., 2015). Our estimates of DOC yield are comparable to, or higher than, previous estimates from high-latitude catchments of similar size that receive high amounts of precipitation and contain extensive organic soils and wetlands (e.g., Naiman, 1982 (DOC yield: $48\,380\,kg\,C\,km^{-2}\,yr^{-1}$); Brooks et al., 1999 (DOC yield: $20\,300\,kg\,C\,km^{-2}\,yr^{-1}$); Ågren et al., 2007 (DOC yield: $32\,043\,kg\,C\,km^{-2}\,yr^{-1}$)). However, many of these catchments represent low- (first- or second-) order headwater streams that drain to higher-order stream reaches, rather than directly to the ocean. Although headwater streams have been shown to export up to 90 % of the total annual carbon in stream systems (Leach et al., 2016), significant processing and loss typically occurs during downstream transit (Battin et al., 2008).

Over much of the incised outer coast of the CTR, small rainfall-dominated catchments contribute high amounts of freshwater runoff to the coastal ocean (Royer, 1982; Morrison et al., 2012; Carmack et al., 2015). Small mountainous watersheds that discharge directly to the ocean can exhibit disproportionately high fluxes of carbon relative to watershed size and in aggregate may deliver more than 50 % of total carbon flux from terrestrial systems to the ocean (Milliman and Syvitski, 1992; Masiello and Druffel, 2001). Extrapolating our estimate of annual DOC yield from Calvert and Hecate island watersheds to the entire hypermaritime subregion of British Columbia's CTR ($29\,935\,km^2$) generates an estimated annual DOC flux of $0.997\,Tg\,C\,yr^{-1}$ (0.721 to $1.305\,Tg\,C\,yr^{-1}$ for our lowest to highest yielding watersheds, respectively), with the caveat that this estimate is rudimentary and does not account for spatial heterogeneity in controlling factors such as wetland extent, topography, and watershed size. Regional comparisons estimate that Southeast Alaska ($104\,000\,km^2$), at the northern range of the CTR, exports approximately $1.25\,Tg\,C\,yr^{-1}$ (Stackpoole et al., 2016), while south of the perhumid CTR, the wet northwestern United States and its associated coastal temperate rainforests export less than $0.153\,Tg\,C\,yr^{-1}$ as DOC (reported as TOC; Butman et al., 2016). This suggests that the hypermaritime coast of British Columbia plays an important role in the export of DOC from coastal temperate rainforest ecosystems of western North America, in a region that is already expected to contribute high quantities of DOC to the coastal ocean.

4.2 DOM composition

The composition of stream water DOM exported from Calvert and Hecate island watersheds is mainly terrestrial, indicating that the production and overall supply of terrestrial material is sufficient to exceed microbial demand, and thus a relatively abundant supply of terrestrial DOM is available for export. Values for δ^{13}C-DOC suggest that terrestrial carbon sources from C3 plants and soils were the dominant input to catchment stream water DOM (Finlay and Kendall, 2007). Measures of S_R and SUVA$_{254}$ were typical of environments that export large quantities of high molecular-weight, highly aromatic DOM such as some tropical rivers (e.g., Lambert et al., 2016; Mann et al., 2014), streams draining wetlands (e.g., Ågren et al., 2008; Austnes et al., 2010), or streams draining small undisturbed catchments comprised of mixed forest and wetlands (e.g., Wickland et al., 2007; Fellman et al., 2009a; Spencer et al., 2010; Yamashita et al., 2011). This suggests that the majority of the DOM pool is comprised of larger molecules that have not been extensively chemically or biologically degraded through processes such as microbial utilization or photodegradation and therefore are potentially more biologically available (Amon and Benner, 1996).

Biological utilization of DOM is influenced by its composition (e.g., Judd et al., 2006; Fasching et al., 2014); therefore, differences in DOM can alter the downstream fate and ecological role of freshwater-exported DOM. For example, the majority of the fluorescent DOM pool was comprised of C1, which is described as humic-like, less processed terrestrial soil and plant material (see Table 2). In addition, although the tryptophan-like component C6, represents a minor proportion of total fluorescence, even a small proteinaceous fraction of the overall DOM pool can play a major role in overall bioavailability and bacterial utilization of DOM (Berggren et al., 2010; Guillamette and Giorgio, 2011). These contributions of stream-exported DOM may represent a relatively fresh, seasonally consistent contribution of terrestrial subsidy from streams to the coastal ecosystem, which in this region is relatively lower in carbon and nutrients throughout much of the year (Whitney et al., 2005; Johannessen et al., 2008).

4.3 DOC and DOM export: sources and seasonal variability

On Calvert and Hecate islands, the relationship between DOC concentration and discharge varied by watershed (see Supplement Fig. S6.1), as might be expected given the known influence of watershed characteristics (e.g., lake area, wetland area, soils, etc.) on DOC concentration and export. However, overall DOC concentrations increased in all watersheds with both discharge and temperature indicating that the overarching drivers of DOC export are the hydrologic coupling of precipitation and runoff from the landscape with the seasonal production and availability of DOC (Fasching et al., 2016).

Precipitation is a well-established driver of stream DOC export (Alvarez-Cobelas et al., 2012), particularly in systems containing organic soils and wetlands (Olefeldt et al., 2013; Wallin et al., 2015; Leach et al., 2016). Frequent, high-intensity precipitation events and short residence times are expected to result in pulsed exports of stream DOC that are rapidly shunted downstream, thus reducing time for in-stream processing (Raymond et al., 2016). Flashy stream hydrographs indicate that hydrologic response times for Calvert and Hecate island watersheds are rapid, presumably as a result of small catchment size, high drainage density, and relatively shallow soils with high hydraulic conductivity (Gibson et al., 2000; Fitzgerald et al., 2003). Rapid runoff is presumably accompanied by rapid increases in water tables and lateral movement of water through shallow soil layers rich in organic matter (Fellman et al., 2009b; D'Amore et al., 2015b). It appears that on Calvert and Hecate islands, the combination of high rainfall, rapid runoff, and abundant sources of DOC from organic-rich wetlands and forests results in high DOC fluxes.

The relationship between DOC, stream temperature, and discharge indicates that seasonal dynamics play an important role in the variability of DOC exported from these systems. For example, DOC concentrations decrease in all watersheds during the wet period of the year; these decreases are associated with clear changes in DOM composition, such as increasing δ^{13}C-DOC and SUVA$_{254}$, and decreasing S_R. This is in contrast with patterns observed during the dry period, when DOC concentrations gradually increase, while δ^{13}C-DOC, SUVA$_{254}$ decrease. Fluctuations in DOC and DOM composition occur throughout the wet and the dry season, suggesting that temperature and runoff – and perhaps other seasonal drivers – are important year-round controls on DOC concentration as well as certain measures of DOM composition, such as δ^{13}C-DOC and S_R.

The process of "DOC flushing" has been shown to increase stream water DOC during higher flows in coastal and temperate watersheds (e.g., Sanderman et al., 2009; Deirmendjian et al., 2017). Flushing can occur through various mechanisms. For example, Boyer et al. (1996) observed that during drier periods, DOC pools can increase in soils and are then flushed to streams when water tables rise. Rising water tables can establish strong hydraulic gradients that initiate and sustain prolonged increases in metrics like SUVA$_{254}$, until the progressive drawdown of upland water tables constrains flow paths (Lambert et al., 2013). DOC concentrations can vary during flushing in response to changing flow paths, which can shift sources of DOC within the soil profile from older material in deeper soil horizons to more recently produced material in shallow horizons (Sanderman et al., 2009) or from changes in the production mechanism of DOC (Lambert et al., 2013). For example, Sanderman et al. (2009), observed distinct relationships between discharge and both δ^{13}C-DOC and SUVA$_{254}$ and postulated that during the rainy season, hillslope flushing shifts DOM sources to more aged soil organic material. In addition, instream production can also provide a source of DOC and therefore affect seasonal variation in DOC concentration and composition (Lambert et al., 2013). The extent of these effects can shift seasonally; relationships between flow paths and DOC export in rain-dominated catchments can vary within and between hydrologic periods depending on factors such as the degree of soil saturation, the duration of previous drying and rewetting cycles, soil chemistry, and DOM source-pool availability (Lambert et al., 2013).

Our observations of changes in DOC and DOM related to discharge and stream temperature suggest that a variety of mechanisms may be important for controlling dynamics of seasonal export in Pacific hypermaritime watersheds. We observed elevated DOC concentrations during precipitation events following extended dry periods, suggesting DOC may accumulate during dry periods and be flushed to streams during runoff events. Increased discharge was significantly related to δ^{13}C-DOC and S_R, with higher discharge resulting in more terrestrial-like DOM. One possible explanation is that hydrologic connectivity increases during higher discharge as soil conditions become more saturated, therefore promoting the mobilization of DOM from across a wider range of the soil profile (McKnight et al., 2001; Kalbitz et al., 2002). In addition, the mechanisms of DOC production and sources of DOC appear to shift seasonally. Relationships between increased temperature and lower values of δ^{13}C-DOC and higher values of the freshness index, C1, and C4 suggest that warmer conditions result in a fresh supply of DOM exported from terrestrial sources (Fellman et al., 2009a; Fasching et al., 2016). This may represent a shift in the source of DOM and/or increased contributions from less aromatic, lower molecular-weight material, such as DOM derived from increased terrestrial primary production (Berggren et al., 2010). Further, fine-scaled investigation into the mechanistic underpinnings of the relationship between discharge, stream temperature, and DOM represents a clear priority for future research in this region.

4.4 Relationships between watershed attributes and exported DOM

Previous studies have implicated wetlands as a major driver of DOM composition (e.g., Xenopoulos et al., 2003; Ågren et al., 2008; Creed et al., 2008); however, the analysis of relationships between Calvert and Hecate island landscape attributes and variation in DOM composition suggests that controls on DOM composition are more nuanced than being solely driven by the extent of wetlands. Ågren et al. (2008) found that when wetland area comprised > 10 % of total catchment area, wetland DOM was the most significant driver of stream DOM composition during periods of high hydrologic connectivity. Although wetlands comprise an av-

erage of 37 % of our study area, they do not appear to be the single leading driver of variability in DOC concentration and DOM composition. Other factors, such as watershed slope, the depth of organic and mineral soil materials, and the presence of lakes also appear to be influence DOC concentration and DOM composition. The presence of cryptic wetlands (Creed et al., 2003) and limitations of the wetland mapping method could also weaken the link between wetland extent, DOC, and DOM.

In these watersheds, soils with pronounced accumulations of organic matter are not restricted to wetland ecosystems. Peat accumulation in wetland ecosystems results in the formation of organic soils (Hemists), where mobile fractions of DOM accumulate under saturated soil conditions and limited drainage, resulting in the enrichment of poorly biodegradable, more stable humic acids (Stevenson, 1994; Marschner and Kalbitz, 2003). Although Hemist soils comprise 27.8 % of our study area, Folic Histosols, which form under more freely drained conditions, such as steeper slopes, occur over an additional 25.7 % of the area (Supplement S1.2). In freely drained organic soils, high rates of respiration can result in further enrichment of aromatic and more complex molecules, and this material may be rapidly mobilized and exported to streams (Glatzel et al., 2003). This suggests the importance of widely distributed, alternative soil DOM source pools, such as Folic Histosols and associated Podzols with thick forest floors on hillslopes, available to contribute high amounts of terrestrial carbon for export.

Although lakes make up a relatively small proportion of the total landscape area, their influence on DOM export appears to be important. The proportion of lake area can be a good predictor of organic carbon loss from a catchment since lakes often increase hydrologic residence times and thus increase opportunities for biogeochemical processing (Algesten et al., 2004; Tranvik et al., 2009). In our study, watersheds with a larger percentage of lake area exhibited a slower response following rain events (Fig. S2.2) and lower DOC yields, and lake area was correlated with parameters that represent greater autochthonous DOM production or microbial processing such as higher freshness index, S_R, and fluorescence index and a higher proportions of component C6. In contrast, watersheds with a high percentage of wetlands contributed DOM that was more allochthonous in composition. Lakes are known to be important landscape predictors of DOC, as increased residence time enables removal via respiration, thus reducing downstream exports from lake outlets (Larson et al., 2007). The proximity of wetlands and lakes to the watershed outlet can also play an important role in the composition of DOM exports (Martin et al., 2006).

5 Conclusions

Previous work has demonstrated that freshwater discharge is substantial along the coastal margin of the North Pacific temperate rainforest and plays an important role in processes such as ocean circulation (Royer, 1982; Eaton and Moore, 2010). Our finding that small catchments in this region contribute high yields of terrestrial DOC to coastal waters suggests that freshwater inputs may also influence ocean biogeochemistry and food web processes through terrestrial organic-matter subsidies. Our findings also suggest that this region may be currently underrepresented in terms of its role in global carbon cycling. Currently, there is no region-wide carbon flux model for the Pacific coastal temperate rainforest or the greater Gulf of Alaska, which would quantify the importance of this region within the global carbon budget. Our estimates point to the importance of the hypermaritime outer-coast zone of the CTR, where subdued terrain, high rainfall, ocean-moderated temperatures and poor bedrock have generated a distinctive bog-forest landscape mosaic within the greater temperate rainforest (Banner et al., 2005). However, even within our geographically limited study area, we observed a range of DOC yields across watersheds. To quantify regional-scale fluxes of rainforest carbon to the coastal ocean, further research will be needed to estimate DOC yields across complex spatial gradients of topography, climate, hydrology, soils, and vegetation. Long-term changes in DOC flux have been observed in many places (e.g., Worrall et al., 2004; Borken et al., 2011; Lepistö et al., 2014; Tank et al., 2016), and continued monitoring of this system will allow us to better understand the underlying drivers of export and evaluate future patterns in DOC yields. Coupled with current studies investigating the fate of terrestrial material in ocean food webs, this work will improve our understanding of coastal carbon patterns and increase capacity for predictions regarding the ecological impacts of climate change.

Author contributions. AAO prepared the paper with contributions from all authors, designed analysis protocols, analyzed samples and performed the modelling and analysis for dissolved organic carbon fluxes, parallel factor analysis of dissolved organic-matter composition, and all remaining statistical analyses. SET assisted with designing the study and overseeing laboratory analyses, crafting the scope of the paper, and determining the analytical approach. IG led the initial DOC sampling design, helped coordinate the research team, oversaw routine sampling and data management, and led the watershed characterization. MCK developed the rating curves and conducted the statistical analysis of discharge measurement uncertainties and rating curve uncertainties. WCF led the hydrology component of this project, selected site locations, installed and designed the hydrometric stations, and developed the rating curves and final discharge calculations. CB and PS collected and analyzed soil field data and prepared the digital soils map of the watersheds. KPL conceived of and co-led the overall study of which this paper is a component, helped assemble and guide the team of researchers who carried out this work and provided input to each stage of the study.

Competing interests. The authors declare that they have no conflict of interest.

Acknowledgements. This work was funded by the Tula Foundation and the Hakai Institute. The authors would like to thank many individuals for their support, including Skye McEwan, Bryn Fedje, Lawren McNab, Nelson Roberts, Adam Turner, Emma Myers, David Norwell, and Chris Coxson for sample collection and data management, Clive Dawson and North Road Analytical for sample processing and data management, Keith Holmes for creating our maps, Matt Foster for database development and support, Shawn Hateley for sensor network maintenance, Jason Jackson, Colby Owen, James McPhail, and the entire staff at Hakai Energy Solutions for installing and maintaining the sensors and telemetry network, and Stewart Butler and Will McInnes for field support. Thanks to Santiago Gonzalez Arriola for generating the watershed summaries and associated data products and Ray Brunsting for overseeing the design and implementation of the sensor network and the data management system at Hakai. Additional thanks to Lori Johnson and Amelia Galuska for soil mapping field assistance, and Francois Guillamette for PARAFAC consultation. Thanks to Dave D'Amore for inspiring the Hakai project to investigate aquatic fluxes at the coastal margin and for technical guidance. Lastly, thanks to Eric Peterson and Christina Munck, who provided significant guidance throughout the process of designing and implementing this study.

Edited by: Steven Bouillon

References

Ågren, A., Buffam, I., Jansson, M., and Laudon, H.: Importance of seasonality and small streams for the landscape regulation of dissolved organic carbon export, J. Geophys. Res.-Biogeosci., 112, https://doi.org/10.1029/2006JG000381, 2007.

Ågren, A., Buffam, I., Berggren, M., Bishop, K., Jansson, M., and Laudon, H.: Dissolved organic carbon characteristics in boreal streams in a forest-wetland gradient during the transition between winter and summer, J. Geophys. Res.-Biogeosci., 113, https://doi.org/10.1029/2007JG000674, 2008.

Akaike, H.: Likelihood of a model and information criteria, J. Econometrics, 16, 3–14, https://doi.org/10.1016/0304-4076(81)90071-3, 1981.

Aitkenhead, J. A. and McDowell, W. H.: Soil C:N ratio as a predictor of annual riverine DOC flux at local and global scales, Global Biogeochem. Cy., 14, 127–138, https://doi.org/10.1029/1999GB900083, 2000.

Alaback, P. B.: Biodiversity patterns in relation to climate: The coastal temperate rainforests of North America, Ecol. Stud., 116, 105–133, https://doi.org/10.1007/978-1-4612-3970-3_7, 1996.

Algesten, G., Sobek, S., Bergström, A., Ågren, A., Tranvik, L., and Jansson, M.: Role of lakes for organic carbon cycling in the boreal zone, Glob. Change Biol., 10, 141–147, https://doi.org/10.1111/j.1365-2486.2003.00721.x, 2004.

Alvarez-Cobelas, M., Angeler, D., Sánchez-Carrillo, S., and Almendros, G.: A worldwide view of organic carbon export from catchments, Biogeochemistry, 107, 275–293, https://doi.org/10.1007/s10533-010-9553-z, 2012.

Amon, R. M. W. and Benner, R.: Bacterial utilization of different size classes of dissolved organic matter, Limnol. Oceanogr., 41, 41–51, 1996.

Aufdenkampe, A., Mayorga, E., Raymond, P., Melack, J., Doney, S., Alin, S., Aalto, R., and Yoo, K.: Riverine coupling of biogeochemical cycles between land, oceans, and atmosphere, Front. Ecol. Environ., 9, 53–60, https://doi.org/10.1890/100014, 2011.

Austnes, K., Evans, C. D., Eliot-Laize, C., Naden, P. S., and Old, G. H.: Effects of storm events on mobilisation and in-stream processing of dissolved organic matter (DOM) in a Welsh peatland catchment, Biogeochemical, 99, 157–173, https://doi.org/10.1007/s10533-009-9399-4, 2010.

Banner, A., LePage, P., Moran, J., and de Groot, A. (Eds.): The HyP3 Project: pattern, process, and productivity in hypermaritime forests of coastal British Columbia – a synthesis of 7-year results, Special Report 10, Res. Br., British Columbia Ministry Forests, Victoria, British Columbia, 142 pp., available at: http://www.for.gov.bc.ca/hfd/pubs/Docs/Srs/Srs10.htm (last access: 11 August 2017), 2005.

Battin, T. J., Kaplan, L. A., Findlay, S., Hopkinson, C. S., Marti, E., Packman, A. I., Newbold, D., and Sabater, F.: Biophysical controls on organic carbon fluxes in fluvial networks, Nat. Geosci., 1, 95–100, 2008.

Bauer, J. E., Cai, W. J., Raymond, P. A., T. S., Bianchi, Hopkinson, C. S., and Regnier, P. A. G.: The changing carbon cycle of the coastal ocean, Nature, 504, 61–70, https://doi.org/10.1038/nature12857, 2013.

Berggren, M., Laudon, H., Haei, M., Ström, L., and Jansson, M.: Efficient aquatic bacterial metabolism of dissolved low-molecular-weight compounds from terrestrial sources, ISME J., 4, 408–416, https://doi.org/10.1038/ismej.2009.120, 2010.

Boehme, J. and Coble, P.: Characterization of Colored Dissolved Organic Matter Using High-Energy Laser Fragmentation, Environ. Sci. Technol., 34, 3283–3290, https://doi.org/10.1021/es9911263, 2000.

Borken, W., Ahrens, B., Schultz, C., and Zimmermann, L.: Site-to-site variability and temporal trends of DOC concentrations and fluxes in temperate forest soils, Glob. Change Biol., 17: 2428–2443, https://doi.org/10.1111/j.1365-2486.2011.02390.x, 2011.

Borges, A. V., Darchambeau, F., Teodoru, C. R., Marwick, T. R., Tamooh, F., Geeraert, N., Omengo, F. O., Guérin, F., Lambert, T., Morana, C., Okuku, E., and Bouillon, S.: Globally significant greenhouse-gas emissions form African inland waters, Nat. Geosci., 8, 637–642, https://doi.org/10.1038/ngeo2486, 2015.

Boyer, E. W., Hornberger, G. M., Bencala, K. E., and McKnight, D.: Overview of a simple model describing variation of dissolved organic carbon in an upland catchment, Ecol. Modell., 86, 183–188, 1996.

Burnham, K. P. and Anderson, D. R.: Model selection and multimodel inference, 2nd Edn., Springer, New York, 2002.

Carmack, E., Winsor, P., and William, W.: The contiguous panarctic Riverine Coastal Domain: A unifying concept, Prog. Oceanogr., 139, 13–23, https://doi.org/10.1016/j.pocean.2015.07.014, 2015.

Castillo, M. M., Allan, J. D., Sinsabaugh, R. L., and Kling, G. W.: Seasonal and interannual variation of bacterial production in lowland rivers of the Orinoco basin, Freshwater Biol., 49, 1400–1414, https://doi.org/10.1111/j.1365-2427.2004.01277.x, 2004.

Clark, J. M., Lane, S. N., Chapman, P. J., and Adamson, J. K.: Export of dissolved organic carbon from an upland peatland during storm events: Implications for flux estimates, J. Hydrol., 347, 438–447, https://doi.org/10.1016/j.jhydrol.2007.09.030, 2007.

Coble, P., Castillo, C., and Avril, B.: Distribution and optical properties of CDOM in the Arabian Sea during the 1995 Southwest Monsoon, Deep-Sea Res. Pt. II, 45, 2195–2223, https://doi.org/10.1016/S0967-0645(98)00068-X, 1998.

Cole, J., Prairie, Y., Caraco, N., McDowell, W., Tranvik, L., Striegl, R., Duarte, C., Kortelainen, P., Downing, J., Middelburg, J., and Melack, J.: Plumbing the Global Carbon Cycle: Integrating Inland Waters into the Terrestrial Carbon Budget, Ecosystems, 10, 172–185, https://doi.org/10.1007/s10021-006-9013-8, 2007.

Cory, R. M. and McKnight, D. M.: Fluorescence spectroscopy reveals ubiquitous presence of oxidized and reduced quinines in dissolved organic matter, Environ. Sci. Technol., 39, 8142–8149, https://doi.org/10.1021/es0506962, 2005.

Creed, I. F., Beall, F. D., Clair, T. A., Dillon, P. J., and Hesslein, R. H.: Predicting export of dissolved organic carbon from forested catchments in glaciated landscapes with shallow soils, Glob. Biogeochem. Cy., 22, GB4024, https://doi.org/10.1029/2008GB003294, 2008.

Creed, I. F., Sanford, S. E., Beall, F. D., Molot, L. A., and Dillon, P. J.: Cryptic wetlands: integrating hidden wetlands in regression models of the export of dissolved organic carbon from forested landscapes, Hydrol. Process., 17, 3629–3648, 2003.

D'Amore, D. V., Edwards, R. T., and Biles, F. E.: Biophysical controls on dissolved organic carbon concentrations of Alaskan coastal temperate rainforest streams, Aquat. Sci., 2, 381–393, https://doi.org/10.1007/s00027-015-0441-4, 2015a.

D'Amore, D. V., Edwards, R. T., Herendeen, P. A., Hood, E., and Fellman, J. B.: Dissolved organic carbon fluxes from hydropedologic units in Alaskan coastal temperate rainforest watersheds, Soil Sci. Soc. Am. J., 79, 378–388, https://doi.org/10.2136/sssaj2014.09.0380, 2015b.

D'Amore, D. V., Biles, F. E., Nay, M., and Rupp, T. S.: Watershed carbon budgets in the southeastern Alaskan coastal forest region, in: Baseline and projected future carbon storage and greenhouse-gas fluxes in ecosystems of Alaska, US Geological Survey Professional Paper, 1826, 196 pp., 2016.

Dai, M., Yin, Z., Meng, F., Liu, Q., and Cai, W.J.: Spatial distribution of riverine DOC inputs to the ocean: an updated global synthesis, Curr. Opin. Sust., 4, 170–178, https://doi.org/10.1016/j.cosust.2012.03.003, 2012.

Deirmendjian, L., Loustau, D., Augusto, L., Lafont, S., Chipeaux, C., Poirier, D., and Abril, G.: Hydrological and ecological controls on dissolved carbon concentrations in groundwater and carbon export to surface waters in a temperate pine forest watershed, Biogeosciences Discuss., https://doi.org/10.5194/bg-2017-90, in review, 2017.

DellaSala, D. A.: Temperate and Boreal Rainforests of the World, Island Press, Washington, DC, 2011.

Emili, L. and Price, J.: Biogeochemical processes in the soil-groundwater system of a forest-peatland complex, north coast British Columbia, Canada, Northwest Sci., 88, 326–348, https://doi.org/10.3955/046.087.0406, 2013.

Fasching, C., Behounek, B., Singer, G., and Battin, T.: Microbial degradation of terrigenous dissolved organic matter and potential consequences for carbon cycling in brown-water streams, Sci. Rep., 4, 4981, https://doi.org/10.1038/srep04981, 2014.

Fasching, C., Ulseth, A., Schelker, J., Steniczka, G., and Battin, T.: Hydrology controls dissolved organic matter export and composition in an Alpine stream and its hyporheic zone, Limnol. Oceanogr., 61, 558–571, https://doi.org/10.1002/lno.10232, 2016.

Fellman, J., Hood, E., D'Amore, D., Edwards, R., and White, D.: Seasonal changes in the chemical quality and biodegradability of dissolved organic matter exported from soils to streams in coastal temperate rainforest watersheds, Biogeochemistry, 95, 277–293, https://doi.org/10.1007/s10533-009-9336-6, 2009a.

Fellman, J., Hood, E., Edwards, R., and D'Amore, D.: Changes in the concentration, biodegradability, and fluorescent properties of dissolved organic matter during stormflows in coastal temperate watersheds, J. Geophys. Res.-Biogeo., 114, https://doi.org/10.1029/2008JG000790, 2009b.

Fellman, J., Hood, E., and Spencer, R.: Fluorescence spectroscopy opens new windows into dissolved organic matter dynamics in freshwater ecosystems: A review, Limnol. Oceanogr., 55, 2452–2462, https://doi.org/10.4319/lo.2010.55.6.2452, 2010.

Fellman, J., Nagorski, S., Pyare, S., Vermilyea, A. W., Scott, D., and Hood, E.: Stream temperature response to variable glacier cover in coastal watersheds of Southeast Alaska, Hydrol. Process., 28, 2062–2073, https://doi.org/10.1002/hyp.9742, 2014

Finlay, J. C. and Kendall, C.: Stable isotope tracing of temporal and spatial variability in organic matter sources and variability in organic matter sources to freshwater ecosytems, in: Stable Isotopes in Ecology and Environmental Science, edited by: Michener, R. and Lajtha, K., Blackwell Publishing Ltd, Oxford, UK, 2, 283–324, 2007.

Fitzgerald, D., Price, J., and Gibson, J.: Hillslope-swamp interactions and flow pathways in a hypermaritime rainforest, British Columbia, Hydrol. Process., 17, 3005–3022, https://doi.org/10.1002/hyp.1279, 2003.

Gibson, J. J., Price, J. S., Aravena, R., Fitzgerald, D. F., and Maloney, D.: Runoff generation in a hypermaritime bog-forest upland, Hydrol. Process., 14, 2711–2730, https://doi.org/10.1002/1099-1085(20001030)14:15<2711::AID-HYP88>3.0.CO;2-2, 2000.

Glatzel, S., Kalbitz, K., Dalva, M., and Moore, T.: Dissolved organic matter properties and their relationship to carbon dioxide efflux from restored peat bogs, Geoderma, 113, 397–411, 2003.

Gonzalez Arriola S., Frazer, G. W., and Giesbrecht, I.: LiDAR-derived watersheds and their metrics for Calvert Island, Hakai Institute, https://doi.org/10.21966/1.15311, 2015.

Gorham, E., Lehman, C., Dyke, A., Clymo, D., and Janssens, J.: Long-term carbon sequestration in North American peatlands, Quaternary Sci. Rev., 58, 77–82, 2012.

Graeber, D., Gelbrecht, J., Pusch, M., Anlanger, C., and von Schiller, D.: Agriculture has changed the amount and composition of dissolved organic matter in Central European headwater streams, Sci. Total Environ., 438, 435–446, https://doi.org/10.1016/j.scitotenv.2012.08.087, 2012.

Green, R. N.: Reconnaissance level terrestrial ecosystem mapping of priority landscape units of the coast EBM planning area: Phase 3, Prepared for British Columbia Ministry Forests, Lands and Natural Resource Ops., Blackwell and Associates, Vancouver, Canada, 2014.

Guillemette, F. and Giorgio, P.: Reconstructing the various facets of dissolved organic carbon bioavailability in freshwater ecosystems, Limnol. Oceanogr., 56, 734–748, https://doi.org/10.4319/lo.2011.56.2.0734, 2011.

Hansen, A. M., Kraus, T. E. C., Pellerin, B. A., Fleck, J. A., Downing, B. D., and Bergamaschi, B. A.: Optical properties of dissolved organic matter (DOM): Effects of biological and photolytic degradation, Limnol. Oceanogr., 61, 1015–1032, https://doi.org/10.1002/lno.10270, 2016.

Harrell, F. E. and Dupont, C.: Hmisc: Harrell Miscellaneous. R package version 4.0-2. https://CRAN.R-project.org/package=Hmisc, 2016.

Harrison, J., Caraco, N., and Seitzinger, S.: Global patterns and sources of dissolved organic matter export to the coastal zone: Results from a spatially explicit, global model, Global Biogeochem. Cy., 19, https://doi.org/10.1029/2005gb002480, 2005.

Helms, J., Stubbins, A., Ritchie, J., Minor, E., Kieber, D., and Mopper, K.: Absorption spectral slopes and slope ratios as indicators of molecular weight, source, and photobleaching of chromophoric dissolved organic matter, Limnol. Oceanogr., 53, 955–969, https://doi.org/10.4319/lo.2008.53.3.0955, 2008.

Helton, A., Wright, M., Bernhardt, E., Poole, G., Cory, R., and Stanford, J.: Dissolved organic carbon lability increases with water residence time in the alluvial aquifer of a river floodplain ecosystem, J. Geophys. Res.-Biogeo., 120, 693–706, https://doi.org/10.1002/2014JG002832, 2015.

Hoffman, K. M., Gavin, D. G., Lertzman, K. P., Smith, D. J., and Starzomski, B. M.: 13 000 years of fire history derived from soil charcoal in a British Columbia coastal temperate rain forest, Ecosphere, 7, e01415, https://doi.org/10.1002/ecs2.1415, 2016.

Hope, D., Billett, M. F., and Cresser, M. S.: A review of the export of carbon in river water: Fluxes and processes, Environ. Pollut., 84, 301–324, https://doi.org/10.1016/0269-7491(94)90142-2, 1994.

Hopkinson, C. S., Buffam, I., Hobbie, J., Vallino, J., and Perdue, M.: Terrestrial inputs of organic matter to coastal ecosystems: An intercomparison of chemical characteristics and bioavailability, Biogeochemistry, 43, 211–234, 1998.

Hudson, N., Baker, A., and Reynolds, D.: Fluorescence analysis of dissolved organic matter in natural, waste and polluted waters-a review, River Res. Appl., 23, 631–649, https://doi.org/10.1002/rra.1005, 2007.

Hurvich, C. M. and Tsai, C.: Regression and time series model selection in small samples, Biometrika, 76, 297–307, https://doi.org/10.2307/2336663, 1989.

IUSS Working Group WRB: World Reference Base for Soil Resources, International soil classification system for naming soils and creating legends for soil maps, World Soil Resources Reports No. 106, Food and Agricultural Organization of the United Nations, Rome, Italy, 2015.

ISO Standard 9196: Liquid flow measurement in open channels – Flow measurements under ice conditions, International Organization for Standardization, available online at: www.iso.org (last access: 1 November 2016), 1992.

ISO Standard 748: Hydrometry – Measurement of liquid flow in open channels using current-meters or floats, International Organization for Standardization, available online at: www.iso.org (last access: 1 November 2016), 2007.

Johannessen, S. C., Potentier, G., Wright, C. A., Masson, D., and Macdonald, R. W.: Water column organic carbon in a Pacific marginal sea (Strait of Georgia, Canada), Mar. Environ. Res., 66, S49–S61, https://doi.org/10.1016/j.marenvres.2008.07.008, 2008.

Johnson, M., Couto, E., Abdo, M., and Lehmann, J.: Fluorescence index as an indicator of dissolved organic carbon quality in hydrologic flowpaths of forested tropical watersheds, Biogeochemistry, 105, 149–157, https://doi.org/10.1007/s10533-011-9595-x, 2011.

Johnson, P. C. D.: Extension of Nakagawa and Schielzeth's R^2_{GLMM} to random slopes models, Methods Ecol. Evol., 5, 944–946, https://doi.org/10.1111/2041-210X.12225, 2014.

Judd, K., Crump, B., and Kling, G.: Variation in dissolved organic matter controls bacterial production and community composition, Ecology, 87, 2068–2079, https://doi.org/10.1890/0012-9658(2006)87[2068:VIDOMC]2.0.CO;2, 2006.

Kalbitz, K., Schmerwitz, J., Schwesig, D., and Matzner, E.: Biodegradation of soil-derived dissolved organic matter as related to its properties, Geoderma, 113, 273–291, https://doi.org/10.1016/S0016-7061(02)00365-8, 2003.

Kling, G., Kipphut, G., Miller, M., and O'Brien, W.: Integration of lakes and streams in a landscape perspective: the importance of material processing on spatial patterns and temporal coherence, Freshwater Biol., 43, 477–497, https://doi.org/10.1046/j.1365-2427.2000.00515.x, 2000.

Koehler, A.-K., Murphy, K., Kiely, G., and Sottocornola, M.: Seasonal variation of DOC concentration and annual loss of DOC from an Atlantic blanket bog in South Western Ireland, Biogeochemistry, 95, 231–242, https://doi.org/10.1007/s10533-009-9333-9, 2009.

Lakowicz, J. R.: Principles of Fluorescence Spectroscopy, 2, Kluwer Academic, New York, 1999.

Larson, J. H., Frost, P. C., Zheng, Z., Johnston, C. A., Bridgham, S. D., Lodge, D. M., and Lamberti, G. A.: Effects of upstream lakes on dissolved organic matter in streams, Limnol. Oceanogr., 52, 60–69, https://doi.org/10.4319/lo.2007.52.1.0060, 2007.

Leighty, W. W., Hamburg, S. P., and Caouette, J.: Effects of management on carbon sequestration in forest biomass in Southeast Alaska, Ecosystems, 9, 1051, https://doi.org/10.1007/s10021-005-0028-3, 2006.

Lalonde, K., Middlestead, P., Gélinas, Y.: Automation of 13C/12C ratio measurement for freshwater and seawater DOC using high temperature combustion, Limnol. Oceanogr.-Meth., 12, 816–829, https://doi.org/10.4319/lom.2014.12.816, 2014.

Lambert, T., Bouillon, S., Darchambeau, F., Massicotte, P., and Borges, A. V.: Shift in the chemical composition of dissolved organic matter in the Congo River network, Biogeosciences, 13, 5405–5420, https://doi.org/10.5194/bg-13-5405-2016, 2016.

Leach, J., Larsson, A., Wallin, M., Nilsson, M., and Laudon, H.: Twelve year interannual and seasonal variability of stream carbon export from a boreal peatland catchment, J. Geophys. Res. 121, 1851–1866, https://doi.org/10.1002/2016JG003357, 2016.

Legendre, P. and Durand, S.: rdaTest, Canonical redundancy analysis, R package version 1.11, available at: http://adn.biol.umontreal.ca/~numericalecology/Rcode/ (last access: 1 January 2017), 2014.

Lepistö, A., Futter, M .N., and Kortelainen, P.: Almost 50 years of monitoring shows that climate, not forestry, controls long-term organic carbon fluxes in a large boreal watershed, Glob. Change Biol., 20, 1225–1237, https://doi.org/10.1111/gcb.12491, 2014.

Liaw, A. and Wiener, M.: Classification and Regression by randomForest, R News, 2, 18–22, 2002.

Lochmuller, C. H. and Saavedra, S. S.: Conformational changes in a soil fulvic acid measured by time dependent fluorescence depolarization, Anal. Chem., 38, 1978–1981, 1986.

Lorenz, D., Runkel, R., and De Cicco, L.: rloadest, River Load Estimation, R package version 0.4.2, available at: https://github.com/USGS-R/rloadest, 2015.

Ludwig, W., Probst, J., and Kempe, S.: Predicting the oceanic input of organic carbon by continental erosion, Global Biogeochem. Cy., 10, 23–41, https://doi.org/10.1029/95GB02925, 1996.

Mann, P. J., Spencer, R. G. M., Dinga, B. J., Poulsen, J. R., Hernes, P. J., Fiske, G., Salter, M. E., Wang, Z. A., Hoering, K. A., Six, J., and Holmes, R. M.: The biogeochemistry of carbon across a gradient of sreams and rivers within the Congo Basin, J. Geophys. Res.-Biogeo., 119, 687–702, https://doi.org/10.1002/2013JG002442, 2014.

Marschner, B. and Kalbitz, K.: Controls on bioavailability and biodegradability of dissolved organic matter in soils, Geoderma, 113, 211–235, 2003.

Martin, S. L. and Soranno, P. A.: Lake landscape position: Relationships to hydrologic connectivity and landscape features, Limnol. Oceanogr., 51, 801–814, https://doi.org/10.4319/lo.2006.51.2.0801, 2006.

Masiello, C. A. and Druffel, E. R. M.: Carbon isotope geochemistry of the Santa Clara River, Global Biogeochem. Cy., 15, 407–416, https://doi.org/10.1029/2000GB001290, 2001.

Mayorga, E., Seitzinger, S., Harrison, J., Dumont, E., Beusen, A., Bouwman, A. F., Fekete, B., Kroeze, C., and Drecht, G.: Global Nutrient Export from WaterSheds 2 (NEWS 2): Model development and implementation, Environ. Model. Softw., 25, 837–853, https://doi.org/10.1016/j.envsoft.2010.01.007, 2010.

McClelland, J., Townsend-Small, A., Holmes, R., Pan, F., Stieglitz, M., Khosh, M., and Peterson, B.: River export of nutrients and organic matter from the North Slope of Alaska to the Beaufort Sea, Water Resour. Res., 50, 1823–1839, https://doi.org/10.1002/2013WR014722, 2014.

McKnight, D., Boyer, E., Westerhoff, P., Doran, P., Kulbe, T., and Andersen, D.: Spectrofluorometric characterization of dissolved organic matter for indication of precursor organic material and aromaticity, Limnol. Oceanogr., 46, 38–48, https://doi.org/10.4319/lo.2001.46.1.0038, 2001.

McLaren, D., Fedje, D., Hay, M. B., Mackie, Q., Walker, I. J., Shugar, D. H., Eamer, J. B. R., Lian, O. B., and Neudorf, C.: A post-glacial sea level hinge on the central Pacific coast of Canada, Quaternary Sci. Rev.., 97, 148–169, 2014.

Meybeck, M.: Carbon, nitrogen, and phosphorus transport by world rivers, Am. J. Sci., 282, 401–450, available from: http://earth.geology.yale.edu/~ajs/1982/04.1982.01.Maybeck.pdf (last access: 11 August 2017), 1982.

Milliman, J. D. and Syvitski J. P. M.: Geomorphic tectonic control of sediment discharge to the ocean: The importance of small mountainous rivers, J. Geol., 100, 525–544, 1992.

Moore, R. D.: Introduction to salt dilution gauging for streamflow measurement part III: Slug injection using salt in solution, Streamline Watershed Management Bulletin, 8, 1–6, 2005.

Morrison, J., Foreman, M. G. G., and Masson, D.: A method for estimating monthly freshwater discharge affecting British Columbia coastal waters, Atmos.-Ocean, 50, 1–8, https://doi.org/10.1080/07055900.2011.637667, 2012.

Mulholland, P. and Watts, J.: Transport of organic carbon to the oceans by rivers of North America: a synthesis of existing data, Tellus, 34, 176–186, https://doi.org/10.1111/j.2153-3490.1982.tb01805.x, 1982.

Murphy, K., Stedmon, C., Graeber, D., and Bro, R.: Fluorescence spectroscopy and multi-way techniques. PARAFAC, Anal. Methods, 5, 6557–6566, https://doi.org/10.1039/C3AY41160E, 2013.

Murphy, K., Stedmon, C., Wenig, P., and Bro, R.: Open-Fluor – A spectral database of auto-fluorescence by organic compounds in the environment, Anal. Methods, 6, 658–661, https://doi.org/10.1039/C3AY41935E, 2014.

Naiman, R. J.: Characteristics of sediment and organic carbon export from pristine boreal forest watersheds, Can. J. Fish. Aquat. Sci., 39, 1699–1718, https://doi.org/10.1139/f82-226, 1982.

Nakagawa, S. and Schielzeth, H.: A general and simple method for obtaining R^2 from generalized linear mixed-effects models, Methods Ecol. Evol., 4, 133–142, https://doi.org/10.1111/j.2041-210x.2012.00261.x, 2013.

Olefeldt, D., Roulet, N., Giesler, R., and Persson, A.: Total waterborne carbon export and DOC composition from ten nested subarctic peatland catchments-importance of peatland cover, groundwater influence, and inter-annual variability of precipitation patterns, Hydrol. Process., 27, 2280–2294, https://doi.org/10.1002/hyp.9358, 2013.

Oliver, A. A., Tank, S. E., Giesbrecht, I., Korver, M. C., Floyd, W. C., Sanborn, P., Bulmer, C., and Lertzman, K. P.: Aquatic carbon flux data package, https://doi.org/10.21966/1.321324, 2017

Pinheiro, J., Bates, D., DebRoy, S., Sarkar, D., and R Core Team: nlme: Linear and Nonlinear Mixed Effects Models, R package version 3.1-128, 2016.

Pojar, J., Klinka, K., and Demarchi, D. A.: Chapter 6, Coastal Western Hemlock Zone, in: Special Report Series 6, Ecosystems of British Columbia, edited by: Meidiner, D. and Pojar, J., Ministry of Forests, British Columbia, Victoria, 330 pp., 1991.

Poulin, B., Ryan, J., and Aiken, G.: Effects of iron on optical properties of dissolved organic matter, Environ. Sci. Technol., 48, 10098–106, https://doi.org/10.1021/es502670r, 2014.

R Core Team, R: A language and environment for statistical computing, R Foundation for Statistical Computing, Vienna, Austria, http://www.R-project.org/ (last access: 11 August 2017), 2013.

Raymond, P., Saiers, J., and Sobczak, W.: Hydrological and biogeochemical controls on watershed dissolved organic matter transport: pulse-shunt concept, Ecology, 97, 5–16, https://doi.org/10.1890/14-1684.1, 2016.

Regnier, P., Friedlingstein, P., Ciais, P., Mackenzie, F., Gruber, N., Janssens, I., Laruelle, G., Lauerwald, R., Luyssaert, S., Andersson, A., Arndt, S., Arnosti, C., Borges, A., Dale, A., Gallego-Sala, A., Goddéris, Y., Goossens, N., Hartmann, J., Heinze, C., Ilyina, T., Joos, F., LaRowe, D., Leifeld, J., Meysman, F., Munhoven, G., Raymond, P., Spahni, R., Suntharalingam, P., and Thullner, M.: Anthropogenic perturbation of the carbon fluxes from land to ocean, Nat. Geosci., 6, 597–607, https://doi.org/10.1038/ngeo1830, 2013.

Roddick, J. R.: Geology, Rivers Inlet-Queens Sound, British Columbia, Open File 3278, Geological Survey of Canada, Ottawa, Canada, 1996.

Royer, T. C.: Coastal fresh water discharge in the northeast, Pacific, J. Geophys. Res., 87, 2017–2021, 1982.

Runkel, R. L., Crawford, C. G., and Cohn, T. A.: Load Estimator (LOADEST): A FORTRAN program for estimating constituent loads in streams and rivers, U.S. Geological Survey Techniques and Methods Book 4, Chapter A5, 65 pp., 2004.

Sanderman, J., Lohse, K. A., Baldock, J. A., and Amundson, R.: Linking soils and streams: Sources and chemistry of dissolved organic matter in a small coastal watershed, Water Resourc. Res., 45, W03418, https://doi.org/10.1029/2008WR006977, 2009.

Spencer, R., Butler, K., and Aiken, G.: Dissolved organic carbon and chromophoric dissolved organic matter properties of rivers in the USA, J. Geophys. Res.-Biogeo., 117, G03001, https://doi.org/10.1029/2011JG001928, 2012.

Spencer, R. G., Hernes, P. J., Ruf, R., Baker, A., Dyda, R. Y., Stubbins, A., and Six, J.: Temporal controls on dissolved organic matter and lignin biogeochemistry in a pristine tropical river, Democratic Republic of Congo, J. Geophys. Res., 115, G03013, https://doi.org/10.1029/2009JG001180, 2010.

Stackpoole, S. M., Butman, D. E., Clow, D. W., Verdin, K. L., Gaglioti, B., and Striegl, R.: Carbon burial, transport, and emission from inland aquatic ecosystems in Alaska, in: Baseline and projected future carbon storage and greenhouse-gas fluxes in ecosystems of Alaska, edited by: Zhiliang, Z., and David, A., US Geological Survey Professional Paper, 1826, 196 pp., 2016.

Stackpoole, S. M., Butman, D. E., Clow, D. W., Verdin, K. L., Gaglioti, B. V., Genet, H., and Striegl, R. G.: Inland waters and their role in the carbon cycle of Alaska, Ecol. Appl., 27, 1403–1420, https://doi.org/10.1002/eap.1552, 2017.

Stedmon, C. and Bro, R.: Characterizing dissolved organic matter fluorescence with parallel factor analysis: a tutorial, Limnol. Oceanogr.-Meth., 6, 572–579, https://doi.org/10.4319/lom.2008.6.572b, 2008.

Stedmon, C. and Markager, S.: Tracing the production and degradation of autochthonous fractions of dissolved organic matter by fluorescence analysis, Limnol. Oceanogr., 50, 1415–1426, https://doi.org/10.4319/lo.2005.50.5.1415, 2005.

Stedmon, C., Markager, S., Bro, R., Stedmon, C., Markager, S., and Bro, R.: Tracing dissolved organic matter in aquatic environments using a new approach to fluorescence spectroscopy, Mar. Chem., 82, 239–254, https://doi.org/10.1016/S0304-4203(03)00072-0, 2003.

Stevenson, F. J.: Humus Chemistry: Genesis, Composition, Reactions, 2, Jon Wiley and Sons Inc., New York, United States of America, 1994.

Symonds, M. R. E. and Moussalli, A.: A brief guide to model selection, multimodel inference, and model averaging in behavioural ecology using Akaike's information criterion, Behav. Ecol. Sociobiol., 65, 13–21, https://doi.org/10.1007/s00265-010-1037-6, 2011.

Tallis, H.: Kelp and rivers subsidize rocky intertidal communities in the Pacific Northwest (USA), Mar. Ecol.-Prog. Ser., 389, 8596, https://doi.org/10.3354/meps08138, 2009.

Tank, S., Raymond, P., Striegl, R., McClelland, J., Holmes, R., Fiske, G., and Peterson, B.: A land-to-ocean perspective on the magnitude, source and implication of DIC flux from major Arctic rivers to the Arctic Ocean, Global Biogeochem. Cy., 26, GB4018, https://doi.org/10.1029/2011GB004192, 2012.

Tank, S., Striegl, R. G., McClelland, J. W., and Kokelij, S. V.: Multi-decadal increases in dissolved organic carbon and alkalinity flux from the Mackenzie drainage basin to the Arctic Ocean, Environ. Res. Lett., 11, https://doi.org/10.1088/1748-9326/11/5/054015, 2016.

Thompson, S. D., Nelson, T. A., Giesbrecht, I., Frazer, G., and Saunders, S. C.: Data-driven regionalization of forested and non-forested ecosystems in coastal British Columbia with LiDAR and RapidEye imagery, Appl. Geogr., 69, 35–50, https://doi.org/10.1016/j.apgeog.2016.02.002, 2016.

Trant, A. J., Niijland, W., Hoffman, K. M., Mathews, D. L., McLaren, D., Nelson, T. A., and Starzomski, B. M.: Intertidal resource use over millennia enhances forest productivity, Nat. Commun., 7, 12491, https://doi.org/10.1038/ncomms12491, 2016.

van Hees, P., Jones, D., Finlay, R., Godbold, D., and Lundström, U.: The carbon we do not see-the impact of low molecular weight compounds on carbon dynamics and respiration in forest soils: a review, Soil Biol. Biochem., 37, 1–13, https://doi.org/10.1016/j.soilbio.2004.06.010, 2005.

Wallin, M., Weyhenmeyer, G., Bastviken, D., Chmiel, H., Peter, S., Sobek, S., and Klemedtsson, L.: Temporal control on concentration, character, and export of dissolved organic carbon in two hemiboreal headwater streams draining contrasting catchments, J. Geophys. Res.-Biogeo., 120, 832–846, https://doi.org/10.1002/2014jg002814, 2015.

Wang, T., Hamann, A., Spittlehouse, D. L., and Murdock, T. Q.: ClimateWNA- High resolution spatial climate data for Western North America, J. Appl. Meterol. Climatol., 51, 16–29, https://doi.org/10.1175/JAMC-D-11-043.1, 2012.

Weishaar, J. L., Aiken, G. R., Bergamaschi, B. A., Fram, M. S., Fujii, R., and Mopper, K.: Evaluation of specific ultraviolet absorbance as an indicator of the chemical composition and reactivity of dissolved organic carbon, Environ. Sci. Technol., 37, 4702–4708, https://doi.org/10.1021/es030360x, 2003.

Whitney, F. A., Crawford, W. R., and Harrison, P. J.: Physical processes that enhance nutrient transport and primary productivity in the coastal and open ocean of the subarctic NE Pacific, Deep-Sea Res. Pt. II, 52, 681–706, 2005.

Wickland, K., Neff, J., and Aiken, G.: Dissolved Organic Carbon in Alaskan Boreal Forest: Sources, Chemical Characteristics, and Biodegradability, Ecosystems, 10, 1323–1340, 2007.

Wilson, H. F. and Xenopoulos, M. A.: Effects of agricultural land use on the composition of fluvial dissolved organic matter, Nat. Geosci., 2, 37–41, https://doi.org/10.1038/ngeo391, 2009.

Wolf, E. C., Mitchell, A. P., and Schoonmaker, P. K.: The Rain Forests of Home: An Atlas of People and Place, Ecotrust, Pacific GIS, Inforain, and Conservation International, Portland, Oregon, 24 pp., available at: http://www.inforain.org/pdfs/ctrf_atlas_orig.pdf, 1995.

Worrall, F., Burt, T., and Adamson, J.: Can climate change explain increases in DOC flux from upland peat catchements?, Sci. Total. Environ., 326, 95–112, https://doi.org/10.1016/j.scitotenv.2003.11.022, 2004.

Xenopoulos, M. A., Lodge, D. M., Frentress, J., Kreps, T. A., Bridgham, S. D., Grossman, E., and Jackson, C. J.: Regional comparisons of watershed determinants of dissolved organic carbon in temperate lakes from the Upper Great Lakes region and selected regions globally, Limnol. Oceanogr., 48, 2321–2334, 2003.

Yamashita, Y. and Jaffei, R.: Characterizing the Interactions between Trace Metals and Dissolved Organic Matter Using Excitation–Emission Matrix and Parallel Factor Analysis, Environ. Sci. Technol., 42, 7374–7379, https://doi.org/10.1021/es801357h, 2008.

Yamashita, Y., Kloeppel, B., Knoepp, J., Zausen, G., and Jaffé, R.: Effects of Watershed History on Dissolved Organic Matter Characteristics in Headwater Streams, Ecosystems, 14, 1110–1122, https://doi.org/10.1007/s10021-011-9469-z, 2011.

Permissions

All chapters in this book were first published in BG, by Copernicus Publications; hereby published with permission under the Creative Commons Attribution License or equivalent. Every chapter published in this book has been scrutinized by our experts. Their significance has been extensively debated. The topics covered herein carry significant findings which will fuel the growth of the discipline. They may even be implemented as practical applications or may be referred to as a beginning point for another development.

The contributors of this book come from diverse backgrounds, making this book a truly international effort. This book will bring forth new frontiers with its revolutionizing research information and detailed analysis of the nascent developments around the world.

We would like to thank all the contributing authors for lending their expertise to make the book truly unique. They have played a crucial role in the development of this book. Without their invaluable contributions this book wouldn't have been possible. They have made vital efforts to compile up to date information on the varied aspects of this subject to make this book a valuable addition to the collection of many professionals and students.

This book was conceptualized with the vision of imparting up-to-date information and advanced data in this field. To ensure the same, a matchless editorial board was set up. Every individual on the board went through rigorous rounds of assessment to prove their worth. After which they invested a large part of their time researching and compiling the most relevant data for our readers.

The editorial board has been involved in producing this book since its inception. They have spent rigorous hours researching and exploring the diverse topics which have resulted in the successful publishing of this book. They have passed on their knowledge of decades through this book. To expedite this challenging task, the publisher supported the team at every step. A small team of assistant editors was also appointed to further simplify the editing procedure and attain best results for the readers.

Apart from the editorial board, the designing team has also invested a significant amount of their time in understanding the subject and creating the most relevant covers. They scrutinized every image to scout for the most suitable representation of the subject and create an appropriate cover for the book.

The publishing team has been an ardent support to the editorial, designing and production team. Their endless efforts to recruit the best for this project, has resulted in the accomplishment of this book. They are a veteran in the field of academics and their pool of knowledge is as vast as their experience in printing. Their expertise and guidance has proved useful at every step. Their uncompromising quality standards have made this book an exceptional effort. Their encouragement from time to time has been an inspiration for everyone.

The publisher and the editorial board hope that this book will prove to be a valuable piece of knowledge for researchers, students, practitioners and scholars across the globe.

List of Contributors

Madhavan Girijakumari Keerthi, Vallivattathillam Parvathi, Iyyappan Suresh, Valiya Parambil Akhil and Pillathu Moolayil Muraleedharan
Physical Oceanography Division, CSIR-National Institute of Oceanography (CSIR-NIO), Goa, India

Marina Levy, Jerome Vialard, Christian Ethé and Olivier Aumont
Sorbonne Universités (UPMC, Univ Paris 06)-CNRS-IRD-MNHN, LOCEAN Laboratory, IPSL, Paris, France

Clément de Boyer Montégut
IFREMER, Univ. Brest, CNRS, IRD, Laboratoire d'Océanographie Physique et Spatiale, IUEM, 29280, Brest, France

Matthieu Lengaigne
Sorbonne Universités (UPMC, Univ Paris 06)-CNRS-IRD-MNHN, LOCEAN Laboratory, IPSL, Paris, France
Indo-French Cell for Water Sciences, IISc-NIO-IITM-IRD Joint International Laboratory, NIO, Goa, India

Ben Wang and Xin Jia
Yanchi Research Station, School of Soil and Water Conservation, Beijing Forestry University, Beijing 100083, PR China
Key Laboratory of State Forestry Administration on Soil and Water Conservation, Beijing Forestry University, Beijing, China
School of Forest Sciences, University of Eastern Finland, P.O. Box 111, 80101 Joensuu, Finland

Jin Nan Gong and Heli Peltola
School of Forest Sciences, University of Eastern Finland, P.O. Box 111, 80101 Joensuu, Finland

Charles Bourque
Faculty of Forestry and Environmental Management, University of New Brunswick, P.O. Box 4400, 28 Dineen Drive, Fredericton, New Brunswick, E3B 5A3, Canada

Tian Shan Zha, Wei Feng, Yun Tian, Bin Wu and Yu Qing Zhang
Yanchi Research Station, School of Soil and Water Conservation, Beijing Forestry University, Beijing 100083, PR China
Key Laboratory of State Forestry Administration on Soil and Water Conservation, Beijing Forestry University, Beijing, China

Jeffrey J. Kelleway
Climate Change Cluster, School of Life Sciences, University of Technology Sydney, Ultimo, NSW 2007, Australia
Department of Environmental Sciences, Macquarie University, Sydney, NSW 2109, Australia

Neil Saintilan
Department of Environmental Sciences, Macquarie University, Sydney, NSW 2109, Australia
School of Life and Environmental Sciences, Centre for Integrative Ecology, Deakin University, Victoria 3216, Australia

Jeffrey A. Baldock
CSIRO Agriculture and Food, Glen Osmond, SA 5064, Australia

Peter I. Macreadie
Climate Change Cluster, School of Life Sciences, University of Technology Sydney, Ultimo, NSW 2007, Australia
School of Life and Environmental Sciences, Centre for Integrative Ecology, Deakin University, Victoria 3216, Australia

Peter J. Ralph
Climate Change Cluster, School of Life Sciences, University of Technology Sydney, Ultimo, NSW 2007, Australia

Francisco Machín
Instituto de Oceanografía y Cambio Global, Grupo QUIMA, Universidad de Las Palmas de Gran Canaria, 35017, Las Palmas de Gran Canaria, Spain
Departamento de Física, Universidad de Las Palmas de Gran Canaria, 35017, Las Palmas de Gran Canaria, Spain

Melchor González-Dávila and J. Magdalena Santana Casiano
Instituto de Oceanografía y Cambio Global, Grupo QUIMA, Universidad de Las Palmas de Gran Canaria, 35017, Las Palmas de Gran Canaria, Spain

David Helman and Itamar M. Lensky
Department of Geography and Environment, Bar-Ilan University, Ramat Gan 52900, Israel

Yagil Osem
Department of Natural Resources, Agricultural Research Organization, Volcani Center, Bet Dagan 50250, Israel

Shani Rohatyn, Eyal Rotenberg and Dan Yakir
Earth and Planetary Sciences, Weizmann Institute of Science, Rehovot 76100, Israel

Yit Arn Teh
Institute of Biological and Environmental Sciences, University of Aberdeen, Aberdeen, UK

Wayne A. Murphy, Juan-Carlos Berrio, Arnoud Boom and Susan E. Page
Department of Geography, University of Leicester, Leicester, UK

Goulven G. Laruelle, Jean-Louis Tison and Pierre Regnier
Department Geoscience, Environment & Society (DGES), Université Libre de Bruxelles, Bruxelles, Belgium

Peter Landschützer
Max Planck Institute for Meteorology, Bundesstr. 53, Hamburg, Germany

Nicolas Gruber
Environmental Physics, Institute of Biogeochemistry and Pollutant Dynamics, ETH Zürich, Zürich, Switzerland

Bruno Delille
Unité d'Oceanographie Chimique, Astrophysics, Geophysics and Oceanography department, University of Liège, Liège, Belgium

TaraW. Hudiburg
Department of Forest, Rangeland, and Fire Sciences, University of Idaho, 875 Perimeter Dr., Moscow, ID 83844-1133, USA

Philip E. Higuera
Department of Ecosystem and Conservation Sciences, University of Montana, 32 Campus Dr.,Missoula, MT 59812, USA

Jeffrey A. Hicke
Department of Geography, University of Idaho, 875 Perimeter Dr.,Moscow, ID 83844-3021, USA

Yao Gao, Tiina Markkanen, Mika Aurela, Tea Thum, Aki Tsuruta and Tuula Aalto
Finnish Meteorological Institute, Helsinki, P.O. Box 503, 00101, Finland

Ivan Mammarella
Department of Physics, University of Helsinki, Helsinki, P.O. Box 48, 00014, Finland

Huiyi Yang
Institute for Climate and Atmospheric Science, School of Earth and Environment, University of Leeds, Leeds, LS2 9JT, UK

Lichao Liu, Peng Zhang, Guang Song, Rong Hui and Zengru Wang
Shapotou Desert Research & Experiment Station, Northwest Institute of Eco-Environment and Resources, Chinese Academy of Sciences, Lanzhou, 730000, China

Yubing Liu and Jin Wang
Shapotou Desert Research & Experiment Station, Northwest Institute of Eco-Environment and Resources, Chinese Academy of Sciences, Lanzhou, 730000, China

Key Laboratory of Stress Physiology and Ecology in Cold and Arid Regions of Gansu Province, Northwest Institute of Eco-Environment and Resources, Chinese Academy of Sciences, Lanzhou, 730000, China

Jörg Schwinger, Jerry Tjiputra and Nadine Goris
Uni Research Climate, Bjerknes Centre for Climate Research, Bergen, Norway

Katharina D. Six and Tatiana Ilyina
Max Planck Institute for Meteorology, Hamburg, Germany

Alf Kirkevåg and Øyvind Seland
Norwegian Meteorological Institute, Oslo, Norway

Christoph Heinze
Geophysical Institute, University of Bergen, Bjerknes Centre for Climate Research, Bergen, Norway
Uni Research Climate, Bjerknes Centre for Climate Research, Bergen, Norway

Magnus Lund and Mikkel P. Tamstorf
Department of Biosciences, Arctic Research Center, Aarhus University, Frederiksborgvej 399, 4000 Roskilde, Denmark

Mathew Williams
School of GeoSciences, University of Edinburgh, Edinburgh, EH93FF, UK

AndreasWestergaard-Nielsen and Birger U. Hansen
Center for Permafrost (CENPERM), Department of Geosciences and Natural Resource Management, University of Copenhagen, Oester Voldgade 10, 1350 Copenhagen, Denmark

Jean-François Exbrayat
School of GeoSciences, University of Edinburgh, Edinburgh, EH93FF, UK

National Centre for Earth Observation, University of Edinburgh, Edinburgh, EH93FF, UK

Torben R. Christensen
Department of Biosciences, Arctic Research Center, Aarhus University, Frederiksborgvej 399, 4000 Roskilde, Denmark
Department of Physical Geography and Ecosystem Science, Lund University, Sölvegatan 12, 223 62 Lund, Sweden

Efrén López-Blanco
Department of Biosciences, Arctic Research Center, Aarhus University, Frederiksborgvej 399, 4000 Roskilde, Denmark
School of GeoSciences, University of Edinburgh, Edinburgh, EH93FF, UK

Barbara Marcolla
Fondazione Edmund Mach, IASMA Research and Innovation Centre, Sustainable Agro-ecosystems and Bioresources Department, 38010 San Michele all'Adige, Trento, Italy

Christian Rödenbeck
Max Planck Institute for Biogeochemistry, 07745 Jena, Germany

Alessandro Cescatti
European Commission, Joint Research Centre, Directorate for Sustainable Resources, 21027 Ispra (VA), Italy

Maartje C. Korver
Hakai Institute, Tula Foundation, P.O. Box 309, Heriot Bay, BC, V0P 1H0, Canada

Chuck Bulmer
BC Ministry of Forests Lands and Natural Resource Operations, 3401 Reservoir Rd, Vernon, BC, V1B 2C7, Canada

Allison A. Oliver and Suzanne E. Tank
University of Alberta, Department of Biological Sciences, CW 405, Biological Sciences Bldg., University of Alberta, Edmonton, AB, T6G 2E9, Canada
Hakai Institute, Tula Foundation, P.O. Box 309, Heriot Bay, BC, V0P 1H0, Canada

Ian Giesbrecht and Ken P. Lertzman
Hakai Institute, Tula Foundation, P.O. Box 309, Heriot Bay, BC, V0P 1H0, Canada
School of Resource and Environmental Management, Simon Fraser University, TASC 1 – Room 8405, 8888 University Drive, Burnaby, BC, V5A 1S6, Canada

Paul Sanborn
Ecosystem Science and Management Program, University of Northern British Columbia, 3333 University Way, Prince George, BC, V2N 4Z9, Canada
Hakai Institute, Tula Foundation, P.O. Box 309, Heriot Bay, BC, V0P 1H0, Canada

William C. Floyd
Ministry of Forests, Lands and Natural Resource Operations, 2100 Labieux Rd, Nanaimo, BC, V9T 6E9, Canada
Vancouver Island University, 900 Fifth Street, Nanaimo, BC, V9R 5S5, Canada
Hakai Institute, Tula Foundation, P.O. Box 309, Heriot Bay, BC, V0P 1H0, Canada

Index